Undergraduate Topics in Computer Science

'Undergraduate Topics in Computer Science' (UTiCS) delivers high-quality instructional content for undergraduates studying in all areas of computing and information science. From core foundational and theoretical material to final-year topics and applications, UTiCS books take a fresh, concise, and modern approach and are ideal for self-study or for a one- or two-semester course. The texts are all authored by established experts in their fields, reviewed by an international advisory board, and contain numerous examples and problems, many of which include fully worked solutions.

The UTiCS concept relies on high-quality, concise books in softback format, and generally a maximum of 275–300 pages. For undergraduate textbooks that are likely to be longer, more expository, Springer continues to offer the highly regarded Texts in Computer Science series, to which we refer potential authors.

More information about this series at https://link.springer.com/bookseries/7592

John Vince

Mathematics for Computer Graphics

Sixth Edition

 Springer

John Vince
Breinton, UK

ISSN 1863-7310   ISSN 2197-1781   (electronic)
Undergraduate Topics in Computer Science
ISBN 978-1-4471-7519-3   ISBN 978-1-4471-7520-9   (eBook)
https://doi.org/10.1007/978-1-4471-7520-9

This Springer imprint is published by the registered company Springer-Verlag London Ltd. part of Springer Nature.
The registered company address is: The Campus, 4 Crinan Street, London, N1 9XW, United Kingdom

This book is dedicated to my wife, Heidi.

Preface

The first edition of this book began life as part of Springer's *Essential* series and contained ten chapters and approximately 220 pages. This sixth and last edition has twenty chapters and approximately 600 pages. Over the intervening editions, I have revised and extended previous descriptions and introduced new chapters on subjects that I believe are relevant to computer graphics, such as differential calculus and interpolation, and new subjects that I had to learn about, such as quaternions and geometric algebra. Hopefully, this edition explores enough mathematical ideas to satisfy most people working in computer graphics.

Although the first edition of this book was produced on a humble PC using WORD, subsequent editions were produced on an Apple iMac using LATEX. I recommend to any budding authors that they should learn LATEX and use Springer's templates to create their first manuscript. Furthermore, today's computers are so fast that I often compile the entire book for the sake of changing a single character—it only takes 5 or 6 seconds!

I have used colour in the text to emphasise the patterns behind certain numbers and in the illustrations to clarify the mathematics.

It is extremely difficult to ensure that there are no spelling mistakes, missing brackets, spurious punctuation marks and, above all, mathematical errors. I truly have done my best to correct the text and associated equations, but if I have missed some, then I apologise now.

In all of my books, I try to mention the names of important mathematicians associated with an invention or discovery and the period over which they were alive. In this book, I mention 50 such people, and the relevant dates are attached to the first citation.

Whilst writing this book I have borne in mind what it was like for me when I was studying different areas of mathematics for the first time. In spite of reading and rereading an explanation several times, it could take days before 'the penny dropped' and a concept became apparent. Hopefully, the reader will find the following explanations useful in developing their understanding of these specific areas of mathematics and enjoy the sound of various pennies dropping!

I would like to thank Helen Desmond, Editor for Computer Science, for allowing me to give up holidays and hobbies in order to complete another book!

Breinton, UK   John Vince
May 2022

Contents

1	**Introduction**	1
	1.1 Mathematics for Computer Graphics	1
	1.2 Understanding Mathematics	1
	1.3 What Makes Mathematics Difficult?	2
	1.4 Background to This Book	2
	1.5 How to Use This Book	2
	1.6 Symbols and Notation	3
	References	4
2	**Numbers**	5
	2.1 Introduction	5
	2.2 Background	5
	2.3 Counting	5
	2.4 Sets of Numbers	6
	2.5 Zero	7
	2.6 Negative Numbers	8
	2.6.1 The Arithmetic of Positive and Negative Numbers	9
	2.7 Observations and Axioms	10
	2.7.1 Commutative Law	10
	2.7.2 Associative Law	10
	2.7.3 Distributive Law	11
	2.8 The Base of a Number System	11
	2.8.1 Background	11
	2.8.2 Octal Numbers	12
	2.8.3 Binary Numbers	13
	2.8.4 Hexadecimal Numbers	13
	2.8.5 Adding Binary Numbers	16
	2.8.6 Subtracting Binary Numbers	18

	2.9	Types of Numbers	18
		2.9.1 Natural Numbers	19
		2.9.2 Integers	19
		2.9.3 Rational Numbers	19
		2.9.4 Irrational Numbers	20
		2.9.5 Real Numbers	20
		2.9.6 Algebraic and Transcendental Numbers	20
		2.9.7 Imaginary Numbers	21
		2.9.8 Complex Numbers	24
		2.9.9 Transcendental and Algebraic Numbers	26
		2.9.10 Infinity	27
	2.10	Summary	28
	2.11	Worked Examples	28
		2.11.1 Algebraic Expansion	28
		2.11.2 Binary Subtraction	28
		2.11.3 Complex Numbers	28
		2.11.4 Complex Rotation	29
3	**Algebra**		**31**
	3.1	Introduction	31
	3.2	Background	32
		3.2.1 Solving the Roots of a Quadratic Equation	33
	3.3	Indices	37
		3.3.1 Laws of Indices	38
	3.4	Logarithms	38
	3.5	Further Notation	40
	3.6	Functions	40
		3.6.1 Explicit and Implicit Equations	41
		3.6.2 Function Notation	41
		3.6.3 Intervals	42
		3.6.4 Function Domains and Ranges	43
		3.6.5 Odd and Even Functions	44
		3.6.6 Power Functions	46
	3.7	Summary	46
	3.8	Worked Examples	46
		3.8.1 Algebraic Manipulation	46
		3.8.2 Solving a Quadratic Equation	47
		3.8.3 Factorising	49
	References		49
4	**Trigonometry**		**51**
	4.1	Introduction	51
	4.2	Background	51
	4.3	Units of Angular Measurement	51

4.4		The Trigonometric Ratios	53
	4.4.1	Domains and Ranges	55
4.5		Inverse Trigonometric Ratios	56
4.6		Trigonometric Identities	58
4.7		The Sine Rule	58
4.8		The Cosine Rule	59
4.9		Compound-Angle Identities	60
	4.9.1	Double-Angle Identities	61
	4.9.2	Multiple-Angle Identities	62
	4.9.3	Half-Angle Identities	63
4.10		Perimeter Relationships	63
4.11		Summary	64
		References	64

5 Coordinate Systems   **65**

5.1		Introduction	65
5.2		Background	65
5.3		The Cartesian Plane	66
5.4		Function Graphs	66
5.5		Shape Representation	66
	5.5.1	2D Polygons	67
	5.5.2	Area of a Shape	67
5.6		Theorem of Pythagoras in 2D	68
5.7		3D Cartesian Coordinates	69
	5.7.1	Theorem of Pythagoras in 3D	70
5.8		Polar Coordinates	70
5.9		Spherical Polar Coordinates	71
5.10		Cylindrical Coordinates	72
5.11		Summary	73
5.12		Worked Examples	73
	5.12.1	Area of a Shape	73
	5.12.2	Distance Between Two Points	73
	5.12.3	Polar Coordinates	74
	5.12.4	Spherical Polar Coordinates	74
	5.12.5	Cylindrical Coordinates	75
		Reference	75

6 Determinants   **77**

6.1		Introduction	77
6.2		Linear Equations with Two Variables	78
6.3		Linear Equations with Three Variables	81
	6.3.1	Sarrus's Rule	88
6.4		Mathematical Notation	88
	6.4.1	Matrix	88
	6.4.2	Order of a Determinant	89

	6.4.3	Value of a Determinant	89
	6.4.4	Properties of Determinants	91
6.5	Summary	....	91
6.6	Worked Examples	....	92
	6.6.1	Determinant Expansion	92
	6.6.2	Complex Determinant	92
	6.6.3	Simple Expansion	93
	6.6.4	Simultaneous Equations	93

7 Vectors **95**
7.1	Introduction	....	95
7.2	Background	....	95
7.3	2D Vectors	....	96
	7.3.1	Vector Notation	96
	7.3.2	Graphical Representation of Vectors	97
	7.3.3	Magnitude of a Vector	98
7.4	3D Vectors	....	99
	7.4.1	Vector Manipulation	100
	7.4.2	Scaling a Vector	100
	7.4.3	Vector Addition and Subtraction	101
	7.4.4	Position Vectors	102
	7.4.5	Unit Vectors	102
	7.4.6	Cartesian Vectors	103
	7.4.7	Products	103
	7.4.8	Scalar Product	104
	7.4.9	The Dot Product in Lighting Calculations	105
	7.4.10	The Scalar Product in Back-Face Detection	106
	7.4.11	The Vector Product	107
	7.4.12	The Right-Hand Rule	112
7.5	Deriving a Unit Normal Vector for a Triangle	112
7.6	Surface Areas	113
	7.6.1	Calculating 2D Areas	114
7.7	Summary	114
7.8	Worked Examples	115
	7.8.1	Position Vector	115
	7.8.2	Unit Vector	115
	7.8.3	Vector Magnitude	115
	7.8.4	Angle Between Two Vectors	116
	7.8.5	Vector Product	116
References	....	117

8 Matrix Algebra **119**
| 8.1 | Introduction | .... | 119 |
| 8.2 | Background | .... | 119 |

	8.3	Matrix Notation	122
	8.3.1	Matrix Dimension or Order	122
	8.3.2	Square Matrix	122
	8.3.3	Column Vector	123
	8.3.4	Row Vector	123
	8.3.5	Null Matrix	123
	8.3.6	Unit Matrix	123
	8.3.7	Trace	124
	8.3.8	Determinant of a Matrix	125
	8.3.9	Transpose	125
	8.3.10	Symmetric Matrix	126
	8.3.11	Antisymmetric Matrix	128
	8.4	Matrix Addition and Subtraction	130
	8.4.1	Scalar Multiplication	130
	8.5	Matrix Products	130
	8.5.1	Row and Column Vectors	131
	8.5.2	Row Vector and a Matrix	131
	8.5.3	Matrix and a Column Vector	132
	8.5.4	Square Matrices	133
	8.5.5	Rectangular Matrices	134
	8.6	Inverse Matrix	134
	8.6.1	Inverting a Pair of Matrices	141
	8.7	Orthogonal Matrix	141
	8.8	Diagonal Matrix	142
	8.9	Summary	143
	8.10	Worked Examples	143
	8.10.1	Matrix Inversion	143
	8.10.2	Identity Matrix	144
	8.10.3	Solving Two Equations Using Matrices	144
	8.10.4	Solving Three Equations Using Matrices	145
	8.10.5	Solving Two Complex Equations	146
	8.10.6	Solving Three Complex Equations	147
	8.10.7	Solving Two Complex Equations	148
	8.10.8	Solving Three Complex Equations	149

9 Complex Numbers 151
	9.1	Introduction	151
	9.2	Definition of a Complex Number	151
	9.2.1	Addition and Subtraction of Complex Numbers	152
	9.2.2	Multiplying a Complex Number by a Scalar	153
	9.2.3	Product of Complex Numbers	153
	9.2.4	Square of a Complex Number	154
	9.2.5	Norm of a Complex Number	154
	9.2.6	Complex Conjugate of a Complex Number	154
	9.2.7	Quotient of Complex Numbers	155

	9.2.8	Inverse of a Complex Number	156
	9.2.9	Square-Root of $\pm i$	156
9.3	Ordered Pairs		158
	9.3.1	Addition and Subtraction of Ordered Pairs	158
	9.3.2	Multiplying an Ordered Pair by a Scalar	159
	9.3.3	Product of Ordered Pairs	159
	9.3.4	Square of an Ordered Pair	160
	9.3.5	Norm of an Ordered Pair	161
	9.3.6	Complex Conjugate of an Ordered Pair	161
	9.3.7	Quotient of an Ordered Pair	161
	9.3.8	Inverse of an Ordered Pair	162
	9.3.9	Square-Root of $\pm i$	163
9.4	Matrix Representation of a Complex Number		164
	9.4.1	Adding and Subtracting Complex Numbers	165
	9.4.2	Product of Two Complex Numbers	166
	9.4.3	Norm Squared of a Complex Number	166
	9.4.4	Complex Conjugate of a Complex Number	167
	9.4.5	Inverse of a Complex Number	167
	9.4.6	Quotient of a Complex Number	168
	9.4.7	Square-Root of $\pm i$	169
9.5	Summary		170
9.6	Worked Examples		170
	9.6.1	Adding and Subtracting Complex Numbers	170
	9.6.2	Product of Complex Numbers	171
	9.6.3	Multiplying a Complex Number by i	172
	9.6.4	The Norm of a Complex Number	173
	9.6.5	The Complex Conjugate of a Complex Number	173
	9.6.6	The Quotient of Two Complex Numbers	174
	9.6.7	Divide a Complex Number by i	175
	9.6.8	Divide a Complex Number by $-i$	176
	9.6.9	The Inverse of a Complex Number	177
	9.6.10	The Inverse of i	178
	9.6.11	The Inverse of $-i$	178
Reference			179
10	**Geometric Transforms**		181
10.1	Introduction		181
10.2	Background		181
10.3	2D Transforms		182
	10.3.1	Translation	182
	10.3.2	Scaling	182
	10.3.3	Reflection	183
10.4	Transforms as Matrices		184
	10.4.1	Systems of Notation	184

10.5	Homogeneous Coordinates	....	184
	10.5.1	2D Translation	....	186
	10.5.2	2D Scaling	....	186
	10.5.3	2D Reflections	....	187
	10.5.4	2D Shearing	....	189
	10.5.5	2D Rotation	....	190
	10.5.6	2D Scaling	....	192
	10.5.7	2D Reflection	....	193
	10.5.8	2D Rotation About an Arbitrary Point	....	194
10.6	3D Transforms	....	194
	10.6.1	3D Translation	....	195
	10.6.2	3D Scaling	....	195
	10.6.3	3D Rotation	....	196
	10.6.4	Gimbal Lock	....	199
	10.6.5	Rotating About an Axis	....	200
	10.6.6	3D Reflections	....	202
10.7	Change of Axes	....	202
	10.7.1	2D Change of Axes	....	202
	10.7.2	Direction Cosines	....	204
	10.7.3	3D Change of Axes	....	205
10.8	Positioning the Virtual Camera	....	205
	10.8.1	Direction Cosines	....	206
	10.8.2	Euler Angles	....	208
10.9	Rotating a Point About an Arbitrary Axis	....	211
	10.9.1	Matrices	....	211
10.10	Transforming Vectors	....	218
10.11	Determinants	....	220
10.12	Perspective Projection	....	222
10.13	Summary	....	224
10.14	Worked Examples	....	224
	10.14.1	2D Scaling Transform	....	224
	10.14.2	2D Scale and Translate	....	224
	10.14.3	3D Scaling Transform	....	225
	10.14.4	2D Rotation	....	225
	10.14.5	2D Rotation About a Point	....	226
	10.14.6	Determinant of the Rotate Transform	....	227
	10.14.7	Determinant of the Shear Transform	....	227
	10.14.8	Yaw, Pitch and Roll Transforms	....	227
	10.14.9	3D Rotation About an Axis	....	228
	10.14.10	3D Rotation Transform Matrix	....	228
	10.14.11	2D Change of Axes	....	229
	10.14.12	3D Change of Axes	....	230
	10.14.13	Rotate a Point About an Axis	....	231
	10.14.14	Perspective Projection	....	232

11 Quaternion Algebra 233
    11.1   Introduction 233
    11.2   Some History 233
    11.3   Defining a Quaternion 237
            11.3.1   The Quaternion Units 239
            11.3.2   Example of Quaternion Products 241
    11.4   Algebraic Definition 241
    11.5   Adding and Subtracting Quaternions 241
    11.6   Real Quaternion 242
    11.7   Multiplying a Quaternion by a Scalar 242
    11.8   Pure Quaternion 243
    11.9   Unit Quaternion 244
    11.10   Additive Form of a Quaternion 245
    11.11   Binary Form of a Quaternion 245
    11.12   The Complex Conjugate of a Quaternion 246
    11.13   Norm of a Quaternion 247
    11.14   Normalised Quaternion 248
    11.15   Quaternion Products 248
            11.15.1   Product of Pure Quaternions 249
            11.15.2   Product of Unit-Norm Quaternions 249
            11.15.3   Square of a Quaternion 250
            11.15.4   Norm of the Quaternion Product 251
    11.16   Inverse Quaternion 252
    11.17   Matrices 253
            11.17.1   Orthogonal Matrix 254
    11.18   Quaternion Algebra 254
    11.19   Summary 255
            11.19.1   Summary of Definitions 255
    11.20   Worked Examples 256
            11.20.1   Adding and Subtracting Quaternions 256
            11.20.2   Norm of a Quaternion 257
            11.20.3   Unit-norm Quaternions 257
            11.20.4   Quaternion Product 257
            11.20.5   Square of a Quaternion 258
            11.20.6   Inverse of a Quaternion 258
    References 258

12 Quaternions in Space 261
    12.1   Introduction 261
    12.2   Some History 261
    12.3   Quaternion Products 261
            12.3.1   Special Case 263
            12.3.2   General Case 266
            12.3.3   Double Angle 268

12.4   Quaternions in Matrix Form 271
        12.4.1   Vector Method 272
        12.4.2   Geometric Verification 274
12.5   Multiple Rotations 276
12.6   Rotating About an Off-Set Axis 277
12.7   Converting a Rotation Matrix to a Quaternion 279
12.8   Summary 280
        12.8.1   Summary of Definitions 281
12.9   Worked Examples 281
        12.9.1   Special Case Quaternion 281
        12.9.2   Rotating a Vector Using a Quaternion 282
        12.9.3   Evaluate qpq^{-1} 282
        12.9.4   Evaluate qpq^{-1} Using a Matrix 282
References 283

13 Interpolation 285
    13.1   Introduction 285
    13.2   Background 285
    13.3   Linear Interpolation 286
    13.4   Non-Linear Interpolation 288
            13.4.1   Trigonometric Interpolation 288
            13.4.2   Cubic Interpolation 289
    13.5   Interpolating Vectors 294
    13.6   Interpolating Quaternions 297
    13.7   Summary 299

14 Curves and Patches 301
    14.1   Introduction 301
    14.2   Background 301
    14.3   The Circle 302
    14.4   The Ellipse 302
    14.5   Bézier Curves 303
            14.5.1   Bernstein Polynomials 303
            14.5.2   Quadratic Bézier Curves 306
            14.5.3   Cubic Bernstein Polynomials 307
    14.6   A Recursive Bézier Formula 310
    14.7   Bézier Curves Using Matrices 311
            14.7.1   Linear Interpolation 312
    14.8   B-Splines 315
            14.8.1   Uniform B-Splines 315
            14.8.2   Continuity 317
            14.8.3   Non-uniform B-Splines 318
            14.8.4   Non-uniform Rational B-Splines 319

14.9   Surface Patches 319
        14.9.1   Planar Surface Patch 319
        14.9.2   Quadratic Bézier Surface Patch 320
        14.9.3   Cubic Bézier Surface Patch 322
14.10   Summary 324

15 Analytic Geometry 325
15.1   Introduction 325
15.2   Background 325
        15.2.1   Angles 325
        15.2.2   Intercept Theorems 326
        15.2.3   Golden Section 327
        15.2.4   Triangles 327
        15.2.5   Centre of Gravity of a Triangle 328
        15.2.6   Isosceles Triangle 328
        15.2.7   Equilateral Triangle 329
        15.2.8   Right Triangle 329
        15.2.9   Theorem of Thales 329
        15.2.10   Theorem of Pythagoras 329
        15.2.11   Quadrilateral 330
        15.2.12   Trapezoid 330
        15.2.13   Parallelogram 331
        15.2.14   Rhombus 331
        15.2.15   Regular Polygon 332
        15.2.16   Circle 332
15.3   2D Analytic Geometry 334
        15.3.1   Equation of a Straight Line 334
        15.3.2   The Hessian Normal Form 335
        15.3.3   Space Partitioning 337
        15.3.4   The Hessian Normal Form from Two Points 337
15.4   Intersection Points 338
        15.4.1   Intersecting Straight Lines 338
        15.4.2   Intersecting Line Segments 339
15.5   Point Inside a Triangle 341
        15.5.1   Area of a Triangle 341
        15.5.2   Hessian Normal Form 343
15.6   Intersection of a Circle with a Straight Line 345
15.7   3D Geometry 347
        15.7.1   Equation of a Straight Line 347
        15.7.2   Intersecting Two Straight Lines 348
15.8   Equation of a Plane 351
        15.8.1   Cartesian Form of the Plane Equation 351
        15.8.2   General Form of the Plane Equation 353
        15.8.3   Parametric Form of the Plane Equation 354

	15.8.4	Converting from the Parametric to the General Form	355
	15.8.5	Plane Equation from Three Points	357
15.9	Intersecting Planes		359
	15.9.1	Intersection of Three Planes	363
	15.9.2	Angle Between Two Planes	365
	15.9.3	Angle Between a Line and a Plane	366
	15.9.4	Intersection of a Line with a Plane	368
15.10	Summary		370

16 Barycentric Coordinates 371
16.1   Introduction 371
16.2   Background 371
16.3   Ceva's Theorem 372
16.4   Ratios and Proportion 373
16.5   Mass Points 374
16.6   Linear Interpolation 380
16.7   Convex Hull Property 387
16.8   Areas 387
16.9   Volumes 396
16.10 Bézier Curves and Patches 398
16.11 Summary 399
Reference 399

17 Geometric Algebra 401
17.1   Introduction 401
17.2   Background 401
17.3   Symmetric and Antisymmetric Functions 402
17.4   Trigonometric Foundations 403
17.5   Vectorial Foundations 405
17.6   Inner and Outer Products 405
17.7   The Geometric Product in 2D 407
17.8   The Geometric Product in 3D 409
17.9   The Outer Product of Three 3D Vectors 411
17.10 Axioms 412
17.11 Notation 413
17.12 Grades, Pseudoscalars and Multivectors 413
17.13 Redefining the Inner and Outer Products 415
17.14 The Inverse of a Vector 415
17.15 The Imaginary Properties of the Outer Product 417
17.16 Duality 419
17.17 The Relationship Between the Vector Product and the Outer Product 420
17.18 The Relationship Between Quaternions and Bivectors 421

17.19   Reflections and Rotations 422
        17.19.1   2D Reflections 422
        17.19.2   3D Reflections 423
        17.19.3   2D Rotations 424
17.20   Rotors 426
17.21   Applied Geometric Algebra 429
17.22   Summary 435
References 435

18   Calculus: Derivatives 437
18.1   Introduction 437
18.2   Background 437
18.3   Small Numerical Quantities 437
18.4   Equations and Limits 439
        18.4.1   Quadratic Function 439
        18.4.2   Cubic Equation 440
        18.4.3   Functions and Limits 442
        18.4.4   Graphical Interpretation of the Derivative 444
        18.4.5   Derivatives and Differentials 445
        18.4.6   Integration and Antiderivatives 445
18.5   Function Types 447
18.6   Differentiating Groups of Functions 448
        18.6.1   Sums of Functions 448
        18.6.2   Function of a Function 450
        18.6.3   Function Products 454
        18.6.4   Function Quotients 458
18.7   Differentiating Implicit Functions 460
18.8   Differentiating Exponential and Logarithmic Functions 463
        18.8.1   Exponential Functions 463
        18.8.2   Logarithmic Functions 465
18.9   Differentiating Trigonometric Functions 467
        18.9.1   Differentiating tan 467
        18.9.2   Differentiating csc 468
        18.9.3   Differentiating sec 469
        18.9.4   Differentiating cot 470
        18.9.5   Differentiating arcsin, arccos and arctan 470
        18.9.6   Differentiating arccsc, arcsec and arccot 471
18.10   Differentiating Hyperbolic Functions 472
        18.10.1   Differentiating sinh, cosh and tanh 474
18.11   Higher Derivatives 475
18.12   Higher Derivatives of a Polynomial 475
18.13   Identifying a Local Maximum or Minimum 477

18.14   Partial Derivatives 480
          18.14.1   Visualising Partial Derivatives 483
          18.14.2   Mixed Partial Derivatives 485
18.15   Chain Rule 486
18.16   Total Derivative 488
18.17   Summary 489
         Reference 490

19   Calculus: Integration 491
19.1   Introduction 491
19.2   Indefinite Integral 491
19.3   Integration Techniques 492
          19.3.1   Continuous Functions 492
          19.3.2   Difficult Functions 493
          19.3.3   Trigonometric Identities 493
          19.3.4   Exponent Notation 495
          19.3.5   Completing the Square 497
          19.3.6   The Integrand Contains a Derivative 498
          19.3.7   Converting the Integrand into a Series
                    of Fractions 500
          19.3.8   Integration by Parts 501
          19.3.9   Integration by Substitution 505
          19.3.10   Partial Fractions 510
19.4   Area Under a Graph 512
19.5   Calculating Areas 513
19.6   Positive and Negative Areas 521
19.7   Area Between Two Functions 523
19.8   Areas with the y-Axis 524
19.9   Area with Parametric Functions 525
19.10   The Riemann Sum 527
19.11   Summary 529

20   Worked Examples 531
20.1   Introduction 531
20.2   Area of Regular Polygon 531
20.3   Area of Any Polygon 532
20.4   Dihedral Angle of a Dodecahedron 533
20.5   Vector Normal to a Triangle 534
20.6   Area of a Triangle Using Vectors 535
20.7   General Form of the Line Equation from Two Points 535
20.8   Angle Between Two Straight Lines 536
20.9   Test if Three Points Lie on a Straight Line 537
20.10   Position and Distance of the Nearest Point on a Line
        to a Point 538

20.11   Position of a Point Reflected in a Line 540
20.12   Intersection of a Line and a Sphere 543
20.13   Sphere Touching a Plane 547
20.14   Summary 549

Appendix A: Limit of $(\sin \theta)/\theta$ 551

Appendix B: Integrating $\cos^n \theta$ 555

Index 557

Chapter 1
Introduction

1.1 Mathematics for Computer Graphics

Computer graphics contains many areas of specialism such as data visualisation, computer animation, film special effects, computer games and virtual reality. Fortunately, not everyone working in computer graphics requires a knowledge of mathematics, but those that do, often look for a book that introduces them to some basic ideas of mathematics, without turning them into mathematicians. This is the objective of this book. Over the following chapters I introduce the reader to some useful mathematical topics that will help them understand the software they work with, and how to solve a wide variety of geometric and algebraic problems. These topics include numbers systems, algebra, trigonometry, 2D and 3D geometry, vectors, equations, matrices, complex numbers, determinants, transforms, quaternions, interpolation, curves, patches and calculus. I have written about some of these topics to a greater level of detail in other books, which you may be interested in exploring.

1.2 Understanding Mathematics

One of the problems with mathematics is its incredible breadth and depth. It embraces everything from geometry, calculus, topology, statistics, complex functions to number theory and propositional calculus. All of these subjects can be studied superficially or to a mind-numbing complexity. Fortunately, no one is required to understand everything, which is why mathematicians tend to specialise in one or two areas and develop a specialist knowledge. If it's any comfort, even Einstein asked friends and colleagues to explain branches of mathematics to help him with his theories.

© Springer-Verlag London Ltd., part of Springer Nature 2022
J. Vince, *Mathematics for Computer Graphics*, Undergraduate Topics
in Computer Science, https://doi.org/10.1007/978-1-4471-7520-9_1

1.3   What Makes Mathematics Difficult?

'What makes mathematics difficult?' is a difficult question to answer, but one that has to be asked and answered. There are many answers to this question, and I believe that problems begin with mathematical notation and how to read it; how to analyse a problem and express a solution using mathematical statements. Unlike learning a foreign language—which I find very difficult—mathematics is a language that needs to be learned by discovering facts and building upon them to discover new facts. Consequently, a good memory is always an advantage, as well as a sense of logic.

Mathematics can be difficult for anyone, including mathematicians. For example, when the idea of $\sqrt{-1}$ was originally proposed, it was criticised and looked down upon by mathematicians, mainly because its purpose was not fully understood. Eventually, it transformed the entire mathematical landscape, including physics. Similarly, when the German mathematician Georg Cantor (1845–1919), published his papers on set theory and transfinite sets, some mathematicians hounded him in a disgraceful manner. The German mathematician Leopold Kronecker (1823–1891), called Cantor a 'scientific charlatan', a 'renegade', and a 'corrupter of youth', and did everything to hinder Cantor's academic career [1]. Similarly, the French mathematician and physicist Henri Poincaré (1854–1912), called Cantor's ideas a 'grave disease' [2], whilst the Austrian-British philosopher and logician Ludwig Wittgenstein (1889–1951), complained that mathematics is 'ridden through and through with the pernicious idioms of set theory' [3]. How wrong they all were. Today, set theory is a major branch of mathematics and has found its way into every math curriculum. So don't be surprised to discover that some mathematical ideas are initially difficult to understand—you are in good company.

1.4   Background to This Book

During my working life in computer animation I came across a wide range of students with an equally wide range of mathematical knowledge. Some students possessed a rudimentary background in mathematics, while others had been taught calculus and supporting subjects. Teaching such a cohort the mathematics of computer graphics was a challenge, to say the least, but somehow I did. By the end of a three-year undergraduate course they were competent programmers and could program a wide variety of mathematical techniques. The first-edition of this book employed much of my teaching material and has been revised and extended.

1.5   How to Use This Book

Initially, I'd recommend to any reader to start at the beginning and start reading chapters on subjects with which they are familiar. One never knows what may be learnt from reading about a familiar subject by a non-mathematician. For those readers with a good background in mathematics, should quick read chapters on topics covered else-where, and settle down on new topics. However you approach this book, I sincerely hope that you discover something new that increases your knowledge of the subject.

1.6 Symbols and Notation

One of the reasons why many people find mathematics inaccessible is due to its symbols and notation. Let's look at symbols first. The English alphabet possesses a reasonable range of familiar character shapes:

$$a,b,c,d,e,f,g,h,i,j,k,l,m,n,o,p,q,r,s,t,u,v,w,x,y,z$$
$$A,B,C,D,E,F,G,H,I,J,K,L,M,N,O,P,Q,R,S,T,U,V,W,X,Y,Z$$

which find their way into every branch of mathematics and physics, and permit us to write equations such as

$$E = mc^2$$

and

$$A = \pi r^2.$$

It is important that when we see an equation, we are able to read it as part of the text. In the case of $E = mc^2$, this is read as 'E equals m, c squared', where E stands for energy, m for mass, c the speed of light, which is multiplied by itself. In the case of $A = \pi r^2$, this is read as 'A equals pi, r squared', where A stands for area, π the ratio of a circle's circumference to its diameter, and r the circle's radius. Greek symbols, which happen to look nice and impressive, have also found their way into many equations, and often disrupt the flow of reading, simply because we don't know their English names. For example, the English theoretical physicist Paul Dirac (1902–1984), derived an equation for a moving electron using the symbols α_i and β, which are 4 × 4 matrices, where

$$\alpha_i \beta + \beta \alpha_i = 0$$

and is read as

'the sum of the products alpha-i beta, and beta alpha-i, equals zero.'
Although we do not come across moving electrons in this book, we do have to be familiar with the following Greek symbols:

α	alpha	ν	nu
β	beta	ξ	xi

γ	gamma	o	omicron
δ	delta	π	pi
ϵ	epsilon	ρ	rho
ζ	zeta	σ	sigma
η	eta	τ	tau
θ	theta	υ	upsilon
ι	iota	ϕ	phi
κ	kappa	χ	chi
λ	lambda	ψ	psi
μ	mu	ω	omega

and some upper-case symbols:

Γ	Gamma	Σ	Sigma
Δ	Delta	Υ	Upsilon
Θ	Theta	Φ	Phi
Λ	Lambda	Ψ	Psi
Ξ	Xi	Ω	Omega
Π	Pi		

Being able to read an equation does not mean that we understand it—but we are a little closer than just being able to stare at a jumble of symbols! Therefore, in future, when I introduce a new mathematical object, I will tell you how it should be read.

References

1. Dauben JW (1979) Georg Cantor his mathematics and philosophy of the infinite. Princeton University Press, Princeton
2. Dauben JW (2004) Georg Cantor and the battle for transfinite set theory (PDF). In: Proceedings of the 9th ACMS conference (Westmont College, Santa Barbara, Calif.), pp 1–22
3. Rodych V (2007) Wittgenstein's philosophy of mathematics. In: Zalta EN (ed) The Stanford encyclopedia of philosophy. Metaphysics Research Lab, Stanford University

Chapter 2
Numbers

2.1 Introduction

This chapter revises some basic ideas about counting and number systems, and how they are employed in the context of mathematics for computer graphics. Omit this chapter, if you are familiar with the subject.

2.2 Background

Over the centuries mathematicians have realised that in order to progress, they must give precise definitions to their discoveries, ideas and concepts, so that they can be built upon and referenced by new mathematical inventions. In the event of any new discovery, these rrrdefinitions have to be occasionally changed or extended. For example, once upon a time integers, rational and irrational numbers, satisfied all the needs of mathematicians, until imaginary quantities were invented. Today, complex numbers have helped shape the current number system hierarchy. Consequently, there must be clear definitions for numbers, and the operators that act upon them. Therefore, we need to identify the types of numbers that exist, what they are used for, and any problems that arise when they are stored in a computer.

2.3 Counting

Our brain's visual cortex possesses some incredible image processing features. For example, children know instinctively when they are given less sweets than another child, and adults know instinctively when they are short-changed by a Parisian taxi driver, or driven around the Arc de Triumph several times, on the way to the airport! Intuitively, we can assess how many donkeys are in a field without counting them,

© Springer-Verlag London Ltd., part of Springer Nature 2022
J. Vince, *Mathematics for Computer Graphics*, Undergraduate Topics
in Computer Science, https://doi.org/10.1007/978-1-4471-7520-9_2

and generally, we seem to know within a second or two, whether there are just a few, dozens, or hundreds of something. But when accuracy is required, one can't beat counting. But what is counting?

Well normally, we are taught to count by our parents by memorising first, the counting words 'one, two, three, four, five, six, seven, eight, nine, ten, ..' and second, associating them with our fingers, so that when asked to count the number of donkeys in a picture book, each donkey is associated with a counting word. When each donkey has been identified, the number of donkeys equals the last word mentioned. However, this still assumes that we know the meaning of 'one, two, three, four, ..' etc. Memorising these counting words is only part of the problem—getting them in the correct sequence is the real challenge. The incorrect sequence 'one, two, five, three, nine, four, ..' etc., introduces an element of randomness into any calculation, but practice makes perfect, and it's useful to master the correct sequence before going to university!

2.4   Sets of Numbers

A *set* is a collection of arbitrary objects called its *elements* or *members*. For example, each system of number belongs to a set with given a name, such as $\mathbb{N}$ for the natural numbers, $\mathbb{R}$ for real numbers, and $\mathbb{Q}$ for rational numbers. When we want to indicate that something is whole, real or rational, etc., we use the notation:

$$n \in \mathbb{N}$$

which reads 'n is a member of ($\in$) the set $\mathbb{N}$', i.e. n is a whole number. Similarly:

$$x \in \mathbb{R}$$

stands for 'x is a real number.'

A *well-ordered set* possesses a unique order, such as the natural numbers $\mathbb{N}$. Therefore, if P is the well-ordered set of prime numbers and $\mathbb{N}$ is the well-ordered set of natural numbers, we can write:

$$P = \{2, 3, 5, 7, 11, 13, 17, 19, 23, 29, 31, 37, 41, 43, 47, \ldots\}$$
$$\mathbb{N} = \{1, 2, 3, 4, 5, 6, 7, 8, 9, 10, 11, 12, 13, 14, 15, 16, 17, \ldots\}.$$

By pairing the prime numbers in P with the numbers in $\mathbb{N}$, we have:

$$\{\{2, 1\}, \{3, 2\}, \{5, 3\}, \{7, 4\}, \{11, 5\}, \{13, 6\}, \{17, 7\}, \{19, 8\}, \{23, 9\}, \ldots\}$$

and we can reason that 2 is the 1st prime, and 3 is the 2nd prime, etc. However, we still have to declare what we mean by 1, 2, 3, 4, 5, ... etc., and without getting too philosophical, I like the idea of defining them as follows. The word 'one', represented by 1, stands for 'oneness' of anything: one finger, one house, one tree, one donkey, etc. The word 'two', represented by 2, is 'one more than one'. The word 'three', represented by 3, is 'one more than two', and so on.

We are now in a position to associate some mathematical notation with our numbers by introducing the $+$ and $=$ signs. We know that $+$ means add, but it also can stand for 'more'. We also know that $=$ means equal, and it can also stand for 'is the same as'. Thus the statement:

$$2 = 1 + 1$$

is read as 'two is the same as one more than one.'

We can also write:

$$3 = 1 + 2$$

which is read as 'three is the same as one more than two.' But as we already have a definition for 2, we can write

$$3 = 1 + 2$$
$$= 1 + 1 + 1.$$

Developing this idea, and including some extra combinations, we have:

$$2 = 1 + 1$$
$$3 = 1 + 2$$
$$4 = 1 + 3 = 2 + 2$$
$$5 = 1 + 4 = 2 + 3$$
$$6 = 1 + 5 = 2 + 4 = 3 + 3$$
$$7 = 1 + 6 = 2 + 5 = 3 + 4$$
$$\text{etc.}$$

and can be continued without limit. These numbers, 1, 2, 3, 4, 5, 6, etc., are called *natural numbers*, and are the set $\mathbb{N}$.

2.5 Zero

The concept of zero has a well-documented history, which shows that it has been used by different cultures over a period of two-thousand years or more. It was the Indian mathematician and astronomer Brahmagupta (598-c.–670), who argued that zero was just as valid as any natural number, with the definition: *the result of subtracting any number from itself.* However, even today, there is no universal agreement as to whether zero belongs to the set $\mathbb{N}$, consequently, the set $\mathbb{N}^0$ stands for the set of natural numbers including zero.

In today's positional decimal system, which is a *place value system*, the digit 0 is a placeholder. For example, 203 stands for: two hundreds, no tens and three units. Although $0 \in \mathbb{N}^0$, it does have special properties that distinguish it from other members of the set, and Brahmagupta also gave rules showing this interaction.

If $x \in \mathbb{N}^0$, then the following rules apply:

$$\begin{aligned}
\text{addition:} \quad & x + 0 = x \\
\text{subtraction:} \quad & x - 0 = x \\
\text{multiplication:} \quad & x \times 0 = 0 \times x = 0 \\
\text{division:} \quad & 0/x = 0 \\
\text{undefined division:} \quad & x/0.
\end{aligned}$$

The expression $0/0$ is called an *indeterminate form*, as it is possible to show that under different conditions, especially limiting conditions, it can equal anything. So for the moment, we will avoid using it until we cover calculus.

2.6 Negative Numbers

When negative numbers were first proposed, they were not accepted with open arms, as it was difficult to visualise -5 of something. For instance, if there are 5 donkeys in a field, and they are all stolen to make salami, the field is now empty, and there is nothing we can do in the arithmetic of donkeys to create a field of -5 donkeys. However, in applied mathematics, numbers have to represent all sorts of quantities such as temperature, displacement, angular rotation, speed, acceleration, etc., and we also need to incorporate ideas such as left and right, up and down, before and after, forwards and backwards, etc. Fortunately, negative numbers are perfect for representing all of the above quantities and ideas.

Consider the expression $4 - x$, where $x \in \mathbb{N}^0$. When x takes on certain values, we have

$$\begin{aligned}
4 - 1 &= 3 \\
4 - 2 &= 2 \\
4 - 3 &= 1 \\
4 - 4 &= 0
\end{aligned}$$

and unless we introduce negative numbers, we are unable to express the result of $4 - 5$. Consequently, negative numbers are visualised as shown in Fig. 2.1, where the *number line* shows negative numbers to the left of the natural numbers, which are *positive*, although the $+$ sign is omitted for clarity.

Moving from left to right, the number line provides a numerical continuum from large negative numbers, through zero, towards large positive numbers. In any

Fig. 2.1 The number line showing negative and positive numbers

calculations, we could agree that angles above the horizon are positive, and angles below the horizon, negative. Similarly, a movement forwards is positive, and a movement backwards is negative. So now we are able to write:

$$4 - 5 = -1$$
$$4 - 6 = -2$$
$$4 - 7 = -3$$
$$\text{etc.,}$$

without worrying about creating impossible conditions.

2.6.1 The Arithmetic of Positive and Negative Numbers

Once again, Brahmagupta compiled all the rules, Tables 2.1 and 2.2, supporting the addition, subtraction, multiplication and division of positive and negative numbers. The real fly in the ointment, being negative numbers, which cause problems for children, math teachers and occasional accidents for mathematicians. Perhaps, the one rule we all remember from our school days is that *two negatives make a positive*.

Another problem with negative numbers arises when we employ the square-root function. As the product of two positive or negative numbers results in a positive result, the square-root of a positive number gives rise to a positive **and** a negative answer. For example, $\sqrt{4} = \pm 2$. This means that the square-root function only applies to positive numbers. Nevertheless, it did not stop the invention of the *imaginary* object i, where $i^2 = -1$. However, i is not a number, but behaves like an operator, and is described later.

Table 2.1 Rules for adding and subtracting positive and negative numbers

$+$	b	$-b$	$-$	b	$-b$
a	$a + b$	$a - b$	a	$a - b$	$a + b$
$-a$	$b - a$	$-(a + b)$	$-a$	$-(a + b)$	$b - a$

Table 2.2 Rules for multiplying and dividing positive and negative numbers

$\times$	b	$-b$	$/$	b	$-b$
a	ab	$-ab$	a	a/b	$-a/b$
$-a$	$-ab$	ab	$-a$	$-a/b$	a/b

2.7   Observations and Axioms

The following *axioms* or laws provide a formal basis for mathematics, and in the following descriptions a *binary operation* is an arithmetic operation such as $+, -, \times, /$ which operate on two operands.

2.7.1   Commutative Law

The *commutative law* in algebra states that when two elements are linked through some binary operation, the result is independent of the order of the elements.

The commutative law of addition is

$$a + b = b + a$$
$$\text{e.g. } 1 + 2 = 2 + 1.$$

The commutative law of multiplication is

$$a \times b = b \times a$$
$$\text{e.g. } 1 \times 2 = 2 \times 1.$$

Note that subtraction is not commutative:

$$a - b \neq b - a$$
$$\text{e.g. } 1 - 2 \neq 2 - 1.$$

2.7.2   Associative Law

The *associative law* in algebra states that when three or more elements are linked together through a binary operation, the result is independent of how each pair of elements is grouped.

The associative law of addition is

$$a + (b + c) = (a + b) + c$$
$$\text{e.g. } 1 + (2 + 3) = (1 + 2) + 3.$$

The associative law of multiplication is

$$a \times (b \times c) = (a \times b) \times c$$

$$\text{e.g. } 1 \times (2 \times 3) = (1 \times 2) \times 3.$$

However, note that subtraction is not associative:

$$a - (b - c) \neq (a - b) - c$$
$$\text{e.g. } 1 - (2 - 3) \neq (1 - 2) - 3,$$

which may seem surprising, but at the same time confirms the need for clear axioms.

2.7.3 Distributive Law

The *distributive law* in algebra describes an operation, which when performed on a combination of elements is the same as performing the operation on the individual elements. The distributive law does not work in all cases of arithmetic. For example, multiplication over addition holds:

$$a(b + c) = ab + ac$$
$$\text{e.g. } 2(3 + 4) = 6 + 8$$

whereas addition over multiplication does not:

$$a + (b \times c) \neq (a + b) \times (a + c)$$
$$\text{e.g. } 3 + (4 \times 5) \neq (3 + 4) \times (3 + 5).$$

Although these laws are natural for numbers, they do not necessarily apply to all mathematical objects. For instance, the vector product, which multiplies two vectors together, is not commutative. The same applies for matrix multiplication.

2.8 The Base of a Number System

2.8.1 Background

Over recent millennia, mankind has invented and discarded many systems for representing number. People have counted on their fingers and toes, used pictures (hieroglyphics), cut marks on clay tablets (cuneiform symbols), employed Greek symbols (Ionic system) and struggled with, and abandoned Roman numerals (I, V, X, L, C, D, M, etc.), until we reach today's decimal place system, which has Hindu-Arabic and Chinese origins. And since the invention of computers we have witnessed the emergence of binary, octal and hexadecimal number systems, where 2, 8 and 16 respectively, replace the 10 in our decimal system.

The decimal number 23 stands for 'two tens and three units', and in English is written 'twenty-three', in French 'vingt-trois' (twenty-three), and in German 'dreiundzwanzig' (three and twenty). Let's investigate the algebra behind the decimal system and see how it can be used to represent numbers to any base. The expression:

$$a \times 1000 + b \times 100 + c \times 10 + d \times 1$$

where a, b, c, d take on any value between 0 and 9, describes any whole number between 0 and 9999. By including

$$e \times 0.1 + f \times 0.01 + g \times 0.001 + h \times 0.0001$$

where e, f, g, h take on any value between 0 and 9, any decimal number between 0 and 9999.9999 can be represented.

Indices bring the notation alive and reveal the true underlying pattern:

$$\ldots a10^3 + b10^2 + c10^1 + d10^0 + e10^{-1} + f10^{-2} + g10^{-3} + h10^{-4} \ldots$$

Remember that any number raised to the power 0 equals 1. By adding extra terms both left and right, any number can be accommodated.

In this example, 10 is the base, which means that the values of a to h range between 0 and 9, 1 less than the base. Therefore, by substituting B for the base we have

$$\ldots aB^3 + bB^2 + cB^1 + dB^0 + eB^{-1} + fB^{-2} + gB^{-3} + hB^{-4} \ldots$$

where the values of a to h range between 0 and $B - 1$.

2.8.2 Octal Numbers

The octal number system has $B = 8$, and a to h range between 0 and 7:

$$\ldots a8^3 + b8^2 + c8^1 + d8^0 + e8^{-1} + f8^{-2} + g8^{-3} + h8^{-4} \ldots$$

and the first 17 octal numbers are:

$$1_8, 2_8, 3_8, 4_8, 5_8, 6_8, 7_8, 10_8, 11_8, 12_8, 13_8, 14_8, 15_8, 16_8, 17_8, 20_8, 21_8.$$

The subscript 8 reminds us that although we may continue to use the words 'twenty-one', it is an octal number, and not a decimal. But what is 14_8 in decimal? Well, it stands for:

$$1 \times 8^1 + 4 \times 8^0 = 12.$$

Thus 356.4_8 is converted to decimal as follows:

$$(3 \times 8^2) + (5 \times 8^1) + (6 \times 8^0) + (4 \times 8^{-1})$$
$$(3 \times 64) + (5 \times 8) + (6 \times 1) + (4 \times 0.125)$$
$$(192 + 40 + 6) + (0.5)$$
$$238.5.$$

Counting in octal appears difficult, simply because we have never been exposed to it, like the decimal system. If we had evolved with 8 fingers, instead of 10, we would be counting in octal!

2.8.3 Binary Numbers

The binary number system has $B = 2$, and a to h are 0 or 1:

$$\ldots a2^3 + b2^2 + c2^1 + d2^0 + e2^{-1} + f2^{-2} + g2^{-3} + h2^{-4} \ldots$$

and the first 13 binary numbers are:

$1_2, 10_2, 11_2, 100_2, 101_2, 110_2, 111_2, 1000_2, 1001_2, 1010_2, 1011_2, 1100_2, 1101_2.$

Thus 11011.11_2 is converted to decimal as follows:

$$\left(1 \times 2^4\right) + \left(1 \times 2^3\right) + \left(0 \times 2^2\right) + \left(1 \times 2^1\right) + \left(1 \times 2^0\right) + \left(1 \times 2^{-1}\right) + \left(1 \times 2^{-2}\right)$$
$$(1 \times 16) + (1 \times 8) + (0 \times 4) + (1 \times 2) + (1 \times 0.5) + (1 \times 0.25)$$
$$(16 + 8 + 2) + (0.5 + 0.25)$$
$$26.75.$$

The reason why computers work with binary numbers—rather than decimal—is due to the difficulty of designing electrical circuits that can store decimal numbers in a stable fashion. A switch, where the open state represents 0, and the closed state represents 1, is the simplest electrical component to emulate. No matter how often it is used, or how old it becomes, it will always behave like a switch. The main advantage of electrical circuits is that they can be switched on and off trillions of times a second, and the only disadvantage is that the encoded binary numbers and characters contain a large number of bits, and humans are not familiar with binary.

2.8.4 Hexadecimal Numbers

The hexadecimal number system has $B = 16$, and a to h can be 0 to 15, which presents a slight problem, as we don't have 15 different numerical characters. Consequently, we use 0 to 9, and the letters A, B, C, D, E, F to represent 10, 11, 12, 13, 14, 15 respectively:

$$\ldots a16^3 + b16^2 + c16^1 + d16^0 + e16^{-1} + f16^{-2} + g16^{-3} + h16^{-4} \ldots$$

and the first 17 hexadecimal numbers are:

$$1_{16}, 2_{16}, 3_{16}, 4_{16}, 5_{16}, 6_{16}, 7_{16}, 8_{16}, 9_{16}, A_{16}, B_{16}, C_{16}, D_{16}, E_{16}, F_{16}, 10_{16}, 11_{16}.$$

Thus $1E.8_{16}$ is converted to decimal as follows:

$$(1 \times 16) + (E \times 1) + (8 \times 16^{-1})$$
$$(16 + 14) + (8/16)$$
$$30.5.$$

Although it is not obvious, binary, octal and hexadecimal numbers are closely related, which is why they are part of a programmer's toolkit. Even though computers work with binary, it's the last thing a programmer wants to use. So to simplify the man-machine interface, binary is converted into octal or hexadecimal. To illustrate this, let's convert the 16-bit binary code 1101011000110001 into octal.

Using the following general binary integer

$$a2^8 + b2^7 + c2^6 + d2^5 + e2^4 + f2^3 + g2^2 + h2^1 + i2^0$$

we group the terms into threes, starting from the right, because $2^3 = 8$:

$$\left(a2^8 + b2^7 + c2^6\right) + \left(d2^5 + e2^4 + f2^3\right) + \left(g2^2 + h2^1 + i2^0\right).$$

Simplifying:

$$2^6\left(a2^2 + b2^1 + c2^0\right) + 2^3\left(d2^2 + e2^1 + f2^0\right) + 2^0\left(g2^2 + h2^1 + i2^0\right)$$
$$8^2\left(a2^2 + b2^1 + c2^1\right) + 8^1\left(d2^2 + e2^1 + f2^0\right) + 8^0\left(g2^2 + h2^1 + i2^0\right)$$
$$8^2 R + 8^1 S + 8^0 T$$

where

$$R = a2^2 + b2^1 + c$$
$$S = d2^2 + e2^1 + f$$
$$T = g2^2 + h2^1 + i$$

and the values of R, S, T vary between 0 and 7. Therefore, given 1101011000 110001, we divide the binary code into groups of three, starting at the right, and adding two leading zeros:

$$(001)(101)(011)(000)(110)(001).$$

For each group, multiply the zeros and ones by 4, 2, 1, right to left:

$$(0 + 0 + 1)(4 + 0 + 1)(0 + 2 + 1)(0 + 0 + 0)(4 + 2 + 0)(0 + 0 + 1)$$
$$(1)(5)(3)(0)(6)(1)$$
$$153061_8.$$

Therefore, $11010110001100001_2 \equiv 153061_8$, ($\equiv$ stands for 'equivalent to') which is much more compact. The secret of this technique is to memorise these patterns:

$$000_2 \equiv 0_8$$
$$001_2 \equiv 1_8$$
$$010_2 \equiv 2_8$$
$$011_2 \equiv 3_8$$
$$100_2 \equiv 4_8$$
$$101_2 \equiv 5_8$$
$$110_2 \equiv 6_8$$
$$111_2 \equiv 7_8.$$

Here are a few more examples, with the binary digits grouped in threes:

$$111_2 \equiv 7_8$$
$$101\ 101_2 \equiv 55_8$$
$$100\ 000_2 \equiv 40_8$$
$$111\ 000\ 111\ 000\ 111_2 \equiv 70707_8.$$

It's just as easy to reverse the process, and convert octal into binary. Here are some examples:

$$567_8 \equiv 101\ 110\ 111_2$$
$$23_8 \equiv 010\ 011_2$$
$$1741_8 \equiv 001\ 111\ 100\ 001_2.$$

A similar technique is used to convert binary to hexadecimal, but this time we divide the binary code into groups of four, because $2^4 = 16$, starting at the right, and adding leading zeros, if necessary. To illustrate this, let's convert the 16-bit binary code 1101 0110 0011 0001 into hexadecimal.

Using the following general binary integer number

$$a2^{11} + b2^{10} + c2^9 + d2^8 + e2^7 + f2^6 + g2^5 + h2^4 + i2^3 + j2^2 + k2^1 + l2^0$$

from the right, we divide the binary code into groups of four:

$$\left(a2^{11} + b2^{10} + c2^9 + d2^8\right) + \left(e2^7 + f2^6 + g2^5 + h2^4\right) + \left(i2^3 + j2^2 + k2^1 + l2^0\right).$$

Simplifying:

$$2^8 \left(a2^3 + b2^2 + c2^1 + d2^0\right) + 2^4 \left(e2^3 + f2^2 + g2^1 + h2^0\right) + 2^0 \left(i2^3 + j2^2 + k2^1 + l2^0\right)$$
$$16^2 \left(a2^3 + b2^2 + c2^1 + d\right) + 16^1 \left(e2^3 + f2^2 + g2^1 + h\right) + 16^0 \left(i2^3 + j2^2 + k2^1 + l\right)$$
$$16^2 R + 16^1 S + 16^0 T$$

where

$$R = a2^3 + b2^2 + c2^1 + d$$
$$S = e2^3 + f2^2 + g2^1 + h$$
$$T = i2^3 + j2^2 + k2^1 + l$$

and the values of R, S, T vary between 0 and 15. Therefore, given 1101011000110001_2, we divide the binary code into groups of fours, starting at the right:

$$(1101)(0110)(0011)(0001)$$

For each group, multiply the zeros and ones by 8, 4, 2, 1 respectively, right to left:

$$(8 + 4 + 0 + 1)(0 + 4 + 2 + 0)(0 + 0 + 2 + 1)(0 + 0 + 0 + 1)$$
$$(13)(6)(3)(1)$$
$$D631_{16}.$$

Therefore, $1101\ 0110\ 0011\ 0001_2 \equiv D631_{16}$, which is even more compact than its octal value 153061_8.

I have deliberately used whole numbers in the above examples, but they can all be extended to include a fractional part. For example, when converting a binary number such as 11.1101_2 to octal, the groups are formed about the binary point:

$$(011).(110)(100) \equiv 3.64_8.$$

Similarly, when converting a binary number such as 101010.100110_2 to hexadecimal, the groups are also formed about the binary point:

$$(0010)(1010).(1001)(1000) \equiv 2A.98_{16}.$$

Table 2.3 shows the first twenty decimal, binary, octal and hexadecimal numbers.

2.8.5   Adding Binary Numbers

When we are first taught the addition of integers containing several digits, we are advised to solve the problem digit by digit, working from right to left. For example, to add 254 to 561 we write:

Table 2.3 The first twenty decimal, binary, octal, and hexadecimal numbers

decimal	binary	octal	hex	decimal	binary	octal	hex
1	1	1	1	11	1011	13	B
2	10	2	2	12	1100	14	C
3	11	3	3	13	1101	15	D
4	100	4	4	14	1110	16	E
5	101	5	5	15	1111	17	F
6	110	6	6	16	10000	20	10
7	111	7	7	17	10001	21	11
8	1000	10	8	18	10010	22	12
9	1001	11	9	19	10011	23	13
10	1010	12	A	20	10100	24	14

Table 2.4 Addition of two decimal integers showing the *carry*

+	0	1	2	3	4	5	6	7	8	9
0	0	1	2	3	4	5	6	7	8	9
1	1	2	3	4	5	6	7	8	9	^{10}
2	2	3	4	5	6	7	8	9	^{10}	^{11}
3	3	4	5	6	7	8	9	^{10}	^{11}	^{12}
4	4	5	6	7	8	9	^{10}	^{11}	^{12}	^{13}
5	5	6	7	8	9	^{10}	^{11}	^{12}	^{13}	^{14}
6	6	7	8	9	^{10}	^{11}	^{12}	^{13}	^{14}	^{15}
7	7	8	9	^{10}	^{11}	^{12}	^{13}	^{14}	^{15}	^{16}
8	8	9	^{10}	^{11}	^{12}	^{13}	^{14}	^{15}	^{16}	^{17}
9	9	^{10}	^{11}	^{12}	^{13}	^{14}	^{15}	^{16}	^{17}	^{18}

Table 2.5 Addition of two binary integers showing the *carry*

+	0	1
0	0	1
1	1	^{10}

$$561$$
$$254$$
$$\overline{815}$$

where $4 + 1 = 5, 5 + 6 = 1$ with a *carry* $= 1, 2 + 5 + carry = 8$.

Table 2.4 shows all the arrangements for adding two digits with the *carry* shown as $^{carry}n$. However, when adding binary numbers, the possible arrangements collapse to the four shown in Table 2.5, which greatly simplifies the process.

For example, to add 124 to 188 as two 16-bit binary integers, we write, showing the status of the *carry* bit:

$$
\begin{array}{ll}
0000000011111000 & carry \\
0000000010111100 & = 188 \\
0000000001111100 & = 124 \\
\hline
0000000100111000 & = 312
\end{array}
$$

Such addition is easily undertaken by digital electronic circuits, and instead of having separate circuitry for subtraction, it is possible to perform subtraction using the technique of *two's complement*.

2.8.6   Subtracting Binary Numbers

Two's complement is a technique for converting a binary number into a form such that when it is added to another binary number, it results in a subtraction. There are two stages to the conversion: inversion, followed by the addition of 1. For example, 24 in binary is 0000000000110000, and is inverted by switching every 1 to 0, and *vice versa*: 1111111111100111. Next, we add 1: 1111111111101000, which now represents −24. If this is added to binary 36: 0000000000100100, we have

$$
\begin{array}{ll}
0000000000100100 & = +36 \\
1111111111101000 & = -24 \\
\hline
0000000000001100 & = +12
\end{array}
$$

Note that the last high-order addition creates a *carry* of 1, which is ignored. Here is another example, 100 − 30:

$$
\begin{array}{lll}
& 0000000000011110 & = +30 \\
inversion & 1111111111100001 & \\
add\ 1 & 0000000000000001 & \\
\hline
& 1111111111100010 & = -30 \\
add\ 100 & 0000000001100100 & = +100 \\
\hline
& 0000000001000110 & = +70
\end{array}
$$

2.9   Types of Numbers

As mathematics evolved, mathematicians introduced different types of numbers to help classify equations and simplify the language employed to describe their work. These are the various types and their set names.

2.9.1 Natural Numbers

The *natural numbers* $\{1, 2, 3, 4, \ldots\}$ are used for counting, ordering and labelling and represented by the set $\mathbb{N}$. When zero is included, $\mathbb{N}^0$ or $\mathbb{N}_0$ is used:

$$\mathbb{N}^0 = \mathbb{N}_0 = \{0, 1, 2, \ldots\}.$$

Note that negative numbers are not included. Natural numbers are used to subscript a quantity to distinguish one element from another, e.g. $x_1, x_2, x_3, x_4, \ldots$.

2.9.2 Integers

Integer numbers include the natural numbers, both positive and negative, and zero, and are represented by the set $\mathbb{Z}$:

$$\mathbb{Z} = \{\ldots, -2, -1, 0, 1, 2, 3, \ldots\}.$$

The reason for using $\mathbb{Z}$ is because the German for whole number is *ganzen Zahlen*. Leopold Kronecker apparently criticised Georg Cantor for his work on set theory with the jibe: '*Die ganzen Zahlen hat der liebe Gott gemacht, alles andere ist Menschenwerk*', which translates: '*God made the integers, and all the rest is man's work*', implying that the rest are artificial. However, Cantor's work on set theory and transfinite numbers proved to be far from artificial.

2.9.3 Rational Numbers

Any number that equals the quotient of one integer divided by another non-zero integer, is a *rational number*, and represented by the set $\mathbb{Q}$. For example, 2, $\sqrt{16}$, 0.25 are rational numbers because

$$2 = 4/2$$
$$\sqrt{16} = \pm 4 = \pm 8/2$$
$$0.25 = 1/4.$$

Some rational numbers can be stored accurately inside a computer, but many others can only be stored approximately. For example, 4/3 produces an infinite sequence of threes $1.333333\ldots$ and is truncated when stored as a binary number.

2.9.4   *Irrational Numbers*

An *irrational number* cannot be expressed as the quotient of two integers. Irrational numbers never terminate, nor contain repeated sequences of digits, consequently, they are always subject to a small error when stored within a computer. Examples are:

$$\sqrt{2} = 1.41421356\ldots$$
$$\phi = 1.61803398\ldots\text{(golden section)}$$
$$e = 2.71828182\ldots$$
$$\pi = 3.14159265\ldots$$

2.9.5   *Real Numbers*

Rational and irrational numbers comprise the set of *real numbers* $\mathbb{R}$. Examples are 1.5, 0.004, 12.999 and 23.0.

2.9.6   *Algebraic and Transcendental Numbers*

Polynomial equations with rational coefficients have the form:

$$f(x) = ax^n + bx^{n-1} + cx^{n-2} \ldots + C$$

such as

$$y = 3x^2 + 2x - 1$$

and their roots belong to the set of *algebraic numbers* $\mathbb{A}$. A consequence of this definition implies that all rational numbers are algebraic, since if

$$x = \frac{p}{q}$$

then

$$qx - p = 0$$

which is a polynomial. Numbers that are not roots to polynomial equations are *transcendental numbers* and include most irrational numbers, but not $\sqrt{2}$, since if

$$x = \sqrt{2}$$

then

$$x^2 - 2 = 0$$

which is a polynomial.

2.9.7 Imaginary Numbers

Imaginary numbers were invented to resolve problems where an equation such as $x^2 + 16 = 0$, has no real solution (roots). The simple idea of declaring the existence of a quantity i, such that $i^2 = -1$, permits the solution to be expressed as

$$x = \pm 4i.$$

For example, if $x = 4i$ we have

$$\begin{aligned} x^2 + 16 &= 16i^2 + 16 \\ &= -16 + 16 \\ &= 0 \end{aligned}$$

and if $x = -4i$ we have

$$\begin{aligned} x^2 + 16 &= 16i^2 + 16 \\ &= -16 + 16 \\ &= 0. \end{aligned}$$

But what is i? In 1637, the French mathematician René Descartes (1596–1650), published *La Géométrie*, in which he stated that numbers incorporating $\sqrt{-1}$ were '*imaginary*', and for centuries this label has stuck. Unfortunately, it was a derogatory remark, as there is nothing '*imaginary*' about i—it *simply* is an object that when introduced into various algebraic expressions, reveals some amazing underlying patterns. i is not a number in the accepted sense, it is a mathematical object or construct that squares to -1. In some respects it is like time, which probably does not really exist, but is useful in describing the universe. However, i does lose its mystery when interpreted as a rotational operator, which we investigate below.

As $i^2 = -1$ then it must be possible to raise i to other powers. For example,

$$i^4 = i^2 i^2 = 1$$

and

$$i^5 = i i^4 = i.$$

Table 2.6 shows the sequence up to i^6.

Table 2.6 Increasing powers of i

i^0	i^1	i^2	i^3	i^4	i^5	i^6
1	i	-1	$-i$	1	i	-1

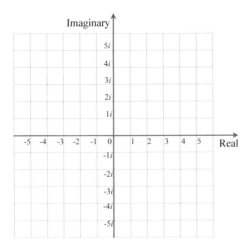

Fig. 2.2 The complex plane

This cyclic pattern is quite striking, and reminds one of a similar pattern:

$$(x, \ y, \ -x, \ -y, \ x, \ldots)$$

that arises when rotating around the Cartesian axes in a anticlockwise direction. Such a similarity cannot be ignored, for when the real number line is combined with a vertical imaginary axis, it creates the *complex plane*, as shown in Fig. 2.2.

The above sequence is summarised as

$$i^{4n} = 1$$
$$i^{4n+1} = i$$
$$i^{4n+2} = -1$$
$$i^{4n+3} = -i$$

where $n \in \mathbb{N}^0$.

But what about negative powers? Well they, too, are also possible. Consider i^{-1}, which is evaluated as follows:

$$i^{-1} = \frac{1}{i} = \frac{1(-i)}{i(-i)} = \frac{-i}{1} = -i.$$

Table 2.7 Decreasing powers of i

i^0	i^{-1}	i^{-2}	i^{-3}	i^{-4}	i^{-5}	i^{-6}
1	$-i$	-1	i	1	$-i$	-1

Similarly,

$$i^{-2} = \frac{1}{i^2} = \frac{1}{-1} = -1$$

and

$$i^{-3} = i^{-1}i^{-2} = -i(-1) = i.$$

Table 2.7 shows the sequence down to i^{-6}.

This time the cyclic pattern is reversed and is similar to the pattern

$$(x, \ -y, \ -x, \ y, \ x, \ldots)$$

that arises when rotating around the Cartesian axes in a clockwise direction.

Now let's investigate how a real number behaves when it is repeatedly multiplied by i. Starting with the number 3, we have:

$$i \times 3 = 3i$$
$$i \times 3i = -3$$
$$i \times (-3) = -3i$$
$$i \times (-3)i = 3.$$

So the cycle is $(3, \ 3i, \ -3, \ -3i, \ 3, \ 3i, \ -3, \ -3i, \ 3, \ldots)$, which has four steps, as shown in Fig. 2.3.

Fig. 2.3 The cycle of points created by repeatedly multiplying 3 by i

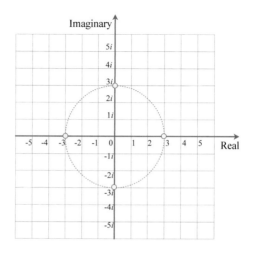

Imaginary objects occur for all sorts of reasons. For example, consider the statements

$$AB = -BA$$
$$BA = -AB$$

where A and B are two undefined objects that obey the associative law, but not the commutative law, and $A^2 = B^2 = 1$. The operation $(AB)^2$ reveals

$$(AB)(AB) = A(BA)B$$
$$= -A(AB)B$$
$$= -(A^2)(B^2)$$
$$= -1$$

which means that the product AB is imaginary. Such objects, which can be matrices, are useful in describing the behaviour of sub-atomic particles.

2.9.8 Complex Numbers

A *complex number* has a real and imaginary part: $z = a + ib$, and represented by the set $\mathbb{C}$:

$$z = a + bi \qquad z \in \mathbb{C}, \quad a, b \in \mathbb{R}, \quad i^2 = -1.$$

Some examples are

$$z = 1 + i$$
$$z = 3 - 2i$$
$$z = -23 + \sqrt{23}i.$$

Complex numbers obey all the normal laws of algebra. For example, if we multiply $(a + bi)$ by $(c + di)$ we have

$$(a + bi)(c + di) = ac + adi + bci + bdi^2.$$

Collecting up like terms and substituting -1 for i^2 we get

$$(a + bi)(c + di) = ac + (ad + bc)i - bd$$

which simplifies to

$$(a + bi)(c + di) = ac - bd + (ad + bc)i$$

which is another complex number.

Something interesting happens when we multiply a complex number by its *complex conjugate*, which is the same complex number but with the sign of the imaginary part reversed:

$$(a + bi)(a - bi) = a^2 - abi + bai - b^2i^2.$$

Collecting up like terms and simplifying we obtain

$$(a + bi)(a - bi) = a^2 + b^2$$

which is a real number, as the imaginary part has been cancelled out by the action of the complex conjugate.

Figure 2.4 shows how complex numbers are represented graphically using the complex plane.

For example, the complex number $P = 4 + 3i$ in Fig. 2.4 is rotated 90° to Q by multiplying it by i. Let's do this, and remember that $i^2 = -1$:

$$i(4 + 3i) = 4i + 3i^2$$
$$= 4i - 3$$
$$= -3 + 4i.$$

The point $Q = -3 + 4i$ is rotated 90° to R by multiplying it by i:

$$i(-3 + 4i) = -3i + 4i^2$$
$$= -3i - 4$$
$$= -4 - 3i.$$

Fig. 2.4 The complex plane showing four complex numbers

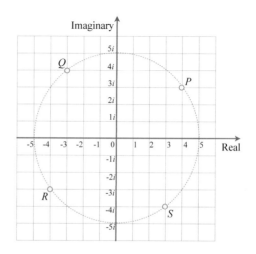

The point $R = -4 - 3i$ is rotated $90°$ to S by multiplying it by i:

$$i(-4 - 3i) = -4i - 3i^2$$
$$= -4i + 3$$
$$= 3 - 4i.$$

Finally, the point $S = 3 - 4i$ is rotated $90°$ back to P by multiplying it by i:

$$i(3 - 4i) = 3i - 4i^2$$
$$= 3i + 4$$
$$= 4 + 3i.$$

As you can see, complex numbers are intimately related to Cartesian coordinates, in that the ordered pair $(x, y) \equiv (x + yi)$.

2.9.9   Transcendental and Algebraic Numbers

Given a polynomial built from integers, for example

$$y = 3x^3 - 4x^2 + x + 23,$$

if the result is an integer, it is called an *algebraic number*, otherwise it is a *transcendental number*. Familiar examples of the latter being $\pi = 3.141\ 592\ 653\ldots$, and $e = 2.718\ 281\ 828\ldots$, which can be represented as various continued fractions:

$$\pi = \cfrac{4}{1 + \cfrac{1^2}{2 + \cfrac{3^2}{2 + \cfrac{5^2}{2 + \cfrac{7^2}{2 + \ldots}}}}}$$

$$e = 2 + \cfrac{1}{1 + \cfrac{1}{2 + \cfrac{1}{1 + \cfrac{1}{1 + \cfrac{1}{4 + \ldots}}}}}$$

2.9.10 Infinity

The term *infinity* is used to describe the size of unbounded systems. For example, there is no end to prime numbers: i.e. they are infinite; so, too, are the sets of other numbers. Consequently, no matter how we try, it is impossible to visualise the size of infinity. Nevertheless, this did not stop Georg Cantor from showing that one infinite set could be infinitely larger than another.

Cantor distinguished between those infinite number sets that could be 'counted', and those that could not. For Cantor, counting meant the one-to-one correspondence of a natural number with the members of another infinite set. If there is a clear correspondence, without leaving any gaps, then the two sets shared a common infinite size, called its *cardinality* using the first letter of the Hebrew alphabet aleph: $\aleph$. The cardinality of the natural numbers $\mathbb{N}$ is $\aleph_0$, called aleph-zero.

Cantor discovered a way of representing the rational numbers as a grid, which is traversed diagonally, back and forth, as shown in Fig. 2.5. Some ratios appear several times, such as $\frac{2}{2}, \frac{3}{3}$ etc., which are not counted. Nevertheless, the one-to-one correspondence with the natural numbers means that the cardinality of rational numbers is also $\aleph_0$.

A real surprise was that there are infinitely more transcendental numbers than natural numbers. Furthermore, there are an infinite number of cardinalities rising to $\aleph_\aleph$. Cantor had been alone working in this esoteric area, and as he published his results, he shook the very foundations of mathematics, which is why he was treated so badly by his fellow mathematicians.

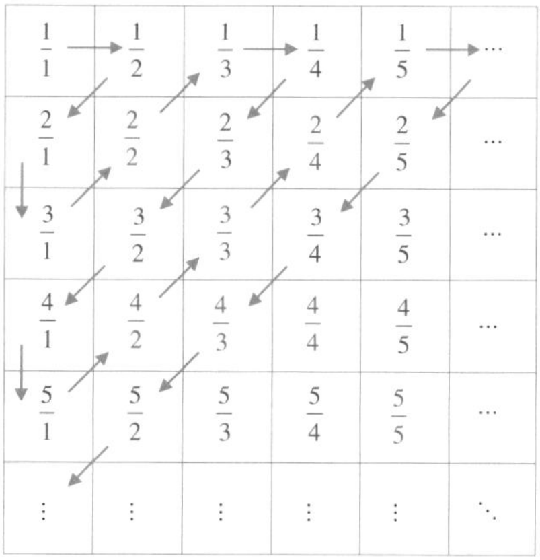

Fig. 2.5 Rational number grid

2.10 Summary

Apart from the natural numbers, integers, rational, irrational, prime, real and complex numbers, there are also Fermat, Mersenne, amicable, chromic, cubic, Fibonacci, pentagonal, perfect, random, square and tetrahedral numbers, which although equally interesting, don't concern us in this text.

Now that we know something about some important number sets, let's revise some ideas behind algebra.

2.11 Worked Examples

2.11.1 Algebraic Expansion

Expand $(a + b)(c + d)$, $(a - b)(c + d)$, and $(a - b)(c - d)$.

$$(a + b)(c + d) = a(c + d) + b(c + d)$$
$$= ac + ad + bc + bd.$$
$$(a - b)(c + d) = a(c + d) - b(c + d)$$
$$= ac + ad - bc - bd.$$
$$(a - b)(c - d) = a(c - d) - b(c - d)$$
$$= ac - ad - bc + bd.$$

2.11.2 Binary Subtraction

Using two's complement, subtract 12 from 50.

	0000000000001100	$= +12$
inversion	1111111111110011	
add 1	0000000000000001	
	1111111111110100	$= -12$
add 50	0000000000110010	$= +50$
	0000000000100110	$= +38$

2.11.3 Complex Numbers

Compute $(3 + 2i) + (2 + 2i) + (5 - 3i)$ and $(3 + 2i)(2 + 2i)(5 - 3i)$.

$$(3 + 2i) + (2 + 2i) + (5 - 3i) = 10 + i.$$

$$
\begin{aligned}
(3 + 2i)(2 + 2i)(5 - 3i) &= (3 + 2i)(10 - 6i + 10i + 6) \\
&= (3 + 2i)(16 + 4i) \\
&= 48 + 12i + 32i - 8 \\
&= 40 + 44i.
\end{aligned}
$$

2.11.4 Complex Rotation

Rotate the complex point $(3 + 2i)$ by $\pm 90°$ and $\pm 180°$.
To rotate $+90°$ (anticlockwise) multiply by i.

$$i(3 + 2i) = (3i - 2) = (-2 + 3i).$$

To rotate $-90°$ (clockwise) multiply by $-i$.

$$-i(3 + 2i) = (-3i + 2) = (2 - 3i).$$

To rotate $+180°$ (anticlockwise) multiply by -1.

$$-1(3 + 2i) = (-3 - 2i).$$

To rotate $-180°$ (clockwise) multiply by -1.

$$-1(3 + 2i) = (-3 - 2i).$$

Chapter 3
Algebra

3.1 Introduction

Some people, including me, find learning a foreign language a real challenge; one of the reasons being the inconsistent rules associated with its syntax. For example, why is a table feminine in French, 'la table', and a bed masculine, 'le lit'? They both have four legs! The rules governing natural language are continuously being changed by each generation, whereas mathematics appears to be logical and consistent. The reason for this consistency is due to the rules associated with numbers and the way they are combined together and manipulated at an abstract level. Such rules, or *axioms*, generally make our life easy, however, as we saw with the invention of negative numbers, extra rules have to be introduced, such as 'two negatives make a positive', which is easily remembered. However, as we explore mathematics, we discover all sorts of inconsistencies, such as there is no real value associated with the square-root of a negative number. It's forbidden to divide a number by zero. Zero divided by zero gives inconsistent results. Nevertheless, such conditions are easy to recognise and avoided. At least in mathematics, we don't have to worry about masculine and feminine numbers!

As a student, I discovered *Principia Mathematica* [1], a three-volume work written by the British philosopher, logician, mathematician and historian Bertrand Russell (1872–1970), and the British mathematician and philosopher Alfred North White-head (1861–1947), in which the authors attempt to deduce all of mathematics using the axiomatic system developed by the Italian mathematician Giuseppe Peano (1858–1932). The first volume established type theory, the second was devoted to numbers, and the third to higher mathematics. The authors did intend a fourth volume on geometry, but it was too much effort to complete. It made extremely intense reading. In fact, I never managed to get pass the first page! It took the authors almost 100 pages of deep logical analysis in the second volume to prove that $1 + 1 = 2$!

© Springer-Verlag London Ltd., part of Springer Nature 2022
J. Vince, *Mathematics for Computer Graphics*, Undergraduate Topics
in Computer Science, https://doi.org/10.1007/978-1-4471-7520-9_3

Russell wrote in *The Principles of Mathematics* [2]:

> The fact that all Mathematics is Symbolic Logic is one of the greatest discoveries of our age; and when this fact has been established, the remainder of the principles of mathematics consists in the analysis of Symbolic Logic itself.

Unfortunately, this dream cannot be realised, for in 1931, the Austrian-born, and later American logician and mathematician Kurt Gödel (1906–1978), showed that even though mathematics is based upon a formal set of axioms, there will always be statements involving natural numbers that cannot be proved or disproved. Furthermore, a consistent axiomatic system cannot demonstrate its own consistency. These theorems are known as Gödel's *incompleteness theorems*.

Even though we start off with some simple axioms, it does not mean that everything discovered in mathematics is provable, which does not mean that we cannot continue our every-day studies using algebra to solve problems. So let's examine the basic rules of algebra and prepare ourselves for the following chapters.

3.2 Background

Modern algebraic notation has evolved over thousands of years where different civilisations developed ways of annotating mathematical and logical problems. The word 'algebra' comes from the Arabic 'al-jabr w'al-muqabal' meaning 'restoration and reduction'. In retrospect, it does seem strange that centuries passed before the 'equals' sign (=) was invented, and concepts such as 'zero' (CE 876) were introduced, especially as they now seem so important. But we are not at the end of this evolution, because new forms of annotation and manipulation will continue to emerge as new mathematical objects are invented.

One fundamental concept of algebra is the idea of giving a name to an unknown quantity. For example, m is often used to represent the slope of a 2D line, and c is the line's y-coordinate where it intersects the y-axis. René Descartes formalised the idea of using letters from the beginning of the alphabet $(a, b, c, \ldots)$ to represent arbitrary quantities, and letters at the end of the alphabet $(p, q, r, s, t, \ldots, x, y, z)$ to represent quantities such as pressure (p), time (t) and coordinates (x, y, z).

With the aid of the basic arithmetic operators: $+, -, \times, /$ we can develop expressions that describe the behaviour of a physical process or a logical computation. For example, the expression $ax + by - d$ equals zero for a straight line. The variables x and y are the coordinates of any point on the line and the values of a, b and d determine the position and orientation of the line. The $=$ sign permits the line equation to be expressed as a self-evident statement:

$$0 = ax + by - d.$$

Such a statement implies that the expressions on the left- and right-hand sides of the $=$ sign are 'equal' or 'balanced', and in order to maintain equality or balance, whatever is done to one side, must also be done to the other. For example, adding d to both sides, the straight-line equation becomes

$$d = ax + by.$$

Similarly, we could double or treble both expressions, divide them by 4, or add 6, without disturbing the underlying relationship. When we are first taught algebra, we are often given the task of rearranging a statement to make different variables the subject. For example, (3.1) can be rearranged such that x is the subject:

$$y = \frac{x + 4}{2 - \dfrac{1}{z}} \tag{3.1}$$

$$y\left(2 - \frac{1}{z}\right) = x + 4$$

$$x = y\left(2 - \frac{1}{z}\right) - 4.$$

Making z the subject requires more effort:

$$y = \frac{x + 4}{2 - \dfrac{1}{z}}$$

$$y\left(2 - \frac{1}{z}\right) = x + 4$$

$$2y - \frac{y}{z} = x + 4$$

$$2y - x - 4 = \frac{y}{z}$$

$$z = \frac{y}{2y - x - 4}.$$

Parentheses are used to isolate part of an expression in order to select a sub-expression that is manipulated in a particular way. For example, the parentheses in $c(a + b) + d$ ensure that the variables a and b are added together before being multiplied by c, and finally added to d.

3.2.1 *Solving the Roots of a Quadratic Equation*

Problem solving is greatly simplified if one has solved it before, and having a good memory is always an advantage. In mathematics, we keep coming across problems that have been encountered before, apart from different numbers. For example,

$(a + b)(a - b)$ always equals $a^2 - b^2$, therefore factorising the following is a trivial exercise:

$$a^2 - 16 = (a + 4)(a - 4)$$
$$x^2 - 49 = (x + 7)(x - 7)$$
$$x^2 - 2 = \left(x + \sqrt{2}\right)\left(x - \sqrt{2}\right).$$

A perfect square has the form:

$$a^2 + 2ab + b^2 = (a + b)^2.$$

Consequently, factorising the following is also a trivial exercise:

$$a^2 + 4ab + 4b^2 = (a + 2b)^2$$
$$x^2 + 14x + 49 = (x + 7)^2$$
$$x^2 - 20x + 100 = (x - 10)^2.$$

Now let's solve the roots of the quadratic equation $ax^2 + bx + c = 0$, i.e. those values of x that make the equation equal zero. As the equation involves an x^2 term, we will exploit any opportunity to factorise it. We begin with the quadratic where $a \neq 0$:

$$ax^2 + bx + c = 0.$$

Step 1: Subtract c from both sides to begin the process of creating a perfect square:

$$ax^2 + bx = -c.$$

Step 2: Divide both sides by a to create an x^2 term:

$$x^2 + \frac{b}{a}x = -\frac{c}{a}.$$

Step 3: Add $b^2/4a^2$ to both sides to create a perfect square on the left side:

$$x^2 + \frac{b}{a}x + \frac{b^2}{4a^2} = \frac{b^2}{4a^2} - \frac{c}{a}.$$

Step 4: Factorise the left side:

$$\left(x + \frac{b}{2a}\right)^2 = \frac{b^2}{4a^2} - \frac{c}{a}.$$

Step 5: Make $4a^2$ the common denominator for the right side:

$$\left(x + \frac{b}{2a}\right)^2 = \frac{b^2 - 4ac}{4a^2}.$$

Step 6: Take the square root of both sides:

$$x + \frac{b}{2a} = \frac{\pm\sqrt{b^2 - 4ac}}{2a}.$$

Step 7: Subtract $b/2a$ from both sides:

$$x = \frac{\pm\sqrt{b^2 - 4ac}}{2a} - \frac{b}{2a}.$$

Step 8: Rearrange the right side:

$$x = \frac{-b \pm \sqrt{b^2 - 4ac}}{2a}$$

which provides the roots for any quadratic equation.

The discriminant $\sqrt{b^2 - 4ac}$ may be positive, negative or zero. A positive value reveals two real roots:

$$x_1 = \frac{-b + \sqrt{b^2 - 4ac}}{2a}, \quad x_2 = \frac{-b - \sqrt{b^2 - 4ac}}{2a}. \tag{3.2}$$

A negative value reveals two complex roots:

$$x_1 = \frac{-b + i\sqrt{|b^2 - 4ac|}}{2a}, \quad x_2 = \frac{-b - i\sqrt{|b^2 - 4ac|}}{2a}.$$

And a zero value reveals a single root:

$$x = \frac{-b}{2a}.$$

For example, Fig. 3.1 shows the graph of $y = x^2 + x - 2$, where we can see that $y = 0$ at two points: $x = -2$ and $x = 1$. In this equation

$$a = 1$$
$$b = 1$$
$$c = -2$$

Fig. 3.1 Graph of
$y = x^2 + x - 2$

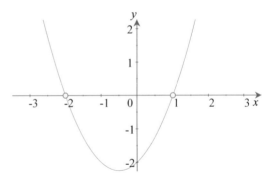

Fig. 3.2 Graph of
$y = x^2 + x + 1$

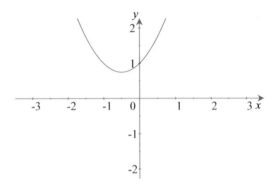

which when plugged into (3.2) confirms the graph:

$$x_1 = \frac{-1 + \sqrt{1 + 8}}{2} = 1$$

$$x_2 = \frac{-1 - \sqrt{1 + 8}}{2} = -2.$$

Figure 3.2 shows the graph of $y = x^2 + x + 1$, where at no point does $y = 0$. In this equation

$$a = 1$$
$$b = 1$$
$$c = 1$$

which when plugged into (3.2) confirms the graph by giving complex roots:

$$x_1 = \frac{-1 + \sqrt{1-4}}{2} = -\frac{1}{2} + i\frac{\sqrt{3}}{2}$$

$$x_2 = \frac{-1 - \sqrt{1-4}}{2} = -\frac{1}{2} - i\frac{\sqrt{3}}{2}.$$

Let's show that x_1 satisfies the original equation:

$$y = x_1^2 + x_1 + 1$$
$$= \left(-\frac{1}{2} + i\frac{\sqrt{3}}{2}\right)^2 - \frac{1}{2} + i\frac{\sqrt{3}}{2} + 1$$
$$= \frac{1}{4} - i\frac{\sqrt{3}}{2} - \frac{3}{4} - \frac{1}{2} + i\frac{\sqrt{3}}{2} + 1$$
$$= 0.$$

x_2 also satisfies the same equation.

Algebraic expressions also contain a wide variety of functions, such as

$$\sqrt{x} = \text{square root of } x$$
$$\sqrt[n]{x} = n\text{th root of } x$$
$$x^n = x \text{ to the power } n$$
$$\sin\theta = \text{sine of } \theta$$
$$\cos\theta = \text{cosine of } \theta$$
$$\tan\theta = \text{tangent of } \theta$$
$$\log x = \text{logarithm of } x$$
$$\ln x = \text{natural logarithm of } x.$$

Trigonometric functions are factorised as follows:

$$\sin^2\theta - \cos^2\theta = (\sin\theta + \cos\theta)(\sin\theta - \cos\theta)$$
$$\sin^2\theta - \tan^2\theta = (\sin\theta + \tan\theta)(\sin\theta - \tan\theta)$$
$$\sin^2\theta + 4\sin\theta\cos\theta + 4\cos^2\theta = (\sin\theta + 2\cos\theta)^2$$
$$\sin^2\theta - 6\sin\theta\cos\theta + 9\cos^2\theta = (\sin\theta - 3\cos\theta)^2.$$

3.3 Indices

Indices are used to imply repeated multiplication and create a variety of situations where laws are required to explain how the result is to be computed.

3.3.1 Laws of Indices

The laws of indices are expressed as follows:

$$a^m \times a^n = a^{m+n}$$

$$\frac{a^m}{a^n} = a^{m-n}$$

$$\left(a^m\right)^n = a^{mn}$$

and are verified using some simple examples:

$$2^3 \times 2^2 = 2^5 = 32$$

$$\frac{2^4}{2^2} = 2^2 = 4$$

$$\left(2^2\right)^3 = 2^6 = 64.$$

From the above laws, it is evident that

$$a^0 = 1$$

$$a^{-p} = \frac{1}{a^p}$$

$$a^{\frac{1}{q}} = \sqrt[q]{a}$$

$$a^{\frac{p}{q}} = \sqrt[q]{a^p}.$$

3.4 Logarithms

Two people are associated with the invention of logarithms: the Scottish theologian and mathematician John Napier (1550–1617), and the Swiss clockmaker and mathematician Joost Bürgi (1552–1632). Both men were frustrated by the time they spent multiplying numbers together, and both realised that multiplication could be replaced by addition using logarithms. Logarithms exploit the addition and subtraction of indices shown above, and are always associated with a base. For example, if $a^x = n$, then $\log_a n = x$, where a is the base. Where no base is indicated, it is assumed to be 10. Two examples bring the idea to life:

$$10^2 = 100 \quad \text{then} \quad \log 100 = 2$$
$$10^3 = 1000 \quad \text{then} \quad \log 1000 = 3$$

which is interpreted as '10 has to be raised to the power (index) 2 to equal 100.' The log operation finds the power of the base for a given number. Thus a multiplication

Fig. 3.3 Graph of log x

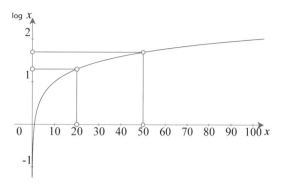

Fig. 3.4 Graph of ln x

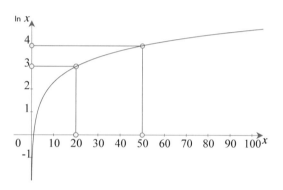

is translated into an addition using logs. Figure 3.3 shows the graph of log x, up to $x = 100$, where we see that log $20 \approx 1.3$ and log $50 \approx 1.7$. Therefore, given suitable software, logarithm tables, or a calculator with a log function, we can compute the product 20×50 as follows:

$$\log(20 \times 50) = \log 20 + \log 50 \approx 1.3 + 1.7 = 3$$
$$10^3 = 1000.$$

In general, the two bases used in calculators and software are 10 and $e = 2.718\,281\,846\ldots$. To distinguish one type of logarithm from the other, a logarithm to the base 10 is written as log, and a natural logarithm to the base e is written ln.

Figure 3.4 shows the graph of ln x, up to $x = 100$, where we see that ln $20 \approx 3$ and ln $50 \approx 3.9$. Therefore, given suitable software, a set of natural logarithm tables or a calculator with a ln function, we can compute the product 20×50 as follows:

$$\ln(20 \times 50) = \ln 20 + \ln 50 \approx 3 + 3.9 = 6.9$$
$$e^{6.9} \approx 1000.$$

From the above notation, it is evident that

$$\log(ab) = \log a + \log b$$
$$\log\left(\frac{a}{b}\right) = \log a - \log b$$
$$\log(a^n) = n \log a.$$

3.5   Further Notation

All sorts of symbols are used to stand in for natural language expressions; here are some examples:

$<$ less than
$>$ greater than
$\leq$ less than or equal to
$\geq$ greater than or equal to
$\approx$ approximately equal to
$\equiv$ equivalent to
$\neq$ not equal to
$|x|$ absolute value of x.

For example, $0 \leq t \leq 1$ is interpreted as: t is greater than or equal to 0, and is less than or equal to 1. Basically, this means t varies between 0 and 1.

3.6   Functions

The theory of *functions* is a large subject, and at this point in the book, I will only touch upon some introductory ideas that will help you understand the following chapters.

The German mathematician Gottfried von Leibniz (1646–1716) is credited with an early definition of a function, based upon the slope of a graph. However, it was the Swiss mathematician Leonhard Euler (1707–1783) who provided a definition along the lines: 'A function is a variable quantity, whose value depends upon one or more independent variables.' Other mathematicians have introduced more rigorous definitions, which are examined later on in the chapter on calculus.

3.6.1 *Explicit and Implicit Equations*

The equation

$$y = 3x^2 + 2x + 4$$

associates the value of y with different values of x. The directness of the equation: '$y =$', is why it is called an *explicit equation*, and their explicit nature is extremely useful. However, simply by rearranging the terms, creates an *implicit equation*:

$$4 = y - 3x^2 - 2x$$

which implies that certain values of x and y combine to produce the result 4. Another implicit form is

$$0 = y - 3x^2 - 2x - 4$$

which means the same thing, but expresses the relationship in a slightly different way.

An implicit equation can be turned into an explicit equation using algebra. For example, the implicit equation

$$4x + 2y = 12$$

has the explicit form:

$$y = 6 - 2x$$

where it is clear what y equals.

3.6.2 *Function Notation*

The explicit equation

$$y = 3x^2 + 2x + 4$$

tells us that the value of y depends on the value of x, and not the other way around. For example, when $x = 1$, $y = 9$; and when $x = 2$, $y = 20$. As y depends upon the value of x, it is called the *dependent variable*; and as x is independent of y, it is called the *independent variable*.

We can also say that y is a function of x, which can be written as

$$y = f(x)$$

where the letter 'f' is the name of the function, and the independent variable is enclosed in brackets. We could have also written $y = g(x)$, $y = h(x)$, etc.

Eventually, we have to identify the nature of the function, which in this case is

$$f(x) = 3x^2 + 2x + 4.$$

Nothing prevents us from writing

$$y = f(x) = 3x^2 + 2x + 4$$

which means: y equals the value of the function $f(x)$, which is determined by the independent variable x using the expression $3x^2 + 2x + 4$.

An equation may involve more than one independent variable, such as the volume of a cylinder:

$$V = \pi r^2 h$$

where r is the radius, and h, the height, and is written:

$$V(r, h) = \pi r^2 h.$$

3.6.3 Intervals

An *interval* is a continuous range of numerical values associated with a variable, which can include or exclude the upper and lower values. For example, a variable such as x is often subject to inequalities like $x \geq a$ and $x \leq b$, which can also be written as

$$a \leq x \leq b$$

and implies that x is located in the *closed interval* $[a, b]$, where the square brackets indicate that the interval includes a and b. For example,

$$1 \leq x \leq 10$$

means that x is located in the closed interval $[1, 10]$, which includes 1 and 10.

When the boundaries of the interval are not included, then we would state $x > a$ and $x < b$, which is written

$$a < x < b$$

and means that x is located in the *open interval* $]a, b[$, where the reverse square brackets indicate that the interval excludes a and b. For example,

$$1 < x < 10$$

means that x is located in the open interval $]1, 10[$, which excludes 1 and 10.

Fig. 3.5 Closed, open and
half-open intervals. The
filled circles indicate that *a*
or *b* are included in the
interval

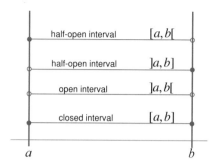

The filled circles indicate that *a* or *b* are included in the interval.

Closed and open intervals may be combined as follows. If $x \geq a$ and $x < b$ then

$$a \leq x < b$$

and means that *x* is located in the *half-open interval* [a, b[. For example,

$$1 \leq x < 10$$

means that *x* is located in the half-open interval [1, 10[, which includes 1, but not 10.
Similarly, if

$$1 < x \leq 10$$

means that *x* is located in the half-open interval]1, 10], which includes 10, but not 1.
An alternative notation employs parentheses instead of reversed brackets:

$$]a, \ b[= (a, \ b)$$
$$[a, \ b[= [a, \ b)$$
$$]a, \ b] = (a, \ b].$$

Figure 3.5 shows open, closed and half-open intervals diagrammatically.

3.6.4 Function Domains and Ranges

The following descriptions of domains and ranges only apply to functions with one
independent variable: $f(x)$.
Returning to the above function:

$$y = f(x) = 3x^2 + 2x + 4$$

the independent variable *x*, can take on any value from $-\infty$ to ∞, which is called
the *domain* of the function. In this case, the domain of $f(x)$ is the set of real numbers

ℝ. The notation used for intervals, is also used for domains, which in this case is

$$] - \infty, \ \infty[$$

and is open, as there are no precise values for $-\infty$ and ∞.

As the independent variable takes on different values from its domain, so the dependent variable, y or $f(x)$, takes on different values from its *range*. Therefore, if the domain of the linear function $f(x) = 3x + 4$ is $[-4, \ 4]$, the range of $f(x)$ is calculated by finding $f(-4)$ and $f(4)$:

$$f(-4) = -12 + 4 = -8$$
$$f(4) = 12 + 4 = 16$$

and the range is $[-4, \ 4]$.

Although calculating the range of linear functions is simple, other types of functions require a knowledge of calculus.

The domain of $\log x$ is

$$]0, \ \infty[$$

which is open, because $x \neq 0$. Whereas, the range of $\log x$ is

$$] - \infty, \ \infty[.$$

The domain of $\sqrt{x}$ is

$$[0, \ \infty[$$

which is half-open, because $\sqrt{0} = 0$, and ∞ has no precise value. Similarly, the range of $\sqrt{x}$ is

$$[0, \ \infty[.$$

Sometimes, a function is sensitive to one specific number. For example, in the function

$$y = f(x) = \frac{1}{x - 1},$$

when $x = 1$, there is a divide by zero, which is meaningless. Consequently, the domain of $f(x)$ is the set of real numbers ℝ, apart from 1.

3.6.5 Odd and Even Functions

An *odd function* satisfies the condition:

$$f(-x) = -f(x)$$

where x is located in a valid domain. Consequently, the graph of an odd function is symmetrical relative to the origin. For example, $\sin(\theta)$ is odd because

$$\sin(-\theta) = -\sin\theta$$

as illustrated in Fig. 3.6. Other odd functions include:

$$f(x) = ax$$
$$f(x) = ax^3.$$

An *even function* satisfies the condition:

$$f(-x) = f(x)$$

where x is located in a valid domain. Consequently, the graph of an even function is symmetrical relative to the $f(x)$ axis. For example, $\cos\theta$ is even because

$$\cos(-\theta) = \cos\theta$$

as illustrated in Fig. 3.7. Other even functions include:

$$f(x) = ax^2$$
$$f(x) = ax^4.$$

Fig. 3.6 The sine function is an odd function

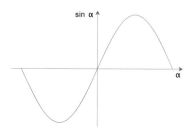

Fig. 3.7 The cosine function is an even function

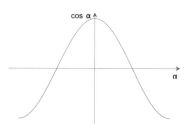

3.6.6 Power Functions

Functions of the form $f(x) = x^n$ are called *power functions of degree n* and are either odd or even. If n is an odd natural number, then the power function is odd, else if n is an even natural number, then the power function is even.

3.7 Summary

The above description of algebra should be sufficient for the reader to understand the following chapters. However, one should remember that this is only the beginning of a very complex subject.

3.8 Worked Examples

3.8.1 Algebraic Manipulation

Rearrange the following equations to make y the subject.

$$7 = \frac{x+4}{3-y}, \quad 23 = \frac{x+68}{3+\dfrac{1}{e^y}}, \quad 23 = \frac{x+68}{3-\sin y}.$$

$$7 = \frac{x+4}{3-y}$$

$$3 - y = \frac{x+4}{7}$$

$$y = 3 - \frac{x+4}{7} = \frac{17-x}{7}.$$

$$23 = \frac{x+68}{3+\dfrac{1}{e^y}}$$

$$3 + \frac{1}{e^y} = \frac{x+68}{23}$$

$$\frac{1}{e^y} = \frac{x+68}{23} - 3$$

$$= \frac{x-1}{23}$$

$$e^y = \frac{23}{x-1}$$

$$y = \ln\left(\frac{23}{x-1}\right).$$

$$23 = \frac{x+68}{3-\sin y}$$

$$3-\sin y = \frac{x+68}{23}$$

$$\sin y = 3 - \frac{x+68}{23}$$

$$= \frac{1-x}{23}$$

$$y = \arcsin\left(\frac{1-x}{23}\right).$$

3.8.2 Solving a Quadratic Equation

Solve the following quadratic equations, and test the answers.

$$0 = x^2 + 4x + 1, \quad 0 = 2x^2 + 4x + 2, \quad 0 = 2x^2 + 4x + 4.$$

$0 = x^2 + 4x + 1$

$$x = \frac{-b \pm \sqrt{b^2 - 4ac}}{2a}$$

$$= \frac{-4 \pm \sqrt{16-4}}{2}$$

$$= \frac{-4 \pm \sqrt{12}}{2}$$

$$= -2 \pm \sqrt{3}.$$

Test with $x = -2 + \sqrt{3}$.

$$x^2 + 4x + 1 = \left(-2 + \sqrt{3}\right)^2 + 4\left(-2 + \sqrt{3}\right) + 1$$
$$= 4 - 4\sqrt{3} + 3 - 8 + 4\sqrt{3} + 1$$
$$= 0.$$

Test with $x = -2 - \sqrt{3}$.

$$x^2 + 4x + 1 = \left(-2 - \sqrt{3}\right)^2 + 4\left(-2 - \sqrt{3}\right) + 1$$

$$= 4 + 4\sqrt{3} + 3 - 8 - 4\sqrt{3} + 1$$
$$= 0.$$

$0 = 2x^2 + 4x + 2$

$$x = \frac{-b \pm \sqrt{b^2 - 4ac}}{2a}$$
$$= \frac{-4 \pm \sqrt{16 - 16}}{4}$$
$$= \frac{-4}{4}$$
$$= -1.$$

Test with $x = -1$.

$$2x^2 + 4x + 2 = 2 - 4 + 2$$
$$= 0.$$

$0 = 2x^2 + 4x + 4$

$$x = \frac{b \pm \sqrt{b^2 - 4ac}}{2a}$$
$$= \frac{-4 \pm \sqrt{16 - 32}}{4}$$
$$= \frac{-4 \pm \sqrt{-16}}{4}$$
$$= -1 \pm \sqrt{-1}$$
$$= -1 \pm i.$$

Test with $x = -1 + i$.

$$2x^2 + 4x + 4 = 2(-1 + i)^2 + 4(-1 + i) + 4$$
$$= 2(1 - 2i - 1) - 4 + 4i + 4$$
$$= -4i + 4i$$
$$= 0.$$

Test with $x = -1 - i$.

$$2x^2 + 4x + 4 = 2(-1 - i)^2 + 4(-1 - i) + 4$$
$$= 2(1 + 2i - 1) - 4 - 4i + 4$$
$$= 4i - 4i$$
$$= 0.$$

3.8.3   *Factorising*

Factorise the following equations:

$$4 \sin^2 \theta - 4 \cos^2 \theta$$
$$9 \sin^2 \theta + 6 \sin \theta \cos \theta + \cos^2 \theta$$
$$25 \sin^2 \theta + 10 \sin \theta \cos \theta + \cos^2 \theta.$$

$$4 \sin^2 \theta - 4 \cos^2 \theta = (2 \sin \theta + 2 \cos \theta)(2 \sin \theta - 2 \cos \theta)$$
$$9 \sin^2 \theta + 6 \sin \theta \cos \theta + \cos^2 \theta = (3 \sin \theta + \cos \theta)^2$$
$$25 \sin^2 \theta + 10 \sin \theta \cos \theta + \cos^2 \theta = (5 \sin \theta + \cos \theta)^2.$$

References

1. Russell B, Whitehead AN (1903) Principia mathematica. Cambridge University Press
2. Russell B (1938) [First published 1903]. Principles of mathematics, 2nd edn. WW Norton & Company

Chapter 4
Trigonometry

4.1 Introduction

This chapter covers some basic features of trigonometry such as angular measure, trigonometric ratios, inverse ratios, trigonometric identities and various rules, with which the reader should be familiar.

4.2 Background

The word 'trigonometry' divides into three parts: 'tri', 'gon', 'metry', which means the measurement of three-sided polygons, i.e. triangles. It is an ancient subject and is used across all branches of mathematics.

4.3 Units of Angular Measurement

The measurement of angles is at the heart of trigonometry, and today two units of angular measurement are part of modern mathematics: *degrees* and *radians*. The degree (or sexagesimal) unit of measure derives from defining one complete rotation as 360°. Each degree divides into 60 min, and each minute divides into 60 s. The number 60 has survived from Mesopotamian days and appears rather incongruous when used alongside today's decimal system—nevertheless, it is still convenient to work with degrees even though the radian is a natural feature of mathematics.

The radian of angular measure does not depend upon any arbitrary constant, and is often defined as the angle created by a circular arc whose length is equal to the circle's radius. And because the perimeter of a circle is $2\pi r$, 2π rad correspond to one complete rotation. As 360° corresponds to 2π rad, 1 rad equals $180°/\pi$, which is approximately 57.3°. The following relationships between radians and degrees are

© Springer-Verlag London Ltd., part of Springer Nature 2022
J. Vince, *Mathematics for Computer Graphics*, Undergraduate Topics
in Computer Science, https://doi.org/10.1007/978-1-4471-7520-9_4

worth remembering:

$$\frac{\pi}{2} \text{ [rad]} \equiv 90°, \qquad \pi \text{ [rad]} \equiv 180°$$

$$\frac{3\pi}{2} \text{ [rad]} \equiv 270°, \quad 2\pi \text{ [rad]} \equiv 360°.$$

To convert $x°$ to radians:

$$\frac{\pi x°}{180} \text{ [rad]}.$$

To convert x [rad] to degrees:

$$\frac{180x}{\pi} \text{ [degrees]}.$$

For those readers wishing to know the background to radians we need to use power series. We start with the power series for e^θ, $\sin \theta$ and $\cos \theta$:

$$e^\theta = 1 + \frac{\theta^1}{1!} + \frac{\theta^2}{2!} + \frac{\theta^3}{3!} + \frac{\theta^4}{4!} + \frac{\theta^5}{5!} + \frac{\theta^6}{6!} + \frac{\theta^7}{7!} + \frac{\theta^8}{8!} + \frac{\theta^9}{9!} + \cdots$$

$$\sin \theta = \theta - \frac{\theta^3}{3!} + \frac{\theta^5}{5!} - \frac{\theta^7}{7!} + \frac{\theta^9}{9!} + \cdots$$

$$\cos \theta = 1 - \frac{\theta^2}{2!} + \frac{\theta^4}{4!} - \frac{\theta^6}{6!} + \frac{\theta^8}{8!} + \cdots .$$

Euler proved that these three power series are related, and when $\theta = \pi$, $\sin \theta = 0$, and $\cos \theta = -1$. Figure 4.1 shows curves of the sine power series for 3, 5, 7 and 9 terms, and when $\theta = 2\pi$, the graph reaches zero.

Fig. 4.1 The sine power series for different number of terms

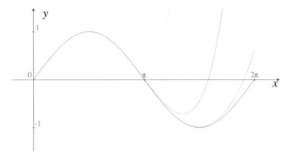

4.4   The Trigonometric Ratios

Ancient civilisations knew that triangles—whatever their size—possessed some inherent properties, especially the ratios of sides and their associated angles. This means that if these ratios are known in advance, problems involving triangles with unknown lengths and angles, can be discovered using these ratios.

Figure 4.2 shows a point P with coordinates (*base*, *height*), on a unit-radius circle rotated through an angle θ. As P is rotated, it moves into the 2nd quadrant, 3rd quadrant, 4th quadrant and returns back to the first quadrant. During the rotation, the sign of *height* and *base* change as follows:

$$
\begin{array}{ll}
\text{1st quadrant:} & \textit{height } (+), \textit{base } (+) \\
\text{2nd quadrant:} & \textit{height } (+), \textit{base } (-) \\
\text{3rd quadrant:} & \textit{height } (-), \textit{base } (-) \\
\text{4th quadrant:} & \textit{height } (-), \textit{base } (+).
\end{array}
$$

Figures 4.3 and 4.4 plot the changing values of *height* and *base* over the four quadrants, respectively. When *radius* = 1, the curves vary between 1 and −1. In the context of triangles, the sides are labelled as follows:

$$
\begin{aligned}
hypotenuse &= radius \\
opposite &= height \\
adjacent &= base.
\end{aligned}
$$

Thus, using the right-angle triangle shown in Fig. 4.5, the trigonometric ratios: sine, cosine and tangent are defined as

$$
\sin\theta = \frac{opposite}{hypotenuse}, \quad \cos\theta = \frac{adjacent}{hypotenuse}, \quad \tan\theta = \frac{opposite}{adjacent}.
$$

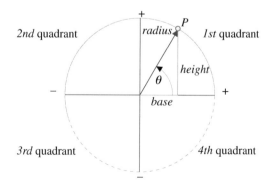

Fig. 4.2 The four quadrants for the trigonometric ratios

Fig. 4.3 The graph of *height* over the four quadrants

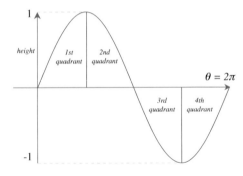

Fig. 4.4 The graph of *base* over the four quadrants

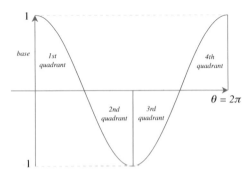

Fig. 4.5 Sides of a right-angle triangle

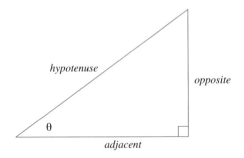

The reciprocals of these functions, cosecant, secant and cotangent are also useful:

$$\csc \theta = \frac{1}{\sin \theta}, \quad \sec \theta = \frac{1}{\cos \theta}, \quad \cot \theta = \frac{1}{\tan \theta}.$$

As an example, Fig. 4.6 shows a triangle where the hypotenuse and an angle are known. The other sides are calculated as follows:

Fig. 4.6 A right-angle
triangle with two unknown
side lengths

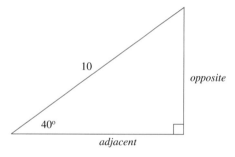

Fig. 4.7 Graph of the
tangent function

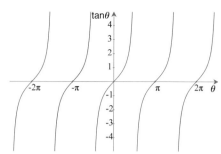

$$\frac{opposite}{10} = \sin 40°$$

$$opposite = 10 \sin 40° \approx 10 \times 0.64278 = 6.4278$$

$$\frac{adjacent}{10} = \cos 40°$$

$$adjacent = 10 \cos 40° \approx 10 \times 0.7660 = 7.660.$$

The theorem of Pythagoras confirms that these lengths are correct:

$$6.4278^2 + 7.660^2 \approx 10^2.$$

Figure 4.7 shows the graph of the tangent function, which, like the sine and cosine functions, is periodic, but with only a period of π radians.

4.4.1 *Domains and Ranges*

The periodic nature of $\sin \theta$, $\cos \theta$ and $\tan \theta$, means that their domains are infinitely large. Consequently, it is customary to confine the domain of $\sin \theta$ to

$$\left[-\frac{\pi}{2}, \frac{\pi}{2} \right]$$

and $\cos\theta$ to

$$[0,\ \pi].$$

The range for both $\sin\theta$ and $\cos\theta$ is

$$[-1,\ 1].$$

The domain for $\tan\theta$ is the open interval

$$\left]-\frac{\pi}{2},\ \frac{\pi}{2}\right[$$

and its range is the open interval:

$$]-\infty,\ \infty[.$$

4.5   Inverse Trigonometric Ratios

The functions $\sin\theta$, $\cos\theta$, $\tan\theta$, $\csc\theta$, $\sec\theta$ and $\cot\theta$ provide different ratios for the angle θ, and the inverse trigonometric functions convert a ratio back into an angle. These are arcsin, arccos, arctan, arccsc, arcsec and arccot, and are sometimes written as $\sin^{-1}$, $\cos^{-1}$, $\tan^{-1}$, $\csc^{-1}$, $\sec^{-1}$ and $\cot^{-1}$. For example, $\sin 30° = 0.5$, therefore, $\arcsin 0.5 = 30°$. Consequently, the domain for arcsin is the range for sin:

$$[-1,\ 1]$$

and the range for arcsin is the domain for sin:

$$\left[-\frac{\pi}{2},\ \frac{\pi}{2}\right]$$

as shown in Fig. 4.8. Similarly, the domain for arccos is the range for cos:

$$[-1,\ 1]$$

and the range for arccos is the domain for cos:

$$[0,\ \pi]$$

as shown in Fig. 4.9.

Fig. 4.8 Graph of the arcsin
function

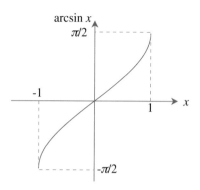

Fig. 4.9 Graph of the arccos
function

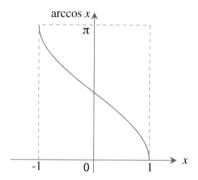

The domain for arctan is the range for tan:

$$] - \infty, \ \infty[$$

and the range for arctan is the domain for tan:

$$\left] -\frac{\pi}{2}, \ \frac{\pi}{2} \right[$$

as shown in Fig. 4.10.

Various programming languages include the atan2 function, which is an arctan function with two arguments: atan2(y, x). The signs of x and y provide sufficient information to locate the quadrant containing the angle, and gives the atan2 function a range of $[0, \ 2\pi]$.

Fig. 4.10 Graph of the
arctan function

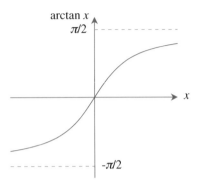

4.6   Trigonometric Identities

The sin and cos curves are identical, apart from being displaced by 90°, and are
related by

$$\cos \theta = \sin \left(\theta + \frac{\pi}{2} \right).$$

Also, simple algebra and the theorem of Pythagoras can be used to derive other
formulae such as

$$\frac{\sin \theta}{\cos \theta} = \tan \theta$$

$$\sin^2 \theta + \cos^2 \theta = 1$$

$$1 + \tan^2 \theta = \sec^2 \theta$$

$$1 + \cot^2 \theta = \csc^2 \theta.$$

4.7   The Sine Rule

Figure 4.11 shows a triangle labeled such that side a is opposite angle A, side b is
opposite angle B, etc. The sine rule states:

$$\frac{a}{\sin A} = \frac{b}{\sin B} = \frac{c}{\sin C}$$

which can be used to compute the length of an unknown length or angle. For example,
if $A = 60°$, $B = 40°$, $C = 80°$, and $b = 10$, then

$$\frac{a}{\sin 60°} = \frac{10}{\sin 40°}$$

Fig. 4.11 An arbitrary triangle

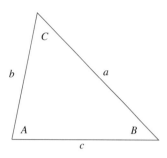

rearranging, we have

$$a = \frac{10 \sin 60°}{\sin 40°} \approx 13.47.$$

Similarly:

$$\frac{c}{\sin 80°} = \frac{10}{\sin 40°}$$

therefore

$$c = \frac{10 \sin 80°}{\sin 40°} \approx 15.32.$$

4.8 The Cosine Rule

The cosine rule expresses the $\sin^2 \theta + \cos^2 \theta = 1$ identity for the arbitrary triangle shown in Fig. 4.11. In fact, there are three versions:

$$a^2 = b^2 + c^2 - 2bc \cos A$$
$$b^2 = c^2 + a^2 - 2ca \cos B$$
$$c^2 = a^2 + b^2 - 2ab \cos C.$$

Three further relationships also hold:

$$a = b \cos C + c \cos B$$
$$b = c \cos A + a \cos C$$
$$c = a \cos B + b \cos A.$$

4.9   Compound-Angle Identities

Trigonometric identities are useful for solving various mathematical problems, but apart from this, their proof often contains a strategy that can be used else where. In the first example, watch out for the technique of multiplying by 1 in the form of a ratio, and swapping denominators. The technique is rather elegant and suggests that the result was known in advance, which probably was the case. Let's begin by finding a way of representing $\sin(\alpha + \beta)$ in terms of $\sin \alpha$, $\cos \alpha$, $\sin \beta$, $\cos \beta$.

With reference to Fig. 4.12:

$$
\begin{aligned}
\sin(\alpha + \beta) &= \frac{FD}{AD} = \frac{BC + ED}{AD} \\
&= \frac{BC}{AD}\frac{AC}{AC} + \frac{ED}{AD}\frac{CD}{CD} \\
&= \frac{BC}{AC}\frac{AC}{AD} + \frac{ED}{CD}\frac{CD}{AD} \\
\sin(\alpha + \beta) &= \sin \alpha \cos \beta + \cos \alpha \sin \beta.
\end{aligned}
\tag{4.1}
$$

To find $\sin(\alpha - \beta)$, reverse the sign of β in (4.1):

$$
\sin(\alpha - \beta) = \sin \alpha \cos \beta - \cos \alpha \sin \beta.
\tag{4.2}
$$

Now let's expand $\cos(\alpha + \beta)$ with reference to Fig. 4.12:

$$
\begin{aligned}
\cos(\alpha + \beta) &= \frac{AE}{AD} = \frac{AB - EC}{AD} \\
&= \frac{AB}{AD}\frac{AC}{AC} - \frac{EC}{AD}\frac{CD}{CD} \\
&= \frac{AB}{AC}\frac{AC}{AD} - \frac{EC}{CD}\frac{CD}{AD} \\
\cos(\alpha + \beta) &= \cos \alpha \cos \beta - \sin \alpha \sin \beta.
\end{aligned}
\tag{4.3}
$$

Fig. 4.12 The geometry to expand $\sin(\alpha + \beta)$

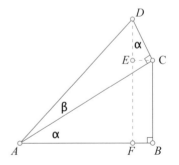

To find $\cos(\alpha - \beta)$, reverse the sign of β in (4.3):

$$\cos(\alpha - \beta) = \cos\alpha\cos\beta + \sin\alpha\sin\beta.$$

To expand $\tan(\alpha + \beta)$, divide (4.1) by (4.3):

$$\frac{\sin(\alpha + \beta)}{\cos(\alpha + \beta)} = \frac{\sin\alpha\cos\beta + \cos\alpha\sin\beta}{\cos\alpha\cos\beta - \sin\alpha\sin\beta}$$

$$= \frac{\dfrac{\sin\alpha\cos\beta}{\cos\alpha\cos\beta} + \dfrac{\cos\alpha\sin\beta}{\cos\alpha\cos\beta}}{\dfrac{\cos\alpha\cos\beta}{\cos\alpha\cos\beta} - \dfrac{\sin\alpha\sin\beta}{\cos\alpha\cos\beta}}$$

$$\tan(\alpha + \beta) = \frac{\tan\alpha + \tan\beta}{1 - \tan\alpha\tan\beta}. \qquad (4.4)$$

To find $\tan(\alpha - \beta)$, reverse the sign of β in (4.4):

$$\tan(\alpha - \beta) = \frac{\tan\alpha - \tan\beta}{1 + \tan\alpha\tan\beta}.$$

4.9.1 Double-Angle Identities

By making $\beta = \alpha$, the three compound-angle identities

$$\sin(\alpha \pm \beta) = \sin\alpha\cos\beta \pm \cos\alpha\sin\beta$$
$$\cos(\alpha \pm \beta) = \cos\alpha\cos\beta \mp \sin\alpha\sin\beta$$
$$\tan(\alpha \pm \beta) = \frac{\tan\alpha \pm \tan\beta}{1 \mp \tan\alpha\tan\beta}$$

provide the starting point for deriving three corresponding double-angle identities:

$$\sin(\alpha \pm \alpha) = \sin\alpha\cos\alpha \pm \cos\alpha\sin\alpha$$
$$\sin(2\alpha) = 2\sin\alpha\cos\alpha.$$

Similarly,

$$\cos(\alpha \pm \alpha) = \cos\alpha\cos\alpha \mp \sin\alpha\sin\alpha$$
$$\cos(2\alpha) = \cos^2\alpha - \sin^2\alpha$$

which can be further simplified using $\sin^2\alpha + \cos^2\alpha = 1$:

$$\cos(2\alpha) = \cos^2 \alpha - \sin^2 \alpha$$
$$\cos(2\alpha) = 2\cos^2 \alpha - 1$$
$$\cos(2\alpha) = 1 - 2\sin^2 \alpha.$$

And for $\tan(2\alpha)$, we have:

$$\tan(\alpha \pm \alpha) = \frac{\tan \alpha \pm \tan \alpha}{1 \mp \tan \alpha \tan \alpha}$$
$$\tan(2\alpha) = \frac{2\tan \alpha}{1 - \tan^2 \alpha}.$$

4.9.2 Multiple-Angle Identities

The French mathematician Abraham de Moivre (1667–1754) published an equation in 1707, which implied

$$\cos \alpha = \tfrac{1}{2}(\cos(n\alpha) + i\sin(n\alpha))^{1/n} + \tfrac{1}{2}(\cos(n\alpha) - i\sin(n\alpha))^{1/n}$$

for all positive, integer values of n. Fifteen years later, de Moivre proved that

$$(\cos \alpha + i\sin \alpha)^n = \cos(n\alpha) + i\sin(n\alpha)$$

which is known as de Moivre's Formula. Euler proved in 1749 that this formula held for $n \in \mathbb{R}$ using his own discovery:

$$\cos \alpha + i\sin \alpha = e^{i\alpha}.$$

Using de Moivre's formula, one can show that

$$\sin(3\alpha) = 3\sin \alpha - 4\sin^3 \alpha$$
$$\cos(3\alpha) = 4\cos^3 \alpha - 3\cos \alpha$$
$$\tan(3\alpha) = \frac{3\tan \alpha - \tan^3 \alpha}{1 - 3\tan^2 \alpha}$$
$$\sin(4\alpha) = 4\sin \alpha \cos \alpha - 8\sin^3 \alpha \cos \alpha$$
$$\cos(4\alpha) = 8\cos^4 \alpha - 8\cos^2 \alpha + 1$$
$$\tan(4\alpha) = \frac{4\tan \alpha - 4\tan^3 \alpha}{1 - 6\tan^2 \alpha + \tan^4 \alpha}$$
$$\sin(5\alpha) = 16\sin^5 \alpha - 20\sin^3 \alpha + 5\sin \alpha$$
$$\cos(5\alpha) = 16\cos^5 \alpha - 20\cos^3 \alpha + 5\cos \alpha$$
$$\tan(5\alpha) = \frac{5\tan \alpha - 10\tan^3 \alpha + \tan^5 \alpha}{1 - 10\tan^2 \alpha + 5\tan^4 \alpha}.$$

4.9.3 Half-Angle Identities

Every now and then, it is necessary to compute the sine, cosine or tangent of a half-angle from the corresponding whole-angle functions. To do this, we rearrange the double-angle identities as follows.

$$\cos(2\alpha) = 1 - 2\sin^2\alpha$$

$$\sin^2\alpha = \frac{1 - \cos(2\alpha)}{2}$$

$$\sin^2\left(\frac{\alpha}{2}\right) = \frac{1 - \cos\alpha}{2}$$

$$\sin\left(\frac{\alpha}{2}\right) = \pm\sqrt{\frac{1 - \cos\alpha}{2}}. \tag{4.5}$$

Similarly,

$$\cos^2\alpha = \frac{1 + \cos(2\alpha)}{2}$$

$$\cos^2\left(\frac{\alpha}{2}\right) = \frac{1 + \cos\alpha}{2}$$

$$\cos\left(\frac{\alpha}{2}\right) = \pm\sqrt{\frac{1 + \cos\alpha}{2}}. \tag{4.6}$$

Dividing (4.5) by (4.6) we have

$$\tan\left(\frac{\alpha}{2}\right) = \pm\sqrt{\frac{1 - \cos\alpha}{1 + \cos\alpha}}.$$

4.10   Perimeter Relationships

Finally, with reference to Fig. 4.11, we come to the relationships that integrate angles with the perimeter of a triangle:

$$s = \tfrac{1}{2}(a + b + c)$$

$$\sin\left(\frac{A}{2}\right) = \pm\sqrt{\frac{(s - b)(s - c)}{bc}}$$

$$\sin\left(\frac{B}{2}\right) = \pm\sqrt{\frac{(s - c)(s - a)}{ca}}$$

$$\sin\left(\frac{C}{2}\right) = \pm\sqrt{\frac{(s-a)(s-b)}{ab}}$$

$$\cos\left(\frac{A}{2}\right) = \pm\sqrt{\frac{s(s-a)}{bc}}$$

$$\cos\left(\frac{B}{2}\right) = \pm\sqrt{\frac{s(s-b)}{ca}}$$

$$\cos\left(\frac{C}{2}\right) = \pm\sqrt{\frac{s(s-c)}{ab}}$$

$$\sin A = \pm\frac{2}{bc}\sqrt{s(s-a)(s-b)(s-c)}$$

$$\sin B = \pm\frac{2}{ca}\sqrt{s(s-a)(s-b)(s-c)}$$

$$\sin C = \pm\frac{2}{ab}\sqrt{s(s-a)(s-b)(s-c)}.$$

4.11 Summary

No derivations have been given for the formulae in this chapter, and the reader who is really interested, will find plenty of books that show their origins. Hopefully, the formulae will be a useful reference when studying the rest of the book, and perhaps will be of some use when solving problems in the future.

I would like to draw the reader's attention to two books I have found a source of information and inspiration [1, 2].

References

1. Harris JW, Stocker H (1998) Handbook of mathematics and computational science. Springer, New York
2. Gullberg J (1997) Mathematics from the birth of numbers. WW Norton & Co., New York

Chapter 5
Coordinate Systems

5.1 Introduction

In this chapter we revise Cartesian coordinates, axial systems, the distance between two points in space, and the area of simple 2D shapes. It also covers polar, spherical polar and cylindrical coordinate systems.

5.2 Background

René Descartes is often credited with the invention of the xy-plane, but the French lawyer and mathematician Pierre de Fermat (1601–1665) was probably the first inventor. In 1636 Fermat was working on a treatise titled *Ad locus planos et solidos isagoge*, which outlined what we now call 'analytic geometry'. Unfortunately, Fermat never published his treatise, although he shared his ideas with other mathematicians such as Blaise Pascal (1623–1662). At the same time, Descartes devised his own system of analytic geometry and in 1637 published his results in the prestigious journal *Géométrie*. In the eyes of the scientific world, the publication date of a technical paper determines when a new idea or invention is released into the public domain. Consequently, ever since this publication Descartes has been associated with the xy-plane, which is why it is called the *Cartesian plane*.

The Cartesian plane is such a simple idea that it is strange that it took so long to be discovered. However, although it is true that René Descartes showed how an orthogonal coordinate system could be used for graphs and coordinate geometry, coordinates had been used by ancient Egyptians, almost 2000 years earlier! If Fermat had been more efficient in publishing his research results, the xy-plane could have been called the Fermatian plane! [1].

© Springer-Verlag London Ltd., part of Springer Nature 2022
J. Vince, *Mathematics for Computer Graphics*, Undergraduate Topics
in Computer Science, https://doi.org/10.1007/978-1-4471-7520-9_5

Fig. 5.1 The Cartesian plane

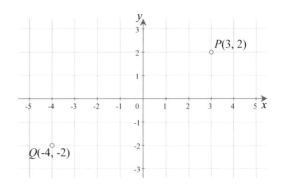

5.3   The Cartesian Plane

The Cartesian plane provides a mechanism for locating points with a unique, ordered pair of numbers (x, y) as shown in Fig. 5.1, where P has coordinates $(3, 2)$ and Q has coordinates $(-4, -2)$. The point $(0, 0)$ is called the *origin*. As previously mentioned, Descartes suggested that the letters x and y should be used to represent variables, and letters at the other end of the alphabet should stand for numbers. Which is why equations such as $y = ax^2 + bx + c$, are written this way.

The axes are said to be *oriented* as the x-axis rotates anticlockwise towards the y-axis. They could have been oriented in the opposite sense, with the y-axis rotating anticlockwise towards the x-axis.

5.4   Function Graphs

When functions such as

$$\text{linear:}\ y = mx + c,$$
$$\text{quadratic:}\ y = ax^2 + bx + c,$$
$$\text{cubic:}\ y = ax^3 + bx^2 + cx + d,$$
$$\text{trigonometric:}\ y = a \sin x,$$

are drawn as graphs, they create familiar shapes that permit the function to be easily identified. Linear functions are straight lines; quadratics are parabolas; cubics, generally, have an 'S' shape; and trigonometric functions often possess a wave-like trace. Figure 5.2 shows examples of each type of function.

5.5   Shape Representation

The Cartesian plane also provides a way to represent 2D shapes numerically, which permits them to be manipulated mathematically. Let's begin with 2D polygons and show how their internal area can be calculated.

Fig. 5.2 Graphs of four
function types

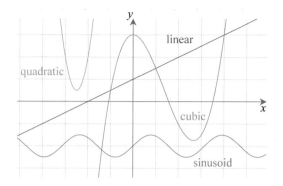

Fig. 5.3 A simple polygon
created by a chain of vertices

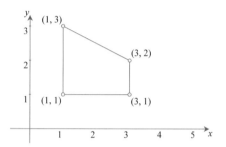

5.5.1 2D Polygons

A polygon is formed from a chain of *vertices* (points) as shown in Fig. 5.3. A straight
line is assumed to connect each pair of neighbouring vertices; intermediate points
on the line are not explicitly stored. There is no convention for starting a chain
of vertices, but software will often dictate whether polygons have a clockwise or
anticlockwise vertex sequence.

We can now subject this list of coordinates to a variety of arithmetic and mathematical operations. For example, if we double the values of x and y and redraw the
vertices, we discover that the shape's geometric integrity is preserved, but its size
is doubled relative to the origin. Similarly, if we divide the values of x and y by 2,
the shape is still preserved, but its size is halved relative to the origin. On the other
hand, if we add 1 to every x-coordinate, and 2 to every y-coordinate, and redraw the
vertices, the shape's size remains the same but is displaced 1 unit horizontally and 2
units vertically.

5.5.2 Area of a Shape

The area of a polygonal shape is readily calculated from its list of coordinates. For
example, using the list of coordinates shown in Table 5.1: the area is computed by

Table 5.1 A polygon's coordinates

x	y
x_0	y_0
x_1	y_1
x_2	y_2
x_3	y_3

$$\text{area} = \tfrac{1}{2}[(x_0 y_1 - x_1 y_0) + (x_1 y_2 - x_2 y_1) + (x_2 y_3 - x_3 y_2) + (x_3 y_0 - x_0 y_3)].$$

You will observe that the calculation sums the results of multiplying an x by the next y, minus the next x by the previous y. When the last vertex is selected, it is paired with the first vertex to complete the process. The result is then halved to reveal the area. As a simple test, let's apply this formula to the shape described in Fig. 5.3:

$$\text{area} = \tfrac{1}{2}[(1 \times 1 - 3 \times 1) + (3 \times 2 - 3 \times 1) + (3 \times 3 - 1 \times 2) + (1 \times 1 - 1 \times 3)]$$
$$\text{area} = \tfrac{1}{2}[-2 + 3 + 7 - 2] = 3.$$

which, by inspection, is the true area. The beauty of this technique is that it works with any number of vertices and any arbitrary shape. The origin of this technique is revealed in Chap. 7.

Another feature of the technique is that if the set of coordinates is clockwise, the area is negative, which means that the calculation computes vertex orientation as well as area. To illustrate this feature, the original vertices are reversed to a clockwise sequence as follows:

$$\text{area} = \tfrac{1}{2}[(1 \times 3 - 1 \times 1) + (1 \times 2 - 3 \times 3) + (3 \times 1 - 3 \times 2) + (3 \times 1 - 1 \times 1)]$$
$$\text{area} = \tfrac{1}{2}[2 - 7 - 3 + 2] = -3.$$

The minus sign confirms that the vertices are in a clockwise sequence.

5.6 Theorem of Pythagoras in 2D

The theorem of Pythagoras is used to calculate the distance between two points. Figure 5.4 shows two arbitrary points $P_1(x_1, \ y_1)$ and $P_2(x_2, \ y_2)$. The distance $\Delta x = x_2 - x_1$ and $\Delta y = y_2 - y_1$. Therefore, the distance d between P_1 and P_2 is given by

$$d = \sqrt{(\Delta x)^2 + (\Delta y)^2}.$$

For example, given $P_1(1, \ 1)$, $P_2(4, \ 5)$, then $d = \sqrt{3^2 + 4^2} = 5$.

Fig. 5.4 Calculating the
distance between two points

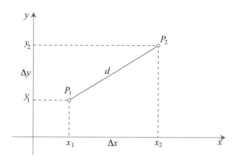

5.7 3D Cartesian Coordinates

Two coordinates are required to locate a point on the 2D Cartesian plane, and three
coordinates are required for 3D space. The corresponding axial system requires three
mutually perpendicular axes; however, there are two ways to add the extra z-axis.
Figure 5.5 shows the two orientations, which are described as *left-* and *right-handed*
axial systems. The left-handed system permits us to align our left hand with the
axes such that the thumb aligns with the x-axis, the first finger aligns with the y-
axis, and the middle finger aligns with the z-axis. The right-handed system permits
the same system of alignment, but using our right hand. The choice between these
axial systems is arbitrary, but one should be aware of the system employed by com-
mercial computer graphics packages. The main problem arises when projecting 3D
points onto a 2D plane, which has an oriented axial system. A right-handed system
is employed throughout this book, as shown in Fig. 5.6, which also shows a point P
with its coordinates. It also worth noting that handedness has no meaning in spaces
with 4 dimensions or more. Also note that the choice of axis as the vertical axis is a
matter of personal preference.

Fig. 5.5 a A left-handed
axial system **b** A
right-handed axial system

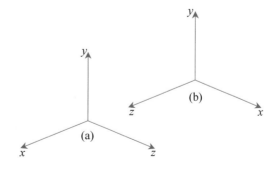

Fig. 5.6 A right-handed
axial system showing the
coordinates of a point P

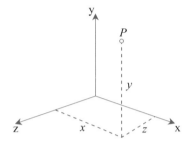

5.7.1   Theorem of Pythagoras in 3D

The theorem of Pythagoras in 3D is a natural extension of the 2D rule. In fact, it
even works in higher dimensions. Given two arbitrary points $P_1(x_1,\ y_1,\ z_1)$ and
$P_2(x_2,\ y_2,\ z_2)$, we compute $\Delta x = x_2 - x_1$, $\Delta y = y_2 - y_1$ and $\Delta z = z_2 - z_1$, from
which the distance d between P_1 and P_2 is given by

$$d = \sqrt{(\Delta x)^2 + (\Delta y)^2 + (\Delta z)^2}$$

and the distance from the origin to a point $P(x,\ y,\ z)$ is simply

$$d = \sqrt{x^2 + y^2 + z^2}.$$

Therefore, the point $(3,\ 4,\ 5)$ is $\sqrt{3^2 + 4^2 + 5^2} \approx 7.07$ from the origin.

5.8   Polar Coordinates

Polar coordinates are used for handling data containing angles, rather than linear
offsets. Figure 5.7 shows the convention used for 2D polar coordinates, where the

Fig. 5.7 2D polar
coordinates

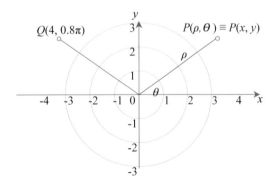

point $P(x,\ y)$ has equivalent polar coordinates $P(\rho,\ \theta)$, where:

$$x = \rho \cos \theta$$
$$y = \rho \sin \theta$$
$$\rho = \sqrt{x^2 + y^2}$$
$$\theta = \arctan\left(\frac{y}{x}\right).$$

For example, the point $Q(4,\ 0.8\pi)$ in Fig. 5.7 has Cartesian coordinates:

$$x = 4\cos(0.8\pi) \approx -3.24$$
$$y = 4\sin(0.8\pi) \approx 2.35$$

and the point $(3,\ 4)$ has polar coordinates:

$$\rho = \sqrt{3^2 + 4^2} = 5$$
$$\theta = \arctan\left(\tfrac{4}{3}\right) \approx 53.13°.$$

These conversion formulae work only for the first quadrant. The atan2 function should be used in a software environment, as it works with all four quadrants.

5.9 Spherical Polar Coordinates

Figure 5.8 shows one convention used for spherical polar coordinates, where the point $P(x,\ y,\ z)$ has equivalent polar coordinates $P(\rho,\ \phi,\ \theta)$, where:

$$x = \rho \sin \phi \cos \theta$$
$$y = \rho \sin \phi \sin \theta$$

Fig. 5.8 Spherical polar coordinates

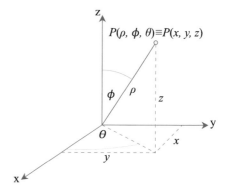

$$z = \rho \cos \phi$$
$$\rho = \sqrt{x^2 + y^2 + z^2}$$
$$\phi = \arccos\left(\frac{z}{\rho}\right)$$
$$\theta = \arctan\left(\frac{y}{x}\right).$$

For example, the point $(3, \ 4, \ 0)$ has spherical polar coordinates $(5, \ 90°, \ 53.13°)$:

$$\rho = \sqrt{3^2 + 4^2 + 0^2} = 5$$
$$\phi = \arccos\left(\frac{0}{5}\right) = 90°$$
$$\theta = \arctan\left(\frac{4}{3}\right) \approx 53.13°.$$

Take great care when using spherical coordinates, as authors often swap ϕ with θ, as well as the alignment of the Cartesian axes; not to mention using a left-handed axial system in preference to a right-handed system!

5.10   Cylindrical Coordinates

Figure 5.9 shows one convention used for cylindrical coordinates, where the point $P(x, \ y, \ z)$ has equivalent cylindrical coordinates $P(\rho, \ \theta, \ z)$, where

$$x = \rho \cos \theta$$
$$y = \rho \sin \theta$$
$$z = z$$
$$\rho = \sqrt{x^2 + y^2}$$
$$\theta = \arctan\left(\frac{y}{x}\right).$$

Fig. 5.9 Cylindrical
coordinates

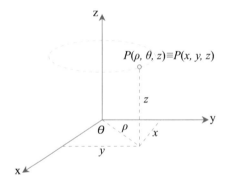

For example, the point $(3, 4, 6)$ has cylindrical coordinates $(5, 53.13°, 6)$:

$$\rho = \sqrt{3^2 + 4^2} = 5$$
$$\theta = \arctan\left(\frac{4}{3}\right) \approx 53.13°$$
$$z = 6.$$

Again, be careful when using cylindrical coordinates to ensure compatibility.

5.11 Summary

All of the above coordinate systems are used in computer graphics. Unfortunately, there are no rigid standards, so be prepared to adjust the formulae used in other books and technical papers.

5.12 Worked Examples

5.12.1 Area of a Shape

Compute the area and orientation of the shape defined by the coordinates in Table 5.2.

$$\text{area} = \frac{1}{2}[(2 \times 2 - 0 \times 2) + (2 \times 2 - 2 \times 1) + (1 \times 1 - 2 \times 1) + (1 \times 1 - 1 \times 0)]$$
$$= \frac{1}{2}(4 + 2 - 1 + 1)$$
$$= 3.$$

The shape is oriented anticlockwise, as the area is positive.

5.12.2 Distance Between Two Points

Find the distance d_{12} between $P_1(1, 1)$ and $P_2(6, 7)$, and d_{34} between $P_3(1, 1, 1)$ and $P_4(7, 8, 9)$.

$$d_{12} = \sqrt{(6 - 1)^2 + (7 - 1)^2} = \sqrt{61} \approx 7.81$$
$$d_{34} = \sqrt{(7 - 1)^2 + (8 - 1)^2 + (9 - 1)^2} = \sqrt{149} \approx 12.21.$$

Table 5.2 Coordinates of a shape

x	0	2	2	1	1	0
y	0	0	2	2	1	1

5.12.3 Polar Coordinates

Convert the 2D polar coordinates $\left(3,\ \frac{\pi}{2}\right)$ to Cartesian form, and the point $(4,\ 5)$ to polar form.

$$\rho = 3$$
$$\theta = \frac{\pi}{2}\,[\text{rad}]$$
$$x = \rho\cos\theta = 3\cos\left(\frac{\pi}{2}\right) = 0$$
$$y = \rho\sin\theta = 3\sin\left(\frac{\pi}{2}\right) = 3$$

therefore, $\left(3,\ \frac{\pi}{2}\right) \equiv (0,\ 3)$.

$$x = 4$$
$$y = 5$$
$$\rho = \sqrt{x^2 + y^2} = \sqrt{4^2 + 5^2} \approx 6.4$$
$$\theta = \arctan\left(\frac{y}{x}\right) = \arctan\left(\frac{5}{4}\right) \approx 51.34°$$

therefore, $(4,\ 5) \approx (6.4,\ 51.34°)$.

5.12.4 Spherical Polar Coordinates

Convert the spherical polar coordinates $\left(10,\ \frac{\pi}{2},\ 45°\right)$ to Cartesian form, and the point $(3,\ 4,\ 5)$ to spherical form.

$$\rho = 10$$
$$\phi = \frac{\pi}{2}\,[\text{rad}] = 90°$$
$$\theta = 45°$$
$$x = \rho\sin\phi\cos\theta = 10\sin 90°\cos 45° = 10\frac{\sqrt{2}}{2} \approx 7.07$$
$$y = \rho\sin\phi\sin\theta = 10\sin 90°\sin 45° = 10\frac{\sqrt{2}}{2} \approx 7.07$$
$$z = \rho\cos\phi = 10\cos 90° = 0$$

therefore, $\left(10,\ \frac{\pi}{2},\ 45°\right) \approx (7.07,\ 7.07,\ 0)$.

$$x = 3$$
$$y = 4$$
$$z = 5$$
$$\rho = \sqrt{x^2 + y^2 + z^2} = \sqrt{3^2 + 4^2 + 5^2} \approx 7.07$$

$$\phi = \arccos\left(\frac{z}{\rho}\right) \approx \arccos\left(\frac{5}{7.07}\right) = 45°$$

$$\theta = \arctan\left(\frac{y}{x}\right) = \arctan\left(\frac{4}{3}\right) \approx 53.13°$$

therefore, $(3, \ 4, \ 5) \approx (7.07, \ 45°, \ 53.13°)$.

5.12.5 Cylindrical Coordinates

Convert the 3D cylindrical coordinates $\left(10, \ \frac{\pi}{2}, \ 5\right)$ to Cartesian form, and the point $(3, \ 4, \ 5)$ to cylindrical form.

$$\rho = 10$$
$$\theta = \frac{\pi}{2}[\text{rad}]$$
$$z = 5$$
$$x = \rho \cos\theta = 10\cos\left(\frac{\pi}{2}\right) = 0$$
$$y = \rho \sin\theta = 10\sin\left(\frac{\pi}{2}\right) = 10$$
$$z = 5$$

therefore, $\left(10, \ \frac{\pi}{2}, \ 5\right) \equiv (0, \ 10, \ 5)$.
  Given the point $(3, \ 4, \ 5)$, then

$$\rho = \sqrt{3^2 + 4^2} = 5$$
$$\theta = \arctan\left(\frac{4}{3}\right) \approx 53.13°$$

therefore, $(3, \ 4, \ 5) \approx (5, \ 53.13°, \ 5)$.

Reference

1. Merzbach UC, Boyer CB (1989) A history of mathematics. Wiley

Chapter 6
Determinants

6.1 Introduction

When patterns of numbers or symbols occur over and over again, mathematicians often devise a way to simplify their description and assign a name to them. For example,

$$\prod_{i=1}^{4} p_i^{\alpha_i}$$

is shorthand for

$$p_1^{\alpha_1} p_2^{\alpha_2} p_3^{\alpha_3} p_4^{\alpha_4}$$

and

$$\sum_{i=1}^{4} p_i^{\alpha_i}$$

is shorthand for

$$p_1^{\alpha_1} + p_2^{\alpha_2} + p_3^{\alpha_3} + p_4^{\alpha_4}.$$

A *determinant* is another example of this process, and is a value derived from a square matrix of terms, often associated with sets of equations. Such problems were studied by the Babylonians around 300 BC and by the Chinese between 200 BC and 100 BC. Since then many mathematicians have been associated with the evolution of determinants and matrices, including Augustin-Louis Cauchy (1789–1857), Arthur Cayley (1821–1895), Girolamo Cardano (1501–1576), Johann Gauss (1777–1855), Pierre-Simon Laplace (1749–1827), Gottfried von Leibniz, Guillaume de L'Hôpital (1661–1704), Takakazu Seki (1642–1708) and Jan de Witt (1625–1672). To understand the rules used to compute a determinant's value, we need to understand their origin, which is in the solution of sets of linear equations.

© Springer-Verlag London Ltd., part of Springer Nature 2022
J. Vince, *Mathematics for Computer Graphics*, Undergraduate Topics
in Computer Science, https://doi.org/10.1007/978-1-4471-7520-9_6

6.2   Linear Equations with Two Variables

Consider the following linear equations where we want to find values of x and y that satisfy both equations:

$$7 = 3x + 2y \tag{6.1}$$
$$10 = 2x + 4y. \tag{6.2}$$

A standard way to resolve this problem is to multiply (6.1) by 2 and subtract (6.2) from (6.1), which removes the y-terms:

$$14 = 6x + 2y$$
$$10 = 2x + 4y$$
$$4 = 4x$$
$$x = 1.$$

Substituting $x = 1$ in (6.1) reveals the value of y:

$$7 = 3 + 2y$$
$$4 = 2y$$
$$y = 2.$$

Therefore, $x = 1$ and $y = 2$, solves (6.1) and (6.2).

The equations must be linearly independent, otherwise we only have one equation. For example, starting with

$$7 = 3x + 2y$$
$$14 = 6x + 4y$$

is a futile exercise, as the second equation is double the first, and does not provide any extra information.

To find a general solution to this problem, we start with

$$d_1 = a_1 x + b_1 y \tag{6.3}$$
$$d_2 = a_2 x + b_2 y. \tag{6.4}$$

Multiply (6.3) by b_2 and (6.4) by b_1:

$$d_1 b_2 = a_1 b_2 x + b_1 b_2 y \tag{6.5}$$
$$b_1 d_2 = b_1 a_2 x + b_1 b_2 y. \tag{6.6}$$

Subtract (6.6) from (6.5):

$$d_1 b_2 - b_1 d_2 = a_1 b_2 x - b_1 a_2 x$$
$$= (a_1 b_2 - b_1 a_2) x$$
$$x = \frac{d_1 b_2 - b_1 d_2}{a_1 b_2 - b_1 a_2}. \tag{6.7}$$

To find y, multiply (6.3) by a_2 and (6.4) by a_1:

$$d_1 a_2 = a_2 a_1 x + b_1 a_2 y \tag{6.8}$$
$$a_1 d_2 = a_2 a_1 x + a_1 b_2 y. \tag{6.9}$$

Subtract (6.8) from (6.9):

$$a_1 d_2 - d_1 a_2 = a_1 b_2 y - b_1 a_2 y$$
$$= (a_1 b_2 - b_1 a_2) y$$
$$y = \frac{a_1 d_2 - d_1 a_2}{a_1 b_2 - b_1 a_2}. \tag{6.10}$$

Observe that both (6.7) and (6.10) share the common denominator: $a_1 b_2 - b_1 a_2$. Furthermore, note the positions of a_1, b_1, a_2 and b_2 in the original equations:

$$\begin{matrix} a_1 & b_1 \\ a_2 & b_2 \end{matrix}$$

and the denominator is formed by cross-multiplying the diagonal terms $a_1 b_2$ and subtracting the other cross-multiplied terms $b_1 a_2$. Placing the four terms between two vertical lines creates a *second-order determinant* whose value equals:

$$\begin{vmatrix} a_1 & b_1 \\ a_2 & b_2 \end{vmatrix} = a_1 b_2 - b_1 a_2.$$

Although the name was originally given by Johann Gauss, it was Augustin-Louis Cauchy who clarified its current modern identity.

If the original equations are linearly related by a factor λ, the determinant equals zero:

$$\begin{vmatrix} a_1 & b_1 \\ \lambda a_1 & \lambda b_1 \end{vmatrix} = a_1 \lambda b_1 - b_1 \lambda a_1 = 0.$$

Observe that the numerators of (6.7) and (6.10) are also second-order determinants:

$$\begin{vmatrix} d_1 & b_1 \\ d_2 & b_2 \end{vmatrix} = d_1 b_2 - b_1 d_2$$

and

$$\begin{vmatrix} a_1 & d_1 \\ a_2 & d_2 \end{vmatrix} = a_1 d_2 - d_1 a_2$$

which means that Eqs. (6.7) and (6.10) can be written using determinants:

$$x = \frac{\begin{vmatrix} d_1 & b_1 \\ d_2 & b_2 \end{vmatrix}}{\begin{vmatrix} a_1 & b_1 \\ a_2 & b_2 \end{vmatrix}}, \qquad y = \frac{\begin{vmatrix} a_1 & d_1 \\ a_2 & d_2 \end{vmatrix}}{\begin{vmatrix} a_1 & b_1 \\ a_2 & b_2 \end{vmatrix}}.$$

And one final piece of algebra permits the solution to be written as

$$\frac{x}{\begin{vmatrix} d_1 & b_1 \\ d_2 & b_2 \end{vmatrix}} = \frac{y}{\begin{vmatrix} a_1 & d_1 \\ a_2 & d_2 \end{vmatrix}} = \frac{1}{\begin{vmatrix} a_1 & b_1 \\ a_2 & b_2 \end{vmatrix}}. \tag{6.11}$$

Observe another pattern in (6.11) where the determinant is

$$\begin{vmatrix} a_1 & b_1 \\ a_2 & b_2 \end{vmatrix}$$

but the d-terms replace the x-coefficients:

$$\begin{vmatrix} d_1 & b_1 \\ d_2 & b_2 \end{vmatrix}$$

and then the y-coefficients

$$\begin{vmatrix} a_1 & d_1 \\ a_2 & d_2 \end{vmatrix}.$$

Returning to the original equations:

$$7 = 3x + 2y$$
$$10 = 2x + 4y$$

and substituting the constants in (6.11), we have

$$\frac{x}{\begin{vmatrix} 7 & 2 \\ 10 & 4 \end{vmatrix}} = \frac{y}{\begin{vmatrix} 3 & 7 \\ 2 & 10 \end{vmatrix}} = \frac{1}{\begin{vmatrix} 3 & 2 \\ 2 & 4 \end{vmatrix}}$$

which, when expanded reveals

$$\frac{x}{28 - 20} = \frac{y}{30 - 14} = \frac{1}{12 - 4}$$

$$\frac{x}{8} = \frac{y}{16} = \frac{1}{8}$$

making $x = 1$ and $y = 2$.

Let's try another example:

$$11 = 4x + y$$
$$5 = x + y$$

and substituting the constants in (6.11), we have

$$\frac{x}{\begin{vmatrix} 11 & 1 \\ 5 & 1 \end{vmatrix}} = \frac{y}{\begin{vmatrix} 4 & 11 \\ 1 & 5 \end{vmatrix}} = \frac{1}{\begin{vmatrix} 4 & 1 \\ 1 & 1 \end{vmatrix}}$$

which, when expanded reveals

$$\frac{x}{11 - 5} = \frac{y}{20 - 11} = \frac{1}{4 - 1}$$

$$\frac{x}{6} = \frac{y}{9} = \frac{1}{3}$$

making $x = 2$ and $y = 3$.

Now let's see how a *third-order* determinant arises from the coefficients of three equations in three unknowns.

6.3 Linear Equations with Three Variables

Consider the following set of three linear equations:

$$13 = 3x + 2y + 2z \qquad (6.12)$$
$$20 = 2x + 3y + 4z \qquad (6.13)$$
$$7 = 2x + y + z. \qquad (6.14)$$

A standard way to resolve this problem is to multiply (6.12) by 2 and subtract (6.13), which removes the z-terms:

$$26 = 6x + 4y + 4z$$
$$20 = 2x + 3y + 4z$$
$$6 = 4x + y \qquad (6.15)$$

leaving (6.15) with two unknowns.

Next, we take (6.13) and (6.14) and remove the z-term by multiplying (6.14) by 4 and subtract (6.13):

$$28 = 8x + 4y + 4z$$
$$20 = 2x + 3y + 4z$$
$$8 = 6x + y \tag{6.16}$$

leaving (6.16) with two unknowns. We are now left with (6.15) and (6.16):

$$6 = 4x + y$$
$$8 = 6x + y$$

which can be solved using (6.11):

$$\frac{x}{\begin{vmatrix} 6 & 1 \\ 8 & 1 \end{vmatrix}} = \frac{y}{\begin{vmatrix} 4 & 6 \\ 6 & 8 \end{vmatrix}} = \frac{1}{\begin{vmatrix} 4 & 1 \\ 6 & 1 \end{vmatrix}}$$

therefore,

$$x = \frac{6 - 8}{4 - 6} = 1$$
$$y = \frac{32 - 36}{4 - 6} = 2.$$

Substituting $x = 1$ and $y = 2$ in (6.12) reveals that $z = 3$.

We can generalise (6.11) for three equations using third-order determinants:

$$\frac{x}{\begin{vmatrix} d_1 & b_1 & c_1 \\ d_2 & b_2 & c_2 \\ d_3 & b_3 & c_3 \end{vmatrix}} = \frac{y}{\begin{vmatrix} a_1 & d_1 & c_1 \\ a_2 & d_2 & c_2 \\ a_3 & d_3 & c_3 \end{vmatrix}} = \frac{z}{\begin{vmatrix} a_1 & b_1 & d_1 \\ a_2 & b_2 & d_2 \\ a_3 & b_3 & d_3 \end{vmatrix}} = \frac{1}{\begin{vmatrix} a_1 & b_1 & c_1 \\ a_2 & b_2 & c_2 \\ a_3 & b_3 & c_3 \end{vmatrix}}. \tag{6.17}$$

Once again, there is an important pattern in (6.17) where the underlying determinant is

$$\begin{vmatrix} a_1 & b_1 & c_1 \\ a_2 & b_2 & c_2 \\ a_3 & b_3 & c_3 \end{vmatrix}$$

but the d-terms replace the x-coefficients:

$$\begin{vmatrix} d_1 & b_1 & c_1 \\ d_2 & b_2 & c_2 \\ d_3 & b_3 & c_3 \end{vmatrix}$$

the d-terms replace the y-coefficients:

$$\begin{vmatrix} a_1 & d_1 & c_1 \\ a_2 & d_2 & c_2 \\ a_3 & d_3 & c_3 \end{vmatrix}$$

and the d-terms replace the z-coefficients:

$$\begin{vmatrix} a_1 & b_1 & d_1 \\ a_2 & b_2 & d_2 \\ a_3 & b_3 & d_3 \end{vmatrix}.$$

We must now find a way of computing the value of a third-order determinant, which requires the following algebraic analysis of three equations in three unknowns. We start with three linear equations:

$$d_1 = a_1 x + b_1 y + c_1 z \qquad (6.18)$$
$$d_2 = a_2 x + b_2 y + c_2 z \qquad (6.19)$$
$$d_3 = a_3 x + b_3 y + c_3 z \qquad (6.20)$$

and derive one equation in two unknowns from (6.18) and (6.19), and another from (6.19) and (6.20).

We multiply (6.18) by c_2, (6.19) by c_1 and subtract them:

$$c_2 d_1 = a_1 c_2 x + b_1 c_2 y + c_1 c_2 z$$
$$c_1 d_2 = c_1 a_2 x + b_2 c_1 y + c_1 c_2 z$$
$$c_2 d_1 - c_1 d_2 = (a_1 c_2 - c_1 a_2)x + (b_1 c_2 - b_2 c_1)y. \qquad (6.21)$$

Next, we multiply (6.19) by c_3, (6.20) by c_2 and subtract them:

$$c_3 d_2 = a_2 c_3 x + b_2 c_3 y + c_2 c_3 z$$
$$c_2 d_3 = a_3 c_2 x + b_3 c_2 y + c_2 c_3 z$$
$$c_3 d_2 - c_2 d_3 = (a_2 c_3 - a_3 c_2)x + (b_2 c_3 - b_3 c_2)y. \qquad (6.22)$$

Simplify (6.21) by letting

$$e_1 = c_2 d_1 - c_1 d_2$$
$$f_1 = a_1 c_2 - c_1 a_2$$
$$g_1 = b_1 c_2 - b_2 c_1$$

therefore,

$$e_1 = f_1 x + g_1 y. \tag{6.23}$$

Simplify (6.22) by letting

$$e_2 = c_3 d_2 - c_2 d_3$$
$$f_2 = a_2 c_3 - a_3 c_2$$
$$g_2 = b_2 c_3 - b_3 c_2$$

therefore,

$$e_2 = f_2 x + g_2 y. \tag{6.24}$$

Now we have two equations in two unknowns:

$$e_1 = f_1 x + g_1 y$$
$$e_2 = f_2 x + g_2 y$$

which are solved using

$$\frac{x}{A} = \frac{y}{B} = \frac{1}{C} \tag{6.25}$$

where

$$A = \begin{vmatrix} e_1 & g_1 \\ e_2 & g_2 \end{vmatrix} = \begin{vmatrix} c_2 d_1 - c_1 d_2 & b_1 c_2 - b_2 c_1 \\ c_3 d_2 - c_2 d_3 & b_2 c_3 - b_3 c_2 \end{vmatrix} \tag{6.26}$$

$$B = \begin{vmatrix} f_1 & e_1 \\ f_2 & e_2 \end{vmatrix} = \begin{vmatrix} a_1 c_2 - c_1 a_2 & c_2 d_1 - c_1 d_2 \\ a_2 c_3 - a_3 c_2 & c_3 d_2 - c_2 d_3 \end{vmatrix} \tag{6.27}$$

$$C = \begin{vmatrix} f_1 & g_1 \\ f_2 & g_2 \end{vmatrix} = \begin{vmatrix} a_1 c_2 - c_1 a_2 & b_1 c_2 - b_2 c_1 \\ a_2 c_3 - a_3 c_2 & b_2 c_3 - b_3 c_2 \end{vmatrix} \tag{6.28}$$

We first compute A, from which we can derive B, because the only difference between (6.26) and (6.27) is that d_1, d_2, d_3 become a_1, a_2, a_3 respectively, and b_1, b_2, b_3 become d_1, d_2, d_3 respectively.

We can derive C from A, as the only difference between (6.26) and (6.28) is that d_1, d_2, d_3 become a_1, a_2, a_3 respectively. Starting with A:

$$A = (c_2d_1 - c_1d_2)(b_2c_3 - b_3c_2) - (b_1c_2 - b_2c_1)(c_3d_2 - c_2d_3)$$
$$= b_2c_2c_3d_1 - b_3c_2^2d_1 - b_2c_1c_3d_2 + b_3c_1c_2d_2$$
$$\quad - b_1c_2c_3d_2 + b_1c_2^2d_3 + b_2c_1c_3d_2 - b_2c_1c_2d_3$$
$$= b_2c_2c_3d_1 - b_3c_2^2d_1 + b_3c_1c_2d_2 - b_1c_2c_3d_2 + b_1c_2^2d_3 - b_2c_1c_2d_3$$
$$= c_2(b_2c_3d_1 - b_3c_2d_1 + b_3c_1d_2 - b_1c_3d_2 + b_1c_2d_3 - b_2c_1d_3)$$
$$A = c_2\Big(d_1(b_2c_3 - c_2b_3) - b_1(d_2c_3 - c_2d_3) + c_1(d_2b_3 - b_2d_3)\Big). \tag{6.29}$$

Using the substitutions described above we can derive B and C from (6.29):

$$B = c_2\Big(a_1(d_2c_3 - c_2d_3) - b_1(a_2c_3 - c_2a_3) + c_1(a_2d_3 - d_2a_3)\Big) \tag{6.30}$$

$$C = c_2\Big(a_1(b_2c_3 - c_2b_3) - b_1(a_2c_3 - c_2a_3) + c_1(a_2b_3 - b_2a_3)\Big). \tag{6.31}$$

We can now rewrite (6.29)–(6.31) using determinant notation. At the same time, we can drop the c_2 terms as they cancel out when computing x, y and z:

$$A = d_1 \begin{vmatrix} b_2 & c_2 \\ b_3 & c_3 \end{vmatrix} - b_1 \begin{vmatrix} d_2 & c_2 \\ d_3 & c_3 \end{vmatrix} + c_1 \begin{vmatrix} d_2 & b_2 \\ d_3 & b_3 \end{vmatrix} \tag{6.32}$$

$$B = a_1 \begin{vmatrix} d_2 & c_2 \\ d_3 & c_3 \end{vmatrix} - d_1 \begin{vmatrix} a_2 & c_2 \\ a_3 & c_3 \end{vmatrix} + c_1 \begin{vmatrix} a_2 & d_2 \\ a_3 & d_3 \end{vmatrix} \tag{6.33}$$

$$C = a_1 \begin{vmatrix} b_2 & c_2 \\ b_3 & c_3 \end{vmatrix} - b_1 \begin{vmatrix} a_2 & c_2 \\ a_3 & c_3 \end{vmatrix} + c_1 \begin{vmatrix} a_2 & b_2 \\ a_3 & b_3 \end{vmatrix}. \tag{6.34}$$

As (6.17) and (6.25) refer to the same x and y, then

$$\begin{vmatrix} d_1 & b_1 & c_1 \\ d_2 & b_2 & c_2 \\ d_3 & b_3 & c_3 \end{vmatrix} = d_1 \begin{vmatrix} b_2 & c_2 \\ b_3 & c_3 \end{vmatrix} - b_1 \begin{vmatrix} d_2 & c_2 \\ d_3 & c_3 \end{vmatrix} + c_1 \begin{vmatrix} d_2 & b_2 \\ d_3 & b_3 \end{vmatrix} \tag{6.35}$$

$$\begin{vmatrix} a_1 & d_1 & c_1 \\ a_2 & d_2 & c_2 \\ a_3 & d_3 & c_3 \end{vmatrix} = a_1 \begin{vmatrix} d_2 & c_2 \\ d_3 & c_3 \end{vmatrix} - d_1 \begin{vmatrix} a_2 & c_2 \\ a_3 & c_3 \end{vmatrix} + c_1 \begin{vmatrix} a_2 & d_2 \\ a_3 & d_3 \end{vmatrix} \tag{6.36}$$

$$\begin{vmatrix} a_1 & b_1 & c_1 \\ a_2 & b_2 & c_2 \\ a_3 & b_3 & c_3 \end{vmatrix} = a_1 \begin{vmatrix} b_2 & c_2 \\ b_3 & c_3 \end{vmatrix} - b_1 \begin{vmatrix} a_2 & c_2 \\ a_3 & c_3 \end{vmatrix} + c_1 \begin{vmatrix} a_2 & b_2 \\ a_3 & b_3 \end{vmatrix}. \tag{6.37}$$

As a consistent algebraic analysis has been pursued to derive (6.35)–(6.37), a consistent pattern has surfaced in Fig. 6.1 which shows how the three determinants are evaluated. This pattern comprises taking each entry in the top row, called a *cofactor*, and multiplying it by the determinant of entries in rows 2 and 3, whilst ignoring the column containing the original term, called a *first minor*. Observe that the second term of the top row is switched negative, called an *inversion correction factor*.

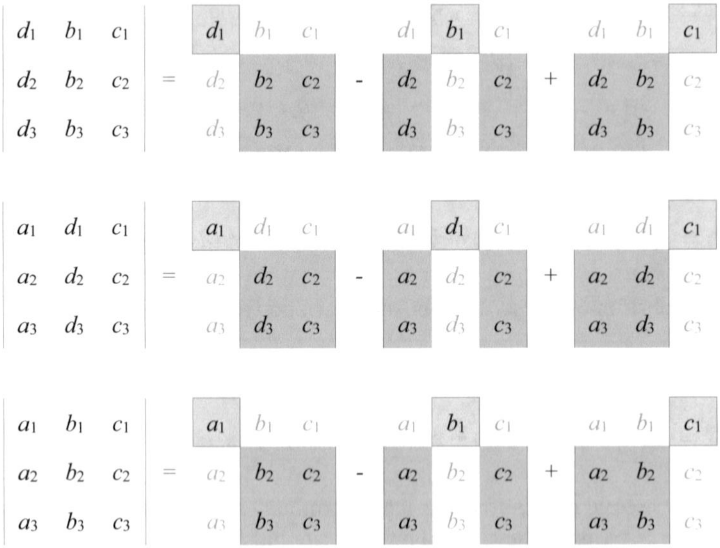

$$\begin{vmatrix} d_1 & b_1 & c_1 \\ d_2 & b_2 & c_2 \\ d_3 & b_3 & c_3 \end{vmatrix} = \begin{vmatrix} d_1 & b_1 & c_1 \\ d_2 & b_2 & c_2 \\ d_3 & b_3 & c_3 \end{vmatrix} - \begin{vmatrix} d_1 & b_1 & c_1 \\ d_2 & b_2 & c_2 \\ d_3 & b_3 & c_3 \end{vmatrix} + \begin{vmatrix} d_1 & b_1 & c_1 \\ d_2 & b_2 & c_2 \\ d_3 & b_3 & c_3 \end{vmatrix}$$

$$\begin{vmatrix} a_1 & d_1 & c_1 \\ a_2 & d_2 & c_2 \\ a_3 & d_3 & c_3 \end{vmatrix} = \begin{vmatrix} a_1 & d_1 & c_1 \\ a_2 & d_2 & c_2 \\ a_3 & d_3 & c_3 \end{vmatrix} - \begin{vmatrix} a_1 & d_1 & c_1 \\ a_2 & d_2 & c_2 \\ a_3 & d_3 & c_3 \end{vmatrix} + \begin{vmatrix} a_1 & d_1 & c_1 \\ a_2 & d_2 & c_2 \\ a_3 & d_3 & c_3 \end{vmatrix}$$

$$\begin{vmatrix} a_1 & b_1 & c_1 \\ a_2 & b_2 & c_2 \\ a_3 & b_3 & c_3 \end{vmatrix} = \begin{vmatrix} a_1 & b_1 & c_1 \\ a_2 & b_2 & c_2 \\ a_3 & b_3 & c_3 \end{vmatrix} - \begin{vmatrix} a_1 & b_1 & c_1 \\ a_2 & b_2 & c_2 \\ a_3 & b_3 & c_3 \end{vmatrix} + \begin{vmatrix} a_1 & b_1 & c_1 \\ a_2 & b_2 & c_2 \\ a_3 & b_3 & c_3 \end{vmatrix}$$

Fig. 6.1 Evaluating the determinants shown in (6.35)–(6.37)

Let's repeat (6.31) again without the c_2 term, as it has nothing to do with the calculation of the determinant.

$$C = a_1(b_2c_3 - c_2b_3) - b_1(a_2c_3 - c_2a_3) + c_1(a_2b_3 - b_2a_3). \tag{6.38}$$

It is possible to arrange the terms of (6.38) as a square matrix such that each row and column sums to C:

$$a_1(b_2c_3 - c_2b_3) - b_1(a_2c_3 - c_2a_3) + c_1(a_2b_3 - b_2a_3)$$
$$-a_2(b_1c_3 - c_1b_3) + b_2(a_1c_3 - c_1a_3) - c_2(a_1b_3 - b_1a_3)$$
$$a_3(b_1c_2 - c_1b_2) - b_3(a_1c_2 - c_1a_2) + c_3(a_1b_2 - b_1a_2)$$

which means that there are six ways to evaluate the determinant C: summing the rows, or summing the columns. Figure 6.2 shows this arrangement with the cofactors in blue, and the first minor determinants in green. Observe how the signs alternate between the terms.

Having discovered the origins of these patterns, let's evaluate the original equations declared at the start of this section using (6.11)

$$13 = 3x + 2y + 2z$$
$$20 = 2x + 3y + 4z$$
$$7 = 2x + y + z.$$

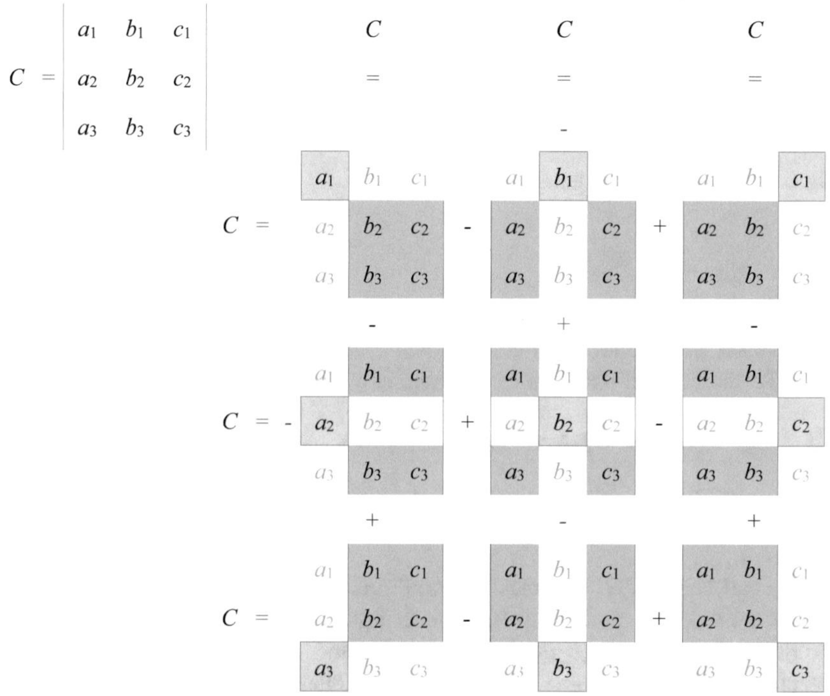

$$C = \begin{vmatrix} a_1 & b_1 & c_1 \\ a_2 & b_2 & c_2 \\ a_3 & b_3 & c_3 \end{vmatrix}$$

Fig. 6.2 The patterns of multipliers with their respective second-order determinants

$$\frac{x}{\begin{vmatrix} d_1 & b_1 & c_1 \\ d_2 & b_2 & c_2 \\ d_3 & b_3 & c_3 \end{vmatrix}} = \frac{y}{\begin{vmatrix} a_1 & d_1 & c_1 \\ a_2 & d_2 & c_2 \\ a_3 & d_3 & c_3 \end{vmatrix}} = \frac{z}{\begin{vmatrix} a_1 & b_1 & d_1 \\ a_2 & b_2 & d_2 \\ a_3 & b_3 & d_3 \end{vmatrix}} = \frac{1}{\begin{vmatrix} a_1 & b_1 & c_1 \\ a_2 & b_2 & c_2 \\ a_3 & b_3 & c_3 \end{vmatrix}}$$

therefore,

$$\frac{x}{\begin{vmatrix} 13 & 2 & 2 \\ 20 & 3 & 4 \\ 7 & 1 & 1 \end{vmatrix}} = \frac{y}{\begin{vmatrix} 3 & 13 & 2 \\ 2 & 20 & 4 \\ 2 & 7 & 1 \end{vmatrix}} = \frac{z}{\begin{vmatrix} 3 & 2 & 13 \\ 2 & 3 & 20 \\ 2 & 1 & 7 \end{vmatrix}} = \frac{1}{\begin{vmatrix} 3 & 2 & 2 \\ 2 & 3 & 4 \\ 2 & 1 & 1 \end{vmatrix}}$$

computing the determinants using the top row entries as cofactors:

$$\frac{x}{-13 + 16 - 2} = \frac{y}{-24 + 78 - 52} = \frac{z}{3 + 52 - 52} = \frac{1}{-3 + 12 - 8}$$

$$\frac{x}{1} = \frac{y}{2} = \frac{z}{3} = \frac{1}{1}$$

therefore, $x = 1$, $y = 2$ and $z = 3$.

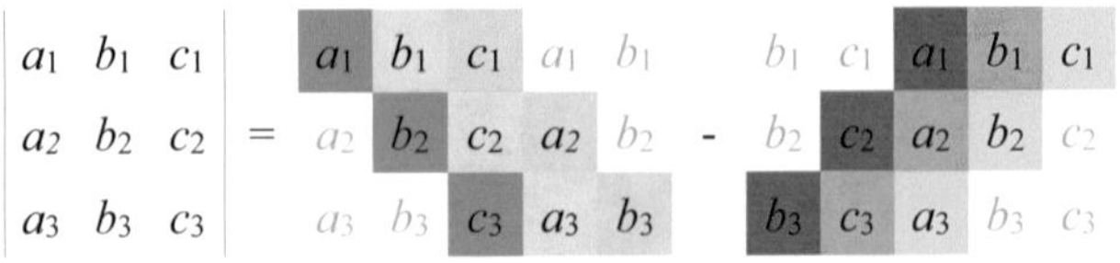

Fig. 6.3 The pattern behind Sarrus's rule

6.3.1 Sarrus's Rule

The French mathematician Pierre Sarrus (1798–1861) discovered another way to compute the value of a third-order determinant, that arises from (6.38):

$$
\begin{aligned}
C &= a_1(b_2c_3 - c_2b_3) - b_1(a_2c_3 - c_2a_3) + c_1(a_2b_3 - b_2a_3) \\
&= a_1b_2c_3 - a_1c_2b_3 - b_1a_2c_3 + b_1c_2a_3 + c_1a_2b_3 - c_1b_2a_3 \\
&= a_1b_2c_3 + b_1c_2a_3 + c_1a_2b_3 - a_1c_2b_3 - b_1a_2c_3 - c_1b_2a_3. \quad (6.39)
\end{aligned}
$$

The pattern in (6.39) becomes clear in Fig. 6.3, where the first two columns of the matrix are repeated, and comprises two diagonal sets of terms: on the left in blue, we have the products $a_1b_2c_3$, $b_1c_2a_3$, $c_1a_2b_3$, and on the right in red and orange, the products $a_1c_2b_3$, $b_1a_2c_3$, $c_1b_2a_3$. These diagonal patterns provide a useful *aide-mémoire* when computing the determinant. Unfortunately, this rule only applies to third-order determinants.

6.4 Mathematical Notation

Having discovered the background of determinants, now let's explore a formal description of their structure and characteristics.

6.4.1 Matrix

In the following definitions, a *matrix* is a square array of entries, with an equal number of rows and columns. The entries may be numbers, vectors, complex numbers or even partial differentials, in the case of a Jacobian. In general, each entry is identified by two subscripts *row col*:

$$a_{row\ col}.$$

A matrix with n rows and m columns has the following entries:

$$\begin{matrix} a_{11} & a_{12} & \dots & a_{1m} \\ a_{21} & a_{22} & \dots & a_{2m} \\ \vdots & \vdots & \ddots & \vdots \\ a_{n1} & a_{n2} & \dots & a_{nm} \end{matrix}$$

The entries lying on the two diagonals are identified as follows: a_{11} and a_{nm} lie on the *main diagonal*, and a_{1m} and a_{n1} lie on the *secondary diagonal*.

6.4.2 Order of a Determinant

The *order* of a square determinant equals the number of rows or columns. For example, a first-order determinant contains a single entry; a second-order determinant has two rows and two columns; and a third-order determinant has three rows and three columns.

6.4.3 Value of a Determinant

A determinant posses a unique, single value derived from its entries. The algorithms used to compute this value must respect the algebra associated with solving sets of linear equations, as discussed above.

Pierre-Simon Laplace developed a way to expand the determinant of any order. The Laplace expansion is the idea described above and shown in Fig. 6.1, where cofactors and first minors or *principal minors* are used. For example, starting with a fourth-order determinant, when any row **and** column are removed, the remaining entries create a third-order determinant, called the *first minor* of the original determinant.

The following equation is used to control the sign of each cofactor:

$$(-1)^{row+col}$$

which, for a fourth-order determinant creates:

$$\begin{vmatrix} + & - & + & - \\ - & + & - & + \\ + & - & + & - \\ - & + & - & + \end{vmatrix}.$$

The Laplace expansion begins by choosing a convenient row or column as the source of cofactors. Any zeros are particularly useful, as they cancel out any contribution by the first minor determinant. It then sums the products of every cofactor in the chosen row or column, with its associated first minor, including an appropriate inversion correction factor to adjust the sign changes. The final result is the determinant's value.

A first-order determinant:

$$\left| a_{11} \right| = a_{11}.$$

A second-order determinant:

$$\begin{vmatrix} a_{11} & a_{12} \\ a_{21} & a_{22} \end{vmatrix} = a_{11}a_{22} - a_{12}a_{21}.$$

A third-order determinant using the Laplace expansion with cofactors from the first row:

$$\begin{vmatrix} a_{11} & a_{12} & a_{13} \\ a_{21} & a_{22} & a_{23} \\ a_{31} & a_{32} & a_{33} \end{vmatrix} = a_{11} \begin{vmatrix} a_{22} & a_{23} \\ a_{32} & a_{33} \end{vmatrix} - a_{12} \begin{vmatrix} a_{21} & a_{23} \\ a_{31} & a_{33} \end{vmatrix} + a_{13} \begin{vmatrix} a_{21} & a_{22} \\ a_{31} & a_{32} \end{vmatrix}.$$

A fourth-order determinant using the Laplace expansion with cofactors from the first row:

$$a_{11} \begin{vmatrix} a_{22} & a_{23} & a_{24} \\ a_{32} & a_{33} & a_{34} \\ a_{42} & a_{43} & a_{44} \end{vmatrix} - a_{12} \begin{vmatrix} a_{21} & a_{23} & a_{24} \\ a_{31} & a_{33} & a_{34} \\ a_{41} & a_{43} & a_{44} \end{vmatrix} +$$

$$a_{13} \begin{vmatrix} a_{21} & a_{22} & a_{24} \\ a_{31} & a_{32} & a_{34} \\ a_{41} & a_{42} & a_{44} \end{vmatrix} - a_{14} \begin{vmatrix} a_{21} & a_{22} & a_{23} \\ a_{31} & a_{32} & a_{33} \\ a_{41} & a_{42} & a_{43} \end{vmatrix} = \begin{vmatrix} a_{11} & a_{12} & a_{13} & a_{14} \\ a_{21} & a_{22} & a_{23} & a_{24} \\ a_{31} & a_{32} & a_{33} & a_{34} \\ a_{41} & a_{42} & a_{43} & a_{44} \end{vmatrix}$$

Sarrus's rule is useful to compute a third-order determinant:

$$\begin{vmatrix} a_{11} & a_{12} & a_{13} \\ a_{21} & a_{22} & a_{23} \\ a_{31} & a_{32} & a_{33} \end{vmatrix} = a_{11}a_{22}a_{33} + a_{12}a_{23}a_{31} + a_{13}a_{21}a_{32} -$$

$$a_{11}a_{23}a_{32} - a_{12}a_{21}a_{33} + a_{13}a_{22}a_{31}$$

The Laplace expansion works with higher-order determinants, as any first minor can itself be expanded using the same expansion.

6.4.4 Properties of Determinants

If a determinant contains a row or column of zeros, the Laplace expansion implies that the value of the determinant is zero.

$$\begin{vmatrix} 3 & 0 & 2 \\ 2 & 0 & 4 \\ 2 & 0 & 1 \end{vmatrix} = 0.$$

If a determinant's rows and columns are interchanged, the Laplace expansion also implies that the value of the determinant is unchanged.

$$\begin{vmatrix} 3 & 12 & 2 \\ 2 & 10 & 4 \\ 2 & 8 & 1 \end{vmatrix} = \begin{vmatrix} 3 & 2 & 2 \\ 12 & 10 & 8 \\ 2 & 4 & 1 \end{vmatrix} = -2.$$

If any two rows, or columns, are interchanged, without changing the order of their entries, the determinant's numerical value is unchanged, but its sign is reversed.

$$\begin{vmatrix} 3 & 12 & 2 \\ 2 & 10 & 4 \\ 2 & 8 & 1 \end{vmatrix} = -2$$

$$\begin{vmatrix} 12 & 3 & 2 \\ 10 & 2 & 4 \\ 8 & 2 & 1 \end{vmatrix} = 2.$$

If the entries of a row or column share a common factor, the entries may be adjusted, and the factor placed outside.

$$\begin{vmatrix} 3 & 12 & 2 \\ 2 & 10 & 4 \\ 2 & 8 & 1 \end{vmatrix} = 2 \begin{vmatrix} 3 & 6 & 2 \\ 2 & 5 & 4 \\ 2 & 4 & 1 \end{vmatrix} = -2.$$

6.5 Summary

This chapter has explored the background of determinants and why they exist. In later chapters we discover their role in matrix algebra.

6.6   Worked Examples

6.6.1   Determinant Expansion

Evaluate this determinant using the Laplace expansion and Sarrus's rule.

$$\begin{vmatrix} 1 & 4 & 7 \\ 2 & 5 & 8 \\ 3 & 6 & 9 \end{vmatrix}.$$

Using the Laplace expansion:

$$\begin{vmatrix} 1 & 4 & 7 \\ 2 & 5 & 8 \\ 3 & 6 & 9 \end{vmatrix} = 1 \begin{vmatrix} 5 & 8 \\ 6 & 9 \end{vmatrix} - 2 \begin{vmatrix} 4 & 7 \\ 6 & 9 \end{vmatrix} + 3 \begin{vmatrix} 4 & 7 \\ 5 & 8 \end{vmatrix}$$

$$= 1(45 - 48) - 2(36 - 42) + 3(32 - 35)$$
$$= -3 + 12 - 9$$
$$= 0.$$

Using Sarrus's rule:

$$\begin{vmatrix} 1 & 4 & 7 \\ 2 & 5 & 8 \\ 3 & 6 & 9 \end{vmatrix} = 1 \times 5 \times 9 + 4 \times 8 \times 3 + 7 \times 2 \times 6 - 7 \times 5 \times 3 - 1 \times 8 \times 6 - 4 \times 2 \times 9$$

$$= 45 + 96 + 84 - 105 - 48 - 72$$
$$= 0.$$

6.6.2   Complex Determinant

Evaluate the complex determinant

$$\begin{vmatrix} 4 + i2 & 1 + i \\ 2 - i3 & 3 + i3 \end{vmatrix}.$$

Using the Laplace expansion:

$$\begin{vmatrix} 4+i2 & 1+i \\ 2-i3 & 3+i3 \end{vmatrix} = (4+i2)(3+i3) - (1+i)(2-i3)$$

$$= (12+i18-6) - (2-i+3)$$
$$= 6+i18-5+i$$
$$= 1+i19.$$

6.6.3 Simple Expansion

Write down the simplest expansion of this determinant with its value:

$$\begin{vmatrix} 1 & 2 & 3 \\ 4 & 5 & 0 \\ 6 & 7 & 0 \end{vmatrix}.$$

Using the Laplace expansion with cofactors from the third column:

$$\begin{vmatrix} 1 & 2 & 3 \\ 4 & 5 & 0 \\ 6 & 7 & 0 \end{vmatrix} = 3 \begin{vmatrix} 4 & 5 \\ 6 & 7 \end{vmatrix} = -6.$$

6.6.4 Simultaneous Equations

Solve the following equations using determinants:

$$3 = 2x + y - z$$
$$12 = x + 2y + z$$
$$8 = 3x - 2y + 2z.$$

Using (6.17):

$$\frac{x}{\begin{vmatrix} 3 & 1 & -1 \\ 12 & 2 & 1 \\ 8 & -2 & 2 \end{vmatrix}} = \frac{y}{\begin{vmatrix} 2 & 3 & -1 \\ 1 & 12 & 1 \\ 3 & 8 & 2 \end{vmatrix}} = \frac{z}{\begin{vmatrix} 2 & 1 & 3 \\ 1 & 2 & 12 \\ 3 & -2 & 8 \end{vmatrix}} = \frac{1}{\begin{vmatrix} 2 & 1 & -1 \\ 1 & 2 & 1 \\ 3 & -2 & 2 \end{vmatrix}}.$$

Therefore,

$$x = \frac{\begin{vmatrix} 3 & 1 & -1 \\ 12 & 2 & 1 \\ 8 & -2 & 2 \end{vmatrix}}{\begin{vmatrix} 2 & 1 & -1 \\ 1 & 2 & 1 \\ 3 & -2 & 2 \end{vmatrix}} = \frac{18 - 16 + 40}{12 + 1 + 8} = \frac{42}{21} = 2,$$

$$y = \frac{\begin{vmatrix} 2 & 3 & -1 \\ 1 & 12 & 1 \\ 3 & 8 & 2 \end{vmatrix}}{\begin{vmatrix} 2 & 1 & -1 \\ 1 & 2 & 1 \\ 3 & -2 & 2 \end{vmatrix}} = \frac{32 + 3 + 28}{12 + 1 + 8} = \frac{63}{21} = 3,$$

$$z = \frac{\begin{vmatrix} 2 & 1 & 3 \\ 1 & 2 & 12 \\ 3 & -2 & 8 \end{vmatrix}}{\begin{vmatrix} 2 & 1 & -1 \\ 1 & 2 & 1 \\ 3 & -2 & 2 \end{vmatrix}} = \frac{80 + 28 - 24}{24 + 1 + 8} = \frac{84}{21} = 4.$$

Chapter 7
Vectors

7.1 Introduction

This chapter provides a comprehensive introduction to vectors. It covers 2D and 3D vectors, unit vectors, position vectors, Cartesian vectors, vector magnitude, vector products, and area calculations. It also shows how vectors are used in lighting calculations and back-face detection.

7.2 Background

Vectors are a relative new invention in the world of mathematics, dating only from the 19th century. They enable us to solve complex geometric problems, the dynamics of moving objects, and problems involving forces and fields.

We often only require a single number to represent quantities used in our daily lives such as height, age, shoe size, waist and chest measurement. The magnitude of these numbers depends on our age and whether we use metric or imperial units. Such quantities are called *scalars*. On the other hand, there are some things that require more than one number to represent them: wind, force, weight, velocity and sound are just a few examples. For example, any sailor knows that wind has a magnitude and a direction. The force we use to lift an object also has a value *and* a direction. Similarly, the velocity of a moving object is measured in terms of its speed (e.g. miles per hour), and a direction such as north-west. Sound, too, has intensity and a direction. Such quantities are called *vectors*.

Complex numbers seemed to be a likely candidate for representing forces, and were being investigated by the Norwegian-Danish mathematician Caspar Wessel (1745–1818), the French amateur mathematician Jean-Robert Argand (1768–1822) and the English mathematician John Warren (1796–1852). At the time, complex numbers were two-dimensional, and their 3D form was being investigated by the Irish mathematician Sir William Rowan Hamilton (1805–1865) who invented them

© Springer-Verlag London Ltd., part of Springer Nature 2022
J. Vince, *Mathematics for Computer Graphics*, Undergraduate Topics
in Computer Science, https://doi.org/10.1007/978-1-4471-7520-9_7

in 1843, calling them *quaternions*. In 1853, Hamilton published his book *Lectures on Quaternions* [1] in which he described terms such as *vector, transvector* and *provector*. Hamilton's vectors were not widely accepted until in 1901, when the American mathematician Edwin Bidwell Wilson (1879–1964) published *Vector Analysis* [2], describing modern vector analysis. This was based upon a series of lectures delivered earlier by the American scientist Josiah Gibbs (1839–1903).

Gibbs was not a fan of the imaginary quantities associated with Hamilton's quaternions, but saw the potential of creating a vectorial system from the imaginary i, j and k into the unit basis vectors **i**, **j** and **k**, which is what we use today.

Some mathematicians were not happy with the direction mathematics had taken. The German mathematician Hermann Gunther Grassmann (1809–1877), believed that his own *geometric calculus* was far superior to Hamilton's quaternions, but he died without managing to convince any of his fellow mathematicians. Fortunately, the English mathematician and philosopher William Kingdon Clifford (1845–1879), recognised the brilliance of Grassmann's ideas, and formalised what today has become known as *geometric algebra*.

With the success of Gibbs' vector analysis, quaternions faded into obscurity, only to be rediscovered in the 1970s when they were employed by the flight simulation community to control the dynamic behaviour of a simulator's motion platform. A decade later they found their way into computer graphics where they are used for rotations about an arbitrary axis.

Now this does not mean that vector analysis is dead—far from it. Vast quantities of scientific software depends upon the vector mathematics developed over a century ago, and will continue to employ it for many years to come. Nevertheless, geometric algebra is destined to emerge as a powerful mathematical framework that could eventually replace vector analysis one day.

Readers interested in the history of vector analysis should read Michael Crowe's book *A History of Vector Analysis* [3].

7.3   2D Vectors

7.3.1   *Vector Notation*

A scalar such as x represents a single numeric quantity. However, as a vector contains two or more numbers, its symbolic name is printed using a **bold** font to distinguish it from a scalar variable. Examples being **n**, **i** and **q**.

When a scalar variable is assigned a value, we use the standard algebraic notation:

$$x = 3.$$

However, a vector has one or more numbers enclosed in brackets, written as a column or as a row—in this text *column vectors* are used:

$$\mathbf{n} = \begin{bmatrix} 3 \\ 4 \end{bmatrix}.$$

The numbers 3 and 4 are the *components* of $\mathbf{n}$, and their sequence within the brackets is important. A *row vector* places the components horizontally:

$$\mathbf{n} = [3 \quad 4].$$

The difference between the two, is appreciated in the context of matrices. Sometimes it is convenient—for presentation purposes—to write a column vector as a row vector, in which case, it is written

$$\mathbf{n} = [3 \quad 4]^{\mathrm{T}},$$

where the superscript ^{T} reminds us that $\mathbf{n}$ is really a *transposed* column vector.

7.3.2 Graphical Representation of Vectors

An arrow is used to represent a vector as it possesses length and direction, as shown in Fig. 7.1. By assigning coordinates to the arrow it is possible to translate the arrow's length and direction into two numbers. For example, in Fig. 7.2 the vector $\mathbf{r}$ has its tail defined by $(x_1, y_1) = (1, 2)$, and its head by $(x_2, y_2) = (3, 4)$. Vector $\mathbf{s}$ has its tail defined by $(x_3, y_3) = (5, 3)$, and its head by $(x_4, y_4) = (3, 1)$. The x- and y-components for $\mathbf{r}$ are computed as follows

$$x_r = x_2 - x_1 = 3 - 1 = 2$$
$$y_r = y_2 - y_1 = 4 - 2 = 2$$

and the components for $\mathbf{s}$ are computed as follows

$$x_s = x_4 - x_3 = 3 - 5 = -2$$
$$y_s = y_4 - y_3 = 1 - 3 = -2.$$

Fig. 7.1 An arrow with magnitude and direction

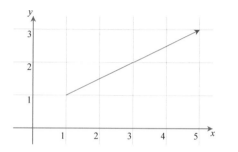

Fig. 7.2 Two vectors **r** and **s** have the same magnitude but opposite directions

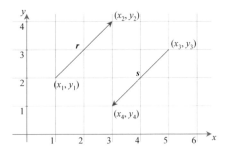

It is the negative value of x_s and y_s that encode the vector's direction. In general, if the coordinates of a vector's head and tail are $(x_h, \ y_h)$ and $(x_t, \ y_t)$ respectively, its components Δx and Δy are given by

$$\Delta x = x_h - x_t$$
$$\Delta y = y_h - y_t.$$

One can readily see from this notation that a vector does not have an absolute position. It does not matter where we place a vector, so long as we preserve its length and orientation, its components are unaltered.

7.3.3   Magnitude of a Vector

The *magnitude* or length of a vector **r** is written $\|\mathbf{r}\|$ and computed using the theorem of Pythagoras:
$$\|\mathbf{r}\| = \sqrt{(\Delta x)^2 + (\Delta y)^2}$$

and used as follows. Consider a vector defined by

$$(x_h, \ y_h) = (4, \ 5)$$
$$(x_t, \ y_t) = (1, \ 1)$$

where the x- and y-components are 3 and 4 respectively. Therefore its magnitude equals $\sqrt{3^2 + 4^2} = 5$. The magnitude of a vector is also written as $|\mathbf{r}|$, with single vertical lines.

Figure 7.3 shows eight vectors, and their geometric properties are listed in Table 7.1.

Fig. 7.3 Eight vectors whose coordinates are shown in Table 7.1

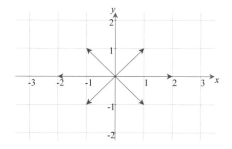

Table 7.1 Values associated with the eight vectors in Fig. 7.3

x_h	y_h	x_t	y_t	Δx	Δy	$\|\mathbf{vector}\|$
2	0	0	0	2	0	2
0	2	0	0	0	2	2
−2	0	0	0	−2	0	2
0	−2	0	0	0	−2	2
1	1	0	0	1	1	$\sqrt{2}$
−1	1	0	0	−1	1	$\sqrt{2}$
−1	−1	0	0	−1	−1	$\sqrt{2}$
1	−1	0	0	1	−1	$\sqrt{2}$

7.4 3D Vectors

The above vector examples are in 2D, but it is easy to extend this notation to embrace an extra dimension. Figure 7.4 shows a 3D vector **r** with its head, tail, components and magnitude annotated. The vector, its components and magnitude are given by

$$\mathbf{r} = [\Delta x \quad \Delta y \quad \Delta z]^{\mathrm{T}}$$

Fig. 7.4 The vector **r** has components Δx, Δy, Δz

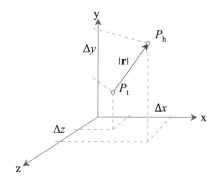

$$\Delta x = x_h - x_t$$
$$\Delta y = y_h - y_t$$
$$\Delta z = z_h - z_t$$
$$\|\mathbf{r}\| = \sqrt{(\Delta x)^2 + (\Delta y)^2 + (\Delta z)^2}.$$

All future examples are three-dimensional.

7.4.1   Vector Manipulation

As vectors are different to scalars, there are rules to control how the two mathematical entities interact with one another. For instance, we need to consider vector addition, subtraction and products, and how a vector is scaled.

7.4.2   Scaling a Vector

Given a vector $\mathbf{n}$, $2\mathbf{n}$ means that the vectors components are scaled by a factor of 2. For example, given

$$\mathbf{n} = \begin{bmatrix} 3 \\ 4 \\ 5 \end{bmatrix}, \quad \text{then} \quad 2\mathbf{n} = \begin{bmatrix} 6 \\ 8 \\ 10 \end{bmatrix}$$

which seems logical. Similarly, if we divide $\mathbf{n}$ by 2, its components are halved. Note that the vector's direction remains unchanged—only its magnitude changes.

In general, given

$$\mathbf{n} = \begin{bmatrix} n_1 \\ n_2 \\ n_3 \end{bmatrix}, \quad \text{then} \quad \lambda\mathbf{n} = \begin{bmatrix} \lambda n_1 \\ \lambda n_2 \\ \lambda n_3 \end{bmatrix}, \quad \text{where} \quad \lambda \in \mathrm{R}.$$

There is no obvious way we can resolve the expression $2 + \mathbf{n}$, for it is not clear which component of $\mathbf{n}$ is to be increased by 2. However, if we can add a scalar to an imaginary (e.g. $2 + 3i$), why can't we add a scalar to a vector (e.g. $2 + \mathbf{n}$)? Well, the answer to this question is two-fold: First, if we change the meaning of 'add' to mean 'associated with', then there is nothing to stop us from 'associating' a scalar with a vector, like complex numbers. Second, the axioms controlling our algebra must be clear on this matter. Unfortunately, the axioms of traditional vector analysis do not support the 'association' of scalars with vectors in this way. However, geometric algebra does! Furthermore, geometric algebra even permits division by a vector, which does sound strange. Consequently, whilst reading the rest of this chapter keep an open mind about what is permitted, and what is not permitted. At the end of the day, virtually anything is possible, so long as we have a well-behaved axiomatic system.

7.4.3 Vector Addition and Subtraction

Given vectors $\mathbf{r}$ and $\mathbf{s}$, $\mathbf{r} \pm \mathbf{s}$ is defined as

$$\mathbf{r} = \begin{bmatrix} x_r \\ y_r \\ z_r \end{bmatrix}, \quad \mathbf{s} = \begin{bmatrix} x_s \\ y_s \\ z_s \end{bmatrix}, \quad \text{then} \quad \mathbf{r} \pm \mathbf{s} = \begin{bmatrix} x_r \pm x_s \\ y_r \pm y_s \\ z_r \pm z_s \end{bmatrix}.$$

Vector addition is commutative:

$$\mathbf{a} + \mathbf{b} = \mathbf{b} + \mathbf{a}$$

$$\text{e.g.} \quad \begin{bmatrix} 1 \\ 2 \\ 3 \end{bmatrix} + \begin{bmatrix} 4 \\ 5 \\ 6 \end{bmatrix} = \begin{bmatrix} 4 \\ 5 \\ 6 \end{bmatrix} + \begin{bmatrix} 1 \\ 2 \\ 3 \end{bmatrix}.$$

However, like scalar subtraction, vector subtraction is not commutative:

$$\mathbf{a} - \mathbf{b} \neq \mathbf{b} - \mathbf{a}$$

$$\text{e.g.} \quad \begin{bmatrix} 4 \\ 5 \\ 6 \end{bmatrix} - \begin{bmatrix} 1 \\ 2 \\ 3 \end{bmatrix} \neq \begin{bmatrix} 1 \\ 2 \\ 3 \end{bmatrix} - \begin{bmatrix} 4 \\ 5 \\ 6 \end{bmatrix}.$$

Let's illustrate vector addition and subtraction with two examples. Figure 7.5 shows the graphical interpretation of adding two vectors $\mathbf{r}$ and $\mathbf{s}$. Note that the tail of vector $\mathbf{s}$ is attached to the head of vector $\mathbf{r}$. The resultant vector $\mathbf{t} = \mathbf{r} + \mathbf{s}$ is defined by adding the corresponding components of $\mathbf{r}$ and $\mathbf{s}$ together. Figure 7.6 shows a graphical interpretation for $\mathbf{r} - \mathbf{s}$. This time, the components of vector $\mathbf{s}$ are reversed

Fig. 7.5 Vector addition $\mathbf{r} + \mathbf{s}$

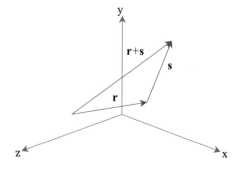

Fig. 7.6 Vector subtraction
$\mathbf{r} - \mathbf{s}$

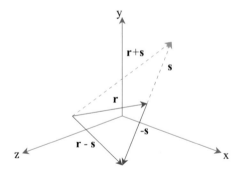

to produce an equal and opposite vector. Then it is attached to $\mathbf{r}$ and added as described above.

7.4.4 Position Vectors

Given any point $P(x, y, z)$, a *position vector* $\mathbf{p}$ is created by assuming that P is the vector's head and the origin is its tail. As the tail coordinates are $(0, 0, 0)$ the vector's components are x, y, z. Consequently, the vector's magnitude $\|\mathbf{p}\|$ equals $\sqrt{x^2 + y^2 + z^2}$.

7.4.5 Unit Vectors

By definition, a *unit vector* has a magnitude of 1. A simple example is $\mathbf{i}$, where

$$\mathbf{i} = [1 \ \ 0 \ \ 0]^{\mathrm{T}}, \quad \text{where} \quad \|\mathbf{i}\| = 1.$$

Unit vectors are extremely useful in the product of two vectors, where their magnitudes are required; and if these are unit vectors, the computation is greatly simplified.

Converting a vector into a unit form is called *normalising*, and is achieved by dividing its components by the vector's magnitude. To formalise this process, consider a vector $\mathbf{r} = [x \ \ y \ \ z]^{\mathrm{T}}$, with magnitude $\|\mathbf{r}\| = \sqrt{x^2 + y^2 + z^2}$. The unit form of $\mathbf{r}$ is given by

$$\hat{\mathbf{r}} = \frac{1}{\|\mathbf{r}\|}[x \ \ y \ \ z]^{\mathrm{T}}$$

This is confirmed by showing that the magnitude of $\hat{\mathbf{r}}$ is 1:

$$\|\hat{\mathbf{r}}\| = \sqrt{\left(\frac{x}{\|\mathbf{r}\|}\right)^2 + \left(\frac{y}{\|\mathbf{r}\|}\right)^2 + \left(\frac{z}{\|\mathbf{r}\|}\right)^2}$$

$$= \frac{1}{\|\mathbf{r}\|}\sqrt{x^2 + y^2 + z^2}$$

$$\|\hat{\mathbf{r}}\| = 1.$$

7.4.6 Cartesian Vectors

A *Cartesian vector* is constructed from three unit vectors: $\mathbf{i}$, $\mathbf{j}$ and $\mathbf{k}$, aligned with the x-, y- and z-axis, respectively:

$$\mathbf{i} = [1 \ \ 0 \ \ 0]^{\mathrm{T}}, \qquad \mathbf{j} = [0 \ \ 1 \ \ 0]^{\mathrm{T}}, \qquad \mathbf{k} = [0 \ \ 0 \ \ 1]^{\mathrm{T}}.$$

Therefore, any vector aligned with the x-, y- or z-axis is a scalar multiple of the associated unit vector. For example, $10\mathbf{i}$ is aligned with the x-axis, with a magnitude of 10. $20\mathbf{k}$ is aligned with the z-axis, with a magnitude of 20. By employing the rules of vector addition and subtraction, we can compose a vector $\mathbf{r}$ by summing three scaled Cartesian unit vectors as follows

$$\mathbf{r} = a\mathbf{i} + b\mathbf{j} + c\mathbf{k}$$

which is equivalent to

$$\mathbf{r} = [a \ \ b \ \ c]^{\mathrm{T}}$$

where the magnitude of $\mathbf{r}$ is

$$\|\mathbf{r}\| = \sqrt{a^2 + b^2 + c^2}.$$

Any pair of Cartesian vectors, such as $\mathbf{r}$ and $\mathbf{s}$, can be combined as follows

$$\mathbf{r} = a\mathbf{i} + b\mathbf{j} + c\mathbf{k}$$
$$\mathbf{s} = d\mathbf{i} + e\mathbf{j} + f\mathbf{k}$$
$$\mathbf{r} \pm \mathbf{s} = (a \pm d)\mathbf{i} + (b \pm e)\mathbf{j} + (c \pm f)\mathbf{k}.$$

7.4.7 Products

The product of two scalars is very familiar: for example, 6×7 or $7 \times 6 = 42$. We often visualise this operation as a rectangular area, where 6 and 7 are the dimensions of a rectangle's sides, and 42 is the area. However, a vector's qualities are its length and orientation, which means that any product must include them in any calculation. The length is easily calculated, but we must know the angle between the two vectors as this reflects their relative orientation. Although the angle can be incorporated within the product in various ways, two particular ways lead to useful results. For example, the product of **r** and **s**, separated by an angle θ could be $\|\mathbf{r}\|\|\mathbf{s}\| \cos\theta$ or $\|\mathbf{r}\|\|\mathbf{s}\| \sin\theta$. It just so happens that $\cos\theta$ forces the product to result in a scalar quantity, and $\sin\theta$ creates a vector. Consequently, there are two products to consider: the *scalar* product, and the *vector* product, which are written as $\mathbf{r}\cdot\mathbf{s}$ and $\mathbf{r}\times\mathbf{s}$ respectively.

7.4.8   Scalar Product

Figure 7.7 shows two vectors **r** and **s** that have been drawn, for convenience, with their tails touching. Taking **s** as the reference vector—which is an arbitrary choice—we compute the projection of **r** on **s**, which takes into account their relative orientation. The length of **r** on **s** is $\|\mathbf{r}\|\cos\beta$. We can now multiply the magnitude of **s** by the projected length of **r**: $\|\mathbf{s}\|\|\mathbf{r}\|\cos\beta$ This scalar product is written

$$\mathbf{r}\cdot\mathbf{s} = \|\mathbf{r}\|\|\mathbf{s}\|\cos\beta. \tag{7.1}$$

Because of the dot symbol '·', the scalar product is also called the *dot* product.

Fortunately, everything is in place to perform this task. To begin with, we define two Cartesian vectors **r** and **s**, and proceed to multiply them together using (7.1):

$$\mathbf{r} = a\mathbf{i} + b\mathbf{j} + c\mathbf{k}$$
$$\mathbf{s} = d\mathbf{i} + e\mathbf{j} + f\mathbf{k}$$
$$\mathbf{r}\cdot\mathbf{s} = (a\mathbf{i} + b\mathbf{j} + c\mathbf{k})\cdot(d\mathbf{i} + e\mathbf{j} + f\mathbf{k})$$
$$= a\mathbf{i}\cdot(d\mathbf{i} + e\mathbf{j} + f\mathbf{k})$$
$$\quad + b\mathbf{j}\cdot(d\mathbf{i} + e\mathbf{j} + f\mathbf{k})$$
$$\quad + c\mathbf{k}\cdot(d\mathbf{i} + e\mathbf{j} + f\mathbf{k})$$
$$= ad\mathbf{i}\cdot\mathbf{i} + ae\mathbf{i}\cdot\mathbf{j} + af\mathbf{i}\cdot\mathbf{k}$$
$$\quad + bd\mathbf{j}\cdot\mathbf{i} + be\mathbf{j}\cdot\mathbf{j} + bf\mathbf{j}\cdot\mathbf{k}$$
$$\quad + cd\mathbf{k}\cdot\mathbf{i} + ce\mathbf{k}\cdot\mathbf{j} + cf\mathbf{k}\cdot\mathbf{k}.$$

Fig. 7.7 The projection of **r** on **s**

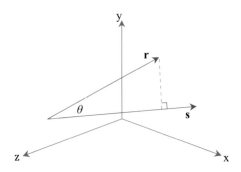

Before we proceed any further, we can see that we have created various dot product terms such as $\mathbf{i} \cdot \mathbf{i}, \mathbf{i} \cdot \mathbf{j}, \ \mathbf{i} \cdot \mathbf{k}$, etc. These terms can be divided into two groups: those that reference the same unit vector, and those that reference different unit vectors.

Using the definition of the dot product (7.1), terms such as $\mathbf{i} \cdot \mathbf{i}, \ \mathbf{j} \cdot \mathbf{j}$ and $\mathbf{k} \cdot \mathbf{k} = 1$, because the angle between $\mathbf{i}$ and $\mathbf{i}$, $\mathbf{j}$ and $\mathbf{j}$, or $\mathbf{k}$ and $\mathbf{k}$, is $0°$; and $\cos 0° = 1$. But as the other vector combinations are separated by $90°$, and $\cos 90° = 0$, all remaining terms collapse to zero, and we are left with

$$\mathbf{r} \cdot \mathbf{s} = ad\mathbf{i} \cdot \mathbf{i} + be\mathbf{j} \cdot \mathbf{j} + cf\mathbf{k} \cdot \mathbf{k}.$$

But as the the magnitude of a unit vector is 1, we can write

$$\mathbf{r} \cdot \mathbf{s} = \|\mathbf{r}\| \|\mathbf{s}\| \cos \theta = ad + be + cf$$

which confirms that the dot product is indeed a scalar quantity.

It is worth pointing out that the angle returned by the dot product ranges between $0°$ and $180°$. This is because, as the angle between two vectors increases beyond $180°$ the returned angle θ is always the smallest angle associated with the geometry.

7.4.9 The Dot Product in Lighting Calculations

Lambert's law states that the intensity of illumination on a diffuse surface is proportional to the cosine of the angle between the surface normal vector and the light source direction. Figure 7.8 shows a scenario where a light source is located at (20, 20, 40), and the illuminated point is (0, 10, 0). In this situation we are interested in calculating $\cos \beta$, which, when multiplied by the light source intensity, gives the incident light intensity on the surface. To begin with, we are given the normal vector $\hat{\mathbf{n}}$ to the surface. In this case $\hat{\mathbf{n}}$ is a unit vector: i.e. $\|\hat{\mathbf{n}}\| = 1$:

$$\hat{\mathbf{n}} = [0 \quad 1 \quad 0]^{\mathsf{T}}$$

The direction of the light source from the surface is defined by the vector $\mathbf{s}$:

$$\mathbf{s} = \begin{bmatrix} 20 - 0 \\ 20 - 10 \\ 40 - 0 \end{bmatrix} = \begin{bmatrix} 20 \\ 10 \\ 40 \end{bmatrix}$$

Fig. 7.8 The geometry associated with Lambert's law

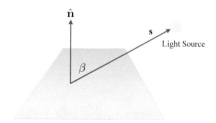

$$\|\mathbf{s}\| = \sqrt{20^2 + 10^2 + 40^2} \approx 45.826$$
$$\|\hat{\mathbf{n}}\| \|\mathbf{s}\| \cos \beta = 0 \times 20 + 1 \times 10 + 0 \times 40 = 10$$
$$1 \times 45.826 \times \cos \beta = 10$$
$$\cos \beta = \frac{10}{45.826} \approx 0.218.$$

Therefore the light intensity at the point $(0, \ 10, \ 0)$ is 0.218 of the original light intensity at $(20, \ 20, \ 40)$, but does not take into account the attenuation due to the inverse-square law of light propagation.

7.4.10 The Scalar Product in Back-Face Detection

A simple way to identify back-facing polygons relative to the virtual camera, is to compute the angle between the polygon's surface normal and the line of sight between the camera and the polygon. If this angle is less than 90°, the polygon is visible; if it equals or exceeds 90°, the polygon is invisible. This geometry is shown in Fig. 7.9. Although it is obvious from Fig. 7.9 that the right-hand polygon is invisible to the camera, let's prove algebraically that this is so.

For example, if the virtual camera is located at $(0, \ 0, \ 0)$ and the polygon's vertex is $(10, \ 10, \ 40)$. The normal vector is $\mathbf{n} = [5 \ \ 5 \ \ -2]^\mathrm{T}$.

$$\mathbf{n} = [5 \ \ 5 \ \ -2]^\mathrm{T}$$
$$\|\mathbf{n}\| = \sqrt{5^2 + 5^2 + (-2)^2} \approx 7.348.$$

The camera vector $\mathbf{c}$ is

$$\mathbf{c} = \begin{bmatrix} 0 - 10 \\ 0 - 10 \\ 0 - 40 \end{bmatrix} = \begin{bmatrix} -10 \\ -10 \\ -40 \end{bmatrix}$$
$$\|\mathbf{c}\| = \sqrt{(-10)^2 + (-10)^2 + (-40)^2} \approx 42.426$$

Fig. 7.9 Back-face detection

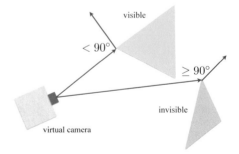

therefore,

$$\|\mathbf{n}\|\,\|\mathbf{c}\|\cos\beta = 5\times(-10)+5\times(-10)+(-2)\times(-40)$$
$$7.348\times42.426\times\cos\beta = -20$$
$$\cos\beta = \frac{-20}{7.348\times42.426}\approx-0.0634$$
$$\beta = \cos^{-1}(-0.0634)\approx93.64°$$

which shows that the polygon is invisible for the camera.

7.4.11 The Vector Product

As mentioned above, the vector product $\mathbf{r}\times\mathbf{s}$ creates a third vector whose magnitude equals $\|\mathbf{r}\|\,\|\mathbf{s}\|\sin\theta$, where θ is the angle between the original vectors. Figure 7.10 reminds us that the area of a parallelogram formed by $\mathbf{r}$ and $\mathbf{s}$ equals $\|\mathbf{r}\|\,\|\mathbf{s}\|\sin\theta$. Because of the cross symbol '×', the vector product is also called the *cross* product.

$$\mathbf{r}\times\mathbf{s} = \mathbf{t} \tag{7.2}$$
$$\|\mathbf{t}\| = \|\mathbf{r}\|\,\|\mathbf{s}\|\sin\theta.$$

We will discover that the vector $\mathbf{t}$ is normal (90°) to the plane containing the vectors $\mathbf{r}$ and $\mathbf{s}$, as shown in Fig. 7.11, which makes it an ideal way of computing the vector normal to a surface. Once again, let's define two vectors and this time multiply them together using (7.2):

$$\mathbf{r} = a\mathbf{i}+b\mathbf{j}+c\mathbf{k}$$
$$\mathbf{s} = d\mathbf{i}+e\mathbf{j}+f\mathbf{k}$$
$$\mathbf{r}\times\mathbf{s} = (a\mathbf{i}+b\mathbf{j}+c\mathbf{k})\times(d\mathbf{i}+e\mathbf{j}+f\mathbf{k})$$
$$= a\mathbf{i}\times(d\mathbf{i}+e\mathbf{j}+f\mathbf{k})$$

Fig. 7.10 The area of the parallelogram formed by $\mathbf{r}$ and $\mathbf{s}$

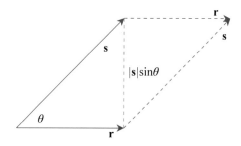

Fig. 7.11 The vector
product

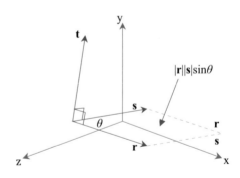

$$+ b\mathbf{j} \times (d\mathbf{i} + e\mathbf{j} + f\mathbf{k})$$
$$c\mathbf{k} \times (d\mathbf{i} + e\mathbf{j} + f\mathbf{k})$$
$$= ad\mathbf{i} \times \mathbf{i} + ae\mathbf{i} \times \mathbf{j} + af\mathbf{i} \times \mathbf{k}$$
$$+ bd\mathbf{j} \times \mathbf{i} + be\mathbf{j} \times \mathbf{j} + bf\mathbf{j} \times \mathbf{k}$$
$$+ cd\mathbf{k} \times \mathbf{i} + ce\mathbf{k} \times \mathbf{j} + cf\mathbf{k} \times \mathbf{k}.$$

As we found with the dot product, there are two groups of vector terms: those that
reference the same unit vector, and those that reference different unit vectors.

Using the definition for the cross product (7.2), operations such as $\mathbf{i} \times \mathbf{i}, \mathbf{j} \times \mathbf{j}$ and
$\mathbf{k} \times \mathbf{k}$ result in a vector whose magnitude is 0. This is because the angle between
the vectors is $0°$, and $\sin 0° = 0$. Consequently these terms disappear and we are left
with

$$\mathbf{r} \times \mathbf{s} = ae\mathbf{i} \times \mathbf{j} + af\mathbf{i} \times \mathbf{k} + bd\mathbf{j} \times \mathbf{i} + bf\mathbf{j} \times \mathbf{k} + cd\mathbf{k} \times \mathbf{i} + ce\mathbf{k} \times \mathbf{j}. \qquad (7.3)$$

Sir William Rowan Hamilton struggled for many years when working on quaternions
to resolve the meaning of a similar result. At the time, he was not using vectors, as
they had yet to be defined, but the imaginary terms i, j and k. Hamilton's problem was to resolve the products ij, jk, ki and their opposites ji, kj and ik. What
did the products mean? He reasoned that $ij = k$, $jk = i$ and $ki = j$, but could not
resolve their opposites. One day in 1843, when he was out walking, thinking about
this problem, he thought the impossible: $ij = k$, but $ji = -k$, $jk = i$, but $kj = -i$,
and $ki = j$, but $ik = -j$. To his surprise, this worked, but it contradicted the commutative multiplication law of scalars where $6 \times 7 = 7 \times 6$. We now accept that the
commutative multiplication law is there to be broken!

Let's continue with Hamilton's rules and reduce the cross product terms of (7.3)
to

$$\mathbf{r} \times \mathbf{s} = ae\mathbf{k} - af\mathbf{j} - bd\mathbf{k} + bf\mathbf{i} + cd\mathbf{j} - ce\mathbf{i}. \qquad (7.4)$$

Equation (7.4) can be tidied up to bring like terms together:

$$\mathbf{r} \times \mathbf{s} = (bf - ce)\mathbf{i} + (cd - af)\mathbf{j} + (ae - bd)\mathbf{k}. \tag{7.5}$$

Now let's repeat the original vector equations to see how Eq. (7.5) is computed:

$$\mathbf{r} = a\mathbf{i} + b\mathbf{j} + c\mathbf{k}$$
$$\mathbf{s} = d\mathbf{i} + e\mathbf{j} + f\mathbf{k}$$
$$\mathbf{r} \times \mathbf{s} = (bf - ce)\mathbf{i} + (cd - af)\mathbf{j} + (ae - bd)\mathbf{k}. \tag{7.6}$$

To compute the $\mathbf{i}$ scalar term we consider the scalars associated with the other two unit vectors, i.e. b, c, e, and f, and cross-multiply and subtract them to form $(bf - ce)$.

To compute the $\mathbf{j}$ scalar term we consider the scalars associated with the other two unit vectors, i.e. a, c, d, and f, and cross-multiply and subtract them to form $(cd - af)$.

To compute the $\mathbf{k}$ scalar term we consider the scalars associated with the other two unit vectors, i.e. a, b, d, and e, and cross-multiply and subtract them to form $(ae - bd)$.

The middle operation seems out of step with the other two, but in fact it preserves a cyclic symmetry often found in mathematics. Nevertheless, some authors reverse the sign of the $\mathbf{j}$ scalar term and cross-multiply and subtract the terms to produce $-(af - cd)$ which maintains a visual pattern for remembering the cross-multiplication. Equation (7.6) now becomes

$$\mathbf{r} \times \mathbf{s} = (bf - ce)\mathbf{i} - (af - cd)\mathbf{j} + (ae - bd)\mathbf{k}. \tag{7.7}$$

However, we now have to remember to introduce a negative sign for the $\mathbf{j}$ scalar term! We can write (7.7) using determinants as follows:

$$\mathbf{r} \times \mathbf{s} = \begin{vmatrix} b & c \\ e & f \end{vmatrix}\mathbf{i} - \begin{vmatrix} a & c \\ d & f \end{vmatrix}\mathbf{j} + \begin{vmatrix} a & b \\ d & e \end{vmatrix}\mathbf{k}.$$

or

$$\mathbf{r} \times \mathbf{s} = \begin{vmatrix} b & c \\ e & f \end{vmatrix}\mathbf{i} + \begin{vmatrix} c & a \\ f & d \end{vmatrix}\mathbf{j} + \begin{vmatrix} a & b \\ d & e \end{vmatrix}\mathbf{k}.$$

Therefore, to derive the cross product of two vectors we first write the vectors in the correct sequence. Remembering that $\mathbf{r} \times \mathbf{s}$ does not equal $\mathbf{s} \times \mathbf{r}$. Second, we compute the three scalar terms and form the resultant vector, which is perpendicular to the plane containing the original vectors.

So far, we have assumed that

$$\mathbf{r} \times \mathbf{s} = \mathbf{t}$$
$$\|\mathbf{t}\| = \|\mathbf{r}\|\|\mathbf{s}\| \sin\theta$$

where θ is the angle between $\mathbf{r}$ and $\mathbf{s}$, and $\mathbf{t}$ is perpendicular to the plane containing $\mathbf{r}$ and $\mathbf{s}$. Now let's prove that this is the case:

$$\mathbf{r} \cdot \mathbf{s} = \|\mathbf{r}\| \|\mathbf{s}\| \cos\theta = x_r x_s + y_r y_s + z_r z_s$$

$$\cos^2\theta = \frac{(x_r x_s + y_r y_s + z_r z_s)^2}{\|\mathbf{r}\|^2 \|\mathbf{s}\|^2}$$

$$\|\mathbf{t}\| = \|\mathbf{r}\| \|\mathbf{s}\| \sin\theta$$

$$\|\mathbf{t}\|^2 = \|\mathbf{r}\|^2 \|\mathbf{s}\|^2 \sin^2\theta$$

$$= \|\mathbf{r}\|^2 \|\mathbf{s}\|^2 \left(1 - \cos^2\theta\right)$$

$$= \|\mathbf{r}\|^2 \|\mathbf{s}\|^2 \left(1 - \frac{(x_r x_s + y_r y_s + z_r z_s)^2}{\|\mathbf{r}\|^2 \|\mathbf{s}\|^2}\right)$$

$$= \|\mathbf{r}\|^2 \|\mathbf{s}\|^2 - (x_r x_s + y_r y_s + z_r z_s)^2$$

$$= \left(x_r^2 + y_r^2 + z_r^2\right)\left(x_s^2 + y_s^2 + z_s^2\right) - (x_r x_s + y_r y_s + z_r z_s)^2$$

$$= x_r^2\left(y_s^2 + z_s^2\right) + y_r^2\left(x_s^2 + z_s^2\right) + z_r^2\left(x_s^2 + y_s^2\right)$$

$$\quad - 2x_r x_s y_r y_s - 2x_r x_s z_r z_s - 2y_r y_s z_r z_s$$

$$= x_r^2 y_s^2 + x_r^2 z_s^2 + y_r^2 x_s^2 + y_r^2 z_s^2 + z_r^2 x_s^2 + z_r^2 y_s^2$$

$$\quad - 2x_r x_s y_r y_s - 2x_r x_s z_r z_s - 2y_r y_s z_r z_s$$

$$= (y_r z_s - z_r y_s)^2 + (z_r x_s - x_r z_s)^2 + (x_r y_s - y_r x_s)^2$$

which in determinant form is

$$\|\mathbf{t}\|^2 = \begin{vmatrix} y_r & z_r \\ y_s & z_s \end{vmatrix}^2 + \begin{vmatrix} z_r & x_r \\ z_s & x_s \end{vmatrix}^2 + \begin{vmatrix} x_r & y_r \\ x_s & y_s \end{vmatrix}^2$$

and confirms that $\mathbf{t}$ is the vector

$$\mathbf{t} = \begin{vmatrix} y_r & z_r \\ y_s & z_s \end{vmatrix} \mathbf{i} + \begin{vmatrix} z_r & x_r \\ z_s & x_s \end{vmatrix} \mathbf{j} + \begin{vmatrix} x_r & y_r \\ x_s & y_s \end{vmatrix} \mathbf{k}.$$

All that remains is to prove that $\mathbf{t}$ is orthogonal (perpendicular) to $\mathbf{r}$ and $\mathbf{s}$, which is achieved by showing that $\mathbf{r} \cdot \mathbf{t} = \mathbf{s} \cdot \mathbf{t} = 0$:

$$\mathbf{r} = x_r \mathbf{i} + y_r \mathbf{j} + z_r \mathbf{k}$$

$$\mathbf{s} = x_s \mathbf{i} + y_s \mathbf{j} + z_s \mathbf{k}$$

$$\mathbf{t} = (y_r z_s - z_r y_s)\mathbf{i} + (z_r x_s - x_r z_s)\mathbf{j} + (x_r y_s - y_r x_s)\mathbf{k}$$

$$\mathbf{r} \cdot \mathbf{t} = x_r(y_r z_s - z_r y_s) + y_r(z_r x_s - x_r z_s) + z_r(x_r y_s - y_r x_s)$$

$$= x_r y_r z_s - x_r y_s z_r + x_s y_r z_r - x_r y_r z_s + x_r y_s z_r - x_s y_r z_r = 0$$

$$\mathbf{s} \cdot \mathbf{t} = x_s(y_r z_s - z_r y_s) + y_s(z_r x_s - x_r z_s) + z_s(x_r y_s - y_r x_s)$$

$$= x_s y_r z_s - x_s y_s z_r + x_s y_s z_r - x_r y_s z_s + x_r y_s z_s - x_s y_r z_s = 0$$

Fig. 7.12 Vector **t** is normal
to the vectors **r** and **s**

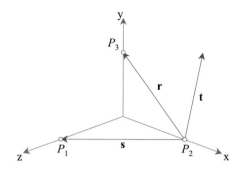

Table 7.2 Coordinates of the vertices used in Fig. 7.12

Vertex	x	y	z
P_1	0	0	1
P_2	1	0	0
P_3	0	1	0

and we have proved that $\mathbf{r} \times \mathbf{s} = \mathbf{t}$, where $\|\mathbf{t}\| = \|\mathbf{r}\|\|\mathbf{s}\| \sin\theta$ and $\mathbf{t}$ is orthogonal to the plane containing **r** and **s**.

Let's now consider two vectors **r** and **s** and compute the normal vector **t**. The vectors are chosen so that we can anticipate approximately the answer. For the sake of clarity, the vector equations include the scalar multipliers 0 and 1. Normally, these are omitted. Figure 7.12 shows the vectors **r** and **s** and the normal vector **t**, and Table 7.2 contains the coordinates of the vertices forming the two vectors which confirms what we expected from Fig. 7.12.

$$\mathbf{r} = [(x_3 - x_2) \quad (y_3 - y_2) \quad (z_3 - z_2)]^{\mathsf{T}}$$
$$\mathbf{s} = [(x_1 - x_2) \quad (y_1 - y_2) \quad (z_1 - z_2)]^{\mathsf{T}}$$
$$P_1 = (0, \ 0, \ 1)$$
$$P_2 = (1, \ 0, \ 0)$$
$$P_3 = (0, \ 1, \ 0)$$
$$\mathbf{r} = -1\mathbf{i} + 1\mathbf{j} + 0\mathbf{k}$$
$$\mathbf{s} = -1\mathbf{i} + 0\mathbf{j} + 1\mathbf{k}$$
$$\mathbf{r} \times \mathbf{s} = [1 \times 1 - 0 \times 0]\mathbf{i}$$
$$\qquad - [-1 \times 1 - (-1) \times 0]\mathbf{j}$$
$$\qquad + [-1 \times 0 - (-1) \times 1]\mathbf{k}$$
$$\mathbf{t} = \mathbf{i} + \mathbf{j} + \mathbf{k}$$

Now let's reverse the vectors to illustrate the importance of vector sequence.

$$\mathbf{s} = -1\mathbf{i} + 0\mathbf{j} + 1\mathbf{k}$$
$$\mathbf{r} = -1\mathbf{i} + 1\mathbf{j} + 0\mathbf{k}$$
$$\mathbf{s} \times \mathbf{r} = [0 \times 0 - 1 \times 1]\mathbf{i}$$
$$- [-1 \times 0 - (-1) \times 1]\mathbf{j}$$
$$+ [-1 \times 1 - (-1) \times 0]\mathbf{k}$$
$$\mathbf{t} = -\mathbf{i} - \mathbf{j} - \mathbf{k}$$

which is in the opposite direction to $\mathbf{r} \times \mathbf{s}$ and confirms that the vector product is non-commutative.

7.4.12 The Right-Hand Rule

The *right-hand rule* is an *aide mémoire* for working out the orientation of the cross product vector. Given the operation $\mathbf{r} \times \mathbf{s}$, if the right-hand thumb is aligned with $\mathbf{r}$, the first finger with $\mathbf{s}$, and the middle finger points in the direction of $\mathbf{t}$. However, we must remember that this only holds in 3D. In 4D and above, it makes no sense.

7.5 Deriving a Unit Normal Vector for a Triangle

Figure 7.13 shows a triangle with vertices defined in an anticlockwise sequence from its visible side. This is the side from which we want the surface normal to point. Using the following information we will compute the surface normal using the cross product and then convert it to a unit normal vector.

Fig. 7.13 The normal vector **t** is derived from the cross product **r** × **s**

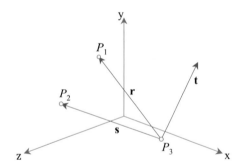

Create vector $\mathbf{r}$ between P_3 and P_1, and vector $\mathbf{s}$ between P_3 and P_2:

$$\mathbf{r} = -1\mathbf{i} + 1\mathbf{j} + 0\mathbf{k}$$
$$\mathbf{s} = -1\mathbf{i} + 0\mathbf{j} + 2\mathbf{k}$$
$$\mathbf{r} \times \mathbf{s} = (1 \times 2 - 0 \times 0)\mathbf{i}$$
$$- (-1 \times 2 - 0 \times -1)\mathbf{j}$$
$$+ (-1 \times 0 - 1 \times -1)\mathbf{k}$$
$$\mathbf{t} = 2\mathbf{i} + 2\mathbf{j} + 1\mathbf{k}$$
$$\|\mathbf{t}\| = \sqrt{2^2 + 2^2 + 1^2} = \sqrt{5}$$
$$\hat{\mathbf{t}}_u = \tfrac{2}{\sqrt{5}}\mathbf{i} + \tfrac{2}{\sqrt{5}}\mathbf{j} + \tfrac{1}{\sqrt{5}}\mathbf{k}.$$

The unit vector $\hat{\mathbf{t}}_u$ can now be used for illumination calculations in computer graphics, and as it has unit length, dot product calculations are simplified.

7.6 Surface Areas

Figure 7.14 shows two vectors $\mathbf{r}$ and $\mathbf{s}$, where the height $h = |\mathbf{s}| \sin \theta$. Therefore the area of the associated parallelogram is

$$\text{area} = \|\mathbf{r}\| \, h = \|\mathbf{r}\|\|\mathbf{s}\| \sin \theta.$$

But this is the magnitude of the cross product vector $\mathbf{t}$. Thus when we calculate $\mathbf{r} \times \mathbf{s}$, the length of the normal vector $\mathbf{t}$ equals the area of the parallelogram formed by $\mathbf{r}$ and $\mathbf{s}$; which means that the triangle formed by halving the parallelogram is half the area.

$$\text{area of parallelogram} = \|\mathbf{t}\|$$
$$\text{area of triangle} = \tfrac{1}{2}\|\mathbf{t}\|.$$

Fig. 7.14 The area of the parallelogram formed by two vectors $\mathbf{r}$ and $\mathbf{s}$

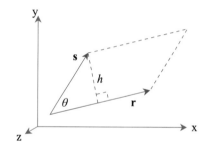

Fig. 7.15 The area of the
triangle formed by the
vectors **r** and **s**

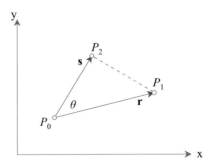

This makes it relatively easy to calculate the surface area of an object constructed
from triangles or parallelograms. In the case of a triangulated surface, we simply
sum the magnitudes of the normals and halve the result.

7.6.1   Calculating 2D Areas

Figure 7.15 shows a triangle with vertices $P_0(x_0, y_0)$, $P_1(x_1, y_1)$ and $P_2(x_2, y_2)$
formed in an anticlockwise sequence. The vectors **r** and **s** are computed as follows:

$$\mathbf{r} = (x_1 - x_0)\mathbf{i} + (y_1 - y_0)\mathbf{j}$$
$$\mathbf{s} = (x_2 - x_0)\mathbf{i} + (y_2 - y_0)\mathbf{j}$$
$$\|\mathbf{r} \times \mathbf{s}\| = (x_1 - x_0)(y_2 - y_0) - (x_2 - x_0)(y_1 - y_0)$$
$$= x_1(y_2 - y_0) - x_0(y_2 - y_0) - x_2(y_1 - y_0) + x_0(y_1 - y_0)$$
$$= x_1 y_2 - x_1 y_0 - x_0 y_2 + x_0 y_0 - x_2 y_1 + x_2 y_0 + x_0 y_1 - x_0 y_0$$
$$= x_1 y_2 - x_1 y_0 - x_0 y_2 - x_2 y_1 + x_2 y_0 + x_0 y_1$$
$$= (x_0 y_1 - x_1 y_0) + (x_1 y_2 - x_2 y_1) + (x_2 y_0 - x_0 y_2).$$

But the area of the triangle formed by the three vertices is $\frac{1}{2}\|\mathbf{r} \times \mathbf{s}\|$. Therefore

$$\text{area} = \tfrac{1}{2}[(x_0 y_1 - x_1 y_0) + (x_1 y_2 - x_2 y_1) + (x_2 y_0 - x_0 y_2)]$$

which is the formula disclosed in Chap. 5!

7.7   Summary

Vectors are extremely useful and relatively easy to use. They are vital to rendering
algorithms and shaders, and most of the time we only need to use the scalar and cross
products.

I have tried to prepare you for an alternative algebra for vectors: geometric algebra. As we shall see later on, geometric algebra shows that mathematics may have taken the wrong direction when it embraced Gibbs' vector analysis. Hermann Grassmann could have been right all along. If the mathematicians of the day had adopted Grassmann's ideas, today we would be familiar with vectors, bivectors, trivectors, quaternions, etc. But we are where we are, and we must prepare ourselves for some new ideas.

Even if you already knew something about vectors, I hope that this chapter has introduced some new ideas and illustrated the role vectors play in computer graphics.

7.8 Worked Examples

7.8.1 Position Vector

Calculate the magnitude of the position vector $\mathbf{p}$, for the point $P(4,\ 5,\ 6)$:

$$\mathbf{p} = [4\quad 5\quad 6]^{\mathrm{T}}, \quad \text{therefore,} \quad \|\mathbf{p}\| = \sqrt{4^2 + 5^2 + 6^2} \approx 8.77.$$

7.8.2 Unit Vector

Convert $\mathbf{r}$ to a unit vector.

$$\mathbf{r} = [1\quad 2\quad 3]^{\mathrm{T}}$$
$$\|\mathbf{r}\| = \sqrt{1^2 + 2^2 + 3^2} = \sqrt{14}$$
$$\hat{\mathbf{r}} = \tfrac{1}{\sqrt{14}}[1\quad 2\quad 3]^{\mathrm{T}} \approx [0.267\quad 0.535\quad 0.802]^{\mathrm{T}}.$$

7.8.3 Vector Magnitude

Compute the magnitude of $\mathbf{r} + \mathbf{s}$.

$$\mathbf{r} = 2\mathbf{i} + 3\mathbf{j} + 4\mathbf{k}$$
$$\mathbf{s} = 5\mathbf{i} + 6\mathbf{j} + 7\mathbf{k}$$
$$\mathbf{r} + \mathbf{s} = 7\mathbf{i} + 9\mathbf{j} + 11\mathbf{k}$$
$$\|\mathbf{r} + \mathbf{s}\| = \sqrt{7^2 + 9^2 + 11^2} \approx 15.84.$$

7.8.4 Angle Between Two Vectors

Find the angle between **r** and **s**.

$$\mathbf{r} = \begin{bmatrix} 2 & 0 & 4 \end{bmatrix}^{\mathrm{T}}$$
$$\mathbf{s} = \begin{bmatrix} 5 & 6 & 10 \end{bmatrix}^{\mathrm{T}}$$
$$\|\mathbf{r}\| = \sqrt{2^2 + 0^2 + 4^2} \approx 4.472$$
$$\|\mathbf{s}\| = \sqrt{5^2 + 6^2 + 10^2} \approx 12.689.$$

Therefore,

$$\|\mathbf{r}\|\|\mathbf{s}\| \cos\theta = 2 \times 5 + 0 \times 6 + 4 \times 10 = 50$$
$$12.689 \times 4.472 \times \cos\theta = 50$$
$$\cos\theta = \frac{50}{12.689 \times 4.472} \approx 0.8811$$
$$\theta = \arccos 0.8811 \approx 28.22°.$$

The angle between the two vectors is approximately $28.22°$.

7.8.5 Vector Product

To show that the vector product works with the unit vectors **i**, **j** and **k**. We start with

$$\mathbf{r} = 1\mathbf{i} + 0\mathbf{j} + 0\mathbf{k}$$
$$\mathbf{s} = 0\mathbf{i} + 1\mathbf{j} + 0\mathbf{k}$$

and then compute (7.7):

$$\mathbf{r} \times \mathbf{s} = (0 \times 0 - 0 \times 1)\mathbf{i} - (1 \times 0 - 0 \times 0)\mathbf{j} + (1 \times 1 - 0 \times 0)\mathbf{k}.$$

The **i** scalar and **j** scalar terms are both zero, but the **k** scalar term is 1, which makes $\mathbf{i} \times \mathbf{j} = \mathbf{k}$.

Let's see what happens when we reverse the vectors. This time we start with

$$\mathbf{r} = 0\mathbf{i} + 1\mathbf{j} + 0\mathbf{k}$$
$$\mathbf{s} = 1\mathbf{i} + 0\mathbf{j} + 0\mathbf{k}$$

and then compute (7.7)

$$\mathbf{r} \times \mathbf{s} = (1 \times 0 - 0 \times 0)\mathbf{i} - (0 \times 0 - 0 \times 1)\mathbf{j} + (0 \times 0 - 1 \times 1)\mathbf{k}.$$

The **i** scalar and **j** scalar terms are both zero, but the **k** scalar term is -1, which makes $\mathbf{j} \times \mathbf{i} = -\mathbf{k}$. So we see that the vector product is *antisymmetric*, i.e. there is a sign reversal when the vectors are reversed. Similarly, it can be shown that

$$\mathbf{j} \times \mathbf{k} = \mathbf{i}$$
$$\mathbf{k} \times \mathbf{i} = \mathbf{j}$$
$$\mathbf{k} \times \mathbf{j} = -\mathbf{i}$$
$$\mathbf{i} \times \mathbf{k} = -\mathbf{j}.$$

References

1. Hamilton WR (1853) Lectures on quaternions. Macmillan & Co., Cambridge
2. Wilson EB (1901) Vector analysis. Yale University Press, New Haven
3. Crowe MJ (1994) A history of vector analysis. Dover Publications Inc., New York

Chapter 8
Matrix Algebra

8.1 Introduction

This chapter introduces matrix algebra, which is a notation widely used in computer graphics. Matrices are used to scale, translate, reflect, shear and rotate 2D shapes and 3D objects, and like determinants, have their background in algebra and offer another way to represent and manipulate equations. Matrices can be added, subtracted and multiplied together, and even inverted, however, they must give the same result obtained through traditional algebraic techniques. Once you have understood the idea behind matrix notation, feel free to go to the next chapter and study their role in geometric transforms, and come back to the more advanced ideas in this chapter.

8.2 Background

Matrix notation was researched by Arthur Cayley around 1858. Cayley formalised matrix algebra, along with the American mathematicians Charles Peirce (1839–1914) and his father, Benjamin Peirce (1809–1880). Previously, Johann Gauss had shown that transforms were not always commutative, i.e. $\mathbf{T}_1\mathbf{T}_2 \neq \mathbf{T}_2\mathbf{T}_1$, (where $\mathbf{T}_1$ and $\mathbf{T}_2$ are transforms) and matrix notation clarified such observations.

Consider the linear transform $\mathbf{T}_1$, where x and y are transformed into x' and y' respectively:

$$\mathbf{T}_1 = \begin{cases} x' = ax + by \\ y' = cx + dy \end{cases} \tag{8.1}$$

and a second linear transform $\mathbf{T}_2$, where x' and y' are transformed into x'' and y'' respectively:

$$\mathbf{T}_2 = \begin{cases} x'' = Ax' + By' \\ y'' = Cx' + Dy' \end{cases} . \tag{8.2}$$

© Springer-Verlag London Ltd., part of Springer Nature 2022
J. Vince, *Mathematics for Computer Graphics*, Undergraduate Topics
in Computer Science, https://doi.org/10.1007/978-1-4471-7520-9_8

Substituting (8.1) in (8.2) we get

$$\mathbf{T_3} = \begin{cases} x'' = A(ax + by) + B(cx + dy) \\ y'' = C(ax + by) + D(cx + dy) \end{cases}$$

which simplifies to

$$\mathbf{T_3} = \begin{cases} x'' = (Aa + Bc)x + (Ab + Bd)y \\ y'' = (Ca + Dc)x + (Cb + Dd)y \end{cases} \cdot \tag{8.3}$$

Having derived the algebra for $\mathbf{T_3}$, let's examine matrix notation, where constants are separated from the variables. For example, the transform (8.4)

$$\begin{aligned} x' &= ax + by \\ y' &= cx + dy \end{aligned} \tag{8.4}$$

can be written in matrix form as:

$$\begin{bmatrix} x' \\ y' \end{bmatrix} = \begin{bmatrix} a & b \\ c & d \end{bmatrix} \begin{bmatrix} x \\ y \end{bmatrix} \tag{8.5}$$

where (8.5) contains two different structures: two single-column matrices or column vectors

$$\begin{bmatrix} x' \\ y' \end{bmatrix} \quad \text{and} \quad \begin{bmatrix} x \\ y \end{bmatrix},$$

and a 2×2 matrix:

$$\begin{bmatrix} a & b \\ c & d \end{bmatrix}.$$

Algebraically, (8.4) and (8.5) are identical, which dictates the way (8.5) is converted to (8.4). Therefore, using (8.5) we have x' followed by the '=' sign, and the sum of the products of the top row of constants a and b with the x and y in the last column vector:

$$x' = ax + by.$$

Next, we have y' followed by the '=' sign, and the sum of the products of the bottom row of constants c and d with the x and y in the last column vector:

$$y' = cx + dy.$$

As an example,

$$\begin{bmatrix} x' \\ y' \end{bmatrix} = \begin{bmatrix} 3 & 4 \\ 5 & 6 \end{bmatrix} \begin{bmatrix} x \\ y \end{bmatrix}$$

is equivalent to

$$x' = 3x + 4y$$
$$y' = 5x + 6y.$$

We can now write T_1 and T_2 using matrix notation:

$$T_1 = \begin{bmatrix} x' \\ y' \end{bmatrix} = \begin{bmatrix} a & b \\ c & d \end{bmatrix} \begin{bmatrix} x \\ y \end{bmatrix} \qquad (8.6)$$

$$T_2 = \begin{bmatrix} x'' \\ y'' \end{bmatrix} = \begin{bmatrix} A & B \\ C & D \end{bmatrix} \begin{bmatrix} x' \\ y' \end{bmatrix} \qquad (8.7)$$

and substituting (8.6) in (8.7) we have

$$T_3 = \begin{bmatrix} x'' \\ y'' \end{bmatrix} = \begin{bmatrix} A & B \\ C & D \end{bmatrix} \begin{bmatrix} a & b \\ c & d \end{bmatrix} \begin{bmatrix} x \\ y \end{bmatrix}. \qquad (8.8)$$

But we have already computed T_3 (8.3), which in matrix form is:

$$T_3 = \begin{bmatrix} x'' \\ y'' \end{bmatrix} = \begin{bmatrix} Aa + Bc & Ab + Bd \\ Ca + Dc & Cb + Dd \end{bmatrix} \begin{bmatrix} x \\ y \end{bmatrix} \qquad (8.9)$$

which implies that

$$\begin{bmatrix} A & B \\ C & D \end{bmatrix} \begin{bmatrix} a & b \\ c & d \end{bmatrix} = \begin{bmatrix} Aa + Bc & Ab + Bd \\ Ca + Dc & Cb + Dd \end{bmatrix}$$

and demonstrates how matrices must be multiplied. Here are the rules for matrix multiplication:

$$\begin{bmatrix} A & B \\ \cdots & \cdots \end{bmatrix} \begin{bmatrix} a & \cdots \\ c & \cdots \end{bmatrix} = \begin{bmatrix} Aa + Bc & \cdots \\ \cdots & \cdots \end{bmatrix}.$$

1: The top left-hand corner element $Aa + Bc$ is the product of the top row of the first matrix by the left column of the second matrix.

$$\begin{bmatrix} A & B \\ \cdots & \cdots \end{bmatrix} \begin{bmatrix} \cdots & b \\ \cdots & d \end{bmatrix} = \begin{bmatrix} \cdots & Ab + Bd \\ \cdots & \cdots \end{bmatrix}.$$

2: The top right-hand element $Ab + Bd$ is the product of the top row of the first matrix by the right column of the second matrix.

$$\begin{bmatrix} \cdots & \cdots \\ C & D \end{bmatrix} \begin{bmatrix} a & \cdots \\ c & \cdots \end{bmatrix} = \begin{bmatrix} \cdots & \cdots \\ Ca + Dc & \cdots \end{bmatrix}.$$

3: The bottom left-hand element $Ca + Dc$ is the product of the bottom row of the first matrix by the left column of the second matrix.

$$
\begin{bmatrix} \cdots \cdots \\ C \ D \end{bmatrix} \begin{bmatrix} \cdots\ b \\ \cdots\ d \end{bmatrix} = \begin{bmatrix} \cdots & \cdots \\ \cdots & Cb + Dd \end{bmatrix}.
$$

4: The bottom right-hand element $Cb + Dd$ is the product of the bottom row of the first matrix by the right column of the second matrix.

Let's multiply the following matrices together:

$$
\begin{bmatrix} 2\ 4 \\ 6\ 8 \end{bmatrix} \begin{bmatrix} 3\ 5 \\ 7\ 9 \end{bmatrix} = \begin{bmatrix} (2 \times 3 + 4 \times 7)\ (2 \times 5 + 4 \times 9) \\ (6 \times 3 + 8 \times 7)\ (6 \times 5 + 8 \times 9) \end{bmatrix} = \begin{bmatrix} 34\ 46 \\ 74\ 102 \end{bmatrix}.
$$

8.3   Matrix Notation

Having examined the background to matrices, we can now formalise their notation.

A matrix is an array of numbers (real, imaginary, complex, etc.) organised in m rows and n columns, where each entry a_{ij} belongs to the ith row and jth column:

$$
A = \begin{bmatrix}
a_{11} & a_{12} & a_{13} & \cdots & a_{1n} \\
a_{21} & a_{22} & a_{23} & \cdots & a_{2n} \\
a_{31} & a_{32} & a_{33} & \cdots & a_{3n} \\
\vdots & \vdots & \vdots & \ddots & \vdots \\
a_{m1} & a_{m2} & a_{m3} & \cdots & a_{mn}
\end{bmatrix}.
$$

It is also convenient to express the above definition as

$$
A = [a_{ij}]_{m\ n}.
$$

8.3.1   Matrix Dimension or Order

The *dimension* or *order* of a matrix is the expression $m \times n$ where m is the number of rows, and n is the number of columns.

8.3.2   Square Matrix

A *square matrix* has the same number of rows as columns:

$$
A = [a_{ij}]_{n\ n} = \begin{bmatrix}
a_{11} & a_{12} & \cdots & a_{1n} \\
a_{21} & a_{22} & \cdots & a_{2n} \\
\vdots & \vdots & \ddots & \vdots \\
a_{n1} & a_{n2} & \cdots & a_{nn}
\end{bmatrix}, \quad \text{e.g.} \quad \begin{bmatrix} 1 & -2 & 4 \\ 6 & 5 & 7 \\ 4 & 3 & 1 \end{bmatrix}.
$$

8.3.3 Column Vector

A *column vector* is a matrix with a single column:

$$\begin{bmatrix} a_{11} \\ a_{21} \\ \vdots \\ a_{m1} \end{bmatrix}, \quad \text{e.g.} \quad \begin{bmatrix} 2 \\ 3 \\ 23 \end{bmatrix}.$$

8.3.4 Row Vector

A *row vector* is a matrix with a single row:

$$\begin{bmatrix} a_{11} \ a_{12} \cdots a_{1n} \end{bmatrix}, \quad \text{e.g.} \quad \begin{bmatrix} 2 \ 3 \ 5 \end{bmatrix}.$$

8.3.5 Null Matrix

A *null matrix* has all its elements equal to zero:

$$\theta_n = [a_{ij}]_{n \, n} = \begin{bmatrix} 0 \ 0 \cdots 0 \\ 0 \ 0 \cdots 0 \\ \vdots \ \vdots \ \ddots \ \vdots \\ 0 \ 0 \cdots 0 \end{bmatrix}, \quad \text{e.g.} \quad \theta_3 = \begin{bmatrix} 0 \ 0 \ 0 \\ 0 \ 0 \ 0 \\ 0 \ 0 \ 0 \end{bmatrix}.$$

The null matrix behaves like zero when used with numbers, where we have, $0 + n = n + 0 = n$ and $0 \times n = n \times 0 = 0$, and similarly, $\theta + \mathbf{A} = \mathbf{A} + \theta = \mathbf{A}$ and $\theta \mathbf{A} = \mathbf{A} \theta = \theta$. For example,

$$\begin{bmatrix} 0 \ 0 \ 0 \\ 0 \ 0 \ 0 \\ 0 \ 0 \ 0 \end{bmatrix} \begin{bmatrix} 1 \ 2 \ 3 \\ 4 \ 5 \ 6 \\ 7 \ 8 \ 9 \end{bmatrix} = \begin{bmatrix} 1 \ 2 \ 3 \\ 4 \ 5 \ 6 \\ 7 \ 8 \ 9 \end{bmatrix} \begin{bmatrix} 0 \ 0 \ 0 \\ 0 \ 0 \ 0 \\ 0 \ 0 \ 0 \end{bmatrix} = \begin{bmatrix} 0 \ 0 \ 0 \\ 0 \ 0 \ 0 \\ 0 \ 0 \ 0 \end{bmatrix}.$$

8.3.6 Unit Matrix

A *unit matrix* $\mathbf{I}_n$, is a square matrix with the elements on its diagonal a_{11} to a_{nn} equal to 1:

$$\mathbf{I}_n = [a_{ij}]_{n \, n} = \begin{bmatrix} 1 \ 0 \cdots 0 \\ 0 \ 1 \cdots 0 \\ \vdots \ \vdots \ \ddots \ \vdots \\ 0 \ 0 \cdots 1 \end{bmatrix}, \quad \text{e.g.} \quad \mathbf{I}_3 = \begin{bmatrix} 1 \ 0 \ 0 \\ 0 \ 1 \ 0 \\ 0 \ 0 \ 1 \end{bmatrix}.$$

The unit matrix behaves like the number 1 in a conventional product, where we have, $1 \times n = n \times 1 = n$, and similarly, $\mathbf{IA} = \mathbf{AI} = \mathbf{A}$. For example,

$$\begin{bmatrix} 1\,0\,0 \\ 0\,1\,0 \\ 0\,0\,1 \end{bmatrix} \begin{bmatrix} 1\,2\,3 \\ 4\,5\,6 \\ 7\,8\,9 \end{bmatrix} = \begin{bmatrix} 1\,2\,3 \\ 4\,5\,6 \\ 7\,8\,9 \end{bmatrix} \begin{bmatrix} 1\,0\,0 \\ 0\,1\,0 \\ 0\,0\,1 \end{bmatrix} = \begin{bmatrix} 1\,2\,3 \\ 4\,5\,6 \\ 7\,8\,9 \end{bmatrix}.$$

8.3.7   Trace

The *trace* of a square matrix is the sum of the elements on its diagonal a_{11} to a_{nn}:

$$\mathrm{Tr}(\mathbf{A}) = \sum_{i=1}^{n} a_{ii}.$$

For example, given

$$\mathbf{A} = \begin{bmatrix} 1\,2\,3 \\ 4\,5\,6 \\ 7\,8\,9 \end{bmatrix}, \quad \text{then} \quad \mathrm{Tr}(\mathbf{A}) = 1 + 5 + 9 = 15.$$

The trace of a rotation matrix can be used to compute the angle of rotation. For example, the matrix to rotate a point about the origin is

$$\mathbf{A} = \begin{bmatrix} \cos\theta & -\sin\theta \\ \sin\theta & \cos\theta \end{bmatrix}$$

where

$$\mathrm{Tr}(\mathbf{A}) = 2\cos\theta$$

which means that

$$\theta = \arccos\left(\frac{\mathrm{Tr}(\mathbf{A})}{2} \right).$$

The three matrices for rotating points about the x-, y- and z-axis are respectively:

$$\mathbf{R}_{\alpha,x} = \begin{bmatrix} 1 & 0 & 0 \\ 0 & \cos\alpha & -\sin\alpha \\ 0 & \sin\alpha & \cos\alpha \end{bmatrix}$$

$$\mathbf{R}_{\alpha,y} = \begin{bmatrix} \cos\alpha & 0 & \sin\alpha \\ 0 & 1 & 0 \\ -\sin\alpha & 0 & \cos\alpha \end{bmatrix}$$

$$\mathbf{R}_{\alpha,z} = \begin{bmatrix} \cos\alpha & -\sin\alpha & 0 \\ \sin\alpha & \cos\alpha & 0 \\ 0 & 0 & 1 \end{bmatrix}$$

and it is clear that

$$\text{Tr}(\mathbf{R}_{\alpha,x}) = \text{Tr}(\mathbf{R}_{\alpha,y}) = \text{Tr}(\mathbf{R}_{\alpha,z}) = 1 + 2\cos\alpha$$

therefore,

$$\alpha = \arccos\left(\frac{\text{Tr}(\mathbf{R}_{\alpha,x}) - 1}{2}\right).$$

8.3.8 Determinant of a Matrix

The *determinant* of a matrix is a scalar value computed from the elements of the matrix. The different methods for computing the determinant are described in Chap. 6. For example, using Sarrus's rule:

$$\mathbf{A} = \begin{bmatrix} 1 & 2 & 3 \\ 4 & 5 & 6 \\ 7 & 8 & 9 \end{bmatrix} \quad \text{then,} \quad \det\mathbf{A} = 45 + 84 + 96 - 105 - 48 - 72 = 0.$$

8.3.9 Transpose

The *transpose* of a matrix exchanges all row elements for column elements. The transposition is indicated by the letter 'T' outside the right-hand bracket.

$$\begin{bmatrix} a_{11} & a_{12} & a_{13} \\ a_{21} & a_{22} & a_{23} \\ a_{31} & a_{32} & a_{33} \end{bmatrix}^{\text{T}} = \begin{bmatrix} a_{11} & a_{21} & a_{31} \\ a_{12} & a_{22} & a_{32} \\ a_{13} & a_{23} & a_{33} \end{bmatrix}.$$

For example,

$$\begin{bmatrix} 1 & 2 & 4 \\ 6 & 5 & 7 \\ 4 & 3 & 1 \end{bmatrix}^{\text{T}} = \begin{bmatrix} 1 & 6 & 4 \\ 2 & 5 & 3 \\ 4 & 7 & 1 \end{bmatrix},$$

and

$$\begin{bmatrix} 2 \\ 3 \\ 5 \end{bmatrix}^{\text{T}} = \begin{bmatrix} 2 & 3 & 5 \end{bmatrix}.$$

To prove that $(\mathbf{AB})^T = \mathbf{B}^T\mathbf{A}^T$, we could develop a general proof using $n \times n$ matrices, but for simplicity, let's employ 3×3 matrices and assume the result generalises to higher dimensions. Given

$$\mathbf{A} = \begin{bmatrix} a_{11} & a_{12} & a_{13} \\ a_{21} & a_{22} & a_{23} \\ a_{31} & a_{32} & a_{33} \end{bmatrix}, \quad \mathbf{A}^T = \begin{bmatrix} a_{11} & a_{21} & a_{31} \\ a_{12} & a_{22} & a_{32} \\ a_{13} & a_{23} & a_{33} \end{bmatrix}$$

and

$$\mathbf{B} = \begin{bmatrix} b_{11} & b_{12} & b_{13} \\ b_{21} & b_{22} & b_{23} \\ b_{31} & b_{32} & b_{33} \end{bmatrix}, \quad \mathbf{B}^T = \begin{bmatrix} b_{11} & b_{21} & b_{31} \\ b_{12} & b_{22} & b_{32} \\ b_{13} & b_{23} & b_{33} \end{bmatrix}$$

then,

$$\mathbf{AB} = \begin{bmatrix} a_{11}b_{11}+a_{12}b_{21}+a_{13}b_{31} & a_{11}b_{12}+a_{12}b_{22}+a_{13}b_{32} & a_{11}b_{13}+a_{12}b_{23}+a_{13}b_{33} \\ a_{21}b_{11}+a_{22}b_{21}+a_{23}b_{31} & a_{21}b_{12}+a_{22}b_{22}+a_{23}b_{32} & a_{21}b_{13}+a_{22}b_{23}+a_{23}b_{33} \\ a_{31}b_{11}+a_{32}b_{21}+a_{33}b_{31} & a_{31}b_{12}+a_{32}b_{22}+a_{33}b_{32} & a_{31}b_{13}+a_{32}b_{23}+a_{33}b_{33} \end{bmatrix}$$

$$(\mathbf{AB})^T = \begin{bmatrix} a_{11}b_{11}+a_{12}b_{21}+a_{13}b_{31} & a_{21}b_{11}+a_{22}b_{21}+a_{23}b_{31} & a_{31}b_{11}+a_{32}b_{21}+a_{33}b_{31} \\ a_{11}b_{12}+a_{12}b_{22}+a_{13}b_{32} & a_{21}b_{12}+a_{22}b_{22}+a_{23}b_{32} & a_{31}b_{12}+a_{32}b_{22}+a_{33}b_{32} \\ a_{11}b_{13}+a_{12}b_{23}+a_{13}b_{33} & a_{21}b_{13}+a_{22}b_{23}+a_{23}b_{33} & a_{31}b_{13}+a_{32}b_{23}+a_{33}b_{33} \end{bmatrix}$$

and

$$\mathbf{B}^T\mathbf{A}^T = \begin{bmatrix} b_{11}a_{11}+b_{21}a_{12}+b_{31}a_{13} & b_{11}a_{21}+b_{21}a_{22}+b_{31}a_{23} & b_{11}a_{31}+b_{21}a_{32}+b_{31}a_{33} \\ b_{12}a_{11}+b_{22}a_{12}+b_{32}a_{13} & b_{12}a_{21}+b_{22}a_{22}+b_{32}a_{23} & b_{12}a_{31}+b_{22}a_{32}+b_{32}a_{33} \\ b_{13}a_{11}+b_{23}a_{12}+b_{33}a_{13} & b_{13}a_{21}+b_{23}a_{22}+b_{33}a_{23} & b_{13}a_{31}+b_{23}a_{32}+b_{33}a_{33} \end{bmatrix}$$

which confirms that $(\mathbf{AB})^T = \mathbf{B}^T\mathbf{A}^T$.

8.3.10   Symmetric Matrix

A *symmetric matrix* is a square matrix that equals its transpose: i.e., $\mathbf{A} = \mathbf{A}^T$. For example, $\mathbf{A}$ is a symmetric matrix:

$$\mathbf{A} = \begin{bmatrix} 1 & 2 & 4 \\ 2 & 5 & 3 \\ 4 & 3 & 6 \end{bmatrix} = \begin{bmatrix} 1 & 2 & 4 \\ 2 & 5 & 3 \\ 4 & 3 & 6 \end{bmatrix}^T.$$

In general, a square matrix $\mathbf{A} = \mathbf{S} + \mathbf{Q}$, where $\mathbf{S}$ is a symmetric matrix, and $\mathbf{Q}$ is an antisymmetric matrix. The symmetric matrix is computed as follows. Given a matrix $\mathbf{A}$ and its transpose $\mathbf{A}^T$

$$\mathbf{A} = \begin{bmatrix} a_{11} & a_{12} & \cdots & a_{1n} \\ a_{21} & a_{22} & \cdots & a_{2n} \\ \vdots & \vdots & \ddots & \vdots \\ a_{n1} & a_{n2} & \cdots & a_{nn} \end{bmatrix}, \quad \mathbf{A}^T = \begin{bmatrix} a_{11} & a_{21} & \cdots & a_{n1} \\ a_{12} & a_{22} & \cdots & a_{n2} \\ \vdots & \vdots & \ddots & \vdots \\ a_{1n} & a_{2n} & \cdots & a_{nn} \end{bmatrix}$$

their sum is

$$
A + A^T = \begin{bmatrix} 2a_{11} & a_{12} + a_{21} & \cdots & a_{1n} + a_{n1} \\ a_{12} + a_{21} & 2a_{22} & \cdots & a_{2n} + a_{n2} \\ \vdots & \vdots & \ddots & \vdots \\ a_{1n} + a_{n1} & a_{2n} + a_{n2} & \cdots & 2a_{nn} \end{bmatrix}.
$$

By inspection, $A + A^T$ is symmetric, and if we divide throughout by 2 we have

$$
S = \tfrac{1}{2}\left(A + A^T\right)
$$

which is defined as the symmetric part of A. For example, given

$$
A = \begin{bmatrix} a_{11} & a_{12} & a_{13} \\ a_{21} & a_{22} & a_{23} \\ a_{31} & a_{32} & a_{33} \end{bmatrix}, \quad A^T = \begin{bmatrix} a_{11} & a_{21} & a_{31} \\ a_{12} & a_{22} & a_{32} \\ a_{13} & a_{23} & a_{33} \end{bmatrix}
$$

then

$$
\begin{aligned}
S &= \tfrac{1}{2}\left(A + A^T\right) \\
&= \begin{bmatrix} a_{11} & (a_{12} + a_{21})/2 & (a_{13} + a_{31})/2 \\ (a_{12} + a_{21})/2 & a_{22} & (a_{23} + a_{32})/2 \\ (a_{13} + a_{31})/2 & (a_{23} + a_{32})/2 & a_{33} \end{bmatrix} \\
&= \begin{bmatrix} a_{11} & s_3/2 & s_2/2 \\ s_3/2 & a_{22} & s_1/2 \\ s_2/2 & s_1/2 & a_{33} \end{bmatrix}
\end{aligned}
$$

where

$$
\begin{aligned}
s_1 &= a_{23} + a_{32} \\
s_2 &= a_{13} + a_{31} \\
s_3 &= a_{12} + a_{21}.
\end{aligned}
$$

Using a real example:

$$
A = \begin{bmatrix} 0 & 1 & 4 \\ 3 & 1 & 4 \\ 4 & 2 & 6 \end{bmatrix}, \quad A^T = \begin{bmatrix} 0 & 3 & 4 \\ 1 & 1 & 2 \\ 4 & 4 & 6 \end{bmatrix}
$$

$$
S = \begin{bmatrix} 0 & 2 & 4 \\ 2 & 1 & 3 \\ 4 & 3 & 6 \end{bmatrix}
$$

which equals its own transpose.

8.3.11   Antisymmetric Matrix

An *antisymmetric matrix* is a matrix whose transpose is its own negative:

$$\mathbf{A}^{\mathrm{T}} = -\mathbf{A}$$

and is also known as a *skew-symmetric matrix*.

As the elements of $\mathbf{A}$ and $\mathbf{A}^{\mathrm{T}}$ are related by

$$a_{row,col} = -a_{col,row}.$$

When $k = row = col$:

$$a_{k,k} = -a_{k,k}$$

which implies that the diagonal elements must be zero. For example, this is an antisymmetric matrix

$$\mathbf{A} = \begin{bmatrix} 0 & -2 & 4 \\ 2 & 0 & -3 \\ -4 & 3 & 0 \end{bmatrix} = - \begin{bmatrix} 0 & -2 & 4 \\ 2 & 0 & -3 \\ -4 & 3 & 0 \end{bmatrix}^{\mathrm{T}}.$$

The antisymmetric part is computed as follows. Given a matrix $\mathbf{A}$ and its transpose $\mathbf{A}^{\mathrm{T}}$

$$\mathbf{A} = \begin{bmatrix} a_{11} & a_{12} & \cdots & a_{1n} \\ a_{21} & a_{22} & \cdots & a_{2n} \\ \vdots & \vdots & \ddots & \vdots \\ a_{n1} & a_{n2} & \cdots & a_{nn} \end{bmatrix}, \quad \mathbf{A}^{\mathrm{T}} = \begin{bmatrix} a_{11} & a_{21} & \cdots & a_{n1} \\ a_{12} & a_{22} & \cdots & a_{n2} \\ \vdots & \vdots & \ddots & \vdots \\ a_{1n} & a_{2n} & \cdots & a_{nn} \end{bmatrix}$$

their difference is

$$\mathbf{A} - \mathbf{A}^{\mathrm{T}} = \begin{bmatrix} 0 & a_{12} - a_{21} & \cdots & a_{1n} - a_{n1} \\ -(a_{12} - a_{21}) & 0 & \cdots & a_{2n} - a_{n2} \\ \vdots & \vdots & \ddots & \vdots \\ -(a_{1n} - a_{n1}) & -(a_{2n} - a_{n2}) & \cdots & 0 \end{bmatrix}.$$

It is clear that $\mathbf{A} - \mathbf{A}^{\mathrm{T}}$ is antisymmetric, and if we divide throughout by 2 we have

$$\mathbf{Q} = \tfrac{1}{2} \left(\mathbf{A} - \mathbf{A}^{\mathrm{T}} \right).$$

For example:

$$\mathbf{A} = \begin{bmatrix} a_{11} & a_{12} & a_{13} \\ a_{21} & a_{22} & a_{23} \\ a_{31} & a_{32} & a_{33} \end{bmatrix}, \quad \mathbf{A}^{\mathrm{T}} = \begin{bmatrix} a_{11} & a_{21} & a_{31} \\ a_{12} & a_{22} & a_{32} \\ a_{13} & a_{23} & a_{33} \end{bmatrix}$$

$$\mathbf{Q} = \begin{bmatrix} 0 & (a_{12} - a_{21})/2 & (a_{13} - a_{31})/2 \\ (a_{21} - a_{12})/2 & 0 & (a_{23} - a_{32})/2 \\ (a_{31} - a_{13})/2 & (a_{32} - a_{23})/2 & 0 \end{bmatrix}$$

and if we maintain some symmetry with the subscripts, we have

$$\mathbf{Q} = \begin{bmatrix} 0 & (a_{12} - a_{21})/2 & -(a_{31} - a_{13})/2 \\ -(a_{12} - a_{21})/2 & 0 & (a_{23} - a_{32})/2 \\ (a_{31} - a_{13})/2 & -(a_{23} - a_{32})/2 & 0 \end{bmatrix}$$

$$= \begin{bmatrix} 0 & q_3/2 & -q_2/2 \\ -q_3/2 & 0 & q_1/2 \\ q_2/2 & -q_1/2 & 0 \end{bmatrix}$$

where

$$q_1 = a_{23} - a_{32}$$
$$q_2 = a_{31} - a_{13}$$
$$q_3 = a_{12} - a_{21}.$$

Using a real example:

$$\mathbf{A} = \begin{bmatrix} 0 & 1 & 4 \\ 3 & 1 & 4 \\ 4 & 2 & 6 \end{bmatrix}, \quad \mathbf{A}^T = \begin{bmatrix} 0 & 3 & 4 \\ 1 & 1 & 2 \\ 4 & 4 & 6 \end{bmatrix}$$

$$\mathbf{Q} = \begin{bmatrix} 0 & -1 & 0 \\ 1 & 0 & 1 \\ 0 & -1 & 0 \end{bmatrix}.$$

Furthermore, we have already computed

$$\mathbf{S} = \begin{bmatrix} 0 & 2 & 4 \\ 2 & 1 & 3 \\ 4 & 3 & 6 \end{bmatrix}$$

and

$$\mathbf{S} + \mathbf{Q} = \begin{bmatrix} 0 & 1 & 4 \\ 3 & 1 & 4 \\ 4 & 2 & 6 \end{bmatrix} = \mathbf{A}.$$

8.4   Matrix Addition and Subtraction

As equations can be added and subtracted together, it follows that matrices can also
be added and subtracted, as long as they have the same dimension. For example,
given

$$\mathbf{A} = \begin{bmatrix} 11 & 22 \\ 14 & -15 \\ 27 & 28 \end{bmatrix} \quad \text{and} \quad \mathbf{B} = \begin{bmatrix} 2 & 1 \\ -4 & 5 \\ 1 & 8 \end{bmatrix}$$

then

$$\mathbf{A} + \mathbf{B} = \begin{bmatrix} 13 & 23 \\ 10 & -10 \\ 28 & 36 \end{bmatrix}, \quad \mathbf{A} - \mathbf{B} = \begin{bmatrix} 9 & 21 \\ 18 & -20 \\ 26 & 20 \end{bmatrix}.$$

8.4.1   Scalar Multiplication

As equations can be scaled and factorised, it follows that matrixes can also be scaled
and factorised.

$$\lambda \mathbf{A} = \lambda \begin{bmatrix} a_{11} & a_{12} & \cdots & a_{1n} \\ a_{21} & a_{22} & \cdots & a_{2n} \\ \vdots & \vdots & \ddots & \vdots \\ a_{m1} & a_{m2} & \cdots & a_{mn} \end{bmatrix} = \begin{bmatrix} \lambda a_{11} & \lambda a_{12} & \cdots & \lambda a_{13} \\ \lambda a_{21} & \lambda a_{22} & \cdots & \lambda a_{23} \\ \vdots & \vdots & \ddots & \vdots \\ \lambda a_{m1} & \lambda a_{m2} & \cdots & \lambda a_{mn} \end{bmatrix}.$$

For example,

$$2 \begin{bmatrix} 1 & 2 & 3 \\ 4 & 5 & 6 \end{bmatrix} = \begin{bmatrix} 2 & 4 & 6 \\ 8 & 10 & 12 \end{bmatrix}.$$

8.5   Matrix Products

We have already seen that matrices can be multiplied together employing rules that
maintain the algebraic integrity of the equations they represent. And as matrices
may be vectors, rectangular or square, we need to examine the products that are
permitted. To keep the notation simple, the definitions and examples are restricted
to a dimension of 3 or 3×3.

We begin with row and column vectors.

8.5.1 Row and Column Vectors

Given

$$\mathbf{A} = \begin{bmatrix} a & b & c \end{bmatrix} \quad \text{and} \quad \mathbf{B} = \begin{bmatrix} \alpha \\ \beta \\ \gamma \end{bmatrix}$$

then

$$\mathbf{AB} = \begin{bmatrix} a & b & c \end{bmatrix} \begin{bmatrix} \alpha \\ \beta \\ \gamma \end{bmatrix} = a\alpha + b\beta + c\gamma$$

which is a scalar and equivalent to the dot or scalar product of two vectors.

For example, given

$$\mathbf{A} = \begin{bmatrix} 2 & 3 & 4 \end{bmatrix} \quad \text{and} \quad \mathbf{B} = \begin{bmatrix} 10 \\ 30 \\ 20 \end{bmatrix}$$

then

$$\mathbf{AB} = \begin{bmatrix} 2 & 3 & 4 \end{bmatrix} \begin{bmatrix} 10 \\ 30 \\ 20 \end{bmatrix} = 20 + 90 + 80 = 190.$$

Whereas,

$$\mathbf{BA} = \begin{bmatrix} b_{11} \\ b_{21} \\ b_{31} \end{bmatrix} \begin{bmatrix} a_{11} & a_{12} & a_{13} \end{bmatrix} = \begin{bmatrix} b_{11}a_{11} & b_{11}a_{12} & b_{11}a_{13} \\ b_{21}a_{11} & b_{21}a_{12} & b_{21}a_{13} \\ b_{31}a_{11} & b_{31}a_{12} & b_{31}a_{13} \end{bmatrix}.$$

For example,

$$\mathbf{BA} = \begin{bmatrix} 10 \\ 30 \\ 20 \end{bmatrix} \begin{bmatrix} 2 & 3 & 4 \end{bmatrix} = \begin{bmatrix} 20 & 30 & 40 \\ 60 & 90 & 120 \\ 40 & 60 & 80 \end{bmatrix}.$$

The products $\mathbf{AA}$ and $\mathbf{BB}$ are not permitted.

8.5.2 Row Vector and a Matrix

Given

$$\mathbf{A} = \begin{bmatrix} a_{11} & a_{12} & a_{13} \end{bmatrix} \quad \text{and} \quad \mathbf{B} = \begin{bmatrix} b_{11} & b_{12} & b_{13} \\ b_{21} & b_{22} & b_{23} \\ b_{m1} & b_{m2} & b_{33} \end{bmatrix}$$

then

$$\mathbf{AB} = \begin{bmatrix} a_{11} & a_{12} & a_{13} \end{bmatrix} \begin{bmatrix} b_{11} & b_{12} & b_{13} \\ b_{21} & b_{22} & b_{23} \\ b_{m1} & b_{m2} & b_{33} \end{bmatrix}$$

$$= \begin{bmatrix} (a_{11}b_{11} + a_{12}b_{21} + a_{13}b_{31}) & (a_{11}b_{12} + a_{12}b_{22} + a_{13}b_{32}) & (a_{11}b_{13} + a_{12}b_{23} + a_{13}b_{33}) \end{bmatrix}.$$

The product **BA** is not permitted.

For example, given

$$\mathbf{A} = \begin{bmatrix} 2 & 3 & 4 \end{bmatrix} \quad \text{and} \quad \mathbf{B} = \begin{bmatrix} 1 & 2 & 3 \\ 3 & 4 & 5 \\ 4 & 5 & 6 \end{bmatrix}$$

then

$$\mathbf{AB} = \begin{bmatrix} 2 & 3 & 4 \end{bmatrix} \begin{bmatrix} 1 & 2 & 3 \\ 3 & 4 & 5 \\ 4 & 5 & 6 \end{bmatrix}$$

$$= \begin{bmatrix} (2 + 9 + 16) & (4 + 12 + 20) & (6 + 15 + 24) \end{bmatrix}$$

$$= \begin{bmatrix} 27 & 36 & 45 \end{bmatrix}.$$

8.5.3 *Matrix and a Column Vector*

Given

$$\mathbf{A} = \begin{bmatrix} a_{11} & a_{12} & a_{13} \\ a_{21} & a_{22} & a_{23} \\ a_{31} & a_{32} & a_{33} \end{bmatrix} \quad \text{and} \quad \mathbf{B} = \begin{bmatrix} b_{11} \\ b_{21} \\ b_{31} \end{bmatrix}$$

then

$$\mathbf{AB} = \begin{bmatrix} a_{11} & a_{12} & a_{13} \\ a_{21} & a_{22} & a_{23} \\ a_{31} & a_{32} & a_{33} \end{bmatrix} \begin{bmatrix} b_{11} \\ b_{21} \\ b_{31} \end{bmatrix} = \begin{bmatrix} a_{11}b_{11} + a_{12}b_{21} + a_{13}b_{31} \\ a_{21}b_{11} + a_{22}b_{21} + a_{23}b_{31} \\ a_{31}b_{11} + a_{32}b_{21} + a_{33}b_{31} \end{bmatrix}.$$

The product **BA** is not permitted.

For example, given

$$\mathbf{A} = \begin{bmatrix} 1 & 2 & 3 \\ 3 & 4 & 5 \\ 4 & 5 & 6 \end{bmatrix}, \quad \text{and} \quad \mathbf{B} = \begin{bmatrix} 2 \\ 3 \\ 4 \end{bmatrix}$$

then

$$\mathbf{AB} = \begin{bmatrix} 1\ 2\ 3 \\ 3\ 4\ 5 \\ 4\ 5\ 6 \end{bmatrix} \begin{bmatrix} 2 \\ 3 \\ 4 \end{bmatrix} = \begin{bmatrix} 2+6+12 \\ 6+12+20 \\ 8+15+24 \end{bmatrix} = \begin{bmatrix} 20 \\ 38 \\ 47 \end{bmatrix}.$$

8.5.4 Square Matrices

To clarify the products, lower-case Greek symbols are used with lower-case letters. Here are their names:

$\alpha = $ alpha, $\beta = $ beta, $\gamma = $ gamma,

$\lambda = $ lambda, $\mu = $ mu, $\nu = $ nu,

$\rho = $ rho, $\sigma = $ sigma, $\tau = $ tau.

Given

$$\mathbf{A} = \begin{bmatrix} a\ b\ c \\ p\ q\ r \\ u\ v\ w \end{bmatrix} \quad \text{and} \quad \mathbf{B} = \begin{bmatrix} \alpha\ \beta\ \gamma \\ \lambda\ \mu\ \nu \\ \rho\ \sigma\ \tau \end{bmatrix}$$

then

$$\mathbf{AB} = \begin{bmatrix} a\ b\ c \\ p\ q\ r \\ u\ v\ w \end{bmatrix} \begin{bmatrix} \alpha\ \beta\ \gamma \\ \lambda\ \mu\ \nu \\ \rho\ \sigma\ \tau \end{bmatrix} = \begin{bmatrix} a\alpha + b\lambda + c\rho & a\beta + b\mu + c\sigma & a\gamma + b\nu + c\tau \\ p\alpha + q\lambda + r\rho & p\beta + q\mu + r\sigma & p\gamma + q\nu + r\tau \\ u\alpha + v\lambda + w\rho & u\beta + v\mu + w\sigma & u\gamma + v\nu + w\tau \end{bmatrix}$$

and

$$\mathbf{BA} = \begin{bmatrix} \alpha\ \beta\ \gamma \\ \lambda\ \mu\ \nu \\ \rho\ \sigma\ \tau \end{bmatrix} \begin{bmatrix} a\ b\ c \\ p\ q\ r \\ u\ v\ w \end{bmatrix} = \begin{bmatrix} \alpha a + \beta p + \gamma u & \alpha b + \beta q + \gamma v & \alpha c + \beta r + \gamma w \\ \lambda a + \mu p + \nu u & \lambda b + \mu q + \nu v & \lambda c + \mu r + \nu w \\ \rho a + \sigma p + \tau u & \rho b + \sigma q + \tau v & \rho c + \sigma r + \tau w \end{bmatrix}.$$

For example, given

$$\mathbf{A} = \begin{bmatrix} 1\ 2\ 3 \\ 3\ 4\ 5 \\ 5\ 6\ 7 \end{bmatrix} \quad \text{and} \quad \mathbf{B} = \begin{bmatrix} 2\ 3\ 4 \\ 4\ 5\ 6 \\ 6\ 7\ 8 \end{bmatrix}$$

then

$$\mathbf{AB} = \begin{bmatrix} 1\ 2\ 3 \\ 3\ 4\ 5 \\ 5\ 6\ 7 \end{bmatrix} \begin{bmatrix} 2\ 3\ 4 \\ 4\ 5\ 6 \\ 6\ 7\ 8 \end{bmatrix} = \begin{bmatrix} 28\ 34\ 40 \\ 52\ 64\ 76 \\ 76\ 92\ 112 \end{bmatrix}.$$

and

$$\mathbf{BA} = \begin{bmatrix} 2 & 3 & 4 \\ 4 & 5 & 6 \\ 6 & 7 & 8 \end{bmatrix} \begin{bmatrix} 1 & 2 & 3 \\ 3 & 4 & 5 \\ 5 & 6 & 7 \end{bmatrix} = \begin{bmatrix} 31 & 40 & 49 \\ 49 & 64 & 89 \\ 67 & 88 & 109 \end{bmatrix}.$$

8.5.5   Rectangular Matrices

Given two rectangular matrices $\mathbf{A}$ and $\mathbf{B}$, where $\mathbf{A}$ has a dimension $m \times n$, the product $\mathbf{AB}$ is permitted, if and only if, $\mathbf{B}$ has a dimension $n \times p$. The resulting matrix has a dimension $m \times p$. For example, given

$$\mathbf{A} = \begin{bmatrix} a_{11} & a_{12} \\ a_{21} & a_{22} \\ a_{31} & a_{32} \end{bmatrix} \quad \text{and} \quad \mathbf{B} = \begin{bmatrix} b_{11} & b_{12} & b_{13} & b_{14} \\ b_{21} & b_{22} & b_{23} & b_{24} \end{bmatrix}$$

then

$$\mathbf{AB} = \begin{bmatrix} a_{11} & a_{12} \\ a_{21} & a_{22} \\ a_{31} & a_{32} \end{bmatrix} \begin{bmatrix} b_{11} & b_{12} & b_{13} & b_{14} \\ b_{21} & b_{22} & b_{23} & b_{24} \end{bmatrix}$$

$$= \begin{bmatrix} (a_{11}b_{11} + a_{12}b_{21}) & (a_{11}b_{12} + a_{12}b_{22}) & (a_{11}b_{13} + a_{12}b_{23}) & (a_{11}b_{14} + a_{12}b_{24}) \\ (a_{21}b_{11} + a_{22}b_{21}) & (a_{21}b_{12} + a_{22}b_{22}) & (a_{21}b_{13} + a_{22}b_{23}) & (a_{21}b_{14} + a_{22}b_{24}) \\ (a_{31}b_{11} + a_{32}b_{21}) & (a_{31}b_{12} + a_{32}b_{22}) & (a_{31}b_{13} + a_{32}b_{23}) & (a_{31}b_{14} + a_{32}b_{24}) \end{bmatrix}.$$

8.6   Inverse Matrix

A square matrix $\mathbf{A}_{nn}$ that is *invertible* satisfies the condition:

$$\mathbf{A}_{nn}\mathbf{A}_{nn}^{-1} = \mathbf{A}_{nn}^{-1}\mathbf{A}_{nn} = \mathbf{I}_n,$$

where $\mathbf{A}_{nn}^{-1}$ is unique, and is the *inverse matrix* of $\mathbf{A}_{nn}$. For example, given

$$\mathbf{A} = \begin{bmatrix} 4 & 3 \\ 5 & 4 \end{bmatrix}$$

then

$$\mathbf{A}^{-1} = \begin{bmatrix} 4 & -3 \\ -5 & 4 \end{bmatrix}$$

because

$$\mathbf{A}\mathbf{A}^{-1} = \begin{bmatrix} 4 & 3 \\ 5 & 4 \end{bmatrix} \begin{bmatrix} 4 & -3 \\ -5 & 4 \end{bmatrix} = \begin{bmatrix} 1 & 0 \\ 0 & 1 \end{bmatrix}.$$

A square matrix whose determinant is 0, cannot have an inverse, and is known as a *singular matrix*.

We now require a way to compute $\mathbf{A}^{-1}$, which is rather easy.

Consider two linear equations:

$$\begin{bmatrix} x' \\ y' \end{bmatrix} = \begin{bmatrix} a & b \\ c & d \end{bmatrix} \begin{bmatrix} x \\ y \end{bmatrix}. \tag{8.10}$$

Let the inverse of

$$\begin{bmatrix} a & b \\ c & d \end{bmatrix}$$

be

$$\begin{bmatrix} e & f \\ g & h \end{bmatrix}$$

therefore,

$$\begin{bmatrix} e & f \\ g & h \end{bmatrix} \begin{bmatrix} a & b \\ c & d \end{bmatrix} = \begin{bmatrix} 1 & 0 \\ 0 & 1 \end{bmatrix}. \tag{8.11}$$

From (8.11) we have

$$ae + cf = 1 \tag{8.12}$$
$$be + df = 0 \tag{8.13}$$
$$ag + ch = 0 \tag{8.14}$$
$$bg + dh = 1. \tag{8.15}$$

Multiply (8.12) by d and (8.13) by c, and subtract:

$$ade + cdf = d$$
$$bce + cdf = 0$$
$$ade - bce = d$$

therefore,

$$e = \frac{d}{ad - bc}.$$

Multiply (8.12) by b and (8.13) by a, and subtract:

$$abe + bcf = b$$
$$abe + adf = 0$$
$$adf - bcf = -b$$

therefore,

$$f = \frac{-b}{ad - bc}.$$

Multiply (8.14) by d and (8.15) by c, and subtract:

$$adg + cdh = 0$$
$$bcg + cdh = c$$
$$adg - bcg = -c$$

therefore,

$$g = \frac{-c}{ad - bc}.$$

Multiply (8.14) by b and (8.15) by a, and subtract:

$$abg + bch = 0$$
$$abg + adh = a$$
$$adh - bch = a$$

therefore,

$$h = \frac{a}{ad - bc}.$$

We now have values for e, f, g and h, which are the elements of the inverse matrix. Consequently, given

$$\mathbf{A} = \begin{bmatrix} a & b \\ c & d \end{bmatrix} \quad \text{and} \quad \mathbf{A}^{-1} = \begin{bmatrix} e & f \\ g & h \end{bmatrix},$$

then

$$\mathbf{A}^{-1} = \frac{1}{\det \mathbf{A}} \begin{bmatrix} d & -b \\ -c & a \end{bmatrix}.$$

The inverse matrix permits us to solve a pair of linear equations as follows. Starting with

$$\begin{bmatrix} x' \\ y' \end{bmatrix} = \begin{bmatrix} a & b \\ c & d \end{bmatrix} \begin{bmatrix} x \\ y \end{bmatrix} = \mathbf{A} \begin{bmatrix} x \\ y \end{bmatrix}$$

multiply both sides by the inverse matrix:

$$\mathbf{A}^{-1} \begin{bmatrix} x' \\ y' \end{bmatrix} = \mathbf{A}^{-1} \mathbf{A} \begin{bmatrix} x \\ y \end{bmatrix}$$

$$\mathbf{A}^{-1} \begin{bmatrix} x' \\ y' \end{bmatrix} = \begin{bmatrix} 1 & 0 \\ 0 & 1 \end{bmatrix} \begin{bmatrix} x \\ y \end{bmatrix} = \begin{bmatrix} x \\ y \end{bmatrix}$$

$$\begin{bmatrix} x \\ y \end{bmatrix} = \mathbf{A}^{-1} \begin{bmatrix} x' \\ y' \end{bmatrix}$$

$$\begin{bmatrix} x \\ y \end{bmatrix} = \frac{1}{\det \mathbf{A}} \begin{bmatrix} d & -b \\ -c & a \end{bmatrix} \begin{bmatrix} x' \\ y' \end{bmatrix}.$$

Although the elements of $\mathbf{A}^{-1}$ come from $\mathbf{A}$, the relationship is not obvious. However, if $\mathbf{A}$ is transposed, a pattern is revealed. Given

$$\mathbf{A} = \begin{bmatrix} a & b \\ c & d \end{bmatrix} \quad \text{then} \quad \mathbf{A}^{\mathrm{T}} = \begin{bmatrix} a & c \\ b & d \end{bmatrix}$$

and placing $\mathbf{A}^{-1}$ alongside $\mathbf{A}^{\mathrm{T}}$, we have

$$\mathbf{A}^{-1} = \begin{bmatrix} e & f \\ g & h \end{bmatrix} \quad \text{and} \quad \mathbf{A}^{\mathrm{T}} = \begin{bmatrix} a & c \\ b & d \end{bmatrix}.$$

The elements of $\mathbf{A}^{-1}$ share a common denominator ($\det \mathbf{A}$), which is placed outside the matrix, therefore, the matrix elements are taken from $\mathbf{A}^{\mathrm{T}}$ as follows. For any entry a_{ij} in $\mathbf{A}^{-1}$, mask out the ith row and jth column in $\mathbf{A}^{\mathrm{T}}$, and the remaining entry is copied to the ijth entry in $\mathbf{A}^{-1}$. In the case of e, it is d. For f, it is b, with a sign reversal. For g, it is c, with a sign reversal, and for h, it is a. The sign change is computed by the same formula used with determinants:

$$(-1)^{i+j}.$$

which generates this pattern:

$$\begin{bmatrix} + & - \\ - & + \end{bmatrix}.$$

You may be wondering what happens when a 3×3 matrix is inverted. Well, the same technique is used, but when the ith row and jth column in $\mathbf{A}^{\mathrm{T}}$ is masked out, it leaves behind a 2×2 determinant, whose value is copied to the ijth entry in $\mathbf{A}^{-1}$, with the appropriate sign change. We investigate this later on.

Let's illustrate this with an example. Given

$$42 = 6x + 2y$$
$$28 = 2x + 3y$$

let

$$\mathbf{A} = \begin{bmatrix} 6 & 2 \\ 2 & 3 \end{bmatrix}$$

then $\det \mathbf{A} = 14$, therefore,

$$\begin{bmatrix} x \\ y \end{bmatrix} = \frac{1}{14} \begin{bmatrix} 3 & -2 \\ -2 & 6 \end{bmatrix} \begin{bmatrix} 42 \\ 28 \end{bmatrix}$$

$$= \frac{1}{14} \begin{bmatrix} 70 \\ 84 \end{bmatrix}$$

$$= \begin{bmatrix} 5 \\ 6 \end{bmatrix}.$$

which is the solution.

Now let's investigate how to invert a 3×3 matrix. Given three simultaneous equations in three unknowns:

$$x' = ax + by + cz$$
$$y' = dx + ey + fz$$
$$z' = gx + hy + jz$$

they can be written using matrices as follows:

$$\begin{bmatrix} x' \\ y' \\ z' \end{bmatrix} = \begin{bmatrix} a & b & c \\ d & e & f \\ g & h & j \end{bmatrix} \begin{bmatrix} x \\ y \\ z \end{bmatrix} = \mathbf{A} \begin{bmatrix} x \\ y \\ z \end{bmatrix}.$$

Let

$$\mathbf{A}^{-1} = \begin{bmatrix} l & m & n \\ p & q & r \\ s & t & u \end{bmatrix}$$

therefore,

$$\begin{bmatrix} l & m & n \\ p & q & r \\ s & t & u \end{bmatrix} \begin{bmatrix} a & b & c \\ d & e & f \\ g & h & j \end{bmatrix} = \begin{bmatrix} 1 & 0 & 0 \\ 0 & 1 & 0 \\ 0 & 0 & 1 \end{bmatrix}. \qquad (8.16)$$

From (8.16) we can write:

$$la + md + ng = 1 \qquad (8.17)$$
$$lb + me + nh = 0 \qquad (8.18)$$
$$lc + mf + nj = 0. \qquad (8.19)$$

Multiply (8.17) by e and (8.18) by d, and subtract:

$$ael + dem + egn = e$$
$$bdl + dem + dhn = 0$$
$$ael - bdl + egn - dhn = e$$
$$l(ae - bd) + n(eg - dh) = e. \qquad (8.20)$$

Multiply (8.18) by f and (8.19) by e, and subtract:

$$bfl + efm + fhn = 0$$
$$cel + efm + ejn = 0$$
$$bfl - cel + fhn - ejn = 0$$
$$l(bf - ce) + n(fh - ej) = 0. \qquad (8.21)$$

Multiply (8.20) by $(fh - ej)$ and (8.21) by $(eg - dh)$, and subtract:

$$l(ae - bd)(fh - ej) + n(eg - dh)(fh - ej) = e(fh - ej)$$
$$l(bf - ce)(eg - dh) + n(eg - dh)(fh - ej) = 0$$
$$l(ae - bd)(fh - ej) - l(bf - ce)(eg - dh) = efh - e^2 j$$
$$l(aefh - ae^2 j - bdfh + bdej - befg + bdfh + ce^2 g - cdeh) = efh - e^2 j$$
$$l(aefh - ae^2 j + bdej - befg + ce^2 g - cdeh) = efh - e^2 j$$
$$l(afh + bdj + ceg - aej - cdh - bfg) = fh - ej$$
$$l(aej + bfg + cdh - afh - bdj - ceg) = ej - fh$$

but $(aej + bfg + cdh - afh - bdj - ceg)$ is the Sarrus expansion for det $\mathbf{A}$, therefore

$$l = \frac{ej - fh}{\det \mathbf{A}}.$$

An exhaustive algebraic analysis reveals:

$$l = \frac{ej - fh}{\det \mathbf{A}}, \qquad m = -\frac{bj - ch}{\det \mathbf{A}}, \qquad n = \frac{bf - ce}{\det \mathbf{A}}$$
$$p = -\frac{dj - gf}{\det \mathbf{A}}, \qquad q = \frac{aj - gc}{\det \mathbf{A}}, \qquad r = -\frac{af - dc}{\det \mathbf{A}}$$
$$s = \frac{dh - ge}{\det \mathbf{A}}, \qquad t = -\frac{ah - gb}{\det \mathbf{A}}, \qquad u = \frac{ae - bd}{\det \mathbf{A}}$$

where

$$\mathbf{A}^{-1} = \begin{bmatrix} l & m & n \\ p & q & r \\ s & t & u \end{bmatrix} \qquad \mathbf{A} = \begin{bmatrix} a & b & c \\ d & e & f \\ g & h & j \end{bmatrix}.$$

However, there does not appear to be an obvious way of deriving $\mathbf{A}^{-1}$ from $\mathbf{A}$. But, as we discovered with the 2×2 matrix, the transpose $\mathbf{A}^{\mathrm{T}}$ resolves the problem:

$$\mathbf{A}^{-1} = \begin{bmatrix} l & m & n \\ p & q & r \\ s & t & u \end{bmatrix}, \qquad \mathbf{A}^{\mathrm{T}} = \begin{bmatrix} a & d & g \\ b & e & h \\ c & f & j \end{bmatrix}.$$

The elements for $\mathbf{A}^{-1}$ share a common denominator (det $\mathbf{A}$), which is placed outside the matrix, therefore, the matrix elements are taken from $\mathbf{A}^{\mathrm{T}}$ as follows. For any entry a_{ij} in $\mathbf{A}^{-1}$, mask out the ith row and jth column in $\mathbf{A}^{\mathrm{T}}$, and the remaining elements, in the form of a 2×2 determinant, is copied to the ijth entry in $\mathbf{A}^{-1}$. In the case of l, it is $(ej - hf)$. For m, it is $(bj - hc)$, with a sign reversal, and for n, it is $(bf - ec)$. The sign change is computed by the same formula used with determinants:

$$(-1)^{i+j},$$

which generates the pattern:

$$\begin{bmatrix} + & - & + \\ - & + & - \\ + & - & + \end{bmatrix}.$$

With the above *aide-mémoire*, it is easy to write down the inverse matrix:

$$\mathbf{A}^{-1} = \frac{1}{\det \mathbf{A}} \begin{bmatrix} ej - fh & -(bj - ch) & bf - ce \\ -(dj - gf) & aj - gc & -(af - dc) \\ dh - ge & -(ah - gb) & ae - bd \end{bmatrix}.$$

This technique is known as the *Laplacian expansion* or the *cofactor expansion*, after Pierre-Simon Laplace. The matrix of minor determinants is called the *cofactor matrix* of $\mathbf{A}$, which permits the inverse matrix to be written as:

$$\mathbf{A}^{-1} = \frac{(\text{cofactor matrix of } \mathbf{A})^{\mathrm{T}}}{\det \mathbf{A}}.$$

Let's illustrate this solution with an example. Given

$$18 = 2x + 2y + 2z$$
$$20 = x + 2y + 3z$$
$$7 = y + z$$

therefore,

$$\begin{bmatrix} 18 \\ 20 \\ 7 \end{bmatrix} = \begin{bmatrix} 2 & 2 & 2 \\ 1 & 2 & 3 \\ 0 & 1 & 1 \end{bmatrix} \begin{bmatrix} x \\ y \\ z \end{bmatrix} = \mathbf{A} \begin{bmatrix} x \\ y \\ z \end{bmatrix}.$$

and

$$\det \mathbf{A} = 4 + 2 - 2 - 6 = -2$$

$$\mathbf{A}^{\mathrm{T}} = \begin{bmatrix} 2 & 1 & 0 \\ 2 & 2 & 1 \\ 2 & 3 & 1 \end{bmatrix}$$

therefore,

$$\mathbf{A}^{-1} = -\frac{1}{2} \begin{bmatrix} -1 & 0 & 2 \\ -1 & 2 & -4 \\ 1 & -2 & 2 \end{bmatrix}$$

and

$$\begin{bmatrix} x \\ y \\ z \end{bmatrix} = -\frac{1}{2} \begin{bmatrix} -1 & 0 & 2 \\ -1 & 2 & -4 \\ 1 & -2 & 2 \end{bmatrix} \begin{bmatrix} 18 \\ 20 \\ 7 \end{bmatrix} = \begin{bmatrix} 2 \\ 3 \\ 4 \end{bmatrix}$$

which is the solution.

8.6.1 Inverting a Pair of Matrices

Having seen how to invert a single matrix, let's investigate how to invert of a pair of matrices.

Given two matrices $\mathbf{T}$ and $\mathbf{R}$, the product $\mathbf{TR}$ and its inverse $(\mathbf{TR})^{-1}$ must equal the identity matrix $\mathbf{I}$:

$$(\mathbf{TR})(\mathbf{TR})^{-1} = \mathbf{I}$$

and multiplying throughout by $\mathbf{T}^{-1}$ we have

$$\mathbf{T}^{-1}\mathbf{TR}(\mathbf{TR})^{-1} = \mathbf{T}^{-1}$$
$$\mathbf{c}(\mathbf{TR})^{-1} = \mathbf{T}^{-1}.$$

Multiplying throughout by $\mathbf{R}^{-1}$ we have

$$\mathbf{R}^{-1}\mathbf{c}(\mathbf{TR})^{-1} = \mathbf{R}^{-1}\mathbf{T}^{-1}$$
$$(\mathbf{TR})^{-1} = \mathbf{R}^{-1}\mathbf{T}^{-1}.$$

Therefore, if $\mathbf{T}$ and $\mathbf{R}$ are invertible, then

$$(\mathbf{TR})^{-1} = \mathbf{R}^{-1}\mathbf{T}^{-1}.$$

Generalising this result to a triple product such as $\mathbf{STR}$ we can reason that

$$(\mathbf{STR})^{-1} = \mathbf{R}^{-1}\mathbf{T}^{-1}\mathbf{S}^{-1}.$$

8.7 Orthogonal Matrix

A matrix is *orthogonal* if its transpose is also its inverse, i.e., matrix $\mathbf{A}$ is orthogonal if

$$\mathbf{A}^{\mathrm{T}} = \mathbf{A}^{-1}.$$

For example,

$$\mathbf{A} = \begin{bmatrix} \frac{1}{\sqrt{2}} & -\frac{1}{\sqrt{2}} \\ \frac{1}{\sqrt{2}} & \frac{1}{\sqrt{2}} \end{bmatrix}$$

and

$$\mathbf{A}^\mathrm{T} = \begin{bmatrix} \frac{1}{\sqrt{2}} & \frac{1}{\sqrt{2}} \\ -\frac{1}{\sqrt{2}} & \frac{1}{\sqrt{2}} \end{bmatrix}$$

and

$$\mathbf{A}\mathbf{A}^\mathrm{T} = \begin{bmatrix} \frac{1}{\sqrt{2}} & -\frac{1}{\sqrt{2}} \\ \frac{1}{\sqrt{2}} & \frac{1}{\sqrt{2}} \end{bmatrix} \begin{bmatrix} \frac{1}{\sqrt{2}} & \frac{1}{\sqrt{2}} \\ -\frac{1}{\sqrt{2}} & \frac{1}{\sqrt{2}} \end{bmatrix} = \begin{bmatrix} 1 & 0 \\ 0 & 1 \end{bmatrix}$$

which implies that $\mathbf{A}^\mathrm{T} = \mathbf{A}^{-1}$.

The following matrix is also orthogonal

$$\mathbf{A} = \begin{bmatrix} \cos\beta & -\sin\beta \\ \sin\beta & \cos\beta \end{bmatrix}$$

because

$$\mathbf{A}^\mathrm{T} = \begin{bmatrix} \cos\beta & \sin\beta \\ -\sin\beta & \cos\beta \end{bmatrix}$$

and

$$\mathbf{A}\mathbf{A}^\mathrm{T} = \begin{bmatrix} \cos\beta & -\sin\beta \\ \sin\beta & \cos\beta \end{bmatrix} \begin{bmatrix} \cos\beta & \sin\beta \\ -\sin\beta & \cos\beta \end{bmatrix} = \begin{bmatrix} 1 & 0 \\ 0 & 1 \end{bmatrix}.$$

Orthogonal matrices play an important role in rotations because they leave the origin fixed and preserve all angles and distances. Consequently, an object's geometric integrity is maintained after a rotation, which is why an orthogonal transform is known as a *rigid motion* transform.

8.8 Diagonal Matrix

A *diagonal matrix* is a square matrix whose elements are zero, apart from its diagonal:

$$\mathbf{A} = \begin{bmatrix} a_{11} & 0 & \dots & 0 \\ 0 & a_{22} & \dots & 0 \\ \vdots & \vdots & \ddots & \vdots \\ 0 & 0 & \dots & a_{nn} \end{bmatrix}.$$

The determinant of a diagonal matrix must be

$$\det \mathbf{A} = a_{11} \times a_{22} \times \cdots \times a_{nn}.$$

Here is a diagonal matrix with its determinant

$$\mathbf{A} = \begin{bmatrix} 2 & 0 & 0 \\ 0 & 3 & 0 \\ 0 & 0 & 4 \end{bmatrix}$$

$$\det \mathbf{A} = 2 \times 3 \times 4 = 24.$$

The identity matrix $\mathbf{I}$ is a diagonal matrix with a determinant of 1.

8.9 Summary

This chapter has covered matrix algebra to some depth and should permit the reader to use matrices with confidence. The following chapter illustrates how matrices are used to perform a wide variety of geometric transformations.

8.10 Worked Examples

8.10.1 Matrix Inversion

Invert $\mathbf{A}$ and show that $\mathbf{A}\mathbf{A}^{-1} = \mathbf{I}_2$.

$$\mathbf{A} = \begin{bmatrix} 3 & 5 \\ 2 & 4 \end{bmatrix}.$$

Using

$$\mathbf{A}^{-1} = \frac{1}{\det \mathbf{A}} \begin{bmatrix} d & -b \\ -c & a \end{bmatrix}$$

then $\det \mathbf{A} = 2$, and

$$\mathbf{A}^{-1} = \frac{1}{2} \begin{bmatrix} 4 & -5 \\ -2 & 3 \end{bmatrix}.$$

Calculating $\mathbf{A}\mathbf{A}^{-1}$:

$$\mathbf{A}\mathbf{A}^{-1} = \frac{1}{2} \begin{bmatrix} 3 & 5 \\ 2 & 4 \end{bmatrix} \begin{bmatrix} 4 & -5 \\ -2 & 3 \end{bmatrix} = \frac{1}{2} \begin{bmatrix} 2 & 0 \\ 0 & 2 \end{bmatrix} = \begin{bmatrix} 1 & 0 \\ 0 & 1 \end{bmatrix}.$$

8.10.2   Identity Matrix

Invert **A** and show that $\mathbf{AA}^{-1} = \mathbf{I}_3$.

$$\mathbf{A} = \begin{bmatrix} 2\ 3\ 4 \\ 1\ 2\ 1 \\ 5\ 6\ 7 \end{bmatrix}.$$

Using Sarrus's rule for det **A**:

$$\det \mathbf{A} = 28 + 15 + 24 - 40 - 12 - 21 = -6.$$

Therefore,

$$\mathbf{A}^{\mathrm{T}} = \begin{bmatrix} 2\ 1\ 5 \\ 3\ 2\ 6 \\ 4\ 1\ 7 \end{bmatrix}$$

$$\mathbf{A}^{-1} = -\tfrac{1}{6} \begin{bmatrix} 14 - 6) & -(21 - 24) & 3 - 8 \\ -(7 - 5) & 14 - 20 & -(2 - 4) \\ 6 - 10) & -(12 - 15) & 4 - 3 \end{bmatrix}$$

$$= -\tfrac{1}{6} \begin{bmatrix} 8 & 3 & -5 \\ -2 & -6 & 2 \\ -4 & 3 & 1 \end{bmatrix}$$

and

$$\mathbf{AA}^{-1} = -\tfrac{1}{6} \begin{bmatrix} 2\ 3\ 4 \\ 1\ 2\ 1 \\ 5\ 6\ 7 \end{bmatrix} . \begin{bmatrix} 8 & 3 & -5 \\ -2 & -6 & 2 \\ -4 & 3 & 1 \end{bmatrix}$$

$$= -\tfrac{1}{6} \begin{bmatrix} -6 & 0 & 0 \\ 0 & -6 & 0 \\ 0 & 0 & -6 \end{bmatrix} = \begin{bmatrix} 1\ 0\ 0 \\ 0\ 1\ 0 \\ 0\ 0\ 1 \end{bmatrix}.$$

8.10.3   Solving Two Equations Using Matrices

Solve the following equations using matrices.

$$20 = 2x + 3y$$
$$36 = 7x + 2y.$$

Let

$$\mathbf{A} = \begin{bmatrix} 2 & 3 \\ 7 & 2 \end{bmatrix}$$

therefore, $\det \mathbf{A} = -17$, and

$$\mathbf{A}^{-1} = -\tfrac{1}{17} \begin{bmatrix} 2 & -3 \\ -7 & 2 \end{bmatrix}$$

therefore,

$$\begin{bmatrix} x \\ y \end{bmatrix} = -\tfrac{1}{17} \begin{bmatrix} 2 & -3 \\ -7 & 2 \end{bmatrix} \begin{bmatrix} 20 \\ 36 \end{bmatrix}$$
$$= -\tfrac{1}{17} \begin{bmatrix} 40 - 108 \\ -140 + 72 \end{bmatrix}$$
$$= -\tfrac{1}{17} \begin{bmatrix} -68 \\ -68 \end{bmatrix}$$
$$= \begin{bmatrix} 4 \\ 4 \end{bmatrix}$$

therefore, $x = y = 4$.

8.10.4 Solving Three Equations Using Matrices

Solve the following equations using matrices.

$$10 = 2x + y - z$$
$$13 = -x - y + z$$
$$28 = -x + 2y + z.$$

Let

$$\mathbf{A} = \begin{bmatrix} 2 & 1 & -1 \\ -1 & -1 & 1 \\ -1 & 2 & 1 \end{bmatrix}.$$

Using Sarrus's rule for $\det \mathbf{A}$:

$$\det \mathbf{A} = -2 - 1 + 2 + 1 - 4 + 1 = -3.$$

Therefore,

$$\mathbf{A}^{\mathrm{T}} = \begin{bmatrix} 2 & -1 & -1 \\ 1 & -1 & 2 \\ -1 & 1 & 1 \end{bmatrix}$$

$$\mathbf{A}^{-1} = -\frac{1}{3} \begin{bmatrix} (-1-2) & -(1+2) & (1-1) \\ -(-1+1) & (2-1) & -(2-1) \\ (-2-1) & -(4+1) & (-2+1) \end{bmatrix}$$

$$= -\frac{1}{3} \begin{bmatrix} -3 & -3 & 0 \\ 0 & 1 & -1 \\ -3 & -5 & -1 \end{bmatrix}$$

therefore,

$$\begin{bmatrix} x \\ y \\ z \end{bmatrix} = -\frac{1}{3} \begin{bmatrix} -3 & -3 & 0 \\ 0 & 1 & -1 \\ -3 & -5 & -1 \end{bmatrix} \begin{bmatrix} 10 \\ 13 \\ 28 \end{bmatrix}$$

$$= -\frac{1}{3} \begin{bmatrix} -30 - 39 \\ 13 - 28 \\ -30 - 65 - 28 \end{bmatrix}$$

$$= -\frac{1}{3} \begin{bmatrix} -69 \\ -15 \\ -123 \end{bmatrix}$$

$$= \begin{bmatrix} 23 \\ 5 \\ 41 \end{bmatrix}$$

therefore, $x = 23, \quad y = 5, \quad z = 41$.

8.10.5   Solving Two Complex Equations

Solve the following complex equations using matrices.

$$7 + i8 = 2x + y$$
$$-4 - i = x - 2y.$$

Let

$$\mathbf{A} = \begin{bmatrix} 2 & 1 \\ 1 & -2 \end{bmatrix}$$

therefore, det $\mathbf{A} = -5$, and

$$\mathbf{A}^{\mathrm{T}} = \begin{bmatrix} 2 & 1 \\ 1 & -2 \end{bmatrix}$$

$$\mathbf{A}^{-1} = -\frac{1}{5} \begin{bmatrix} -2 & -1 \\ -1 & 2 \end{bmatrix}$$

therefore,

$$
\begin{bmatrix} x \\ y \end{bmatrix} = -\frac{1}{5} \begin{bmatrix} -2 & -1 \\ -1 & 2 \end{bmatrix} \begin{bmatrix} 7+i8 \\ -4-i \end{bmatrix}
$$

$$
= -\frac{1}{5} \begin{bmatrix} -14-i16+4+i \\ -7-i8-8-i2 \end{bmatrix}
$$

$$
= -\frac{1}{5} \begin{bmatrix} -10-i15 \\ -15-i10 \end{bmatrix}
$$

$$
= \begin{bmatrix} 2+i3 \\ 3+i2 \end{bmatrix}
$$

therefore, $x = 2 + i3, \quad y = 3 + i2.$

8.10.6 Solving Three Complex Equations

Solve the following complex equations using matrices.

$$
0 = x + y - z
$$
$$
3 + i3 = 2x - y + z
$$
$$
-5 - i5 = -x + y - 2z.
$$

Let

$$
A = \begin{bmatrix} 1 & 1 & -1 \\ 2 & -1 & 1 \\ -1 & 1 & -2 \end{bmatrix}
$$

therefore, det $A = 2 - 1 - 2 + 1 - 1 + 4 = 3$, and

$$
A^T = \begin{bmatrix} 1 & 2 & -1 \\ 1 & -1 & 1 \\ -1 & 1 & -2 \end{bmatrix}
$$

$$
A^{-1} = \frac{1}{3} \begin{bmatrix} (2-1) & -(-2+1) & 0 \\ -(-4+1) & (-2-1) & -(1+2) \\ (2-1) & -(1+1) & (-1-2) \end{bmatrix}
$$

therefore,

$$
\begin{bmatrix} x \\ y \\ z \end{bmatrix} = \frac{1}{3} \begin{bmatrix} 1 & 1 & 0 \\ 3 & -3 & -3 \\ 1 & -2 & -3 \end{bmatrix} \begin{bmatrix} 0 \\ 3+i3 \\ -5-i5 \end{bmatrix}
$$

$$= \tfrac{1}{3} \begin{bmatrix} 3 + i3 \\ -9 - i9 + 15 + i15 \\ -6 - i6 + 15 + i15 \end{bmatrix}$$

$$= \begin{bmatrix} 1 + i \\ 2 + i2 \\ 3 + i3 \end{bmatrix}$$

therefore, $x = 1 + i$, $\quad y = 2 + i2$, $\quad z = 3 + i3$.

8.10.7 Solving Two Complex Equations

Solve the following complex equations using matrices.

$$3 + i5 = ix + 2y$$
$$5 + i = 3x - iy.$$

Let

$$\mathbf{A} = \begin{bmatrix} i & 2 \\ 3 & -i \end{bmatrix}$$

therefore, set $\mathbf{A} = 1 - 6 = -5$, and

$$\mathbf{A}^{\mathrm{T}} = \begin{bmatrix} i & 3 \\ 2 & -i \end{bmatrix}$$

$$\mathbf{A}^{-1} = -\tfrac{1}{5} \begin{bmatrix} -i & -2 \\ -3 & i \end{bmatrix}$$

therefore,

$$\begin{bmatrix} x \\ y \end{bmatrix} = -\tfrac{1}{5} \begin{bmatrix} -i & -2 \\ -3 & i \end{bmatrix} \begin{bmatrix} 3 + i5 \\ 5 + i \end{bmatrix}$$

$$= -\tfrac{1}{5} \begin{bmatrix} -i3 + 5 - 10 - i2 \\ -9 - i15 + i5 - 1 \end{bmatrix}$$

$$= -\tfrac{1}{5} \begin{bmatrix} -5 - i5 \\ -10 - i10 \end{bmatrix}$$

$$= \begin{bmatrix} 1 + i \\ 2 + i2 \end{bmatrix}$$

therefore, $x = 1 + i$, $\quad y = 2 + i2$.

8.10.8 *Solving Three Complex Equations*

Solve the following complex equations using matrices.

$$6 + i2 = ix + 2y - iz$$
$$-2 + i6 = 2x - iy + i2z$$
$$2 + i10 = i2x + iy + 2z.$$

Let

$$\mathbf{A} = \begin{bmatrix} i & 2 & -i \\ 2 & -i & i2 \\ i2 & i & 2 \end{bmatrix}$$

therefore, det $\mathbf{A} = 2 - 8 + 2 + i2 + i2 - 8 = -12 + i4$, and

$$\mathbf{A}^{\mathrm{T}} = \begin{bmatrix} i & 2 & i2 \\ 2 & -i & i \\ -i & i2 & 2 \end{bmatrix}$$

$$\mathbf{A}^{-1} = \frac{1}{-12 + i4} \begin{bmatrix} -i2 + 2 & -(4-1) & i4+1 \\ -(4+4) & i2-2 & -(-2+i2) \\ i2-2 & -(-1-i4) & 1-4 \end{bmatrix}$$

$$= \frac{1}{-12 + i4} \begin{bmatrix} 2-i2 & -3 & 1+i4 \\ -8 & -2+i2 & 2-i2 \\ -2+i2 & 1+i4 & -3 \end{bmatrix}$$

therefore,

$$\begin{bmatrix} x \\ y \\ z \end{bmatrix} = \frac{1}{-12 + i4} \begin{bmatrix} 2-i2 & -3 & 1+i4 \\ -8 & -2+i2 & 2-i2 \\ -2+i2 & 1+i4 & -3 \end{bmatrix} \begin{bmatrix} 6+i2 \\ -2+i6 \\ 2+i10 \end{bmatrix}$$

$$= \frac{1}{-12 + i4} \begin{bmatrix} (2-i2)(6+i2) - 3(-2+i6) + (1+i4)(2+i10) \\ -8(6+i2) + (-2+i2)(-2+i6) + (2-i2)(2+i10) \\ (-2+i2)(6+i2) + (1+i4)(-2+i6) - 3(2+i10) \end{bmatrix}$$

$$= \frac{1}{-12 + i4} \begin{bmatrix} 12 + i4 - i12 + 4 + 6 - i18 + 2 + i10 + i8 - 40 \\ -48 - i16 + 4 - i12 - i4 - 12 + 4 + i20 - i4 + 20 \\ -12 - i4 + i12 - 4 - 2 + i6 - i8 - 24 - 6 - i30 \end{bmatrix}$$

$$= \frac{1}{-12 + i4} \begin{bmatrix} -16 - i8 \\ -32 - i16 \\ -48 - i24 \end{bmatrix}$$

multiply by the conjugate of $-12 + i4$:

$$\begin{bmatrix} x \\ y \\ z \end{bmatrix} = \frac{-12 - i4}{160} \begin{bmatrix} -16 - i8 \\ -32 - i16 \\ -48 - i24 \end{bmatrix}$$

therefore,

$$
\begin{aligned}
x &= \tfrac{1}{160}(-12 - i4)(-16 - i8) \\
&= \tfrac{1}{160}(192 + i64 + i96 - 32) \\
&= \tfrac{1}{160}(160 + i160) = 1 + i \\
y &= \tfrac{1}{160}(-12 - i4)(-32 - i16) \\
&= \tfrac{1}{160}(384 + i128 + i192 - 64) \\
&= \tfrac{1}{160}(320 + i320) = 2 + i2 \\
z &= \tfrac{1}{160}(-12 - i4)(-48 - i24) \\
&= \tfrac{1}{160}(576 + i192 + i288 - 96) \\
&= \tfrac{1}{160}(480 + i480) = 3 + i3
\end{aligned}
$$

therefore, $x = 1 + i, \quad y = 2 + i2, \quad z = 3 + i3.$

Chapter 9
Complex Numbers

9.1 Introduction

In this chapter we investigate complex numbers and show how they can be thought of as an ordered pair. We also show how they are represented by a matrix. Many of the qualities associated with quaternions are found in complex numbers, which is why they are worthy of close examination. Readers interested in this subject may want to examine the author's book *Imaginary Mathematics for Computer Science* [1].

9.2 Definition of a Complex Number

By definition, a *complex number* is the combination of a real number and an imaginary number, and is expressed as

$$z = a + bi, \quad a, b \in \mathbb{R}, \quad i^2 = -1.$$

The set of complex numbers is $\mathbb{C}$, which permits us to write $z \in \mathbb{C}$. For example, $3 + 4i$ is a complex number where 3 is the real part and $4i$ is the imaginary part. The following are all complex numbers:

$$3, \quad 3 + 4i, \quad -4 - 6i, \quad 7i, \quad 5.5 + 6.7i.$$

A real number is also a complex number—it just has no imaginary part. This leads to the idea that the set of real numbers is a subset of complex numbers, which is expressed as:

$$\mathbb{R} \subset \mathbb{C}$$

where $\subset$ means *is a subset of*.

© Springer-Verlag London Ltd., part of Springer Nature 2022
J. Vince, *Mathematics for Computer Graphics*, Undergraduate Topics
in Computer Science, https://doi.org/10.1007/978-1-4471-7520-9_9

Although some mathematicians place i before its multiplier: $i4$, others place it after the multiplier: $4i$, which is the convention used in this book. However, when i is associated with trigonometric functions, it is good practice to place it before the function to avoid any confusion with the function's angle. For example, $\sin \alpha i$ could imply that the angle is imaginary, whereas $i \sin \alpha$ implies that the value of $\sin \alpha$ is imaginary.

Therefore, a complex number can be constructed in all sorts of ways:

$$\sin \alpha + i \cos \beta, \quad 2 - i \tan \alpha, \quad 23 + x^2 i.$$

In general, we write a complex number as $a + bi$ and subject it to the normal rules of real algebra. All that we have to remember is that whenever we encounter i^2 it is replaced by -1. For example:

$$(2 + 3i)(3 + 4i) = 2 \times 3 + 2 \times 4i + 3i \times 3 + 3i \times 4i$$
$$= 6 + 8i + 9i + 12i^2$$
$$= 6 + 17i - 12$$
$$= -6 + 17i.$$

9.2.1   Addition and Subtraction of Complex Numbers

Given two complex numbers:

$$z_1 = a_1 + b_1 i$$
$$z_2 = a_2 + b_2 i$$

then,
$$z_1 \pm z_2 = (a_1 \pm a_2) + (b_1 \pm b_2)i$$

where the real and imaginary parts are added or subtracted, respectively. The operations are closed, so long as $a_1, b_1, a_2, b_2 \in \mathbb{R}$.

For example:

$$z_1 = 2 + 3i$$
$$z_2 = 4 + 2i$$
$$z_1 + z_2 = 6 + 5i$$
$$z_1 - z_2 = -2 + i.$$

9.2.2 Multiplying a Complex Number by a Scalar

A complex number is multiplied by a scalar using normal algebraic rules. For example, the complex number $a + bi$ is multiplied by the scalar λ as follows:

$$\lambda(a + bi) = \lambda a + \lambda bi$$

for example:
$$3(2 + 5i) = 6 + 15i.$$

9.2.3 Product of Complex Numbers

Given two complex numbers:

$$z_1 = a_1 + b_1 i$$
$$z_2 = a_2 + b_2 i$$

their product is

$$z_1 z_2 = (a_1 + b_1 i)(a_2 + b_2 i)$$
$$= a_1 a_2 + a_1 b_2 i + b_1 a_2 i + b_1 b_2 i^2$$
$$= (a_1 a_2 - b_1 b_2) + (a_1 b_2 + b_1 a_2)i$$

which is another complex number and confirms that the operation is closed. For example:

$$z_1 = 3 + 4i$$
$$z_2 = 3 - 2i$$
$$z_1 z_2 = (3 + 4i)(3 - 2i)$$
$$= 9 - 6i + 12i - 8i^2$$
$$= 9 + 6i + 8$$
$$= 17 + 6i.$$

Note that the addition, subtraction and multiplication of complex numbers obey the normal axioms of algebra.

9.2.4   Square of a Complex Number

Given a complex number z, its square z^2 is given by:

$$z = a + bi$$
$$z^2 = (a + bi)(a + bi)$$
$$= (a^2 - b^2) + 2abi.$$

For example:

$$z = 4 + 3i$$
$$z^2 = (4 + 3i)(4 + 3i)$$
$$= (4^2 - 3^2) + 2 \times 4 \times 3i$$
$$= 7 + 24i.$$

9.2.5   Norm of a Complex Number

The *norm*, *modulus* or *absolute value* of a complex number z is written $|z|$ and by definition is

$$z = a + bi$$
$$|z| = \sqrt{a^2 + b^2}.$$

For example, the norm of $3 + 4i$ is 5. We'll see why this is so when we cover the polar representation of a complex number.

9.2.6   Complex Conjugate of a Complex Number

The product of two complex numbers, where the only difference between them is the sign of the imaginary part, gives rise to a special result:

$$(a + bi)(a - bi) = a^2 - abi + abi - b^2i^2$$
$$= a^2 + b^2.$$

This type of product *always* results in a real quantity and is used to resolve the quotient of two complex numbers. Because this real value is such an interesting result, $a - bi$ is called the *complex conjugate* of $z = a + bi$, and is written either with a bar $\bar{z}$, or an asterisk z^*, and implies that

$$zz^* = a^2 + b^2 = |z|^2.$$

For example:

$$z = 3 + 4i$$
$$z^* = 3 - 4i$$
$$zz^* = 9 + 16 = 25.$$

9.2.7 Quotient of Complex Numbers

The complex conjugate provides us with a mechanism to divide one complex number by another. For instance, the quotient

$$\frac{a_1 + b_1 i}{a_2 + b_2 i}$$

is resolved by multiplying the numerator and denominator by the denominator's complex conjugate $a_2 - b_2 i$ to create a real denominator:

$$\frac{a_1 + b_1 i}{a_2 + b_2 i} = \frac{(a_1 + b_1 i)(a_2 - b_2 i)}{(a_2 + b_2 i)(a_2 - b_2 i)}$$
$$= \frac{a_1 a_2 - a_1 b_2 i + b_1 a_2 i - b_1 b_2 i^2}{a_2^2 + b_2^2}$$
$$= \left(\frac{a_1 a_2 + b_1 b_2}{a_2^2 + b_2^2} \right) + \left(\frac{b_1 a_2 - a_1 b_2}{a_2^2 + b_2^2} \right) i.$$

For example, to evaluate

$$\frac{4 + 3i}{3 + 4i}.$$

we multiply top and bottom by the complex conjugate $3 - 4i$:

$$\frac{4 + 3i}{3 + 4i} = \frac{(4 + 3i)(3 - 4i)}{(3 + 4i)(3 - 4i)}$$
$$= \frac{12 - 16i + 9i - 12i^2}{25}$$
$$= \frac{24}{25} - \frac{7}{25}i.$$

9.2.8   *Inverse of a Complex Number*

To compute the inverse of $z = a + bi$ we start with

$$z^{-1} = \frac{1}{z}.$$

Multiplying top and bottom by z^* we have

$$z^{-1} = \frac{z*}{zz*}.$$

But we have previously shown that $zz^* = |z|^2$, therefore,

$$z^{-1} = \frac{z^*}{|z|^2}$$

$$= \left(\frac{a}{a^2 + b^2}\right) - \left(\frac{b}{a^2 + b^2}\right)i.$$

As an example, the inverse of $3 + 4i$ is

$$(3 + 4i)^{-1} = \tfrac{3}{25} - \tfrac{4}{25}i.$$

Let's test this result by multiplying $3 + 4i$ by its inverse:

$$(3 + 4i)\left(\tfrac{3}{25} - \tfrac{4}{25}i\right) = \tfrac{9}{25} - \tfrac{12}{25}i + \tfrac{12}{25}i + \tfrac{16}{25} = 1$$

which confirms the correctness of the result.

9.2.9   *Square-Root of $\pm i$*

To find $\sqrt{i}$ we assume that the roots are complex. Therefore, we start with

$$\sqrt{i} = a + bi$$
$$i = (a + bi)(a + bi)$$
$$= a^2 + 2abi - b^2$$
$$= a^2 - b^2 + 2abi$$

and equating real and imaginary parts we have

$$a^2 - b^2 = 0$$
$$2ab = 1.$$

From this we deduce that
$$a = b = \pm\tfrac{\sqrt{2}}{2}.$$

Therefore, the roots are
$$\sqrt{i} = \pm\tfrac{\sqrt{2}}{2}(1+i).$$

Let's test this result by squaring each root to ensure the answer is i:

$$\left(\pm\tfrac{\sqrt{2}}{2}\right)^2 (1+i)(1+i) = \tfrac{1}{2}2i = i.$$

To find $\sqrt{-i}$ we assume that the roots are complex. Therefore, we start with

$$\sqrt{-i} = a + bi$$
$$-i = (a + bi)(a + bi)$$
$$= a^2 + 2abi - b^2$$
$$= a^2 - b^2 + 2abi$$

and equating real and imaginary parts we have

$$a^2 - b^2 = 0$$
$$2ab = -1.$$

From this we deduce that
$$a = b = \pm\tfrac{\sqrt{2}}{2}i.$$

Therefore, the roots are

$$\sqrt{-i} = \pm\tfrac{\sqrt{2}}{2}i(1+i)$$
$$= \pm\tfrac{\sqrt{2}}{2}(-1+i)$$
$$= \pm\tfrac{\sqrt{2}}{2}(1-i).$$

Let's test this result by squaring each root to ensure the answer is $-i$:

$$\left(\pm\tfrac{\sqrt{2}}{2}\right)^2 (1-i)(1-i) = -\tfrac{1}{2}2i = -i.$$

We use these roots in the next chapter to investigate the rotational properties of complex numbers.

9.3 Ordered Pairs

So far, we have chosen to express a complex number as $a + bi$ where we can distinguish between the real and imaginary parts. However, one thing we cannot assume is that the real part is always first, and the imaginary part second, because $bi + a$ is also a complex number. Consequently, two functions are employed to extract the real and imaginary coefficients as follows:

$$\text{Re}(a + bi) = a$$
$$\text{Im}(a + bi) = b$$

and leads us to the idea of representing a complex number by an ordered pair where order is guaranteed:

$$a + bi = (a, \ b)$$

where b follows a to define the order. Thus the set $\mathbb{C}$ of complex numbers is equivalent to the set $\mathbb{R}^2$ of ordered pairs $(a, \ b)$.

Writing a complex number as an ordered pair was a great contribution, and first made by Hamilton in 1833. Such notation is very succinct and free from any imaginary term, which can be added whenever required.

9.3.1 Addition and Subtraction of Ordered Pairs

Given two complex numbers:

$$z_1 = a_1 + b_1 i$$
$$z_2 = a_2 + b_2 i$$

they are written as ordered pairs:

$$z_1 = (a_1, \ b_1)$$
$$z_2 = (a_2, \ b_2)$$

and

$$z_1 \pm z_2 = (a_1 \pm a_2, \ b_1 \pm b_2)$$

where the two parts are added or subtracted, respectively.

For example:

$$z_1 = 2 + 3i = (2, \ 3)$$
$$z_2 = 4 + 2i = (4, \ 2)$$

$$z_1 + z_2 = (6,\ 5)$$
$$z_1 - z_2 = (-2,\ 1).$$

9.3.2 Multiplying an Ordered Pair by a Scalar

We have already seen how a complex number is multiplied by a scalar, which must be the same as ordered pairs:

$$\lambda(a,\ b) = (\lambda a,\ \lambda b).$$

An example is

$$3(2,\ 5) = (6,\ 15).$$

9.3.3 Product of Ordered Pairs

Given two complex numbers:

$$z_1 = a_1 + b_1 i$$
$$z_2 = a_2 + b_2 i$$

their product is

$$z_1 z_2 = (a_1 a_2 - b_1 b_2) + (a_1 b_2 + b_1 a_2)i$$

which must also work with ordered pairs:

$$z_1 = (a_1,\ b_1)$$
$$z_2 = (a_2,\ b_2)$$
$$z_1 z_2 = (a_1,\ b_1)(a_2,\ b_2)$$
$$= (a_1 a_2 - b_1 b_2,\ a_1 b_2 + b_1 a_2).$$

For example:

$$z_1 = (6,\ 2)$$
$$z_2 = (4,\ 3)$$
$$z_1 z_2 = (6,\ 2)(4,\ 3)$$
$$= (24 - 6,\ 18 + 8)$$
$$= (18,\ 26).$$

9.3.4 Square of an Ordered Pair

The square of a complex number is given by:

$$z = a + bi$$
$$z^2 = (a + bi)(a + bi)$$
$$= (a^2 - b^2) + 2abi.$$

Therefore, the square of an ordered pair is:

$$z = (a,\ b)$$
$$z^2 = (a,\ b)(a,\ b)$$
$$= (a^2 - b^2,\ 2ab).$$

For example:

$$z = (4,\ 3)$$
$$z^2 = (4,\ 3)(4,\ 3)$$
$$= (4^2 - 3^2,\ 2 \times 4 \times 3)$$
$$= (7,\ 24).$$

Let's continue to develop an algebra based upon ordered pairs that is identical to the algebra of complex numbers. We start by writing

$$z = (a,\ b)$$
$$= (a,\ 0) + (0,\ b)$$
$$= a(1,\ 0) + b(0,\ 1)$$

which creates the unit ordered pairs $(1,\ 0)$ and $(0,\ 1)$.

Now let's compute the product $(1,\ 0)(1,\ 0)$:

$$(1,\ 0)(1,\ 0) = (1 - 0,\ 0)$$
$$= (1,\ 0)$$

which shows that $(1,\ 0)$ behaves like the real number 1. i.e. $(1,\ 0) = 1$.

Next, let's compute the product $(0,\ 1)(0,\ 1)$:

$$(0,\ 1)(0,\ 1) = (0 - 1,\ 0)$$
$$= (-1,\ 0)$$

which is the real number -1:

$$(0, \ 1)^2 = -1$$

or

$$(0, \ 1) = \sqrt{-1} \quad \text{and is imaginary.}$$

This means that the ordered pair $(a, \ b)$, together with its associated rules, represents a complex number. i.e. $(a, \ b) \equiv a + bi$.

9.3.5 Norm of an Ordered Pair

The *norm, modulus* or *absolute value* of an ordered pair z is written $|z|$ and by definition is

$$z = (a, \ b)$$
$$|z| = \sqrt{a^2 + b^2}.$$

For example, the norm of $(3, \ 4)$ is 5.

9.3.6 Complex Conjugate of an Ordered Pair

The complex conjugate of $z = a + bi$ is defined as $z^* = a - bi$, which in terms of an ordered pair is $z^* = (a, \ -b)$:

$$z = (a, \ b)$$
$$z^* = (a, \ -b)$$
$$zz^* = (a, \ b)(a, \ -b)$$
$$= (a^2 + b^2, \ ba - ab)$$
$$= (a^2 + b^2, \ 0)$$
$$= a^2 + b^2 = |z|^2.$$

9.3.7 Quotient of an Ordered Pair

The technique for resolving z_1/z_2 is to multiply the expression by z_2^*/z_2^*, which using ordered pairs is

$$\frac{z_1}{z_2} = \frac{(a_1, \ b_1)}{(a_2, \ b_2)}$$

$$= \frac{(a_1, \; b_1) \, (a_2, \; -b_2)}{(a_2, \; b_2) \, (a_2, \; -b_2)}$$

$$= \frac{(a_1 a_2 + b_1 b_2, \; b_1 a_2 - a_1 b_2)}{(a_2^2 + b_2^2, \; 0)}$$

$$= \left(\frac{a_1 a_2 + b_1 b_2}{a_2^2 + b_2^2}, \; \frac{b_1 a_2 - a_1 b_2}{a_2^2 + b_2^2} \right).$$

For example, to evaluate

$$\frac{(4, \; 3)}{(3, \; 4)}.$$

we multiply top and bottom by the complex conjugate $(3, \; -4)$:

$$\frac{(4, \; 3)}{(3, \; 4)} = \frac{(4, \; 3)(3, \; -4)}{(3, \; 4)(3, \; -4)}$$

$$= \left(\frac{12 + 12}{25}, \; \frac{9 - 16}{25} \right)$$

$$= \left(\tfrac{24}{25}, \; -\tfrac{7}{25} \right).$$

9.3.8   *Inverse of an Ordered Pair*

We have previously shown that z^{-1} is

$$z^{-1} = \frac{z^*}{zz^*} = \frac{z^*}{|z|^2}$$

which using ordered pairs is

$$z = (a, \; b)$$

$$z^{-1} = \frac{(a, \; -b)}{(a, \; b)(a, \; -b)}$$

$$= \frac{(a, \; -b)}{(a^2 + b^2, \; 0)}$$

$$= \left(\frac{a}{a^2 + b^2}, \; \frac{-b}{a^2 + b^2} \right).$$

As an illustration, the inverse of $(3, \; 4)$ is

$$(3, \; 4)^{-1} = \left(\tfrac{3}{25}, \; -\tfrac{4}{25} \right).$$

Let's test this result by multiplying $(3, 4)$ by its inverse:

$$(3, 4)\left(\tfrac{3}{25}, -\tfrac{4}{25}\right) = \left(\tfrac{9}{25} + \tfrac{16}{25}, -\tfrac{12}{25} + \tfrac{12}{25}\right)$$
$$= (1, 0).$$

9.3.9 Square-Root of $\pm i$

To find $\sqrt{i}$ we assume that the roots are complex. Therefore, we start with

$$\sqrt{i} = (a, b)$$
$$i = (a, b)(a, b)$$
$$(0, 1) = \left(a^2 - b^2, 2ab\right)$$

and equating left and right ordered elements we have

$$a^2 - b^2 = 0$$
$$2ab = 1.$$

From this we deduce that

$$a = b = \pm\tfrac{\sqrt{2}}{2}.$$

Therefore, the roots are

$$\sqrt{i} = \pm\tfrac{\sqrt{2}}{2}(1, 1).$$

Let's test this result by squaring each root to ensure the answer is i:

$$\left(\pm\tfrac{\sqrt{2}}{2}\right)^2 (1, 1)(1, 1) = \tfrac{1}{2}(0, 2) = (0, 1).$$

To find $\sqrt{-i}$ we assume that the roots are complex. Therefore, we start with

$$\sqrt{-i} = (a, b)$$
$$-i = (a, b)(a, b)$$
$$(0, -1) = \left(a^2 - b^2, 2ab\right)$$

and equating left and right ordered elements we have

$$a^2 - b^2 = 0$$
$$2ab = -1.$$

From this we deduce that

$$a = b = \pm\tfrac{\sqrt{2}}{2}i$$
$$= \pm\tfrac{\sqrt{2}}{2}(0,\ 1)(1,\ 1)$$
$$= \pm\tfrac{\sqrt{2}}{2}(-1,\ 1).$$

Therefore, the roots are

$$\sqrt{-i} = \pm\tfrac{\sqrt{2}}{2}(1,\ -1).$$

Let's test this result by squaring each root to ensure the answer is $-i$:

$$\left(\pm\tfrac{\sqrt{2}}{2}\right)^2 (1,\ -1)(1,\ -1) = \tfrac{1}{2}(0,\ -2) = (0,\ -1).$$

It is obvious from the above definitions that ordered pairs provide an alternative notation for expressing complex numbers, where the imaginary feature is embedded within the product axiom. We will also use ordered pairs to define a quaternion with three imaginary terms, which when incorporated within the product axiom remain hidden.

9.4 Matrix Representation of a Complex Number

As quaternions have a matrix representation, perhaps we should investigate the matrix representation for a complex number.

Although I have only hinted that i can be regarded as some sort of rotational operator, this is the perfect way of visualising it. In Chap. 2 we discovered that multiplying a complex number by i effectively rotates the number $90°$ anticlockwise. So for the moment, it can be represented by a rotation matrix of $90°$:

$$i \equiv \begin{bmatrix} \cos 90° & -\sin 90° \\ \sin 90° & \cos 90° \end{bmatrix} = \begin{bmatrix} 0 & -1 \\ 1 & 0 \end{bmatrix}$$

and the 2×2 identity matrix is

$$\begin{bmatrix} 1 & 0 \\ 0 & 1 \end{bmatrix}.$$

This permits us to write a complex number as:

$$a + bi = a\begin{bmatrix} 1 & 0 \\ 0 & 1 \end{bmatrix} + b\begin{bmatrix} 0 & -1 \\ 1 & 0 \end{bmatrix}$$
$$= \begin{bmatrix} a & 0 \\ 0 & a \end{bmatrix} + \begin{bmatrix} 0 & -b \\ b & 0 \end{bmatrix}$$
$$= \begin{bmatrix} a & -b \\ b & a \end{bmatrix}.$$

Note that the matrix representing i squares to -1:

$$\begin{bmatrix} 0 & -1 \\ 1 & 0 \end{bmatrix}\begin{bmatrix} 0 & -1 \\ 1 & 0 \end{bmatrix} = \begin{bmatrix} -1 & 0 \\ 0 & -1 \end{bmatrix}$$
$$= -1\begin{bmatrix} 1 & 0 \\ 0 & 1 \end{bmatrix}.$$

However, we must also remember that $i^2 = (-i)^2 = -1$, which is interpreted as anticlockwise and clockwise rotations in the complex plane. This is confirmed by:

$$\begin{bmatrix} 0 & 1 \\ -1 & 0 \end{bmatrix}\begin{bmatrix} 0 & 1 \\ -1 & 0 \end{bmatrix} = \begin{bmatrix} -1 & 0 \\ 0 & -1 \end{bmatrix}$$
$$= -1\begin{bmatrix} 1 & 0 \\ 0 & 1 \end{bmatrix}.$$

Now let's employ matrix notation for all the arithmetic operations used for complex numbers.

9.4.1 Adding and Subtracting Complex Numbers

Two complex numbers are added or subtracted as follows:

$$z_1 = a_1 + b_1 i$$
$$z_2 = a_2 + b_2 i$$
$$z_1 = \begin{bmatrix} a_1 & -b_1 \\ b_1 & a_1 \end{bmatrix}$$
$$z_2 = \begin{bmatrix} a_2 & -b_2 \\ b_2 & a_2 \end{bmatrix}$$
$$z_1 \pm z_2 = \begin{bmatrix} a_1 & -b_1 \\ b_1 & a_1 \end{bmatrix} \pm \begin{bmatrix} a_2 & -b_2 \\ b_2 & a_2 \end{bmatrix}$$
$$= \begin{bmatrix} a_1 \pm a_2 & -(b_1 \pm b_2) \\ b_1 \pm b_2 & a_1 \pm a_2 \end{bmatrix}.$$

For example:

$$z_1 = 2 + 3i$$
$$z_2 = 4 + 2i$$
$$z_1 = \begin{bmatrix} 2 & -3 \\ 3 & 2 \end{bmatrix}$$

$$z_2 = \begin{bmatrix} 4 & -2 \\ 2 & 4 \end{bmatrix}$$

$$z_1 \pm z_2 = \begin{bmatrix} 2 & -3 \\ 3 & 2 \end{bmatrix} \pm \begin{bmatrix} 4 & -2 \\ 2 & 4 \end{bmatrix}$$

$$z_1 + z_2 = \begin{bmatrix} 6 & -5 \\ 5 & 6 \end{bmatrix} = 6 + 5i$$

$$z_1 - z_2 = \begin{bmatrix} -2 & -1 \\ 1 & -2 \end{bmatrix} = -2 + i.$$

9.4.2 Product of Two Complex Numbers

The product of two complex numbers is computed as follows:

$$z_1 = a_1 + b_1 i$$
$$z_2 = a_2 + b_2 i$$
$$z_1 z_2 = \begin{bmatrix} a_1 & -b_1 \\ b_1 & a_1 \end{bmatrix} \begin{bmatrix} a_2 & -b_2 \\ b_2 & a_2 \end{bmatrix}$$
$$= \begin{bmatrix} a_1 a_2 - b_1 b_2 & -(a_1 b_2 + b_1 a_2) \\ a_1 b_2 + b_1 a_2 & a_1 a_2 - b_1 b_2 \end{bmatrix}.$$

For example:

$$z_1 = 6 + 2i$$
$$z_2 = 4 + 3i$$
$$z_1 z_2 = \begin{bmatrix} 6 & -2 \\ 2 & 6 \end{bmatrix} \begin{bmatrix} 4 & -3 \\ 3 & 4 \end{bmatrix}$$
$$= \begin{bmatrix} 24 - 6 & -(18 + 8) \\ 18 + 8 & 24 - 6 \end{bmatrix}$$
$$= \begin{bmatrix} 18 & -26 \\ 26 & 18 \end{bmatrix}.$$

9.4.3 Norm Squared of a Complex Number

The square of the norm is as the determinant of the matrix:

$$z = a + bi$$

$$= \begin{bmatrix} a & -b \\ b & a \end{bmatrix}$$

$$|z|^2 = a^2 + b^2 = \begin{vmatrix} a & -b \\ b & a \end{vmatrix}.$$

9.4.4 Complex Conjugate of a Complex Number

The complex conjugate of a complex number is

$$z = a + bi = \begin{bmatrix} a & -b \\ b & a \end{bmatrix}$$

$$z^* = a - bi = \begin{bmatrix} a & b \\ -b & a \end{bmatrix}.$$

The product $zz^* = a^2 + b^2$:

$$zz^* = \begin{bmatrix} a & -b \\ b & a \end{bmatrix} \begin{bmatrix} a & b \\ -b & a \end{bmatrix}$$

$$= \begin{bmatrix} a^2 + b^2 & 0 \\ 0 & a^2 + b^2 \end{bmatrix}$$

$$= (a^2 + b^2) \begin{bmatrix} 1 & 0 \\ 0 & 1 \end{bmatrix}.$$

For example:

$$z = 3 + 4i = \begin{bmatrix} 3 & -4 \\ 4 & 3 \end{bmatrix}$$

$$z^* = 3 - 4i = \begin{bmatrix} 3 & 4 \\ -4 & 3 \end{bmatrix}$$

$$zz^* = \begin{bmatrix} 3 & -4 \\ 4 & 3 \end{bmatrix} \begin{bmatrix} 3 & 4 \\ -4 & 3 \end{bmatrix} = \begin{bmatrix} 25 & 0 \\ 0 & 25 \end{bmatrix}$$

$$= 25 \begin{bmatrix} 1 & 0 \\ 0 & 1 \end{bmatrix}.$$

9.4.5 Inverse of a Complex Number

The inverse of 2×2 matrix $\mathbf{A}$ is given by

$$A = \begin{bmatrix} a_{11} & a_{12} \\ a_{21} & a_{22} \end{bmatrix}$$

$$A^{-1} = \frac{1}{a_{11}a_{22} - a_{12}a_{21}} \begin{bmatrix} a_{22} & -a_{12} \\ -a_{21} & a_{12} \end{bmatrix}$$

therefore, the inverse of z is given by

$$z = a + bi$$

$$z = \begin{bmatrix} a & -b \\ b & a \end{bmatrix}$$

$$z^{-1} = \frac{1}{a^2 + b^2} \begin{bmatrix} a & b \\ -b & a \end{bmatrix}.$$

For example:

$$z = 3 + 4i$$

$$z = \begin{bmatrix} 3 & -4 \\ 4 & 3 \end{bmatrix}$$

$$z^{-1} = \tfrac{1}{25} \begin{bmatrix} 3 & 4 \\ -4 & 3 \end{bmatrix}.$$

9.4.6 Quotient of a Complex Number

The quotient of two complex numbers is computed as follows:

$$z_1 = a_1 + b_1 i$$

$$z_2 = a_2 + b_2 i$$

$$\frac{z_1}{z_2} = z_1 z_2^{-1}$$

$$= \begin{bmatrix} a_1 & -b_1 \\ b_1 & a_1 \end{bmatrix} \frac{1}{a_2^2 + b_2^2} \begin{bmatrix} a_2 & b_2 \\ -b_2 & a_2 \end{bmatrix}$$

$$= \frac{1}{a_2^2 + b_2^2} \begin{bmatrix} a_1 a_2 + b_1 b_2 & -(b_1 a_2 - a_1 b_2) \\ b_1 a_2 - a_1 b_2 & a_1 a_2 + b_1 b_2 \end{bmatrix}.$$

For example:

$$z_1 = 4 + 3i$$

$$z_2 = 3 + 4i$$

$$\frac{z_1}{z_2} = z_1 z_2^{-1}$$

$$= \begin{bmatrix} 4 & -3 \\ 3 & 4 \end{bmatrix} \frac{1}{9+16} \begin{bmatrix} 3 & 4 \\ -4 & 3 \end{bmatrix}$$

$$= \tfrac{1}{25} \begin{bmatrix} 24 & 7 \\ -7 & 24 \end{bmatrix}.$$

9.4.7 Square-Root of $\pm i$

To find $\sqrt{i}$ we assume that the roots are complex. Therefore, we start with

$$\sqrt{i} = \begin{bmatrix} a & -b \\ b & a \end{bmatrix}$$

$$i = \begin{bmatrix} a & -b \\ b & a \end{bmatrix} \begin{bmatrix} a & -b \\ b & a \end{bmatrix}$$

$$\begin{bmatrix} 0 & -1 \\ 1 & 0 \end{bmatrix} = \begin{bmatrix} a^2 - b^2 & -2ab \\ 2ab & a^2 - b^2 \end{bmatrix}$$

and equating left and right matrices we have

$$a^2 - b^2 = 0$$
$$2ab = 1.$$

From this we deduce that

$$a = b = \pm\tfrac{\sqrt{2}}{2}.$$

Therefore, the roots are

$$\sqrt{i} = \pm\tfrac{\sqrt{2}}{2} \begin{bmatrix} 1 & -1 \\ 1 & 1 \end{bmatrix}.$$

Let's test this result by squaring each root to ensure the answer is i:

$$\left(\pm\tfrac{\sqrt{2}}{2}\right)^2 \begin{bmatrix} 1 & -1 \\ 1 & 1 \end{bmatrix} \begin{bmatrix} 1 & -1 \\ 1 & 1 \end{bmatrix} = \tfrac{1}{2} \begin{bmatrix} 0 & -2 \\ 2 & 0 \end{bmatrix} = i$$

To find $\sqrt{-i}$ we assume that the roots are complex. Therefore, we start with

$$\sqrt{-i} = \begin{bmatrix} a & -b \\ b & a \end{bmatrix}$$

$$-i = \begin{bmatrix} a & -b \\ b & a \end{bmatrix} \begin{bmatrix} a & -b \\ b & a \end{bmatrix}$$

$$\begin{bmatrix} 0 & 1 \\ -1 & 0 \end{bmatrix} = \begin{bmatrix} a^2 - b^2 & -2ab \\ 2ab & a^2 - b^2 \end{bmatrix}$$

and equating left and right matrices we have

$$a^2 - b^2 = 0$$
$$2ab = -1.$$

From this we deduce that

$$a = b = \pm\tfrac{\sqrt{2}}{2}i.$$

Therefore, the roots are

$$\sqrt{-i} = \pm\tfrac{\sqrt{2}}{2}\begin{bmatrix} 0 & -1 \\ 1 & 0 \end{bmatrix}\begin{bmatrix} 1 & -1 \\ 1 & 1 \end{bmatrix} = \pm\tfrac{\sqrt{2}}{2}\begin{bmatrix} 1 & 1 \\ -1 & 1 \end{bmatrix}.$$

Let's test this result by squaring each root to ensure the answer is i:

$$\left(\pm\tfrac{\sqrt{2}}{2}\right)^2\begin{bmatrix} 1 & 1 \\ -1 & 1 \end{bmatrix}\begin{bmatrix} 1 & 1 \\ -1 & 1 \end{bmatrix} = \tfrac{1}{2}\begin{bmatrix} 0 & 2 \\ -2 & 0 \end{bmatrix} = -i$$

9.5 Summary

We have shown in this chapter that there is a one-to-one correspondence between a complex number and an ordered pair, and that a complex number can be represented as a matrix, which permits us to compute all complex number operations as matrix operations or ordered pairs.

If this the first time you have come across complex numbers you probably will have found them strange on the one hand, and amazing on the other. Simply by declaring the existence of i that squares to -1, opens up a new number system that unifies large areas of mathematics.

9.6 Worked Examples

Here are some worked examples that employ the ideas described above. In some cases a test is included to confirm the result.

9.6.1 Adding and Subtracting Complex Numbers

Add and subtract z_1 and z_2.

Complex Number:

$$z_1 = 12 + 6i$$
$$z_2 = 10 - 4i$$
$$z_1 + z_2 = 22 + 2i$$
$$z_1 - z_2 = 2 + 10i.$$

Ordered Pair:

$$z_1 = (12, \ 6)$$
$$z_2 = (10, \ -4)$$
$$z_1 + z_2 = (12, \ 6) + (10, \ -4) = (22, \ 2)$$
$$z_1 - z_2 = (12, \ 6) - (10, \ -4) = (2, \ 10).$$

Matrix:

$$z_1 = \begin{bmatrix} 12 & -6 \\ 6 & 12 \end{bmatrix}$$
$$z_2 = \begin{bmatrix} 10 & 4 \\ -4 & 10 \end{bmatrix}$$
$$z_1 + z_2 = \begin{bmatrix} 12 & -6 \\ 6 & 12 \end{bmatrix} + \begin{bmatrix} 10 & 4 \\ -4 & 10 \end{bmatrix} = \begin{bmatrix} 22 & -2 \\ 2 & 22 \end{bmatrix}$$
$$z_1 - z_2 = \begin{bmatrix} 12 & -6 \\ 6 & 12 \end{bmatrix} - \begin{bmatrix} 10 & 4 \\ -4 & 10 \end{bmatrix} = \begin{bmatrix} 2 & -10 \\ 10 & 2 \end{bmatrix}.$$

9.6.2 Product of Complex Numbers

Compute the product $z_1 z_2$.

Complex Number:

$$z_1 = 12 + 6i$$
$$z_2 = 10 - 4i$$
$$z_1 z_2 = (12 + 6i)(10 - 4i)$$
$$= 144 + 12i.$$

Ordered Pair:

$$z_1 = (12, \ 6)$$
$$z_2 = (10, \ -4)$$
$$z_1 z_2 = (12, \ 6)(10, \ -4)$$
$$= (120 + 24, \ -48 + 60)$$
$$= (144, \ 12).$$

Matrix:

$$z_1 = \begin{bmatrix} 12 & -6 \\ 6 & 12 \end{bmatrix}$$
$$z_2 = \begin{bmatrix} 10 & 4 \\ -4 & 10 \end{bmatrix}$$
$$z_1 z_2 = \begin{bmatrix} 12 & -6 \\ 6 & 12 \end{bmatrix}\begin{bmatrix} 10 & 4 \\ -4 & 10 \end{bmatrix} = \begin{bmatrix} 144 & -12 \\ 12 & 144 \end{bmatrix}.$$

9.6.3 Multiplying a Complex Number by i

Multiply z_1 by i.

Complex Number:

$$z_1 = 12 + 6i$$
$$z_1 i = (12 + 6i)i$$
$$= -6 + 12i.$$

Ordered Pair:

$$z_1 = (12, \ 6)$$
$$i = (0, \ 1)$$
$$z_1 i = (12, \ 6)(0, \ 1)$$
$$= (-6, \ 12).$$

Matrix:

$$z_1 = \begin{bmatrix} 12 & -6 \\ 6 & 12 \end{bmatrix}$$

$$i = \begin{bmatrix} 0 & -1 \\ 1 & 0 \end{bmatrix}$$

$$z_1 z_2 = \begin{bmatrix} 12 & -6 \\ 6 & 12 \end{bmatrix} \begin{bmatrix} 0 & -1 \\ 1 & 0 \end{bmatrix} = \begin{bmatrix} -6 & -12 \\ 12 & -6 \end{bmatrix}.$$

9.6.4 The Norm of a Complex Number

Compute the norm of z_1.

Complex Number:

$$z_1 = 12 + 6i$$
$$|z_1| = \sqrt{12^2 + 6^2} \approx 13.416.$$

Ordered Pair:

$$z_1 = (12, \ 6)$$
$$|z_1| = \sqrt{12^2 + 6^2} \approx 13.416.$$

Matrix:

$$z_1 = \begin{bmatrix} 12 & -6 \\ 6 & 12 \end{bmatrix}$$
$$|z_1| = \begin{vmatrix} 12 & -6 \\ 6 & 12 \end{vmatrix} = \sqrt{12^2 + 6^2} \approx 13.416.$$

9.6.5 The Complex Conjugate of a Complex Number

Compute the complex conjugate of the following.

Complex Number:

$$(2 + 3i)^* = 2 - 3i$$

$$1^* = 1$$
$$i^* = -i.$$

Ordered Pair:

$$(2, \ 3)^* = (2, \ -3)$$
$$(1, \ 0)^* = (1, \ 0)$$
$$(0, \ 1)^* = (0, \ -1).$$

Matrix:

$$z = \begin{bmatrix} 2 & -3 \\ 3 & 2 \end{bmatrix}$$

$$z^* = \begin{bmatrix} 2 & 3 \\ -3 & 2 \end{bmatrix}$$

$$1 = \begin{bmatrix} 1 & 0 \\ 0 & 1 \end{bmatrix}$$

$$1^* = \begin{bmatrix} 1 & 0 \\ 0 & 1 \end{bmatrix}$$

$$i = \begin{bmatrix} 0 & -1 \\ 1 & 0 \end{bmatrix}$$

$$i^* = \begin{bmatrix} 0 & 1 \\ -1 & 0 \end{bmatrix}.$$

9.6.6   The Quotient of Two Complex Numbers

Compute the quotient $(2 + 3i)/(3 + 4i)$.

Complex Number:

$$\frac{2 + 3i}{3 + 4i} = \frac{(2 + 3i)\,(3 - 4i)}{(3 + 4i)\,(3 - 4i)}$$
$$= \frac{6 - 8i + 9i + 12}{25}$$
$$= \tfrac{18}{25} + \tfrac{1}{25}i.$$

Test:

$$(3 + 4i)\left(\tfrac{18}{25} + \tfrac{1}{25}i\right) = \tfrac{54}{25} + \tfrac{3}{25}i + \tfrac{72}{25}i - \tfrac{4}{25}$$
$$= 2 + 3i.$$

Ordered Pair:

$$\frac{(2,\ 3)}{(3,\ 4)} = \frac{(2,\ 3)\ (3,\ -4)}{(3,\ 4)\ (3,\ -4)}$$
$$= \frac{(6 + 12,\ 1)}{(9 + 16,\ 0)}$$
$$= \left(\tfrac{18}{25},\ \tfrac{1}{25}\right).$$

Matrix:

$$z_1 = \begin{bmatrix} 2 & -3 \\ 3 & 2 \end{bmatrix}$$
$$z_2 = \begin{bmatrix} 3 & -4 \\ 4 & 3 \end{bmatrix}$$
$$\frac{z_1}{z_2} = z_1 z_2^{-1}$$
$$= \tfrac{1}{25} \begin{bmatrix} 2 & -3 \\ 3 & 2 \end{bmatrix} \begin{bmatrix} 3 & 4 \\ -4 & 3 \end{bmatrix}$$
$$= \tfrac{1}{25} \begin{bmatrix} 18 & -1 \\ 1 & 18 \end{bmatrix}.$$

9.6.7 Divide a Complex Number by i

Divide $2 + 3i$ by i.

Complex Number:

$$\frac{2 + 3i}{0 + i} = \frac{(2 + 3i)\ (0 - i)}{(0 + i)\ (0 - i)}$$
$$= \frac{-2i + 3}{1}$$
$$= 3 - 2i.$$

Test:

$$i(3 - 2i) = 2 + 3i.$$

Ordered Pair:

$$\frac{(2, \ 3)}{(0, \ 1)} = \frac{(2, \ 3)}{(0, \ 1)} \frac{(0, \ -1)}{(0, \ -1)}$$
$$= \frac{(3, \ -2)}{(1, \ 0)}$$
$$= (3, \ -2).$$

Matrix:

$$z = \begin{bmatrix} 2 & -3 \\ 3 & 2 \end{bmatrix}$$
$$i = \begin{bmatrix} 0 & -1 \\ 1 & 0 \end{bmatrix}$$
$$i^{-1} = \begin{bmatrix} 0 & 1 \\ -1 & 0 \end{bmatrix}$$
$$zi^{-1} = \begin{bmatrix} 2 & -3 \\ 3 & 2 \end{bmatrix} \begin{bmatrix} 0 & 1 \\ -1 & 0 \end{bmatrix} = \begin{bmatrix} 3 & 2 \\ -2 & 3 \end{bmatrix}.$$

9.6.8   Divide a Complex Number by $-i$

Divide $2 + 3i$ by $-i$.

Complex Number:

$$\frac{2 + 3i}{0 - i} = \frac{(2 + 3i)}{(0 - i)} \frac{(0 + i)}{(0 + i)}$$
$$= \frac{2i - 3}{1}$$
$$= -3 + 2i.$$

Test:

$$-i(-3 + 2i) = 2 + 3i.$$

Ordered Pair:

$$
\begin{aligned}
\frac{(2,\ 3)}{(0,\ -1)} &= \frac{(2,\ 3)}{(0,\ -1)} \frac{(0,\ 1)}{(0,\ 1)} \\
&= \frac{(-3,\ 2)}{1} \\
&= (-3,\ 2).
\end{aligned}
$$

Matrix:

$$
z = \begin{bmatrix} 2 & -3 \\ 3 & 2 \end{bmatrix}
$$

$$
-i = \begin{bmatrix} 0 & 1 \\ -1 & 0 \end{bmatrix}
$$

$$
-i^{-1} = \begin{bmatrix} 0 & -1 \\ 1 & 0 \end{bmatrix}
$$

$$
z\left(-i^{-1}\right) = \begin{bmatrix} 2 & -3 \\ 3 & 2 \end{bmatrix} \begin{bmatrix} 0 & -1 \\ 1 & 0 \end{bmatrix} = \begin{bmatrix} -3 & -2 \\ 2 & -3 \end{bmatrix}.
$$

9.6.9 The Inverse of a Complex Number

Compute the inverse of $2 + 3i$.

Complex Number:

$$
\begin{aligned}
\frac{1}{2 + 3i} &= \frac{1}{(2 + 3i)} \frac{(2 - 3i)}{(2 - 3i)} \\
&= \frac{2 - 3i}{13} \\
&= \tfrac{2}{13} - \tfrac{3}{13}i.
\end{aligned}
$$

Ordered Pair:

$$
\begin{aligned}
\frac{1}{(2,\ 3)} &= \frac{1}{(2,\ 3)} \frac{(2,\ -3)}{(2,\ -3)} \\
&= \frac{(2,\ -3)}{13} \\
&= \left(\tfrac{2}{13},\ -\tfrac{3}{13}\right).
\end{aligned}
$$

Matrix:

$$z = \begin{bmatrix} 2 & -3 \\ 3 & 2 \end{bmatrix}$$

$$z^{-1} = \frac{1}{13} \begin{bmatrix} 2 & 3 \\ -3 & 2 \end{bmatrix}.$$

9.6.10   The Inverse of i

Compute the inverse of i.

Complex Number:

$$\frac{1}{0+i} = \frac{1}{(0+i)} \frac{(0-i)}{(0-i)}$$
$$= \frac{-i}{1} = -i.$$

Ordered Pair:

$$\frac{1}{(0,\ 1)} = \frac{1}{(0,\ 1)} \frac{(0,\ -1)}{(0,\ -1)}$$
$$= \frac{(0,\ -1)}{(1,\ 0)} = (0,\ -1) = -i.$$

Matrix:

$$i = \begin{bmatrix} 0 & -1 \\ 1 & 0 \end{bmatrix}$$

$$i^{-1} = \begin{bmatrix} 0 & 1 \\ -1 & 0 \end{bmatrix} = -i.$$

9.6.11   The Inverse of −i

Compute the inverse of $-i$.

Complex Number:

$$\frac{1}{0 - i} = \frac{1}{(0 - i)} \frac{(0 + i)}{(0 + i)}$$
$$= \frac{i}{1} = i.$$

Ordered Pair:

$$\frac{1}{(0, \ -1)} = \frac{1}{(0, \ -1)} \frac{(0, \ 1)}{(0, \ 1)}$$
$$= \frac{(0, \ 1)}{(1, \ 0)} = (0, \ 1) = i.$$

Matrix:

$$-i = \begin{bmatrix} 0 & 1 \\ -1 & 0 \end{bmatrix}$$
$$-i^{-1} = \begin{bmatrix} 0 & -1 \\ 1 & 0 \end{bmatrix} = i.$$

Reference

1. Vince J (2018) Imaginary mathematics for computer science. Springer

Chapter 10
Geometric Transforms

10.1 Introduction

This chapter shows how matrices are used to scale, translate, reflect, shear and rotate 2D shapes and 3D objects. The reader should try to understand the construction of the various matrices and recognise the role of each matrix element. After a little practice it will be possible to define a wide variety of matrices without thinking about the underlying algebra.

10.2 Background

A point $P(x, y)$ is transformed into $P'(x', y')$ by manipulating the original coordinates x and y using

$$x' = ax + by + e$$
$$y' = cx + dy + f,$$

where a, b, c, d, e and f have assigned values. Similarly, a 3D point $P(x, y, z)$ is transformed into $P'(x', y', z')$ using

$$x' = ax + by + cz + k$$
$$y' = dx + ey + fz + l$$
$$z' = gx + hy + jz + m.$$

The values for a, b, c, ... etc. determine whether the transform translates, shears, scales, reflects or rotates a point.

© Springer-Verlag London Ltd., part of Springer Nature 2022
J. Vince, *Mathematics for Computer Graphics*, Undergraduate Topics
in Computer Science, https://doi.org/10.1007/978-1-4471-7520-9_10

Although transforms have an algebraic origin, it is convenient to express them as matrices, which provide certain advantages for viewing the transform and for interfacing to various types of computer graphics hardware. We begin with an algebraic approach and then introduce matrix notation.

10.3   2D Transforms

10.3.1   Translation

Cartesian coordinates provide a one-to-one relationship between number and shape, such that when we change a shape's coordinates, we change its geometry. For example, if $P(x, \ y)$ is a shape's vertex, when we apply the operation $x' = x + 3$ we create a new point $P'(x', \ y)$ three units to the right. Similarly, the operation $y' = y + 1$ creates a new point $P'(x, \ y')$ displaced one unit vertically. By applying both of these transforms to every vertex on the original shape, the shape is displaced as shown in Fig. 10.1.

10.3.2   Scaling

Shape scaling is effected by multiplying coordinates as follows:

$$x' = 2.5x$$
$$y' = 1.5y.$$

Fig. 10.1 The translated shape results by adding 3 to every x-coordinate, and 1 to every y-coordinate to the original shape

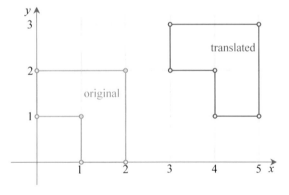

Fig. 10.2 The scaled shape
results by multiplying the
x-coordinates by 2.5 and the
y-coordinates by 1.5

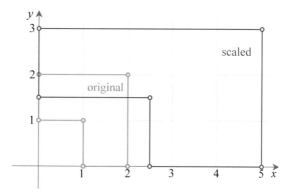

This transform results in a horizontal scaling of 2.5 and a vertical scaling of 1.5 as
illustrated in Fig. 10.2. Note that a point located at the origin does not change its
place, so scaling is relative to the origin.

10.3.3 Reflection

To make a reflection of a shape relative to the y-axis, we simply reverse the sign of
the x-coordinates, leaving the y-coordinates unchanged:

$$x' = -x$$
$$y' = y$$

and to reflect a shape relative to the x-axis we reverse the y-coordinates:

$$x' = x$$
$$y' = -y.$$

Figure 10.3 shows three reflections derived from the original shape by reversing the
signs for the x- and y-coordinates. Note that a shape's vertex order is reversed for
each reflection.

  Before proceeding, we pause to introduce matrix notation so that we can develop
further transforms using algebra and matrix algebra side by side.

Fig. 10.3 The original shape
gives rise to three reflections
simply by reversing the signs
of its coordinates

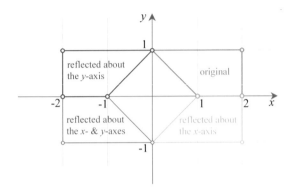

10.4   Transforms as Matrices

10.4.1   Systems of Notation

Over time two systems of matrix notation have evolved: one where the matrix multiplies a column vector, as described above, and another where a *row vector* multiplies the matrix:

$$\begin{bmatrix} x' & y' \end{bmatrix} = \begin{bmatrix} x & y \end{bmatrix} \begin{bmatrix} a & c \\ b & d \end{bmatrix} = \begin{bmatrix} ax + by & cx + dy \end{bmatrix}.$$

Note how the elements of the matrix are transposed to accommodate the algebraic correctness of the transform. There is no preferred system of notation, and you will find technical books and papers supporting both. Personally, I prefer a matrix premultiplying a column vector, as it is very similar to the original algebraic equations. However, the important thing to remember is that the rows and columns of the matrix are transposed when moving between the two systems.

10.5   Homogeneous Coordinates

Chapter 8 showed how a pair of equations such as

$$x' = ax + by$$
$$y' = cx + dy$$

can be written in matrix notation as:

$$\begin{bmatrix} x' \\ y' \end{bmatrix} = \begin{bmatrix} a & b \\ c & d \end{bmatrix} \begin{bmatrix} x \\ y \end{bmatrix}.$$

One immediate problem with this notation is that there is no apparent mechanism to add or subtract a constant such as e or f:

$$x' = ax + by + e$$
$$y' = cx + dy + f.$$

Mathematicians resolved this by using *homogeneous coordinates*, which appeared in the early 19th century where they were independently proposed by the German mathematician August Möbius (1790–1868) (who also is associated with a one-sided curled band, the Möbius strip), and the German mathematician and physicist Julius Plücker (1801–1868). Möbius called them *barycentric coordinates*, and they have also been called *areal coordinates* because of their area-calculating properties.

Basically, homogeneous coordinates define a point in a plane using three coordinates instead of two. Initially, Plücker located a homogeneous point relative to the sides of a triangle, but later revised his notation to the one employed in contemporary mathematics and computer graphics. This states that for a point (x, y) there exists a homogeneous point (xt, yt, t) where t is an arbitrary number. For example, the point $(3, 4)$ has homogeneous coordinates $(6, 8, 2)$, because $3 = 6/2$ and $4 = 8/2$. But the homogeneous point $(6, 8, 2)$ is not unique to $(3, 4)$; $(12, 16, 4)$, $(15, 20, 5)$ and $(300, 400, 100)$ are all possible homogeneous coordinates for $(3, 4)$.

The reason why this coordinate system is called 'homogeneous' is because it is possible to transform functions such as $f(x, y)$ into the form $f(x/t, y/t)$ without disturbing the degree of the curve. To the non-mathematician this may not seem anything to get excited about, but in the field of projective geometry it is a very powerful concept.

Figure 10.4 shows a 3D homogeneous space with axes x, y and h, where a point $(x, y, 1)$ is associated with a projected point (xt, yt, t). The figure shows a triangle on the $h = 1$ plane, and a similar triangle on the plane $h = t$. Thus instead of working in two dimensions, we can work on an arbitrary xy-plane in three dimensions. The h-coordinate of the plane is immaterial because the x- and y-coordinates are eventually divided by t. However, to keep things simple it seems a good idea to choose $t = 1$.

Fig. 10.4 2D homogeneous coordinates can be visualised as a plane in 3D space generally where $h = 1$, for convenience

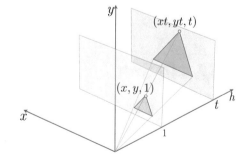

This means that the point (x, y) has homogeneous coordinates $(x, y, 1)$ making scaling superfluous.

If we substitute 3D homogeneous coordinates for traditional 2D Cartesian coordinates we must attach 1 to every (x, y) pair. When a point $(x, y, 1)$ is transformed, it emerges as $(x', y', 1)$, and we discard the 1. This may seem a futile exercise, but it resolves the problem of creating a translation transform.

Consider the following transform on the homogeneous point $(x, y, 1)$:

$$\begin{bmatrix} x' \\ y' \\ 1 \end{bmatrix} = \begin{bmatrix} a & b & e \\ c & d & f \\ 0 & 0 & 1 \end{bmatrix} \begin{bmatrix} x \\ y \\ 1 \end{bmatrix}.$$

This expands to

$$x' = ax + by + e$$
$$y' = cx + dy + f$$
$$1 = 1$$

and solves the above problem of adding a constant. Now let's move on to see how homogeneous coordinates are used in practice.

10.5.1   2D Translation

The algebraic and matrix notation for 2D translation is

$$x' = x + t_x$$
$$y' = y + t_y$$

or using matrices:

$$\begin{bmatrix} x' \\ y' \\ 1 \end{bmatrix} = \begin{bmatrix} 1 & 0 & t_x \\ 0 & 1 & t_y \\ 0 & 0 & 1 \end{bmatrix} \begin{bmatrix} x \\ y \\ 1 \end{bmatrix}.$$

10.5.2   2D Scaling

The algebraic and matrix notation for 2D scaling is

$$x' = s_x x$$
$$y' = s_y y$$

or using matrices:

$$\begin{bmatrix} x' \\ y' \\ 1 \end{bmatrix} = \begin{bmatrix} s_x & 0 & 0 \\ 0 & s_y & 0 \\ 0 & 0 & 1 \end{bmatrix} \begin{bmatrix} x \\ y \\ 1 \end{bmatrix}.$$

The scaling action is relative to the origin, i.e. the point $(0, 0)$ remains unchanged. All other points move away from the origin when $s_x > 1$, or move towards the origin when $s_x < 1$. To scale relative to another point (p_x, p_y) we first subtract (p_x, p_y) from (x, y) respectively. This effectively makes the reference point (p_x, p_y) the new origin. Second, we perform the scaling operation relative to the new origin, and third, add (p_x, p_y) back to the new (x, y) respectively to compensate for the original subtraction. Algebraically this is

$$x' = s_x(x - p_x) + p_x$$
$$y' = s_y(y - p_y) + p_y$$

which simplifies to

$$x' = s_x x + p_x(1 - s_x)$$
$$y' = s_y y + p_y(1 - s_y)$$

or as a homogeneous matrix:

$$\begin{bmatrix} x' \\ y' \\ 1 \end{bmatrix} = \begin{bmatrix} s_x & 0 & p_x(1 - s_x) \\ 0 & s_y & p_y(1 - s_y) \\ 0 & 0 & 1 \end{bmatrix} \begin{bmatrix} x \\ y \\ 1 \end{bmatrix}. \tag{10.1}$$

For example, to scale a shape by 2 relative to the point $(1, 1)$ the matrix is

$$\begin{bmatrix} x' \\ y' \\ 1 \end{bmatrix} = \begin{bmatrix} 2 & 0 & -1 \\ 0 & 2 & -1 \\ 0 & 0 & 1 \end{bmatrix} \begin{bmatrix} x \\ y \\ 1 \end{bmatrix}.$$

10.5.3 2D Reflections

The matrix notation for reflecting about the y-axis is

$$\begin{bmatrix} x' \\ y' \\ 1 \end{bmatrix} = \begin{bmatrix} -1 & 0 & 0 \\ 0 & 1 & 0 \\ 0 & 0 & 1 \end{bmatrix} \begin{bmatrix} x \\ y \\ 1 \end{bmatrix}$$

or about the x-axis:

$$
\begin{bmatrix} x' \\ y' \\ 1 \end{bmatrix} = \begin{bmatrix} 1 & 0 & 0 \\ 0 & -1 & 0 \\ 0 & 0 & 1 \end{bmatrix} \begin{bmatrix} x \\ y \\ 1 \end{bmatrix}.
$$

However, to make a reflection about an arbitrary vertical or horizontal axis we need to introduce some more algebraic deception.

To make a reflection about the vertical axis $x = 1$, we first subtract 1 from the x-coordinate. This effectively makes the $x = 1$ axis coincident with the major y-axis. Next, we perform the reflection by reversing the sign of the modified x-coordinate. And finally, we add 1 to the reflected coordinate to compensate for the original subtraction. Algebraically, the three steps are

$$
\begin{aligned}
x_1 &= x - 1 \\
x_2 &= -(x - 1) \\
x' &= -(x - 1) + 1
\end{aligned}
$$

which simplifies to

$$
\begin{aligned}
x' &= -x + 2 \\
y' &= y
\end{aligned}
$$

or in matrix form:

$$
\begin{bmatrix} x' \\ y' \\ 1 \end{bmatrix} = \begin{bmatrix} -1 & 0 & 2 \\ 0 & 1 & 0 \\ 0 & 0 & 1 \end{bmatrix} \begin{bmatrix} x \\ y \\ 1 \end{bmatrix}.
$$

Figure 10.5 illustrates this process.

To reflect a point about an arbitrary y-axis, $x = a_x$, the following transform is required:

$$
\begin{aligned}
x' &= -(x - a_x) + a_x \quad = -x + 2a_x \\
y' &= y
\end{aligned}
$$

Fig. 10.5 The shape on the right is reflected about the $x = 1$ axis

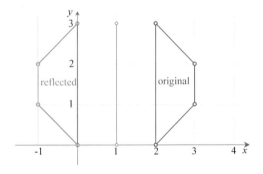

or in matrix form:

$$\begin{bmatrix} x' \\ y' \\ 1 \end{bmatrix} = \begin{bmatrix} -1 & 0 & 2a_x \\ 0 & 1 & 0 \\ 0 & 0 & 1 \end{bmatrix} \begin{bmatrix} x \\ y \\ 1 \end{bmatrix}. \tag{10.2}$$

Similarly, to reflect a point about an arbitrary x-axis $y = a_y$, the following transform is required:

$$x' = x$$
$$y' = -(y - a_y) + a_y \quad = -y + 2a_y$$

or in matrix form:

$$\begin{bmatrix} x' \\ y' \\ 1 \end{bmatrix} = \begin{bmatrix} 1 & 0 & 0 \\ 0 & -1 & 2a_y \\ 0 & 0 & 1 \end{bmatrix} \begin{bmatrix} x \\ y \\ 1 \end{bmatrix}.$$

10.5.4 2D Shearing

A shape is sheared by leaning it over at an angle β. Figure 10.6 illustrates the geometry, and we see that the y-coordinates remain unchanged but the x-coordinates are a function of y and $\tan \beta$.

$$x' = x + y \tan \beta$$
$$y' = y$$

or in matrix form:

Fig. 10.6 The original green, square shape is sheared to the right by an angle β, and the horizontal shear is proportional to $y \tan \beta$

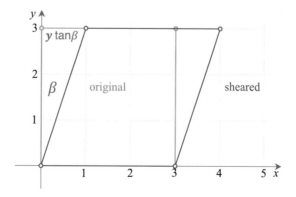

$$\begin{bmatrix} x' \\ y' \\ 1 \end{bmatrix} = \begin{bmatrix} 1 & \tan \beta & 0 \\ 0 & 1 & 0 \\ 0 & 0 & 1 \end{bmatrix} \begin{bmatrix} x \\ y \\ 1 \end{bmatrix}.$$

10.5.5   2D Rotation

Figure 10.7 shows a point $P(x, y)$, distance R from the origin, which is to be rotated by an angle β about the origin to $P'(x', y')$. It can be seen that

$$x' = R\cos(\theta + \beta)$$
$$y' = R\sin(\theta + \beta)$$

and substituting the identities for $\cos(\theta + \beta)$ and $\sin(\theta + \beta)$ we have

$$x' = R(\cos\theta \cos\beta - \sin\theta \sin\beta)$$
$$y' = R(\sin\theta \cos\beta + \cos\theta \sin\beta)$$
$$x' = R\left(\frac{x}{R}\cos\beta - \frac{y}{R}\sin\beta\right)$$
$$y' = R\left(\frac{y}{R}\cos\beta + \frac{x}{R}\sin\beta\right)$$
$$x' = x\cos\beta - y\sin\beta$$
$$y' = x\sin\beta + y\cos\beta$$

or in matrix form:

$$\begin{bmatrix} x' \\ y' \\ 1 \end{bmatrix} = \begin{bmatrix} \cos\beta & -\sin\beta & 0 \\ \sin\beta & \cos\beta & 0 \\ 0 & 0 & 1 \end{bmatrix} \begin{bmatrix} x \\ y \\ 1 \end{bmatrix}.$$

Fig. 10.7 The point $P(x, y)$ is rotated through an angle β to $P'(x', y')$

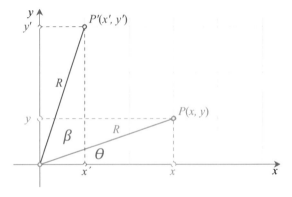

For example, to rotate a point through 90° the matrix is

$$
\begin{bmatrix} x' \\ y' \\ 1 \end{bmatrix} = \begin{bmatrix} 0 & -1 & 0 \\ 1 & 0 & 0 \\ 0 & 0 & 1 \end{bmatrix} \begin{bmatrix} x \\ y \\ 1 \end{bmatrix}.
$$

Thus the point $(1, 0)$ becomes $(0, 1)$. If we rotate through 360° the matrix becomes

$$
\begin{bmatrix} x' \\ y' \\ 1 \end{bmatrix} = \begin{bmatrix} 1 & 0 & 0 \\ 0 & 1 & 0 \\ 0 & 0 & 1 \end{bmatrix} \begin{bmatrix} x \\ y \\ 1 \end{bmatrix}.
$$

Such a matrix has a null effect and is called an *identity matrix*.

To rotate a point (x, y) about an arbitrary point (p_x, p_y) we first, subtract (p_x, p_y) from the coordinates (x, y) respectively. This enables us to perform the rotation about the origin. Second, we perform the rotation, and third, we add (p_x, p_y) to compensate for the original subtraction. Here are the steps:

1. Subtract (p_x, p_y):

$$
x_1 = (x - p_x)
$$
$$
y_1 = (y - p_y).
$$

2. Rotate β about the origin:

$$
x_2 = (x - p_x) \cos \beta - (y - p_y) \sin \beta
$$
$$
y_2 = (x - p_x) \sin \beta + (y - p_y) \cos \beta.
$$

3. Add (p_x, p_y):

$$
x' = (x - p_x) \cos \beta - (y - p_y) \sin \beta + p_x
$$
$$
y' = (x - p_x) \sin \beta + (y - p_y) \cos \beta + p_y.
$$

Simplifying,

$$
x' = x \cos \beta - y \sin \beta + p_x (1 - \cos \beta) + p_y \sin \beta
$$
$$
y' = x \sin \beta + y \cos \beta + p_y (1 - \cos \beta) - p_x \sin \beta
$$

and in matrix form:

$$
\begin{bmatrix} x' \\ y' \\ 1 \end{bmatrix} = \begin{bmatrix} \cos \beta & -\sin \beta & p_x(1 - \cos \beta) + p_y \sin \beta \\ \sin \beta & \cos \beta & p_y(1 - \cos \beta) - p_x \sin \beta \\ 0 & 0 & 1 \end{bmatrix} \begin{bmatrix} x \\ y \\ 1 \end{bmatrix}. \qquad (10.3)
$$

For example, to rotate a point 90° about the point $(1, 1)$ the matrix operation becomes

$$\begin{bmatrix} x' \\ y' \\ 1 \end{bmatrix} = \begin{bmatrix} 0 & -1 & 2 \\ 1 & 0 & 0 \\ 0 & 0 & 1 \end{bmatrix} \begin{bmatrix} x \\ y \\ 1 \end{bmatrix}.$$

A simple test is to substitute the point $(2, \; 1)$ for $(x, \; y)$; which is transformed correctly to $(1, \; 2)$.

The algebraic approach in deriving the above transforms is relatively easy. However, it is also possible to use matrices to derive compound transforms, such as a reflection relative to an arbitrary line and scaling and rotation relative to an arbitrary point. These transforms are called *affine*, as parallel lines remain parallel after being transformed. Furthermore, the word 'affine' is used to imply that there is a strong geometric *affinity* between the original and transformed shape. One can not always guarantee that angles and lengths are preserved, as the scaling transform can alter these when different x and y scaling factors are used. For completeness, we will repeat these transforms from a matrix perspective.

10.5.6   2D Scaling

The strategy used to scale a point $(x, \; y)$ relative to some arbitrary point $(p_x, \; p_y)$ is to first, translate $(-p_x, \; -p_y)$; second, perform the scaling; and third translate $(p_x, \; p_y)$. These three transforms are represented in matrix form as follows:

$$\begin{bmatrix} x' \\ y' \\ 1 \end{bmatrix} = \begin{bmatrix} \text{translate}(p_x, \; p_y) \end{bmatrix} \begin{bmatrix} \text{scale}(s_x, \; s_y) \end{bmatrix} \begin{bmatrix} \text{translate}(-p_x, \; -p_y) \end{bmatrix} \begin{bmatrix} x \\ y \\ 1 \end{bmatrix}$$

which expands to

$$\begin{bmatrix} x' \\ y' \\ 1 \end{bmatrix} = \begin{bmatrix} 1 & 0 & p_x \\ 0 & 1 & p_y \\ 0 & 0 & 1 \end{bmatrix} \begin{bmatrix} s_x & 0 & 0 \\ 0 & s_y & 0 \\ 0 & 0 & 1 \end{bmatrix} \begin{bmatrix} 1 & 0 & -p_x \\ 0 & 1 & -p_y \\ 0 & 0 & 1 \end{bmatrix} \begin{bmatrix} x \\ y \\ 1 \end{bmatrix}.$$

Note the sequence of the transforms, as this often causes confusion. The first transform acting on the point $(x, \; y, \; 1)$ is translate $(-p_x, \; -p_y)$, followed by scale $(s_x, \; s_y)$, followed by translate $(p_x, \; p_y)$. If they are placed in any other sequence, you will discover, like Gauss, that transforms are not commutative!

We can now combine these matrices into a single matrix by multiplying them together. This can be done in any sequence, so long as we preserve the original order. Let's start with scale $(s_x, \; s_y)$ and translate $(-p_x, \; -p_y)$ matrices. This produces

$$\begin{bmatrix} x' \\ y' \\ 1 \end{bmatrix} = \begin{bmatrix} 1 & 0 & p_x \\ 0 & 1 & p_y \\ 0 & 0 & 1 \end{bmatrix} \begin{bmatrix} s_x & 0 & -s_x p_x \\ 0 & s_y & -s_y p_y \\ 0 & 0 & 1 \end{bmatrix} \begin{bmatrix} x \\ y \\ 1 \end{bmatrix}$$

and finally:

$$
\begin{bmatrix} x' \\ y' \\ 1 \end{bmatrix} = \begin{bmatrix} s_x & 0 & p_x(1 - s_x) \\ 0 & s_y & p_y(1 - s_y) \\ 0 & 0 & 1 \end{bmatrix} \begin{bmatrix} x \\ y \\ 1 \end{bmatrix}
$$

which is the same as the previous transform (10.1).

10.5.7   2D Reflection

A reflection about the y-axis is given by

$$
\begin{bmatrix} x' \\ y' \\ 1 \end{bmatrix} = \begin{bmatrix} -1 & 0 & 0 \\ 0 & 1 & 0 \\ 0 & 0 & 1 \end{bmatrix} \begin{bmatrix} x \\ y \\ 1 \end{bmatrix}.
$$

Therefore, using matrices, we can reason that a reflection transform about an arbitrary axis $x = a_x$, parallel with the y-axis, is given by

$$
\begin{bmatrix} x' \\ y' \\ 1 \end{bmatrix} = \begin{bmatrix} \text{translate}(a_x,\ 0) \end{bmatrix} \begin{bmatrix} \text{reflection} \end{bmatrix} \begin{bmatrix} \text{translate}(-a_x,\ 0) \end{bmatrix} \begin{bmatrix} x \\ y \\ 1 \end{bmatrix}
$$

which expands to

$$
\begin{bmatrix} x' \\ y' \\ 1 \end{bmatrix} = \begin{bmatrix} 1 & 0 & a_x \\ 0 & 1 & 0 \\ 0 & 0 & 1 \end{bmatrix} \begin{bmatrix} -1 & 0 & 0 \\ 0 & 1 & 0 \\ 0 & 0 & 1 \end{bmatrix} \begin{bmatrix} 1 & 0 & -a_x \\ 0 & 1 & 0 \\ 0 & 0 & 1 \end{bmatrix} \begin{bmatrix} x \\ y \\ 1 \end{bmatrix}.
$$

We can now combine these matrices into a single matrix by multiplying them together. Let's begin by multiplying the reflection and the translate $(-a_x,\ 0)$ matrices together. This produces

$$
\begin{bmatrix} x' \\ y' \\ 1 \end{bmatrix} = \begin{bmatrix} 1 & 0 & a_x \\ 0 & 1 & 0 \\ 0 & 0 & 1 \end{bmatrix} \begin{bmatrix} -1 & 0 & a_x \\ 0 & 1 & 0 \\ 0 & 0 & 1 \end{bmatrix} \begin{bmatrix} x \\ y \\ 1 \end{bmatrix}
$$

and finally:

$$
\begin{bmatrix} x' \\ y' \\ 1 \end{bmatrix} = \begin{bmatrix} -1 & 0 & 2a_x \\ 0 & 1 & 0 \\ 0 & 0 & 1 \end{bmatrix} \begin{bmatrix} x \\ y \\ 1 \end{bmatrix}
$$

which is the same as the previous transform (10.2).

10.5.8 2D Rotation About an Arbitrary Point

A rotation about the origin is given by

$$\begin{bmatrix} x' \\ y' \\ 1 \end{bmatrix} = \begin{bmatrix} \cos\beta & -\sin\beta & 0 \\ \sin\beta & \cos\beta & 0 \\ 0 & 0 & 1 \end{bmatrix} \begin{bmatrix} x \\ y \\ 1 \end{bmatrix}$$

Therefore, using matrices, we can develop a rotation about an arbitrary point $(p_x, \ p_y)$ as follows:

$$\begin{bmatrix} x' \\ y' \\ 1 \end{bmatrix} = \begin{bmatrix} \text{translate}(p_x, \ p_y) \end{bmatrix} \begin{bmatrix} \text{rotate}\beta \end{bmatrix} \begin{bmatrix} \text{translate}(-p_x, \ -p_y) \end{bmatrix} \begin{bmatrix} x \\ y \\ 1 \end{bmatrix}$$

which expands to

$$\begin{bmatrix} x' \\ y' \\ 1 \end{bmatrix} = \begin{bmatrix} 1 & 0 & p_x \\ 0 & 1 & p_y \\ 0 & 0 & 1 \end{bmatrix} \begin{bmatrix} \cos\beta & -\sin\beta & 0 \\ \sin\beta & \cos\beta & 0 \\ 0 & 0 & 1 \end{bmatrix} \begin{bmatrix} 1 & 0 & -p_x \\ 0 & 1 & -p_y \\ 0 & 0 & 1 \end{bmatrix} \begin{bmatrix} x \\ y \\ 1 \end{bmatrix}.$$

We can now combine these matrices into a single matrix by multiplying them together. Let's begin by multiplying the rotate β and the translate $(-p_x, \ -p_y)$ matrices together. This produces

$$\begin{bmatrix} x' \\ y' \\ 1 \end{bmatrix} = \begin{bmatrix} 1 & 0 & p_x \\ 0 & 1 & p_y \\ 0 & 0 & 1 \end{bmatrix} \begin{bmatrix} \cos\beta & -\sin\beta & -p_x\cos\beta + p_y\sin\beta \\ \sin\beta & \cos\beta & -p_x\sin\beta - p_y\cos\beta \\ 0 & 0 & 1 \end{bmatrix} \begin{bmatrix} x \\ y \\ 1 \end{bmatrix}$$

$$\begin{bmatrix} x' \\ y' \\ 1 \end{bmatrix} = \begin{bmatrix} \cos\beta & -\sin\beta & p_x(1-\cos\beta) + p_y\sin\beta \\ \sin\beta & \cos\beta & p_y(1-\cos\beta) - p_x\sin\beta \\ 0 & 0 & 1 \end{bmatrix} \begin{bmatrix} x \\ y \\ 1 \end{bmatrix}$$

which is the same as the previous transform (10.3).

I hope it is now clear to the reader that one can derive all sorts of transforms either algebraically, or by using matrices—it is just a question of convenience.

10.6 3D Transforms

Now we come to transforms in three dimensions, where we apply the same reasoning as in two dimensions. Scaling and translation are basically the same, but where in 2D we rotated a shape about a point, in 3D we rotate an object about an axis.

10.6.1   3D Translation

The algebra is so simple for 3D translation that we can simply write the homogeneous matrix directly:

$$
\begin{bmatrix} x' \\ y' \\ z' \\ 1 \end{bmatrix} =
\begin{bmatrix} 1 & 0 & 0 & t_x \\ 0 & 1 & 0 & t_y \\ 0 & 0 & 1 & t_z \\ 0 & 0 & 0 & 1 \end{bmatrix}
\begin{bmatrix} x \\ y \\ z \\ 1 \end{bmatrix}.
$$

10.6.2   3D Scaling

The algebra for 3D scaling is

$$
x' = s_x x
$$
$$
y' = s_y y
$$
$$
z' = s_z z
$$

which in matrix form is

$$
\begin{bmatrix} x' \\ y' \\ z' \\ 1 \end{bmatrix} =
\begin{bmatrix} s_x & 0 & 0 & 0 \\ 0 & s_y & 0 & 0 \\ 0 & 0 & s_z & 0 \\ 0 & 0 & 0 & 1 \end{bmatrix}
\begin{bmatrix} x \\ y \\ z \\ 1 \end{bmatrix}.
$$

The scaling is relative to the origin, but we can arrange for it to be relative to an arbitrary point $(p_x,\ p_y,\ p_z)$ using the following algebra:

$$
x' = s_x(x - p_x) + p_x
$$
$$
y' = s_y(y - p_y) + p_y
$$
$$
z' = s_z(z - p_z) + p_z
$$

which in matrix form is

$$
\begin{bmatrix} x' \\ y' \\ z' \\ 1 \end{bmatrix} =
\begin{bmatrix} s_x & 0 & 0 & p_x(1 - s_x) \\ 0 & s_y & 0 & p_y(1 - s_y) \\ 0 & 0 & s_z & p_z(1 - s_z) \\ 0 & 0 & 0 & 1 \end{bmatrix}
\begin{bmatrix} x \\ y \\ z \\ 1 \end{bmatrix}.
$$

10.6.3   3D Rotation

In two dimensions a shape is rotated about a point, whether it be the origin or some other position. In three dimensions an object is rotated about an axis, whether it be the x-, y- or z-axis, or some arbitrary axis. To begin with, let's look at rotating a vertex about one of the three orthogonal axes; such rotations are called *Euler rotations* after Leonhard Euler.

Recall that a general 2D rotation transform is given by

$$\begin{bmatrix} x' \\ y' \\ 1 \end{bmatrix} = \begin{bmatrix} \cos\beta & -\sin\beta & 0 \\ \sin\beta & \cos\beta & 0 \\ 0 & 0 & 1 \end{bmatrix} \begin{bmatrix} x \\ y \\ 1 \end{bmatrix}$$

which in 3D can be visualised as rotating a point $P(x,\ y,\ z)$ on a plane parallel with the xy-plane as shown in Fig. 10.8. In algebraic terms this is written as

$$x' = x\cos\beta - y\sin\beta$$
$$y' = x\sin\beta + y\cos\beta$$
$$z' = z.$$

Therefore, the 3D rotation transform is

$$\begin{bmatrix} x' \\ y' \\ z' \\ 1 \end{bmatrix} = \begin{bmatrix} \cos\beta & -\sin\beta & 0 & 0 \\ \sin\beta & \cos\beta & 0 & 0 \\ 0 & 0 & 1 & 0 \\ 0 & 0 & 0 & 1 \end{bmatrix} \begin{bmatrix} x \\ y \\ z \\ 1 \end{bmatrix}$$

which basically rotates a point about the z-axis.

When rotating about the x-axis, the x-coordinates remain constant whilst the y- and z-coordinates are changed. Algebraically, this is

Fig. 10.8 Rotating the point P, through an angle β, about the z-axis

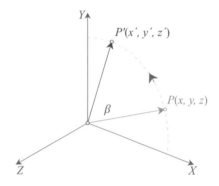

$$x' = x$$
$$y' = y\cos\beta - z\sin\beta$$
$$z' = y\sin\beta + z\cos\beta$$

or in matrix form:
$$\begin{bmatrix} x' \\ y' \\ z' \\ 1 \end{bmatrix} = \begin{bmatrix} 1 & 0 & 0 & 0 \\ 0 & \cos\beta & -\sin\beta & 0 \\ 0 & \sin\beta & \cos\beta & 0 \\ 0 & 0 & 0 & 1 \end{bmatrix} \begin{bmatrix} x \\ y \\ z \\ 1 \end{bmatrix}.$$

When rotating about the y-axis, the y-coordinate remains constant whilst the x- and z-coordinates are changed. Algebraically, this is

$$x' = z\sin\beta + x\cos\beta$$
$$y' = y$$
$$z' = z\cos\beta - x\sin\beta$$

or in matrix form:
$$\begin{bmatrix} x' \\ y' \\ z' \\ 1 \end{bmatrix} = \begin{bmatrix} \cos\beta & 0 & \sin\beta & 0 \\ 0 & 1 & 0 & 0 \\ -\sin\beta & 0 & \cos\beta & 0 \\ 0 & 0 & 0 & 1 \end{bmatrix} \begin{bmatrix} x \\ y \\ z \\ 1 \end{bmatrix}.$$

Note that the matrix terms do not appear to share the symmetry seen in the previous two matrices. Nothing really has gone wrong, it is just the way the axes are paired together to rotate the coordinates.

The above rotations are also known as *yaw*, *pitch* and *roll*, and great care should be taken with these angles when referring to other books and technical papers. Sometimes a left-handed system of axes is used rather than a right-handed set, and the vertical axis may be the y-axis or the z-axis. Consequently, the matrices representing the rotations can vary greatly. In this chapter all Cartesian coordinate systems are right-handed, and the vertical axis is always the y-axis.

I will define the roll, pitch and yaw angles as follows:

- *roll* is the angle of rotation about the z-axis,
- *pitch* is the angle of rotation about the x-axis,
- *yaw* is the angle of rotation about the y-axis.

Figure 10.9 illustrates these rotations and the sign convention. The homogeneous matrices representing these rotations are as follows:

- rotate *roll* about the z-axis:

Fig. 10.9 The convention
for *roll*, *pitch* and *yaw*
angles

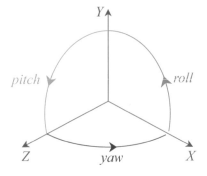

$$\begin{bmatrix} \cos roll & -\sin roll & 0 & 0 \\ \sin roll & \cos roll & 0 & 0 \\ 0 & 0 & 1 & 0 \\ 0 & 0 & 0 & 1 \end{bmatrix}.$$

- rotate *pitch* about the *x*-axis:

$$\begin{bmatrix} 1 & 0 & 0 & 0 \\ 0 & \cos pitch & -\sin pitch & 0 \\ 0 & \sin pitch & \cos pitch & 0 \\ 0 & 0 & 0 & 1 \end{bmatrix}.$$

- rotate *yaw* about the *y*-axis:

$$\begin{bmatrix} \cos yaw & 0 & \sin yaw & 0 \\ 0 & 1 & 0 & 0 \\ -\sin yaw & 0 & \cos yaw & 0 \\ 0 & 0 & 0 & 1 \end{bmatrix}.$$

A common sequence for applying these rotations is *roll, pitch, yaw*, as seen in
the following transform:

$$\begin{bmatrix} x' \\ y' \\ z' \\ 1 \end{bmatrix} = \begin{bmatrix} yaw \end{bmatrix}\begin{bmatrix} pitch \end{bmatrix}\begin{bmatrix} roll \end{bmatrix}\begin{bmatrix} x \\ y \\ z \\ 1 \end{bmatrix}$$

and if a translation is involved,

Fig. 10.10 The $X'Y'Z'$ axial system after a *pitch* of $90°$

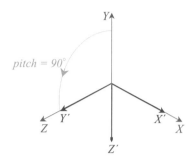

Fig. 10.11 The $X'Y'Z'$ axial system after a *yaw* of $90°$

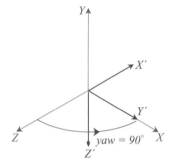

$$\begin{bmatrix} x' \\ y' \\ z' \\ 1 \end{bmatrix} = \begin{bmatrix} translate \end{bmatrix} \begin{bmatrix} yaw \end{bmatrix} \begin{bmatrix} pitch \end{bmatrix} \begin{bmatrix} roll \end{bmatrix} \begin{bmatrix} x \\ y \\ z \\ 1 \end{bmatrix}.$$

When these rotation transforms are applied, the vertex is first rotated about the z-axis (*roll*), followed by a rotation about the x-axis (*pitch*), followed by a rotation about the y-axis (*yaw*). Euler rotations are relative to the fixed frame of reference. This is not always easy to visualise as one's attention is normally with the rotating frame of reference. Let's consider a simple example where an axial system is subjected to a pitch rotation followed by a yaw rotation relative to fixed frame of reference.

We begin with two frames of reference XYZ and $X'Y'Z'$ mutually aligned. Figure 10.10 shows the orientation of $X'Y'Z'$ after it is subjected to a pitch of $90°$ about the X-axis. And Fig. 10.11 shows the final orientation after $X'Y'Z'$ is subjected to a yaw of $90°$ about the Y-axis.

10.6.4 Gimbal Lock

Let's take another example starting from the point where the two axial systems are mutually aligned. Figure 10.12 shows the orientation of $X'Y'Z'$ after it is subjected

Fig. 10.12 The $X'Y'Z'$ axial
system after a *roll* of 45°

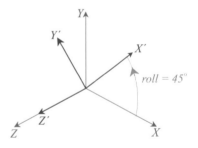

Fig. 10.13 The $X'Y'Z'$ axial
system after a *pitch* of 90°

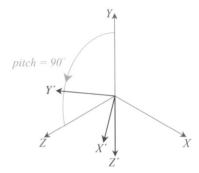

to a roll of 45° about the Z-axis, and Fig. 10.13 shows the orientation of $X'Y'Z'$ after
it is subjected to a pitch of 90° about the X-axis. Now the interesting thing about
this orientation is that if we now performed a yaw of 45° about the Y-axis, it would
rotate the X'-axis towards the X-axis, counteracting the effect of the original roll.
Yaw has become a negative roll rotation, caused by the 90° pitch. This situation is
known as *gimbal lock*, because one degree of rotational freedom has been lost. Quite
innocently, we have stumbled across one of the major weaknesses of Euler angles:
under certain conditions it is only possible to rotate an object about two axes. One
way of preventing this is to create a secondary set of axes constructed from three
orthogonal vectors that are also rotated alongside an object or virtual camera. But
instead of making the rotations relative to the fixed frame of reference, the roll, pitch
and yaw rotations are relative to the rotating frame of reference.

10.6.5   Rotating About an Axis

The above rotations were relative to the x-, y-, z-axis. Now let's consider rotations
about an axis parallel to one of these axes. To begin with, we will rotate about an axis
parallel with the z-axis, as shown in Fig. 10.14. The scenario is very reminiscent of
the 2D case for rotating a point about an arbitrary point, and the general transform
is given by

Fig. 10.14 Rotating a point about an axis parallel with the x-axis

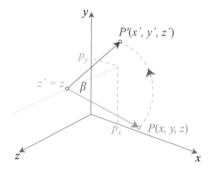

$$\begin{bmatrix} x' \\ y' \\ z' \\ 1 \end{bmatrix} = \big[\, \text{translate}(p_x,\ p_y,\ 0) \,\big]\big[\, \text{rotate}\beta \,\big]\big[\, \text{translate}(-p_x,\ -p_y,\ 0) \,\big] \begin{bmatrix} x \\ y \\ z \\ 1 \end{bmatrix}$$

and the matrix is

$$\begin{bmatrix} x' \\ y' \\ z' \\ 1 \end{bmatrix} = \begin{bmatrix} \cos\beta & -\sin\beta & 0 & p_x(1-\cos\beta)+p_y\sin\beta \\ \sin\beta & \cos\beta & 0 & p_y(1-\cos\beta)-p_x\sin\beta \\ 0 & 0 & 1 & 0 \\ 0 & 0 & 0 & 1 \end{bmatrix} \begin{bmatrix} x \\ y \\ z \\ 1 \end{bmatrix}.$$

I hope you can see the similarity between rotating in 3D and 2D: the x- and y-coordinates are updated while the z-coordinate is held constant. We can now state the other two matrices for rotating about an axis parallel with the x-axis and parallel with the y-axis:

- rotating about an axis parallel with the x-axis:

$$\begin{bmatrix} x' \\ y' \\ z' \\ 1 \end{bmatrix} = \begin{bmatrix} 1 & 0 & 0 & 0 \\ 0 & \cos\beta & -\sin\beta & p_y(1-\cos\beta)+p_z\sin\beta \\ 0 & \sin\beta & \cos\beta & p_z(1-\cos\beta)-p_y\sin\beta \\ 0 & 0 & 0 & 1 \end{bmatrix} \begin{bmatrix} x \\ y \\ z \\ 1 \end{bmatrix}.$$

- rotating about an axis parallel with the y-axis:

$$\begin{bmatrix} x' \\ y' \\ z' \\ 1 \end{bmatrix} = \begin{bmatrix} \cos\beta & 0 & \sin\beta & p_x(1-\cos\beta)-p_z\sin\beta \\ 0 & 1 & 0 & 0 \\ -\sin\beta & 0 & \cos\beta & p_z(1-\cos\beta)+p_x\sin\beta \\ 0 & 0 & 0 & 1 \end{bmatrix} \begin{bmatrix} x \\ y \\ z \\ 1 \end{bmatrix}.$$

10.6.6   3D Reflections

Reflections in 3D occur with respect to a plane, rather than an axis. The matrix giving the reflection relative to the yz-plane is

$$\begin{bmatrix} x' \\ y' \\ z' \\ 1 \end{bmatrix} = \begin{bmatrix} -1 & 0 & 0 & 0 \\ 0 & 1 & 0 & 0 \\ 0 & 0 & 1 & 0 \\ 0 & 0 & 0 & 1 \end{bmatrix} \begin{bmatrix} x \\ y \\ z \\ 1 \end{bmatrix}$$

and the reflection relative to a plane parallel to, and a_x units from the yz-plane is

$$\begin{bmatrix} x' \\ y' \\ z' \\ 1 \end{bmatrix} = \begin{bmatrix} -1 & 0 & 0 & 2a_x \\ 0 & 1 & 0 & 0 \\ 0 & 0 & 1 & 0 \\ 0 & 0 & 0 & 1 \end{bmatrix} \begin{bmatrix} x \\ y \\ z \\ 1 \end{bmatrix}.$$

It is left to the reader to develop similar matrices for the other major axial planes.

10.7   Change of Axes

Points in one coordinate system often have to be referenced in another one. For example, to view a 3D scene from an arbitrary position, a virtual camera is positioned in the world space using a series of transforms. An object's coordinates, which are relative to the world frame of reference, are computed relative to the camera's axial system, and then used to develop a perspective projection. Before explaining how this is achieved in 3D, let's examine the simple case of changing axial systems in two dimensions.

10.7.1   2D Change of Axes

Figure 10.15 shows a point $P(x, y)$ relative to the XY-axes, but we require to know the coordinates relative to the $X'Y'$-axes. To do this, we need to know the relationship between the two coordinate systems, and ideally we want to apply a technique that works in 2D and 3D. If the second coordinate system is a simple translation (t_x, t_y) relative to the reference system, as shown in Fig. 10.15, the point $P(x, y)$ has coordinates relative to the translated system $(x - t_x, y - t_y)$:

Fig. 10.15 The $X'Y'$ axial system is translated (t_x, t_y)

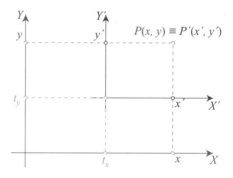

Fig. 10.16 The $X'Y'$ axial system is rotated β

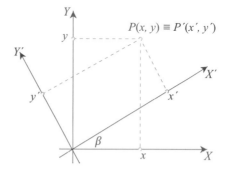

$$\begin{bmatrix} x' \\ y' \\ 1 \end{bmatrix} = \begin{bmatrix} 1 & 0 & -t_x \\ 0 & 1 & -t_y \\ 0 & 0 & 1 \end{bmatrix} \begin{bmatrix} x \\ y \\ 1 \end{bmatrix}.$$

If the $X'Y'$-axes are rotated β relative to the XY-axes, as shown in Fig. 10.16, a point $P(x, y)$ relative to the XY-axes becomes $P'(x', y')$ relative to the rotated axes is given by

$$\begin{bmatrix} x' \\ y' \\ 1 \end{bmatrix} = \begin{bmatrix} \cos(-\beta) & -\sin(-\beta) & 0 \\ \sin(-\beta) & \cos(-\beta) & 0 \\ 0 & 0 & 1 \end{bmatrix} \begin{bmatrix} x \\ y \\ 1 \end{bmatrix}$$

which simplifies to

$$\begin{bmatrix} x' \\ y' \\ 1 \end{bmatrix} = \begin{bmatrix} \cos \beta & \sin \beta & 0 \\ -\sin \beta & \cos \beta & 0 \\ 0 & 0 & 1 \end{bmatrix} \begin{bmatrix} x \\ y \\ 1 \end{bmatrix}.$$

When a coordinate system is rotated and translated relative to the reference system, a point $P(x, y)$ becomes $P'(x', y')$ relative to the new axes given by

Fig. 10.17 If the X'- and Y'-axes are assumed to be unit vectors, their direction cosines form the elements of the rotation matrix

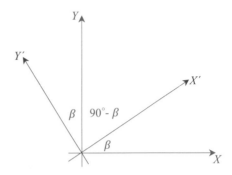

$$\begin{bmatrix} x' \\ y' \\ 1 \end{bmatrix} = \begin{bmatrix} \cos\beta & \sin\beta & 0 \\ -\sin\beta & \cos\beta & 0 \\ 0 & 0 & 1 \end{bmatrix} \begin{bmatrix} 1 & 0 & -t_x \\ 0 & 1 & -t_y \\ 0 & 0 & 1 \end{bmatrix} \begin{bmatrix} x \\ y \\ 1 \end{bmatrix}$$

which simplifies to

$$\begin{bmatrix} x' \\ y' \\ 1 \end{bmatrix} = \begin{bmatrix} \cos\beta & \sin\beta & -t_x\cos\beta - t_y\sin\beta \\ -\sin\beta & \cos\beta & t_x\sin\beta - t_y\cos\beta \\ 0 & 0 & 1 \end{bmatrix} \begin{bmatrix} x \\ y \\ 1 \end{bmatrix}.$$

10.7.2 Direction Cosines

Direction cosines are the cosines of the angles between a vector and the Cartesian axes, and for unit vectors they are the vector's components. Figure 10.17 shows two unit vectors X' and Y', and by inspection the direction cosines for X' are $\cos\beta$ and $\cos(90° - \beta)$, which can be rewritten as $\cos\beta$ and $\sin\beta$, and the direction cosines for Y' are $\cos(90° + \beta)$ and $\cos\beta$, which can be rewritten as $-\sin\beta$ and $\cos\beta$. But these direction cosines $\cos\beta$, $\sin\beta$, $-\sin\beta$ and $\cos\beta$ are the four elements of the rotation matrix used above

$$\begin{bmatrix} \cos\beta & \sin\beta \\ -\sin\beta & \cos\beta \end{bmatrix}.$$

The top row contains the direction cosines for the X'-axis and the bottom row contains the direction cosines for the Y'-axis. This relationship also holds in 3D.

As an example, let's evaluate a simple 2D case where a set of axes is rotated 45° as shown in Fig. 10.18. The appropriate transform is

Fig. 10.18 The vertices of a
unit square relative to the
two axial systems

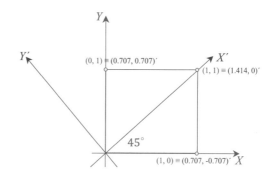

$$
\begin{bmatrix} x' \\ y' \\ 1 \end{bmatrix} = \begin{bmatrix} \cos 45° & \sin 45° & 0 \\ -\sin 45° & \cos 45° & 0 \\ 0 & 0 & 1 \end{bmatrix} \begin{bmatrix} x \\ y \\ 1 \end{bmatrix}
$$

$$
\approx \begin{bmatrix} 0.707 & 0.707 & 0 \\ -0.707 & 0.707 & 0 \\ 0 & 0 & 1 \end{bmatrix} \begin{bmatrix} x \\ y \\ 1 \end{bmatrix}.
$$

The four vertices on a unit square become

$$(0, 0) \rightarrow (0, 0)$$
$$(1, 0) \rightarrow (0.707, -0.707)$$
$$(1, 1) \rightarrow (1.1414, 0)$$
$$(0, 1) \rightarrow (0.707, 0.707)$$

which by inspection of Fig. 10.18 are correct.

10.7.3 3D Change of Axes

The ability to reference a collection of coordinates is fundamental in computer graphics, especially in 3D. And rather than investigate them within this section, let's delay their analysis for the next section, where we see how the technique is used for relating an object's coordinates relative to an arbitrary virtual camera.

10.8 Positioning the Virtual Camera

Four coordinate systems are used in the computer graphics pipeline: *object space*, *world space*, *camera space* and *image space*.

- The object space is a domain where objects are modelled and assembled.
- The world space is where objects are positioned and animated through appropriate transforms. The world space also hosts a virtual camera or observer.
- The camera space is a transform of the world space relative to the camera.
- Finally, the image space is a projection—normally perspective—of the camera space onto an image plane.

  The transforms considered so far are used to manipulate and position objects within the world space. What we will consider next is how a virtual camera or observer is positioned in world space, and the process of converting world coordinates to camera coordinates. The procedure used generally depends on the method employed to define the camera's frame of reference within the world space, which may involve the use of direction cosines, Euler angles or quaternions.

10.8.1 Direction Cosines

A 3D unit vector has three components $[x \quad y \quad z]^T$, which are equal to the cosines of the angles formed between the vector and the three orthogonal axes. These angles are known as *direction cosines* and can be computed taking the dot product of the vector and the Cartesian unit vectors. Figure 10.19 shows the direction cosines and the angles. These direction cosines enable any point $P(x, y, z)$ in one frame of reference to be transformed into $P'(x', y', z')$ in another frame of reference as follows:

$$\begin{bmatrix} x' \\ y' \\ z' \\ 1 \end{bmatrix} = \begin{bmatrix} r_{11} & r_{12} & r_{13} & 0 \\ r_{21} & r_{22} & r_{23} & 0 \\ r_{31} & r_{32} & r_{33} & 0 \\ 0 & 0 & 0 & 1 \end{bmatrix} \begin{bmatrix} x \\ y \\ z \\ 1 \end{bmatrix}$$

where:

- r_{11}, r_{12}, r_{13} are the direction cosines of the secondary x-axis,
- r_{21}, r_{22}, r_{23} are the direction cosines of the secondary y-axis,

Fig. 10.19 The components of a unit vector are equal to the cosines of the angles between the vector and the axes

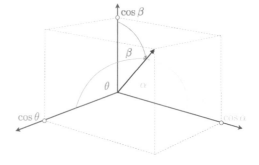

Fig. 10.20 Two axial
systems mutually aligned

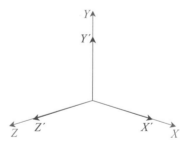

Fig. 10.21 The $X'Y'Z'$ axial
system after a roll of $90°$

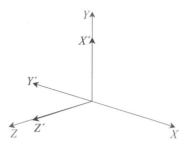

- r_{31}, r_{32}, r_{33} are the direction cosines of the secondary z-axis.

To illustrate this operation, consider the scenario shown in Fig. 10.20 with two axial systems mutually aligned. Evaluating the direction cosines results in the following matrix transformation:

$$\begin{bmatrix} x' \\ y' \\ z' \\ 1 \end{bmatrix} = \begin{bmatrix} 1 & 0 & 0 & 0 \\ 0 & 1 & 0 & 0 \\ 0 & 0 & 1 & 0 \\ 0 & 0 & 0 & 1 \end{bmatrix} \begin{bmatrix} x \\ y \\ z \\ 1 \end{bmatrix}$$

which is the identity matrix and implies that $(x', y', z') = (x, y, z)$.

Figure 10.21 shows another scenario where the axes are rolled $90°$, and the associated transform is

$$\begin{bmatrix} x' \\ y' \\ z' \\ 1 \end{bmatrix} = \begin{bmatrix} 0 & 1 & 0 & 0 \\ -1 & 0 & 0 & 0 \\ 0 & 0 & 1 & 0 \\ 0 & 0 & 0 & 1 \end{bmatrix} \begin{bmatrix} x \\ y \\ z \\ 1 \end{bmatrix}.$$

Substituting $(1, 1, 0)$ for (x, y, z) produces $(1, -1, 0)$ for (x', y', z') in the new frame of reference, which by inspection, is correct.

If the virtual camera is offset by (t_x, t_y, t_z) the transform relating points in world space to camera space is expressed as a compound operation consisting of a translation back to the origin, followed by a change of axial systems. This is expressed as

Fig. 10.22 The secondary axial system is subject to a *yaw* of 180° and an offset of (10, 1, 1)

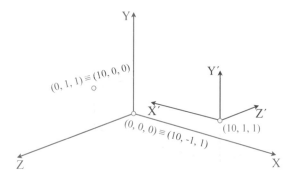

$$\begin{bmatrix} x' \\ y' \\ z' \\ 1 \end{bmatrix} = \begin{bmatrix} r_{11} & r_{12} & r_{13} & 0 \\ r_{21} & r_{22} & r_{23} & 0 \\ r_{31} & r_{32} & r_{33} & 0 \\ 0 & 0 & 0 & 1 \end{bmatrix} \begin{bmatrix} 1 & 0 & 0 & -t_x \\ 0 & 1 & 0 & -t_y \\ 0 & 0 & 1 & -t_z \\ 0 & 0 & 0 & 1 \end{bmatrix} \begin{bmatrix} x \\ y \\ z \\ 1 \end{bmatrix}.$$

To illustrate this, consider the scenario shown in Fig. 10.22. The values of $(t_x,\ t_y,\ t_z)$ are (10, 1, 1), and the direction cosines are as shown in the following matrix operation:

$$\begin{bmatrix} x' \\ y' \\ z' \\ 1 \end{bmatrix} = \begin{bmatrix} -1 & 0 & 0 & 0 \\ 0 & 1 & 0 & 0 \\ 0 & 0 & -1 & 0 \\ 0 & 0 & 0 & 1 \end{bmatrix} \begin{bmatrix} 1 & 0 & 0 & -10 \\ 0 & 1 & 0 & -1 \\ 0 & 0 & 1 & -1 \\ 0 & 0 & 0 & 1 \end{bmatrix} \begin{bmatrix} x \\ y \\ z \\ 1 \end{bmatrix}$$

which simplifies to

$$\begin{bmatrix} x' \\ y' \\ z' \\ 1 \end{bmatrix} = \begin{bmatrix} -1 & 0 & 0 & 10 \\ 0 & 1 & 0 & -1 \\ 0 & 0 & -1 & 1 \\ 0 & 0 & 0 & 1 \end{bmatrix} \begin{bmatrix} x \\ y \\ z \\ 1 \end{bmatrix}.$$

Substituting (0, 0, 0) for $(x,\ y,\ z)$ in the above transform produces (10, −1, 1) for $(x',\ y',\ z')$, which can be confirmed from Fig. 10.22. Similarly, substituting (0, 1, 1) for $(x,\ y,\ z)$ produces (10, 0, 0) for $(x',\ y',\ z')$, which is also correct.

10.8.2 Euler Angles

Another approach for locating the virtual camera involves Euler angles, but we must remember that they suffer from gimbal lock. However, if the virtual camera is located in world space using Euler angles, the transform relating world coordinates to camera coordinates can be derived from the inverse operations. The *yaw, pitch, roll* matrices described above are called *orthogonal matrices*, as the inverse matrix is the transpose of the original rows and columns. Consequently, to rotate through angles $-roll$, $-pitch$ and $-yaw$, we use

- rotate $-roll$ about the z-axis:

$$\begin{bmatrix} \cos roll & \sin roll & 0 & 0 \\ -\sin roll & \cos roll & 0 & 0 \\ 0 & 0 & 1 & 0 \\ 0 & 0 & 0 & 1 \end{bmatrix}.$$

- rotate $-pitch$ about the x-axis:

$$\begin{bmatrix} 1 & 0 & 0 & 0 \\ 0 & \cos pitch & \sin pitch & 0 \\ 0 & -\sin pitch & \cos pitch & 0 \\ 0 & 0 & 0 & 1 \end{bmatrix}.$$

- rotate $-yaw$ about the y-axis:

$$\begin{bmatrix} \cos yaw & 0 & -\sin yaw & 0 \\ 0 & 1 & 0 & 0 \\ \sin yaw & 0 & \cos yaw & 0 \\ 0 & 0 & 0 & 1 \end{bmatrix}.$$

The same result is obtained by substituting $-roll$, $-pitch$, $-yaw$ in the original matrices. As described above, the virtual camera will normally be translated from the origin by (t_x, t_y, t_z), which implies that the transform from the world space to the camera space must be evaluated as follows:

$$\begin{bmatrix} x' \\ y' \\ z' \\ 1 \end{bmatrix} = [-roll][-pitch][-yaw][translate(-t_x, -t_y, -t_z)] \begin{bmatrix} x \\ y \\ z \\ 1 \end{bmatrix}$$

which is represented by a single homogeneous matrix:

$$\begin{bmatrix} x' \\ y' \\ z' \\ 1 \end{bmatrix} = \begin{bmatrix} T_{11} & T_{12} & T_{13} & T_{14} \\ T_{21} & T_{22} & T_{23} & T_{24} \\ T_{31} & T_{32} & T_{33} & T_{34} \\ T_{41} & T_{42} & T_{43} & T_{44} \end{bmatrix} \begin{bmatrix} x \\ y \\ z \\ 1 \end{bmatrix}$$

where

$$T_{11} = \cos(yaw)\cos(roll) + \sin(yaw)\sin(pitch)\sin(roll)$$
$$T_{12} = \cos(pitch)\sin(roll)$$
$$T_{13} = -\sin(yaw)\cos(roll) + \cos(yaw)\sin(pitch)\sin(roll)$$
$$T_{14} = -\left(t_x T_{11} + t_y T_{12} + t_z T_{13}\right)$$
$$T_{21} = -\cos(yaw)\sin(roll) + \sin(yaw)\sin(pitch)\cos(roll)$$
$$T_{22} = \cos(pitch)\cos(roll)$$
$$T_{23} = -\sin(yaw)\sin(roll) + \cos(yaw)\sin(pitch)\cos(roll)$$
$$T_{24} = -\left(t_x T_{21} + t_y T_{22} + t_z T_{23}\right)$$
$$T_{31} = \sin(yaw)\cos(pitch)$$
$$T_{32} = -\sin(pitch)$$
$$T_{33} = \cos(yaw)\cos(pitch)$$
$$T_{34} = -\left(t_x T_{31} + t_y T_{32} + t_z T_{33}\right)$$
$$T_{41} = T_{42} = T_{43} = 0$$
$$T_{44} = 1.$$

For example, consider the scenario shown in Fig. 10.22 where the following conditions prevail:

$$roll = 0°$$
$$pitch = 0°$$
$$yaw = 180°$$
$$t_x = 10$$
$$t_y = 1$$
$$t_z = 1.$$

The transform is

$$\begin{bmatrix} x' \\ y' \\ z' \\ 1 \end{bmatrix} = \begin{bmatrix} -1 & 0 & 0 & 10 \\ 0 & 1 & 0 & -1 \\ 0 & 0 & -1 & 1 \\ 0 & 0 & 0 & 1 \end{bmatrix} \begin{bmatrix} x \\ y \\ z \\ 1 \end{bmatrix}$$

which is identical to the equation used for direction cosines.

Another scenario is shown in Fig. 10.23 where the following conditions prevail:

Fig. 10.23 The secondary axial system is subject to a *roll* of 90°, a *pitch* of 180° and a translation of (0.5, 0.5, 11)

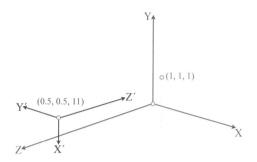

$$roll = 90°$$
$$pitch = 180°$$
$$yaw = 0°$$
$$t_x = 0.5$$
$$t_y = 0.5$$
$$t_z = 11.$$

The transform is

$$\begin{bmatrix} x' \\ y' \\ z' \\ 1 \end{bmatrix} = \begin{bmatrix} 0 & -1 & 0 & 0.5 \\ -1 & 0 & 0 & 0.5 \\ 0 & 0 & -1 & 11 \\ 0 & 0 & 0 & 1 \end{bmatrix} \begin{bmatrix} x \\ y \\ z \\ 1 \end{bmatrix}.$$

Substituting $(1, 1, 1)$ for (x, y, z) produces $(-0.5, -0.5, 10)$ for (x', y', z'). Similarly, substituting $(0, 0, 1)$ for (x, y, z) produces $(0.5, 0.5, 10)$ for (x', y', z'), which can be visually verified from Fig. 10.23.

10.9 Rotating a Point About an Arbitrary Axis

10.9.1 Matrices

Let's consider two ways of developing a matrix for rotating a point about an arbitrary axis. The first approach employs vector analysis and is quite succinct. The second technique is less analytical and relies on matrices and trigonometric evaluation and is rather laborious. Fortunately, they both arrive at the same result!

Figure 10.24 shows a view of the geometry associated with the task at hand. For clarification, Fig. 10.25 shows a cross-section and a plan view of the geometry.

Fig. 10.24 A view of the
geometry associated with
rotating a point about an
arbitrary axis

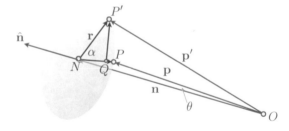

Fig. 10.25 A cross-section
and plan view of the
geometry associated with
rotating a point about an
arbitrary axis

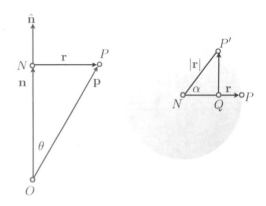

The axis of rotation is given by the unit vector:

$$\hat{\mathbf{n}} = a\mathbf{i} + b\mathbf{j} + c\mathbf{k}.$$

$P(x_p,\ y_p,\ z_p)$ is the point to be rotated by angle α to $P'(x'_p,\ y'_p,\ z'_p)$.

O is the origin, whilst $\mathbf{p}$ and $\mathbf{p}'$ are position vectors for P and P' respectively.
From Figs. 10.24 and 10.25:

$$\mathbf{p}' = \overrightarrow{ON} + \overrightarrow{NQ} + \overrightarrow{QP'}.$$

To find $\overrightarrow{ON}$:

$$\|\mathbf{n}\| = \|\mathbf{p}\| \cos\theta = \hat{\mathbf{n}} \cdot \mathbf{p}$$

therefore,

$$\overrightarrow{ON} = \mathbf{n} = \hat{\mathbf{n}}(\hat{\mathbf{n}} \cdot \mathbf{p}).$$

To find $\overrightarrow{NQ}$:

$$\overrightarrow{NQ} = \frac{NQ}{NP}\mathbf{r} = \frac{NQ}{NP'}\mathbf{r} = \cos\alpha\ \mathbf{r}$$

but

$$\mathbf{p} = \mathbf{n} + \mathbf{r} = \hat{\mathbf{n}}(\hat{\mathbf{n}} \cdot \mathbf{p}) + \mathbf{r}$$

therefore,

$$\mathbf{r} = \mathbf{p} - \hat{\mathbf{n}}(\hat{\mathbf{n}} \cdot \mathbf{p})$$

and

$$\overrightarrow{NQ} = [\mathbf{p} - \hat{\mathbf{n}}(\hat{\mathbf{n}} \cdot \mathbf{p})] \cos \alpha.$$

To find $\overrightarrow{QP'}$:
Let

$$\hat{\mathbf{n}} \times \mathbf{p} = \mathbf{w}$$

where

$$\|\mathbf{w}\| = \|\hat{\mathbf{n}}\| \|\mathbf{p}\| \sin \theta = \|\mathbf{p}\| \sin \theta$$

but

$$\|\mathbf{r}\| = \|\mathbf{p}\| \sin \theta$$

therefore,

$$\|\mathbf{w}\| = \|\mathbf{r}\|.$$

Now

$$\frac{QP'}{NP'} = \frac{QP'}{\|\mathbf{r}\|} = \frac{QP'}{\|\mathbf{w}\|} = \sin \alpha$$

therefore,

$$\overrightarrow{QP'} = \mathbf{w} \sin \alpha = (\hat{\mathbf{n}} \times \mathbf{p}) \sin \alpha$$

then

$$\mathbf{p}' = \hat{\mathbf{n}}(\hat{\mathbf{n}} \cdot \mathbf{p}) + [\mathbf{p} - \hat{\mathbf{n}}(\hat{\mathbf{n}} \cdot \mathbf{p}] \cos \alpha + (\hat{\mathbf{n}} \times \mathbf{p}) \sin \alpha$$

and

$$\mathbf{p}' = \mathbf{p} \cos \alpha + \hat{\mathbf{n}}(\hat{\mathbf{n}} \cdot \mathbf{p})(1 - \cos \alpha) + (\hat{\mathbf{n}} \times \mathbf{p}) \sin \alpha.$$

Let

$$K = 1 - \cos \alpha$$

then

$$\mathbf{p}' = \mathbf{p} \cos \alpha + \hat{\mathbf{n}}(\hat{\mathbf{n}} \cdot \mathbf{p}) K + (\hat{\mathbf{n}} \times \mathbf{p}) \sin \alpha$$

and

$$\mathbf{p}' = (x_p\mathbf{i} + y_p\mathbf{j} + z_p\mathbf{k})\cos\alpha + (a\mathbf{i} + b\mathbf{j} + c\mathbf{k})(ax_p + by_p + cz_p)K$$
$$+ [(bz_p - cy_p)\mathbf{i} + (cx_p - az_p)\mathbf{j} + (ay_p - bx_p)\mathbf{k}]\sin\alpha$$
$$= [x_p\cos\alpha + a(ax_p + by_p + cz_p)K + (bz_p - cy_p)\sin\alpha]\mathbf{i}$$
$$+ [y_p\cos\alpha + b(ax_p + by_p + cz_p)K + (cx_p - az_p)\sin\alpha]\mathbf{j}$$
$$+ [z_p\cos\alpha + c(ax_p + by_p + cz_p)K + (ay_p - bx_p)\sin\alpha]\mathbf{k}$$
$$= [x_p\left(a^2K + \cos\alpha\right) + y_p(abK - c\sin\alpha) + z_p(acK + b\sin\alpha)]\mathbf{i}$$
$$+ [x_p(abK + c\sin\alpha) + y_p\left(b^2K + \cos\alpha\right) + z_p(bcK - a\sin\alpha)]\mathbf{j}$$
$$+ [x_p(acK - b\sin\alpha) + y_p(bcK + a\sin\alpha) + z_p\left(c^2K + \cos\alpha\right)]\mathbf{k}$$

and the transform is:

$$
\begin{bmatrix} x'_p \\ y'_p \\ z'_p \\ 1 \end{bmatrix}
=
\begin{bmatrix}
a^2K + \cos\alpha & abK - c\sin\alpha & acK + b\sin\alpha & 0 \\
abK + c\sin\alpha & b^2K + \cos\alpha & bcK - a\sin\alpha & 0 \\
acK - b\sin\alpha & bcK + a\sin\alpha & c^2K + \cos\alpha & 0 \\
0 & 0 & 0 & 1
\end{bmatrix}
\begin{bmatrix} x_p \\ y_p \\ z_p \\ 1 \end{bmatrix}
$$

where

$$K = 1 - \cos\alpha.$$

Now let's approach the problem using transforms and trigonometric identities. The following is extremely tedious, but it is a good exercise for improving one's algebraic skills!

Figure 10.26 shows a point $P(x, y, z)$ to be rotated through an angle α to $P'(x', y', z')$ about an axis defined by

$$\mathbf{v} = a\mathbf{i} + b\mathbf{j} + c\mathbf{k}$$

where $\|\mathbf{v}\| = 1$.

The transforms to achieve this operation is expressed as follows:

Fig. 10.26 The geometry
associated with rotating a
point about an arbitrary axis

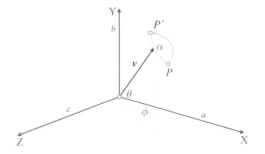

$$\begin{bmatrix} x' \\ y' \\ z' \end{bmatrix} = [\mathbf{T}_5]\ [\mathbf{T}_4]\ [\mathbf{T}_3]\ [\mathbf{T}_2]\ [\mathbf{T}_1] \begin{bmatrix} x \\ y \\ z \end{bmatrix}$$

which aligns the axis of rotation with the x-axis, performs the rotation of P through an angle α about the x-axis, and returns the axis of rotation back to its original position. Therefore,

$$\mathbf{T}_1 \text{ rotates } +\phi \text{ about the } y\text{-axis}$$
$$\mathbf{T}_2 \text{ rotates } -\theta \text{ about the } z\text{-axis}$$
$$\mathbf{T}_3 \text{ rotates } +\alpha \text{ about the } x\text{-axis}$$
$$\mathbf{T}_4 \text{ rotates } +\theta \text{ about the } z\text{-axis}$$
$$\mathbf{T}_5 \text{ rotates } -\phi \text{ about the } y\text{-axis}$$

where

$$\mathbf{T}_1 = \begin{bmatrix} \cos\phi & 0 & \sin\phi \\ 0 & 1 & 0 \\ -\sin\phi & 0 & \cos\phi \end{bmatrix}, \quad \mathbf{T}_2 = \begin{bmatrix} \cos\theta & \sin\theta & 0 \\ -\sin\theta & \cos\theta & 0 \\ 0 & 0 & 1 \end{bmatrix}$$

$$\mathbf{T}_3 = \begin{bmatrix} 1 & 0 & 0 \\ 0 & \cos\alpha & -\sin\alpha \\ 0 & \sin\alpha & \cos\alpha \end{bmatrix}, \quad \mathbf{T}_4 = \begin{bmatrix} \cos\theta & -\sin\theta & 0 \\ \sin\theta & \cos\theta & 0 \\ 0 & 0 & 1 \end{bmatrix}$$

$$\mathbf{T}_5 = \begin{bmatrix} \cos\phi & 0 & -\sin\phi \\ 0 & 1 & 0 \\ \sin\phi & 0 & \cos\phi \end{bmatrix}.$$

Let

$$[\mathbf{T}_5]\ [\mathbf{T}_4]\ [\mathbf{T}_3]\ [\mathbf{T}_2]\ [\mathbf{T}_1] = \begin{bmatrix} E_{11} & E_{12} & E_{13} & 0 \\ E_{21} & E_{22} & E_{23} & 0 \\ E_{31} & E_{32} & E_{33} & 0 \\ 0 & 0 & 0 & 1 \end{bmatrix}$$

where by multiplying the matrices together we find that:

$$E_{11} = \cos^2\phi\cos^2\theta + \cos^2\phi\sin^2\theta\cos\alpha + \sin^2\phi\cos\alpha$$

$$E_{12} = \cos\phi\cos\theta\sin\theta - \cos\phi\sin\theta\cos\theta\cos\alpha - \sin\phi\cos\theta\sin\alpha$$

$$E_{13} = \cos\phi\sin\phi\cos^2\theta + \cos\phi\sin\phi\sin^2\theta\cos\alpha + \sin^2\phi\sin\theta\sin\alpha$$
$$+ \cos^2\phi\sin\theta\sin\alpha - \cos\phi\sin\phi\cos\alpha$$

$$E_{21} = \sin\theta\cos\theta\cos\phi - \cos\theta\sin\theta\cos\phi\cos\alpha + \cos\theta\sin\phi\sin\alpha$$

$$E_{22} = \sin^2\theta + \cos^2\theta\cos\alpha$$

$$E_{23} = \sin\theta\cos\theta\sin\phi - \cos\theta\sin\theta\sin\phi\cos\alpha - \cos\theta\cos\phi\sin\alpha$$

$$E_{31} = \cos\phi\sin\phi\cos^2\theta + \cos\phi\sin\phi\sin^2\theta\cos\alpha - \cos^2\phi\sin\theta\sin\alpha$$
$$= -\cos\phi\sin\phi\cos\alpha$$

$$E_{32} = \sin\phi\cos\theta\sin\theta - \sin\phi\sin\theta\cos\theta\cos\alpha + \cos\phi\cos\theta\sin\alpha$$

$$E_{33} = \sin^2\phi\cos^2\theta + \sin^2\phi\sin^2\theta\cos\alpha - \cos\phi\sin\phi\sin\theta\sin\alpha$$
$$+ \cos\phi\sin\phi\sin\theta\sin\alpha + \cos^2\phi\cos\alpha.$$

From Fig. 10.26 we compute the sin and cos of θ and ϕ in terms of a, b and c, and then compute their equivalent $\sin^2$ and $\cos^2$ values:

$$\cos\theta = \sqrt{1 - b^2} \Rightarrow \cos^2\theta = 1 - b^2$$
$$\sin\theta = b \Rightarrow \sin^2\theta = b^2$$
$$\cos\phi = \frac{a}{\sqrt{1 - b^2}} \Rightarrow \cos^2\phi = \frac{a^2}{1 - b^2}$$
$$\sin\phi = \frac{c}{\sqrt{1 - b^2}} \Rightarrow \sin^2\phi = \frac{c^2}{1 - b^2}.$$

To find E_{11}:

$$E_{11} = \cos^2\phi\cos^2\theta + \cos^2\phi\sin^2\theta\cos\alpha + \sin^2\phi\cos\alpha$$
$$= \frac{a^2}{1 - b^2}(1 - b^2) + \frac{a^2}{1 - b^2}b^2\cos\alpha + \frac{c^2}{1 - b^2}\cos\alpha$$
$$= a^2 + \frac{a^2 b^2}{1 - b^2}\cos\alpha + \frac{c^2}{1 - b^2}\cos\alpha$$
$$= a^2 + \left(\frac{c^2 + a^2 b^2}{1 - b^2}\right)\cos\alpha$$

but

$$a^2 + b^2 + c^2 = 1 \Rightarrow c^2 = 1 - a^2 - b^2$$

substituting c^2 in E_{11}

$$E_{11} = a^2 + \left(\frac{1 - a^2 - b^2 + a^2 b^2}{1 - b^2} \right) \cos \alpha$$

$$= a^2 + \left(\frac{(1 - a^2)(1 - b^2)}{1 - b^2} \right) \cos \alpha$$

$$= a^2 + (1 - a^2) \cos \alpha$$

$$= a^2 (1 - \cos \alpha) + \cos \alpha.$$

Let

$$K = 1 - \cos \alpha$$

then

$$E_{11} = a^2 K + \cos \alpha.$$

To find E_{12}:

$$E_{12} = \cos \phi \cos \theta \sin \theta - \cos \phi \sin \theta \cos \theta \cos \alpha - \sin \phi \cos \theta \sin \alpha$$

$$= \frac{a}{\sqrt{1 - b^2}} \sqrt{1 - b^2} b - \frac{a}{\sqrt{1 - b^2}} b \sqrt{1 - b^2} \cos \alpha - \frac{c}{\sqrt{1 - b^2}} \sqrt{1 - b^2} \sin \alpha$$

$$= ab - ab \cos \alpha - c \sin \alpha$$

$$= ab(1 - \cos \alpha) - c \sin \alpha$$

$$E_{12} = abK - c \sin \alpha.$$

To find E_{13}:

$$E_{13} = \cos \phi \sin \phi \cos^2 \theta + \cos \phi \sin \phi \sin^2 \theta \cos \alpha + \sin^2 \phi \sin \theta \sin \alpha$$

$$\quad + \cos^2 \phi \sin \theta \sin \alpha - \cos \phi \sin \phi \cos \alpha$$

$$= \cos \phi \sin \phi \cos^2 \theta + \cos \phi \sin \phi \sin^2 \theta \cos \alpha + \sin \theta \sin \alpha - \cos \phi \sin \phi \cos \alpha$$

$$= \frac{a}{\sqrt{1 - b^2}} \frac{c}{\sqrt{1 - b^2}} (1 - b^2) + \frac{a}{\sqrt{1 - b^2}} \frac{c}{\sqrt{1 - b^2}} b^2 \cos \alpha + b \sin \alpha$$

$$\quad - \frac{a}{\sqrt{1 - b^2}} \frac{c}{\sqrt{1 - b^2}} \cos \alpha$$

$$= ac + ac \frac{b^2}{(1 - b^2)} \cos \alpha + b \sin \alpha - \frac{ac}{(1 - b^2)} \cos \alpha$$

$$= ac + ac \frac{(b^2 - 1)}{(1 - b^2)} \cos \alpha + b \sin \alpha$$

$$= ac(1 - \cos \alpha) + b \sin \alpha$$

$$E_{13} = acK + b \sin \alpha.$$

Using similar algebraic methods, we discover that:

$$E_{21} = abK + c\sin\alpha$$
$$E_{22} = b^2 K + \cos\alpha$$
$$E_{23} = bcK - a\sin\alpha$$
$$E_{31} = acK - b\sin\alpha$$
$$E_{32} = bcK + a\sin\alpha$$
$$E_{33} = c^2 K + \cos\alpha$$

and our original matrix transform becomes:

$$
\begin{bmatrix} x'_p \\ y'_p \\ z'_p \\ 1 \end{bmatrix} =
\begin{bmatrix}
a^2 K + \cos\alpha & abK - c\sin\alpha & acK + b\sin\alpha & 0 \\
abK + c\sin\alpha & b^2 K + \cos\alpha & bcK - a\sin\alpha & 0 \\
acK - b\sin\alpha & bcK + a\sin\alpha & c^2 K + \cos\alpha & 0 \\
0 & 0 & 0 & 1
\end{bmatrix}
\begin{bmatrix} x_p \\ y_p \\ z_p \\ 1 \end{bmatrix}
$$

where
$$K = 1 - \cos\alpha.$$

which is identical to the transformation derived from the first approach. Now let's test the matrix with a simple example that can be easily verified. We do this by rotating a point $P(10, 5, 0)$, about an arbitrary axis $\mathbf{v} = \mathbf{i} + \mathbf{j} + \mathbf{k}$, through $360°$, which should return it to itself producing $P(10, 5, 0)$.
    Therefore,

$$\alpha = 360°, \quad \cos\alpha = 1, \quad \sin\alpha = 0, \quad K = 0$$

$$a = 1, \quad b = 1, \quad c = 1$$

and

$$
\begin{bmatrix} 10 \\ 5 \\ 0 \\ 1 \end{bmatrix} =
\begin{bmatrix}
1 & 0 & 0 & 0 \\
0 & 1 & 0 & 0 \\
0 & 0 & 1 & 0 \\
0 & 0 & 0 & 1
\end{bmatrix}
\begin{bmatrix} 10 \\ 5 \\ 0 \\ 1 \end{bmatrix}.
$$

As the matrix is an identity matrix $P' = P$.

10.10   Transforming Vectors

The transforms described in this chapter have been used to transform single points. However, a geometric database will not only contain pure vertices, but vectors, which must also be subject to any prevailing transform. A generic transform $\mathbf{Q}$ of a 3D point is represented by

$$\begin{bmatrix} x' \\ y' \\ z' \\ 1 \end{bmatrix} = [\mathbf{Q}] \begin{bmatrix} x \\ y \\ z \\ 1 \end{bmatrix}$$

and as a vector is defined by two points we can write

$$\begin{bmatrix} x' \\ y' \\ z' \\ 1 \end{bmatrix} = [\mathbf{Q}] \begin{bmatrix} x_2 - x_1 \\ y_2 - y_1 \\ z_2 - z_1 \\ 1 - 1 \end{bmatrix}$$

where we see the homogeneous scaling term collapse to zero; which implies that any vector $[x \ \ y \ \ z]^T$ can be transformed using

$$\begin{bmatrix} x' \\ y' \\ z' \\ 0 \end{bmatrix} = [\mathbf{Q}] \begin{bmatrix} x \\ y \\ z \\ 0 \end{bmatrix}$$

Let's put this to the test by using a transform from an earlier example. The problem concerned a change of axial system where a virtual camera was subject to the following:

$$roll = 90°$$
$$pitch = 180°$$
$$yaw = 90°$$
$$t_x = 2$$
$$t_y = 2$$
$$t_z = 0$$

and the transform is

$$\begin{bmatrix} x' \\ y' \\ z' \\ 1 \end{bmatrix} = \begin{bmatrix} 0 & -1 & 0 & 2 \\ 0 & 0 & 1 & 0 \\ -1 & 0 & 0 & 2 \\ 0 & 0 & 0 & 1 \end{bmatrix} \begin{bmatrix} x \\ y \\ z \\ 1 \end{bmatrix}.$$

The point (1, 1, 0) is transformed to (1, 0, 1), as shown in Fig. 10.27. And the vector $[1 \ \ 1 \ \ 0]^T$ is transformed to $[-1 \ \ 0 \ \ -1]^T$, using the following transform

Fig. 10.27 Vector
$[1 \ \ 1 \ \ 0]^T$ is transformed to
$[-1 \ \ 0 \ \ -1]^T$

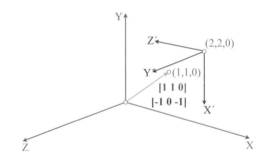

$$
\begin{bmatrix} -1 \\ 0 \\ -1 \\ 0 \end{bmatrix} = \begin{bmatrix} 0 & -1 & 0 & 2 \\ 0 & 0 & 1 & 0 \\ -1 & 0 & 0 & 2 \\ 0 & 0 & 0 & 1 \end{bmatrix} \begin{bmatrix} 1 \\ 1 \\ 0 \\ 0 \end{bmatrix}
$$

which is correct with reference to Fig. 10.27.

10.11 Determinants

Before concluding this chapter, I would like to expand upon the role of the determinant in transforms.

In Chap. 6 we saw that determinants arise in the solution of linear equations. Now let's investigate their graphical significance. Consider the transform:

$$
\begin{bmatrix} x' \\ y' \end{bmatrix} = \begin{bmatrix} a & b \\ c & d \end{bmatrix} \begin{bmatrix} x \\ y \end{bmatrix}.
$$

The determinant of the transform is $ad - cb$. If we subject the vertices of a unit-square to this transform, we create the situation shown in Fig. 10.28. The vertices of the unit-square are transformed as follows:

$$(0, \ 0) \Rightarrow (0, \ 0)$$
$$(1, \ 0) \Rightarrow (a, \ c)$$
$$(1, \ 1) \Rightarrow (a + b, \ c + d)$$
$$(0, \ 1) \Rightarrow (b, \ d).$$

From Fig. 10.28 it can be seen that the area of the transformed unit-square A' is given by

Fig. 10.28 The inner parallelogram is the transformed unit square

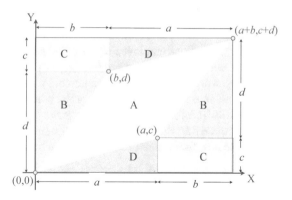

$$\text{area} = (a + b)(c + d) - 2B - 2C - 2D$$
$$= (ac + ad + cb + bd) - bd - 2cb - ac$$
$$= ad - cb$$

which is the determinant of the transform. But as the area of the original unit-square is 1, the determinant of the transform controls the scaling factor applied to the transformed shape.

Let's examine the determinants of two transforms: The first 2D transform encodes a scaling of 2, and results in an overall area scaling of 4:

$$\begin{bmatrix} 2 & 0 \\ 0 & 2 \end{bmatrix}$$

and the determinant is

$$\begin{vmatrix} 2 & 0 \\ 0 & 2 \end{vmatrix} = 4.$$

The second 2D transform encodes a scaling of 3 and a translation of $(3, \ 3)$, and results in an overall area scaling of 9:

$$\begin{bmatrix} 3 & 0 & 3 \\ 0 & 3 & 3 \\ 0 & 0 & 1 \end{bmatrix}$$

and the determinant is

$$3 \begin{vmatrix} 3 & 3 \\ 0 & 1 \end{vmatrix} - 0 \begin{vmatrix} 0 & 3 \\ 0 & 1 \end{vmatrix} + 0 \begin{vmatrix} 0 & 3 \\ 3 & 3 \end{vmatrix} = 9.$$

These two examples demonstrate the extra role played by the elements of a matrix.

10.12   Perspective Projection

Of all the projections employed in computer graphics, the *perspective projection* is one most widely used. There are two stages to its computation: the first involves converting world coordinates to the camera's frame of reference, and the second transforms camera coordinates to the projection plane coordinates. We have already looked at the transforms for locating a camera in world space, and the inverse transform for converting world coordinates to the camera's frame of reference. Let's now investigate how these camera coordinates are transformed into a perspective projection.

We begin by assuming that the camera is directed along the z-axis as shown in Fig. 10.29. Positioned d units along the z-axis is a projection screen, which is used to capture a perspective projection of an object. Figure 10.29 shows that any point $(x_c,\ y_c,\ z_c)$ is transformed to $(x_p,\ y_p,\ d)$. It also shows that the screen's x-axis is pointing in the opposite direction to the camera's x-axis, which can be compensated for by reversing the sign of x_p when it is computed.

Figure 10.30 shows a plan view of the scenario depicted in Figs. 10.29 and 10.31 a side view. Next, we reverse the sign of x_p and state:

Fig. 10.29 The axial system used to produce a perspective view

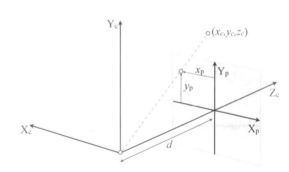

Fig. 10.30 The plan view of the camera's axial system

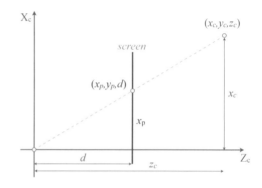

Fig. 10.31 The side view of the camera's axial system

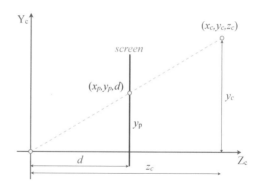

$$\frac{x_c}{z_c} = \frac{-x_p}{d}$$

$$x_p = \frac{-x_c}{z_c/d}$$

and

$$\frac{y_c}{z_c} = \frac{y_p}{d}$$

$$y_p = \frac{y_c}{z_c/d}.$$

This is expressed in matrix form as

$$\begin{bmatrix} x_p \\ y_p \\ z_p \\ w \end{bmatrix} = \begin{bmatrix} -1 & 0 & 0 & 0 \\ 0 & 1 & 0 & 0 \\ 0 & 0 & 1 & 0 \\ 0 & 0 & 1/d & 0 \end{bmatrix} \begin{bmatrix} x_c \\ y_c \\ z_c \\ 1 \end{bmatrix}.$$

At first the transform seems strange, but if we multiply this out we get

$$[x_p \quad y_p \quad z_p \quad w]^\mathrm{T} = [-x_c \quad y_c \quad z_c \quad z_c/d]^\mathrm{T}$$

and if we remember the idea behind homogeneous coordinates, we must divide the terms x_p, y_p, z_p by w to get the scaled terms, which produces

$$x_p = \frac{-x_c}{z_c/d}$$

$$y_p = \frac{y_c}{z_c/d}$$

$$z_p = \frac{z_c}{z_c/d} = d$$

which, after all, is rather elegant. Notice that this transform takes into account the sign change that occurs with the x-coordinate. Some algorithms delay this sign reversal until the mapping is made to screen coordinates.

10.13   Summary

The purpose of this chapter was to introduce the reader to transforms and matrices—I hope this has been achieved. This is not the end of the subject, as one can do so much with matrices. For example, it would be interesting to see how a matrix behaves when some of its elements are changed dynamically.

10.14   Worked Examples

10.14.1   2D Scaling Transform

State the 2D homogeneous matrix to scale by a factor of 2 in the x-direction and 3 in the y-direction.

$$\begin{bmatrix} x' \\ y' \\ 1 \end{bmatrix} = \begin{bmatrix} 2 & 0 & 0 \\ 0 & 3 & 0 \\ 0 & 0 & 1 \end{bmatrix} \begin{bmatrix} x \\ y \\ 1 \end{bmatrix}.$$

10.14.2   2D Scale and Translate

Given matrix $\mathbf{T}_1$ which scales a 2D point by a factor of 2, and $\mathbf{T}_2$ which translates a 2D point by $x = 2$ and $y = 2$, combine them in two possible ways and show that the point $(1,\ 1)$ is transformed to two different places.

$$\mathbf{T}_1 = \begin{bmatrix} x' \\ y' \\ 1 \end{bmatrix} = \begin{bmatrix} 2 & 0 & 0 \\ 0 & 2 & 0 \\ 0 & 0 & 1 \end{bmatrix} \begin{bmatrix} x \\ y \\ 1 \end{bmatrix}$$

$$\mathbf{T}_2 = \begin{bmatrix} x' \\ y' \\ 1 \end{bmatrix} = \begin{bmatrix} 1 & 0 & 2 \\ 0 & 1 & 2 \\ 0 & 0 & 1 \end{bmatrix} \begin{bmatrix} x \\ y \\ 1 \end{bmatrix}$$

$$\mathbf{T}_1 \mathbf{T}_2 = \begin{bmatrix} x' \\ y' \\ 1 \end{bmatrix} = \begin{bmatrix} 2 & 0 & 0 \\ 0 & 2 & 0 \\ 0 & 0 & 1 \end{bmatrix} \begin{bmatrix} 1 & 0 & 2 \\ 0 & 1 & 2 \\ 0 & 0 & 1 \end{bmatrix} \begin{bmatrix} x \\ y \\ 1 \end{bmatrix}$$

$$= \begin{bmatrix} 2 & 0 & 4 \\ 0 & 2 & 4 \\ 0 & 0 & 1 \end{bmatrix} \begin{bmatrix} x \\ y \\ 1 \end{bmatrix}$$

and the point (1, 1) is transformed to (6, 6).

$$\mathbf{T_2T_1} = \begin{bmatrix} x' \\ y' \\ 1 \end{bmatrix} = \begin{bmatrix} 1 & 0 & 2 \\ 0 & 1 & 2 \\ 0 & 0 & 1 \end{bmatrix} \begin{bmatrix} 2 & 0 & 0 \\ 0 & 2 & 0 \\ 0 & 0 & 1 \end{bmatrix} \begin{bmatrix} x \\ y \\ 1 \end{bmatrix}$$

$$= \begin{bmatrix} 2 & 0 & 2 \\ 0 & 2 & 2 \\ 0 & 0 & 1 \end{bmatrix} \begin{bmatrix} x \\ y \\ 1 \end{bmatrix}$$

and the point (1, 1) is transformed to (4, 4).

10.14.3 3D Scaling Transform

Derive the 3D homogeneous matrix to scale by a factor of 2 in the x-direction, 3 in the y-direction and 4 in the z-direction, relative to the point (1, 1, 1), and compute the transformed position of (2, 2, 2).

$$\begin{bmatrix} x' \\ y' \\ z' \\ 1 \end{bmatrix} = \begin{bmatrix} s_x & 0 & 0 & p_x(1-s_x) \\ 0 & s_y & 0 & p_y(1-s_y) \\ 0 & 0 & s_z & p_z(1-s_z) \\ 0 & 0 & 0 & 1 \end{bmatrix} \begin{bmatrix} x \\ y \\ z \\ 1 \end{bmatrix}.$$

Substituting the given values:

$$\begin{bmatrix} x' \\ y' \\ z' \\ 1 \end{bmatrix} = \begin{bmatrix} 2 & 0 & 0 & -1 \\ 0 & 3 & 0 & -2 \\ 0 & 0 & 4 & -3 \\ 0 & 0 & 0 & 1 \end{bmatrix} \begin{bmatrix} x \\ y \\ z \\ 1 \end{bmatrix}.$$

The point (2, 2, 2) is transformed to (3, 4, 5):

$$\begin{bmatrix} 3 \\ 4 \\ 5 \\ 1 \end{bmatrix} = \begin{bmatrix} 2 & 0 & 0 & -1 \\ 0 & 3 & 0 & -2 \\ 0 & 0 & 4 & -3 \\ 0 & 0 & 0 & 1 \end{bmatrix} \begin{bmatrix} 2 \\ 2 \\ 2 \\ 1 \end{bmatrix}.$$

10.14.4 2D Rotation

Compute the coordinates of the unit square in Table 10.1 after a rotation of 90°. The points are rotated as follows:

Table 10.1 Original and rotated coordinates of the unit square

x	y	x'	y'
0	0	0	0
1	0	0	1
1	1	−1	1
0	1	−1	0

$$\begin{bmatrix} x' \\ y' \\ 1 \end{bmatrix} = \begin{bmatrix} \cos\beta & -\sin\beta & 0 \\ \sin\beta & \cos\beta & 0 \\ 0 & 0 & 1 \end{bmatrix} \begin{bmatrix} x \\ y \\ 1 \end{bmatrix}$$

$$= \begin{bmatrix} 0 & -1 & 0 \\ 1 & 0 & 0 \\ 0 & 0 & 1 \end{bmatrix} \begin{bmatrix} x \\ y \\ 1 \end{bmatrix}$$

$$\begin{bmatrix} 0 \\ 0 \\ 1 \end{bmatrix} = \begin{bmatrix} 0 & -1 & 0 \\ 1 & 0 & 0 \\ 0 & 0 & 1 \end{bmatrix} \begin{bmatrix} 0 \\ 0 \\ 1 \end{bmatrix}$$

$$\begin{bmatrix} 0 \\ 1 \\ 1 \end{bmatrix} = \begin{bmatrix} 0 & -1 & 0 \\ 1 & 0 & 0 \\ 0 & 0 & 1 \end{bmatrix} \begin{bmatrix} 1 \\ 0 \\ 1 \end{bmatrix}$$

$$\begin{bmatrix} -1 \\ 1 \\ 1 \end{bmatrix} = \begin{bmatrix} 0 & -1 & 0 \\ 1 & 0 & 0 \\ 0 & 0 & 1 \end{bmatrix} \begin{bmatrix} 1 \\ 1 \\ 1 \end{bmatrix}$$

$$\begin{bmatrix} -1 \\ 0 \\ 1 \end{bmatrix} = \begin{bmatrix} 0 & -1 & 0 \\ 1 & 0 & 0 \\ 0 & 0 & 1 \end{bmatrix} \begin{bmatrix} 0 \\ 1 \\ 1 \end{bmatrix}.$$

10.14.5   2D Rotation About a Point

Derive the 2D homogeneous matrix to rotate 180° about (−1, 0), and compute the transformed position of (0, 0).

$$\begin{bmatrix} x' \\ y' \\ 1 \end{bmatrix} = \begin{bmatrix} \cos\beta & -\sin\beta & p_x(1-\cos\beta)+p_y\sin\beta \\ \sin\beta & \cos\beta & p_y(1-\cos\beta)-p_x\sin\beta \\ 0 & 0 & 1 \end{bmatrix} \begin{bmatrix} x \\ y \\ 1 \end{bmatrix}$$

$$= \begin{bmatrix} \cos 180° & -\sin 180° & -1(1-\cos 180°)+0\sin 180° \\ \sin 180° & \cos 180° & 0(1-\cos 180°)+1\sin 180° \\ 0 & 0 & 1 \end{bmatrix} \begin{bmatrix} x \\ y \\ 1 \end{bmatrix}$$

$$\begin{bmatrix} -2 \\ 0 \\ 1 \end{bmatrix} = \begin{bmatrix} -1 & 0 & -2 \\ 0 & -1 & 0 \\ 0 & 0 & 1 \end{bmatrix} \begin{bmatrix} 0 \\ 0 \\ 1 \end{bmatrix}.$$

The point (0, 0) is rotated to (−2, 0).

10.14.6 Determinant of the Rotate Transform

Using determinants, show that the rotate transform preserves area.
    The determinant of a 2D matrix transform reflects the area change produced by the transform. Therefore, if area is preserved, the determinant must equal 1. Using Sarrus's rule:

$$\left|\begin{bmatrix} \cos\beta & -\sin\beta & 0 \\ \sin\beta & \cos\beta & 0 \\ 0 & 0 & 1 \end{bmatrix}\right| = \cos^2\beta + \sin^2\beta = 1$$

which confirms the role of the determinant.

10.14.7 Determinant of the Shear Transform

Using determinants, show that the shear transform preserves area.
    The determinant of a 2D matrix transform reflects the area change produced by the transform. Therefore, if area is preserved, the determinant must equal 1. Using Sarrus's rule:

$$\left|\begin{bmatrix} 1 & \tan\beta & 0 \\ 0 & 1 & 0 \\ 0 & 0 & 1 \end{bmatrix}\right| = 1$$

which confirms the role of the determinant.

10.14.8 Yaw, Pitch and Roll Transforms

Using the yaw and pitch transforms in the sequence $yaw \times pitch$, compute how the point $(1,\ 1,\ 1)$ is transformed with $yaw = pitch = 90°$.

$$\begin{bmatrix} x' \\ y' \\ 1 \end{bmatrix} = \begin{bmatrix} \cos yaw & 0 & \sin yaw & 0 \\ 0 & 1 & 0 & 0 \\ -\sin yaw & 0 & \cos yaw & 0 \\ 0 & 0 & 0 & 1 \end{bmatrix} \begin{bmatrix} 1 & 0 & 0 & 0 \\ 0 & \cos pitch & -\sin pitch & 0 \\ 0 & \sin pitch & \cos pitch & 0 \\ 0 & 0 & 0 & 1 \end{bmatrix} \begin{bmatrix} x \\ y \\ z \\ 1 \end{bmatrix}$$

$$= \begin{bmatrix} 0 & 0 & 1 & 0 \\ 0 & 1 & 0 & 0 \\ -1 & 0 & 0 & 0 \\ 0 & 0 & 0 & 1 \end{bmatrix} \begin{bmatrix} 1 & 0 & 0 & 0 \\ 0 & 0 & -1 & 0 \\ 0 & 1 & 0 & 0 \\ 0 & 0 & 0 & 1 \end{bmatrix} \begin{bmatrix} x \\ y \\ z \\ 1 \end{bmatrix}$$

$$\begin{bmatrix} 1 \\ -1 \\ -1 \\ 1 \end{bmatrix} = \begin{bmatrix} 0 & 1 & 0 & 0 \\ 0 & 0 & -1 & 0 \\ -1 & 0 & 0 & 0 \\ 0 & 0 & 0 & 1 \end{bmatrix} \begin{bmatrix} 1 \\ 1 \\ 1 \\ 1 \end{bmatrix}$$

therefore, $(1, \ 1, \ 1)$ is transformed to $(1, \ -1, \ -1)$.

10.14.9   3D Rotation About an Axis

Derive a homogeneous matrix to rotate $(-1, \ 1, \ 0)$, $270°$ about an axis parallel to the y-axis, and intersecting $(1, \ 0, \ 0)$.

$$
\begin{bmatrix} x' \\ y' \\ z' \\ 1 \end{bmatrix} = \begin{bmatrix} \cos\beta & 0 & \sin\beta & p_x(1-\cos\beta) - p_z\sin\beta \\ 0 & 1 & 0 & 0 \\ -\sin\beta & 0 & \cos\beta & p_z(1-\cos\beta) + p_x\sin\beta \\ 0 & 0 & 0 & 1 \end{bmatrix} \begin{bmatrix} x \\ y \\ z \\ 1 \end{bmatrix}
$$

$$
= \begin{bmatrix} \cos 270° & 0 & \sin 270° & 1(1-\cos 270°) - 0\sin 270° \\ 0 & 1 & 0 & 0 \\ -\sin 270° & 0 & \cos 270° & 0(1-\cos 270°) + 1\sin 270° \\ 0 & 0 & 0 & 1 \end{bmatrix} \begin{bmatrix} x \\ y \\ z \\ 1 \end{bmatrix}
$$

$$
= \begin{bmatrix} 0 & 0 & -1 & 1(1-0) \\ 0 & 1 & 0 & 0 \\ 1 & 0 & 0 & 0(1-0) - 1 \\ 0 & 0 & 0 & 1 \end{bmatrix} \begin{bmatrix} x \\ y \\ z \\ 1 \end{bmatrix}
$$

$$
= \begin{bmatrix} 0 & 0 & -1 & 1 \\ 0 & 1 & 0 & 0 \\ 1 & 0 & 0 & -1 \\ 0 & 0 & 0 & 1 \end{bmatrix} \begin{bmatrix} x \\ y \\ z \\ 1 \end{bmatrix}
$$

$$
\begin{bmatrix} 1 \\ 1 \\ -2 \\ 1 \end{bmatrix} = \begin{bmatrix} 0 & 0 & -1 & 1 \\ 0 & 1 & 0 & 0 \\ 1 & 0 & 0 & -1 \\ 0 & 0 & 0 & 1 \end{bmatrix} \begin{bmatrix} -1 \\ 1 \\ 0 \\ 1 \end{bmatrix}.
$$

The point $(-1, \ 1, \ 0)$ is rotated to $(1, \ 1, \ -2)$.

10.14.10   3D Rotation Transform Matrix

Show that the matrix for rotating a point about an arbitrary axis corresponds to the three matrices for rotating about the x-, y- and z-axis.

$$
\begin{bmatrix} a^2K + \cos\alpha & abK - c\sin\alpha & acK + b\sin\alpha & 0 \\ abK + c\sin\alpha & b^2K + \cos\alpha & bcK - a\sin\alpha & 0 \\ acK - b\sin\alpha & bcK + a\sin\alpha & c^2K + \cos\alpha & 0 \\ 0 & 0 & 0 & 1 \end{bmatrix}
$$

Pitch about the x-axis: $\hat{\mathbf{n}} = \mathbf{i}$, where $a = 1$ and $b = c = 0$; $K = 1 - \cos\alpha$.

$$\text{pitch} = \begin{bmatrix} 1 & 0 & 0 & 0 \\ 0 & \cos\alpha & -\sin\alpha & 0 \\ 0 & \sin\alpha & \cos\alpha & 0 \\ 0 & 0 & 0 & 1 \end{bmatrix}$$

Yaw about the y-axis: $\hat{\mathbf{n}} = \mathbf{j}$, where $b = 1$ and $a = c = 0$; $K = 1 - \cos\alpha$.

$$\text{yaw} = \begin{bmatrix} \cos\alpha & 0 & \sin\alpha & 0 \\ 0 & 1 & 0 & 0 \\ -\sin\alpha & 0 & \cos\alpha & 0 \\ 0 & 0 & 0 & 1 \end{bmatrix}$$

Roll about the z-axis: $\hat{\mathbf{n}} = \mathbf{k}$, where $c = 1$ and $a = b = 0$; $K = 1 - \cos\alpha$.

$$\text{roll} = \begin{bmatrix} \cos\alpha & -\sin\alpha & 0 & 0 \\ \sin\alpha & \cos\alpha & 0 & 0 \\ 0 & 0 & 1 & 0 \\ 0 & 0 & 0 & 1 \end{bmatrix}.$$

10.14.11 2D Change of Axes

Derive a 2D homogeneous matrix to compute $(1,\ 1)$ in an axial system with direction cosines $\cos\beta = \sqrt{2}/2$ and $\sin\beta = -\sqrt{2}/2$.

$$\begin{bmatrix} x' \\ y' \\ 1 \end{bmatrix} = \begin{bmatrix} \cos\beta & \sin\beta \\ -\sin\beta & \cos\beta \end{bmatrix} \begin{bmatrix} x \\ y \\ 1 \end{bmatrix}$$

$$= \begin{bmatrix} \sqrt{2}/2 & -\sqrt{2}/2 \\ \sqrt{2}/2 & \sqrt{2}/2 \end{bmatrix} \begin{bmatrix} 1 \\ 1 \\ 1 \end{bmatrix}$$

$$\begin{bmatrix} 0 \\ \sqrt{2} \\ 1 \end{bmatrix} = \begin{bmatrix} \sqrt{2}/2 & -\sqrt{2}/2 \\ \sqrt{2}/2 & \sqrt{2}/2 \end{bmatrix} \begin{bmatrix} 1 \\ 1 \\ 1 \end{bmatrix}.$$

The point $(1,\ 1)$ has coordinates $(0,\ \sqrt{2})$ in the rotated axial system.

10.14.12   3D Change of Axes

Derive a 3D homogeneous matrix to compute the positions of $(0, \ 0, \ 0)$ and $(0, \ 1, \ 0)$ in an axial system with $180°$ yaw, $0°$ pitch, $180°$ roll, and translated by $(10, \ 0, \ 0)$.

$$
\begin{bmatrix} x' \\ y' \\ z' \\ 1 \end{bmatrix} = \begin{bmatrix} T_{11} & T_{12} & T_{13} & T_{14} \\ T_{21} & T_{22} & T_{23} & T_{24} \\ T_{31} & T_{32} & T_{33} & T_{34} \\ T_{41} & T_{42} & T_{43} & T_{44} \end{bmatrix} \begin{bmatrix} x \\ y \\ z \\ 1 \end{bmatrix}
$$

where

$$T_{11} = \cos(yaw)\cos(roll) + \sin(yaw)\sin(pitch)\sin(roll)$$
$$T_{12} = \cos(pitch)\sin(roll)$$
$$T_{13} = -\sin(yaw)\cos(roll) + \cos(yaw)\sin(pitch)\sin(roll)$$
$$T_{14} = -(t_x T_{11} + t_y T_{12} + t_z T_{13})$$
$$T_{21} = -\cos(yaw)\sin(roll) + \sin(yaw)\sin(pitch)\cos(roll)$$
$$T_{22} = \cos(pitch)\cos(roll)$$
$$T_{23} = -\sin(yaw)\sin(roll) + \cos(yaw)\sin(pitch)\cos(roll)$$
$$T_{24} = -(t_x T_{21} + t_y T_{22} + t_z T_{23})$$
$$T_{31} = \sin(yaw)\cos(pitch)$$
$$T_{32} = -\sin(pitch)$$
$$T_{33} = \cos(yaw)\cos(pitch)$$
$$T_{34} = -(t_x T_{31} + t_y T_{32} + t_z T_{33})$$
$$T_{41} = T_{42} = T_{43} = 0$$
$$T_{44} = 1.$$

Substituting the above values:

$$T_{11} = \cos 180° \cos 180° + \sin 180° \sin 0° \sin 180° = 1$$
$$T_{12} = \cos 0° \sin 180° = 0$$
$$T_{13} = -\sin 180° \cos 180° + \cos 180° \sin 0° \sin 180° = 0$$
$$T_{14} = -(-10T_{11} + 0T_{12} + 0T_{13}) = 10$$
$$T_{21} = -\cos 180° \sin 180° + \sin 180° \sin 0° \cos 180° = 0$$
$$T_{22} = \cos 0° \cos 180° = -1$$
$$T_{23} = -\sin 180° \sin 180° + \cos 180° \sin 0° \cos 180° = 0$$
$$T_{24} = -(-10T_{21} + 0T_{22} + 0T_{23}) = 0$$
$$T_{31} = \sin 180° \cos 0° = 0$$
$$T_{32} = -\sin 0° = 0$$

$$T_{33} = \cos 180° \cos 0° = -1$$
$$T_{34} = -(-10T_{31} + 0T_{32} + 0T_{33}) = 0$$
$$T_{41} = T_{42} = T_{43} = 0$$
$$T_{44} = 1.$$

Therefore:

$$\begin{bmatrix} 10 \\ 0 \\ 0 \\ 1 \end{bmatrix} = \begin{bmatrix} 1 & 0 & 0 & 10 \\ 0 & -1 & 0 & 0 \\ 0 & 0 & -1 & 0 \\ 0 & 0 & 0 & 0 \end{bmatrix} \begin{bmatrix} 0 \\ 0 \\ 0 \\ 1 \end{bmatrix}$$

and

$$\begin{bmatrix} 10 \\ -1 \\ 0 \\ 1 \end{bmatrix} = \begin{bmatrix} 1 & 0 & 0 & 10 \\ 0 & -1 & 0 & 0 \\ 0 & 0 & -1 & 0 \\ 0 & 0 & 0 & 0 \end{bmatrix} \begin{bmatrix} 0 \\ 1 \\ 0 \\ 1 \end{bmatrix}.$$

The positions of $(0, 0, 0)$ and $(0, 1, 0)$ in the transformed axial system are $(10, 0, 0)$ and $(10, -1, 0)$ respectively.

10.14.13 Rotate a Point About an Axis

Derive a 3D homogeneous matrix to rotate $(1, 0, 0)$, $180°$ about an axis whose parallel vector is $\hat{\mathbf{n}} = 1/\sqrt{2}\mathbf{j} + 1/\sqrt{2}\mathbf{k}$.

Given

$$\begin{bmatrix} x'_p \\ y'_p \\ z'_p \\ 1 \end{bmatrix} = \begin{bmatrix} a^2 K + \cos\alpha & abK - c\sin\alpha & acK + b\sin\alpha & 0 \\ abK + c\sin\alpha & b^2 K + \cos\alpha & bcK - a\sin\alpha & 0 \\ acK - b\sin\alpha & bcK + a\sin\alpha & c^2 K + \cos\alpha & 0 \\ 0 & 0 & 0 & 1 \end{bmatrix} \begin{bmatrix} x_p \\ y_p \\ z_p \\ 1 \end{bmatrix}$$

where

$$K = 1 - \cos\alpha.$$

Therefore,

$$\begin{bmatrix} -1 \\ 0 \\ 0 \\ 1 \end{bmatrix} = \begin{bmatrix} -1 & 0 & 0 & 0 \\ 0 & 0 & 1 & 0 \\ 0 & 1 & 0 & 0 \\ 0 & 0 & 0 & 1 \end{bmatrix} \begin{bmatrix} 1 \\ 0 \\ 0 \\ 1 \end{bmatrix}.$$

The rotated point is $(-1, 0, 0)$.

Table 10.2   Coordinates of a 3D cube

Vertex	x_c	y_c	z_c	x_p	y_p
1	0	0	10	0	0
2	10	0	10	20	0
3	10	10	10	20	20
4	0	10	10	0	20
5	0	0	20	0	0
6	10	0	20	10	0
7	10	10	20	10	10
8	0	10	20	0	10

Fig. 10.32 A perspective
sketch of a 3D cube

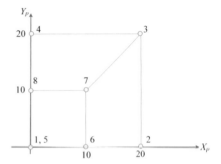

10.14.14 *Perspective Projection*

Compute the perspective coordinates of a 3D cube stored in Table 10.2 with the
projection screen distance $d = 20$. Sketch the result.

   Using the perspective transform:

$$\begin{bmatrix} x_p \\ y_p \\ z_p \\ w \end{bmatrix} = \begin{bmatrix} -1 & 0 & 0 & 0 \\ 0 & 1 & 0 & 0 \\ 0 & 0 & 1 & 0 \\ 0 & 0 & 1/d & 0 \end{bmatrix} \begin{bmatrix} x_c \\ y_c \\ z_c \\ 1 \end{bmatrix}.$$

the perspective coordinates are stored in Table 10.2, and Fig. 10.32 shows a sketch
of the result.

Chapter 11
Quaternion Algebra

11.1 Introduction

This chapter contains some historical background to the invention of quaternions, and covers the evolution of quaternion algebra. I show how quaternion algebra is greatly simplified by treating a quaternion as an ordered pair, and provide examples of addition, subtraction, real, pure and unit quaternions. After defining the complex conjugate, norm, quaternion product, square and inverse, I show how a quaternion is represented by a matrix. The chapter concludes with a summary of the important definitions and several worked examples.

11.2 Some History

A complex number is defined as

$$z = a + ib, \quad a, \ b \in \mathbb{R}, \quad i^2 = -1.$$

Complex numbers can be regarded as a 2D point, which begs the question: is there a complex object for a 3D point? After many years of thinking, Sir Willian Rowan Hamilton found the answer in the form of a *quaternion*.

Hamilton defined a quaternion q, and its associated rules as

$$q = s + ia + jb + kc, \quad s, \ a, \ b, \ c \in \mathbb{R}$$

where,

$$i^2 = j^2 = k^2 = ijk = -1$$

$$ij = k, \quad jk = i, \quad ki = j$$
$$ji = -k, \ kj = -i, \ ik = -j$$

© Springer-Verlag London Ltd., part of Springer Nature 2022
J. Vince, *Mathematics for Computer Graphics*, Undergraduate Topics
in Computer Science, https://doi.org/10.1007/978-1-4471-7520-9_11

[1–3], but we tend to write quaternions as

$$q = s + ai + bj + ck.$$

Observe from Hamilton's rules how the occurrence of ij is replaced by k. The extra imaginary k term is key to the cyclic patterns $ij = k$, $jk = i$, and $ki = j$, which are very similar to the cross product of two unit Cartesian vectors:

$$\mathbf{i} \times \mathbf{j} = \mathbf{k}, \quad \mathbf{j} \times \mathbf{k} = \mathbf{i}, \quad \mathbf{k} \times \mathbf{i} = \mathbf{j}.$$

In fact, this similarity is no coincidence, as Hamilton also invented the scalar and vector products. However, although quaternions provided an algebraic framework to describe vectors, one must acknowledge that vectorial quantities had been studied for many years prior to Hamilton.

Hamilton also saw that the i, j, k terms could represent three Cartesian unit vectors $\mathbf{i}$, $\mathbf{j}$ and $\mathbf{k}$, which had to possess imaginary qualities. i.e. $\mathbf{i}^2 = -1$, etc., which didn't go down well with some mathematicians and scientists who were suspicious of the need to involve so many imaginary terms.

Hamilton's motivation to search for a 3D equivalent of complex numbers was part algebraic, and part geometric. For if a complex number is represented by a couple and is capable of rotating points on the plane by 90°, then perhaps a *triple* rotates points in space by 90°. In the end, a triple had to be replaced by a a quadruple—a quaternion.

One can regard Hamilton's rules from two perspectives. The first, is that they are an algebraic consequence of combining three imaginary terms. The second, is that they reflect an underlying geometric structure of space. The latter interpretation was adopted by the Scottish mathematical physicist Peter Guthrie Tait (1831–1901), and outlined in his book *An Elementary Treatise on Quaternions*. Tait's approach assumes three unit vectors $\mathbf{i}$, $\mathbf{j}$, $\mathbf{k}$ aligned with the x-, y-, z-axes respectively:

> The result of the multiplication of $\mathbf{i}$ into $\mathbf{j}$ or $\mathbf{ij}$ is defined to be the turning of $\mathbf{j}$ through a right angle in the plane perpendicular to $\mathbf{i}$ in the positive direction, in other words, the operation of $\mathbf{i}$ on $\mathbf{j}$ turns it round so as to make it coincide with $\mathbf{k}$; and therefore briefly $\mathbf{ij} = \mathbf{k}$.

> To be consistent it is requisite to admit that if $\mathbf{i}$ instead of operating on $\mathbf{j}$ had operated on any other unit vector perpendicular to $\mathbf{i}$ in the plane yz, it would have turned it through a right-angle in the same direction, so that $\mathbf{ik}$ can be nothing else than $-\mathbf{j}$.

> Extending to other unit vectors the definition which we have illustrated by referring to $\mathbf{i}$, it is evident that $\mathbf{j}$ operating on $\mathbf{k}$ must bring it round to $\mathbf{i}$, or $\mathbf{jk} = \mathbf{i}$. [4]

Tait's explanation is illustrated in Fig. 11.1a–d. Figure 11.1a shows the original alignment of $\mathbf{i}$, $\mathbf{j}$, $\mathbf{k}$. Figure 11.1b shows the effect of turning $\mathbf{j}$ into $\mathbf{k}$. Figure 11.1c shows the turning of $\mathbf{k}$ into $\mathbf{i}$, and Fig. 11.1d shows the turning of $\mathbf{i}$ in to $\mathbf{j}$.

So far, there is no mention of imaginary quantities—we just have:

$$\mathbf{ij} = \mathbf{k}, \quad \mathbf{jk} = \mathbf{i}, \quad \mathbf{ki} = \mathbf{j}$$
$$\mathbf{ji} = -\mathbf{k}, \quad \mathbf{kj} = -\mathbf{i}, \quad \mathbf{ik} = -\mathbf{j}.$$

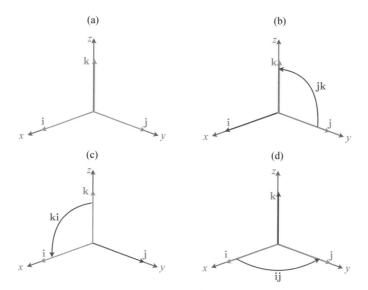

Fig. 11.1 Interpreting the products **jk, ki, ij**

If we assume that these vectors obey the distributive and associative axioms of algebra, their imaginary qualities are exposed. For example:

$$\mathbf{ij} = \mathbf{k}$$

and multiplying throughout by **i**:

$$\mathbf{iij} = \mathbf{ik} = -\mathbf{j}$$

therefore,

$$\mathbf{ii} = \mathbf{i}^2 = -1.$$

Similarly, we can show that $\mathbf{j}^2 = \mathbf{k}^2 = -1$.
    Next:

$$\mathbf{ijk} = \mathbf{i}(\mathbf{jk}) = \mathbf{ii} = \mathbf{i}^2 = -1.$$

Thus, simply by declaring the action of the cross-product, Hamilton's rules emerge, with all of their imaginary features. Tait also made the following observation:

A very curious speculation, due to Servois, and published in 1813 in Gergonne's Annales is the only one, so far has been discovered, in which the slightest trace of an anticipation of Quaternions is contained. Endeavouring to extend to *space* the form $a + b\sqrt{-1}$ for the plane, he is guided by analogy to write a directed unit-line in space the form

$$p \cos \alpha + q \cos \beta + r \cos \gamma,$$

where α, β, γ are its inclinations to the three axes. He perceives easily that p, q, r must be *non-reals* : but, he asks, "seraient-elles imaginaires réductibles à la forme générale $A + B\sqrt{-1}$?" This could not be the answer. In fact they are the **i**, **j**, **k** of the Quaternion Calculus. [4]

So the French mathematician François-Joseph Servois (1768–1847), was another person who came very close to discovering quaternions. Furthermore, both Tait and Hamilton were apparently unaware of a paper on transformation groups published by the French banker and mathematician Olinde Rodrigues (1795–1851) in 1840.

And it doesn't stop there: the brilliant mathematician Carl Friedrich Gauss was extremely cautious, and nervous of publishing anything too revolutionary, just in case he was ridiculed by fellow mathematicians. His diaries reveal that he had anticipated non-euclidean geometry ahead of the Russian mathematician Nikolai Lobachevsky (1792–1856). And in a short note from his diary in 1819 [5] he reveals that he had identified a method of finding the product of two quadruples (a, b, c, d) and $(\alpha, \beta, \gamma, \delta)$ as:

$$\begin{aligned}(A, B, C, D) &= (a, b, c, d)(\alpha, \beta, \gamma, \delta)\\ &= (a\alpha - b\beta - c\gamma - d\delta, \ a\beta + b\alpha - c\delta + d\gamma,\\ &\quad a\gamma + b\delta + c\alpha - d\beta, \ a\delta - b\gamma + c\beta + d\alpha).\end{aligned}$$

At first glance, this result does not look like a quaternion product, but if we transpose the second and third coordinates of the quadruples, and treat them as quaternions, we have:

$$\begin{aligned}(A, B, C, D) &= (a + ci + bj + dk)(\alpha + \gamma i + \beta j + \delta k)\\ &= a\alpha - c\gamma - b\beta - d\delta + a(\gamma i + \beta j + \delta k)\\ &\quad + \alpha(ci + bj + dk), \ (b\delta - d\beta)i + (d\gamma - c\delta)j + (c\beta - b\gamma)k\end{aligned}$$

which is identical to Hamilton's quaternion product! Furthermore, Gauss also realised that the product was non-commutative. However, he did not publish his findings, and it was left to Hamilton to invent quaternions for himself, publish his results and take the credit.

In 1881 and 1884, Josiah Willard Gibbs, at Yale University, printed his lecture notes on vector analysis for his students. Gibbs had cut the 'umbilical cord' between the real and vector parts of a quaternion and raised the 3D vector as an independent object without any imaginary connotations. Gibbs also took on board the ideas of Grassmann, who had been developing his own ideas for a vectorial system since 1832. Gibbs also defined the scalar and vector products using the relevant parts of the quaternion product. Finally, in 1901, a student of Gibbs, Edwin Bidwell Wilson, published Gibbs' notes in book form: *Vector Analysis* [6], which contains the notation in use today.

Quaternion algebra is definitely imaginary, yet simply by isolating the vector part and ignoring the imaginary rules, Gibbs was able to reveal a new branch of mathematics that exploded into vector analysis.

Hamilton and his supporters were unable to persuade their peers that quaternions could represent vectorial quantities, and eventually, Gibbs' notation won the day, and quaternions faded from the scene.

In recent years, quaternions have been rediscovered by the flight simulation industry, and more recently by the computer graphics community, where they are used to rotate vectors about an arbitrary axis. In the intervening years, various people have had the opportunity to investigate the algebra, and propose new ways of harnessing its qualities.

So let's look at three ways of annotating a quaternion q:

$$q = s + xi + yj + zk \qquad\qquad (11.1)$$
$$q = s + \mathbf{v} \qquad\qquad (11.2)$$
$$q = [s, \ \mathbf{v}] \qquad\qquad (11.3)$$
$$\text{where } s, \ x, \ y, \ z \in \mathbb{R}, \quad \mathbf{v} \in \mathbb{R}^3$$
$$\text{and } i^2 = j^2 = k^2 = -1.$$

The difference is rather subtle. In (11.1) we have Hamilton's original definition with its imaginary terms and associated rules. In (11.2) a '+' sign is used to add a scalar to a vector, which seems strange, yet works. In (11.3) we have an ordered pair comprising a scalar and a vector.

Now you may be thinking: How is it possible to have three different definitions for the same object? Well, I would argue that you can call an object whatever you like, so long as they are algebraically identical. For example, matrix notation is used to represent a set of linear equations, and leads to the same results as every-day equations. Therefore, both systems of notation are equally valid.

Although I have employed the notation in (11.1) and (11.2) in other publications, in this book I have used ordered pairs. So what we need to show is that Hamilton's original definition of a quaternion (11.1), with its scalar and three imaginary terms, can be replaced by an ordered pair (11.3) comprising a scalar and a 'modern' vector.

11.3 Defining a Quaternion

Let's start with two quaternions q_a and q_b à la Hamilton:

$$q_a = s_a + x_a i + y_a j + z_a k$$
$$q_b = s_b + x_b i + y_b j + z_b k$$

and the obligatory rules:
$$i^2 = j^2 = k^2 = ijk = -1$$

$$ij = k, \quad jk = i, \quad ki = j$$
$$ji = -k, \ kj = -i, \ ik = -j.$$

Our objective is to show that q_a and q_b can also be represented by the ordered pairs

$$q_a = [s_a, \mathbf{a}]$$
$$q_b = [s_b, \mathbf{b}], \quad s_a, s_b \in \mathbb{R}, \quad \mathbf{a}, \mathbf{b} \in \mathbb{R}^3.$$

The quaternion product $q_a q_b$ expands to

$$
\begin{aligned}
q_a q_b = [s_a, \mathbf{a}][s_b, \mathbf{b}] &= [s_a + x_a i + y_a j + z_a k][s_b + x_b i + y_b j + z_b k] \\
&= [(s_a s_b - x_a x_b - y_a y_b - z_a z_b) \\
&\quad + (s_a x_b + s_b x_a + y_a z_b - y_b z_a)i \\
&\quad + (s_a y_b + s_b y_a + z_a x_b - z_b x_a)j \\
&\quad + (s_a z_b + s_b z_a + x_a y_b - x_b y_a)k].
\end{aligned}
\tag{11.4}
$$

Equation (11.4) takes the form of another quaternion, and confirms that the quaternion product is closed.

At this stage, Hamilton turned the imaginary terms i, j, k into unit Cartesian vectors $\mathbf{i}$, $\mathbf{j}$, $\mathbf{k}$ and transformed (11.4) into a vector form. The problem with this approach is that the vectors retain their imaginary roots. The author Simon Altmann suggests replacing the imaginaries by the ordered pairs:

$$i = [0, \mathbf{i}], \quad j = [0, \mathbf{j}], \quad k = [0, \mathbf{k}]$$

which are themselves quaternions, and called *quaternion units*.

The idea of defining a quaternion in terms of quaternion units is exactly the same as defining a vector in terms of its unit Cartesian vectors. Furthermore, it permits vectors to exist without any imaginary associations.

Let's substitute these quaternion units in (11.4) together with $[1, \mathbf{0}] = 1$:

$$
\begin{aligned}
[s_a, \mathbf{a}][s_b, \mathbf{b}] &= [(s_a s_b - x_a x_b - y_a y_b - z_a z_b)[1, \mathbf{0}] \\
&\quad + (s_a x_b + s_b x_a + y_a z_b - y_b z_a)[0, \mathbf{i}] \\
&\quad + (s_a y_b + s_b y_a + z_a x_b - z_b x_a)[0, \mathbf{j}] \\
&\quad + (s_a z_b + s_b z_a + x_a y_b - x_b y_a)[0, \mathbf{k}]].
\end{aligned}
\tag{11.5}
$$

Next, we expand (11.5) using previously defined rules:

$$
\begin{aligned}
[s_a, \mathbf{a}][s_b, \mathbf{b}] &= [[s_a s_b - x_a x_b - y_a y_b - z_a z_b, \mathbf{0}] \\
&\quad + [0, (s_a x_b + s_b x_a + y_a z_b - y_b z_a)\mathbf{i}] \\
&\quad + [0, (s_a y_b + s_b y_a + z_a x_b - z_b x_a)\mathbf{j}] \\
&\quad + [0, (s_a z_b + s_b z_a + x_a y_b - x_b y_a)\mathbf{k}]].
\end{aligned}
\tag{11.6}
$$

A vertical scan of (11.6) reveals some hidden vectors:

$$[s_a, \ \mathbf{a}][s_b, \ \mathbf{b}] = [[s_a s_b - x_a x_b - y_a y_b - z_a z_b, \ \mathbf{0}]$$
$$+ [0, \ s_a(x_b\mathbf{i} + y_b\mathbf{j} + z_b\mathbf{k}) + s_b(x_a\mathbf{i} + y_a\mathbf{j} + z_a\mathbf{k})$$
$$+ (y_a z_b - y_b z_a)\mathbf{i} + (z_a x_b - z_b x_a)\mathbf{j} + (x_a y_b - x_b y_a)\mathbf{k}]]. \quad (11.7)$$

Equation (11.7) contains two ordered pairs which can now be combined:

$$[s_a, \ \mathbf{a}][s_b, \ \mathbf{b}] = [s_a s_b - x_a x_b - y_a y_b - z_a z_b,$$
$$+ s_a(x_b\mathbf{i} + y_b\mathbf{j} + z_b\mathbf{k}) + s_b(x_a\mathbf{i} + y_a\mathbf{j} + z_a\mathbf{k})$$
$$+ (y_a z_b - y_b z_a)\mathbf{i} + (z_a x_b - z_b x_a)\mathbf{j} + (x_a y_b - x_b y_a)\mathbf{k}]. \quad (11.8)$$

If we make

$$\mathbf{a} = x_a\mathbf{i} + y_a\mathbf{j} + z_a\mathbf{k}$$
$$\mathbf{b} = x_b\mathbf{i} + y_b\mathbf{j} + z_b\mathbf{k}$$

and substitute them in (11.8) we get:

$$[s_a, \ \mathbf{a}][s_b, \ \mathbf{b}] = [s_a s_b - \mathbf{a} \cdot \mathbf{b}, \ s_a\mathbf{b} + s_b\mathbf{a} + \mathbf{a} \times \mathbf{b}] \quad (11.9)$$

which defines the quaternion product.

From now on, we don't have to worry about Hamilton's rules as they are embedded within (11.9). Furthermore, our vectors have no imaginary associations.

Although Rodrigues did not have access to Gibbs' vector notation used in (11.9), he managed to calculate the equivalent algebraic expression, which was some achievement.

11.3.1 The Quaternion Units

Using (11.9) we can check to see if the quaternion units are imaginary by squaring them:

$$i = [0, \ \mathbf{i}]$$
$$i^2 = [0, \ \mathbf{i}][0, \ \mathbf{i}]$$
$$= [\mathbf{i} \cdot \mathbf{i}, \ \mathbf{i} \times \mathbf{i}]$$
$$= [-1, \ \mathbf{0}]$$

which is a *real quaternion* and equivalent to -1, confirming that $[0, \ \mathbf{i}]$ is imaginary. Using a similar expansion we can shown that $[0, \ \mathbf{j}]$ and $[0, \ \mathbf{k}]$ have the same property.

Now let's compute the products $ij, \; jk$ and ki:

$$ij = [0, \; \mathbf{i}][0, \; \mathbf{j}]$$
$$= [-\mathbf{i} \cdot \mathbf{j}, \; \mathbf{i} \times \mathbf{j}]$$
$$= [0, \; \mathbf{k}]$$

which is the quaternion unit k.

$$jk = [0, \; \mathbf{j}][0, \; \mathbf{k}]$$
$$= [-\mathbf{j} \cdot \mathbf{k}, \; \mathbf{j} \times \mathbf{k}]$$
$$= [0, \; \mathbf{i}]$$

which is the quaternion unit i.

$$ki = [0, \; \mathbf{k}][0, \; \mathbf{i}]$$
$$= [-\mathbf{k} \cdot \mathbf{i}, \; \mathbf{k} \times \mathbf{i}]$$
$$= [0, \; \mathbf{j}]$$

which is the quaternion unit j.

Next, let's confirm that $ijk = -1$:

$$ijk = [0, \; \mathbf{i}][0, \; \mathbf{j}][0, \; \mathbf{k}]$$
$$= [0, \; \mathbf{k}][0, \; \mathbf{k}]$$
$$= [-\mathbf{k} \cdot \mathbf{k}, \; \mathbf{k} \times \mathbf{k}]$$
$$= [-1, \; \mathbf{0}]$$

which is a real quaternion equivalent to -1, confirming that $ijk = -1$.

Thus the notation of ordered pairs upholds all of Hamilton's rules. However, the last double product assumes that quaternions are associative. So let's double check to show that $(ij)k = i(jk)$:

$$i(jk) = [0, \; \mathbf{i}][0, \; \mathbf{j}][0, \; \mathbf{k}]$$
$$= [0, \; \mathbf{i}][0, \; \mathbf{i}]$$
$$= [-\mathbf{i} \cdot \mathbf{i}, \; \mathbf{i} \times \mathbf{i}]$$
$$= [-1, \; \mathbf{0}]$$

which is correct.

11.3.2 Example of Quaternion Products

Although we have yet to discover how quaternions are used to rotate vectors, let's concentrate on their algebraic traits by evaluating an example.

$$q_a = [1, \ 2\mathbf{i} + 3\mathbf{j} + 4\mathbf{k}]$$
$$q_b = [2, \ 3\mathbf{i} + 4\mathbf{j} + 5\mathbf{k}]$$

the product $q_a q_b$ is

$$
\begin{aligned}
q_a q_b &= [1, \ 2\mathbf{i} + 3\mathbf{j} + 4\mathbf{k}][2, \ 3\mathbf{i} + 4\mathbf{j} + 5\mathbf{k}] \\
&= [1 \times 2 - (2 \times 3 + 3 \times 4 + 4 \times 5), \\
&\quad 1(3\mathbf{i} + 4\mathbf{j} + 5\mathbf{k}) + 2(2\mathbf{i} + 3\mathbf{j} + 4\mathbf{k}) \\
&\quad + (3 \times 5 - 4 \times 4)\mathbf{i} - (2 \times 5 - 4 \times 3)\mathbf{j} + (2 \times 4 - 3 \times 3)\mathbf{k}] \\
&= [-36, \ 7\mathbf{i} + 10\mathbf{j} + 13\mathbf{k} - \mathbf{i} + 2\mathbf{j} - \mathbf{k}] \\
&= [-36, \ 6\mathbf{i} + 12\mathbf{j} + 12\mathbf{k}]
\end{aligned}
$$

which is another ordered pair representing a quaternion.

Having shown that Hamilton's *imaginary* notation has a vector equivalent, and can be represented as an ordered pair, we continue with this notation and describe other features of quaternions. Note that we can abandon Hamilton's rules as they are embedded within the definition of the quaternion product, and will surface in the following definitions.

11.4 Algebraic Definition

A quaternion is the ordered pair:

$$q = [s, \ \mathbf{v}], \quad s \in \mathbb{R}, \quad \mathbf{v} \in \mathbb{R}^3.$$

If we express $\mathbf{v}$ in terms of its components, we have

$$q = [s, \ x\mathbf{i} + y\mathbf{j} + z\mathbf{k}], \quad s, x, y, z \in \mathbb{R}.$$

11.5 Adding and Subtracting Quaternions

Addition and subtraction employ the following rule:

$$q_a = [s_a, \ \mathbf{a}]$$

$$q_b = [s_b, \ \mathbf{b}]$$
$$q_a \pm q_b = [s_a \pm s_b, \ \mathbf{a} \pm \mathbf{b}].$$

For example:

$$q_a = [0.5, \ 2\mathbf{i} + 3\mathbf{j} - 4\mathbf{k}]$$
$$q_b = [0.1, \ 4\mathbf{i} + 5\mathbf{j} + 6\mathbf{k}]$$
$$q_a + q_b = [0.6, \ 6\mathbf{i} + 8\mathbf{j} + 2\mathbf{k}]$$
$$q_a - q_b = [0.4, \ -2\mathbf{i} - 2\mathbf{j} - 10\mathbf{k}].$$

11.6   Real Quaternion

A *real quaternion* has a zero vector term:

$$q = [s, \ \mathbf{0}].$$

The product of two real quaternions is

$$q_a = [s_a, \ \mathbf{0}]$$
$$q_b = [s_b, \ \mathbf{0}]$$
$$q_a q_b = [s_a, \ \mathbf{0}][s_b, \ \mathbf{0}]$$
$$= [s_a s_b, \ \mathbf{0}]$$

which is another real quaternion, and shows that they behave just like real numbers:

$$[s, \ \mathbf{0}] \equiv s.$$

We have already come across this with complex numbers containing a zero imaginary term:

$$a + bi = a, \quad \text{when } b = 0.$$

11.7   Multiplying a Quaternion by a Scalar

Intuition suggests that multiplying a quaternion by a scalar should obey the rule:

$$q = [s, \ \mathbf{v}]$$
$$\lambda q = \lambda[s, \ \mathbf{v}], \quad \lambda \in \mathbb{R}$$
$$= [\lambda s, \ \lambda \mathbf{v}].$$

For example:

$$q = 3[2, \ 3\mathbf{i} + 4\mathbf{j} + 5\mathbf{k}]$$
$$= [6, \ 9\mathbf{i} + 12\mathbf{j} + 15\mathbf{k}].$$

We can confirm our intuition by multiplying a quaternion by a scalar in the form of a real quaternion:

$$q = [s, \ \mathbf{v}]$$
$$\lambda = [\lambda, \ \mathbf{0}]$$
$$\lambda q = [\lambda, \ \mathbf{0}][s, \ \mathbf{v}]$$
$$= [\lambda s, \ \lambda \mathbf{v}]$$

which is excellent confirmation.

11.8 Pure Quaternion

Hamilton defined a *pure quaternion* as one having a zero scalar term:

$$q = xi + yj + zk$$

and is just a vector, but with imaginary qualities. Simon Altmann, and others, believe that this was a serious mistake on Hamilton's part to call a quaternion with a zero real term, a vector.

The main issue is that there are two types of vectors: *polar* and *axial*, also called a *pseudovector*. Richard Feynman describes polar vectors as '*honest*' vectors [7] and represent the every-day vectors of directed lines. Whereas, axial vectors are computed from polar vectors, such as in a vector product. However, these two types of vector do not behave in the same way when transformed. For example, given two 'honest', polar vectors $\mathbf{a}$ and $\mathbf{b}$, we can compute the axial vector: $\mathbf{c} = \mathbf{a} \times \mathbf{b}$. Next, if we subject $\mathbf{a}$ and $\mathbf{b}$ to an inversion transform through the origin, such that $\mathbf{a}$ becomes $-\mathbf{a}$, and $\mathbf{b}$ becomes $-\mathbf{b}$, and compute their cross product $(-\mathbf{a}) \times (-\mathbf{b})$, we still get $\mathbf{c}$! Which implies that the axial vector $\mathbf{c}$ must not be transformed along with $\mathbf{a}$ and $\mathbf{b}$.

It could be argued that the inversion transform is not a 'proper' transform as it turns a right-handed set of axes into a left-handed set. But in physics, laws of nature are expected to work in either system. Unfortunately, Hamilton was not aware of this distinction, as he had only just invented vectors. However, in the intervening years, it has become evident that Hamilton's quaternion vector is an axial vector, and not a polar vector.

As we will see, in 3D rotations quaternions take the form

$$q = \left[\cos\left(\tfrac{\theta}{2}\right), \ \sin\left(\tfrac{\theta}{2}\right) \mathbf{v} \right]$$

where θ is the angle of rotation and $\mathbf{v}$ is the axis of rotation, and when we set $\theta = 180°$, we get

$$q = [0, \ \mathbf{v}]$$

which remains a quaternion, even though it only contains a vector part.

Consequently, we define a *pure quaternion* as

$$q = [0, \ \mathbf{v}].$$

The product of two pure quaternions is

$$q_a = [0, \ \mathbf{a}]$$
$$q_b = [0, \ \mathbf{b}]$$
$$q_a q_b = [0, \ \mathbf{a}][0, \ \mathbf{b}]$$
$$= [-\mathbf{a} \cdot \mathbf{b}, \ \mathbf{a} \times \mathbf{b}]$$

which is no longer 'pure', as some of the original vector information has 'tunnelled' across into the real part via the dot product.

11.9   Unit Quaternion

Let's pursue this analysis further by introducing some familiar vector notation.

Give vector $\mathbf{v}$, then

$$\mathbf{v} = \lambda \hat{\mathbf{v}}, \quad \text{where } \lambda = \|\mathbf{v}\| \text{ and } \|\hat{\mathbf{v}}\| = 1.$$

Combining this with the definition of a pure quaternion we get:

$$q = [0, \ \mathbf{v}]$$
$$= [0, \ \lambda\hat{\mathbf{v}}]$$
$$= \lambda[0, \ \hat{\mathbf{v}}]$$

and reveals the object $[0, \ \hat{\mathbf{v}}]$ which is called the *unit quaternion* and comprises a zero scalar and a unit vector. It is convenient to identify this unit quaternion as $\hat{q}$:

$$\hat{q} = [0, \ \hat{\mathbf{v}}].$$

So now we have a notation similar to that of vectors where a vector $\mathbf{v}$ is described in terms of its unit form:

$$\mathbf{v} = \lambda\hat{\mathbf{v}}$$

and a quaternion q is also described in terms of its unit form:

$$q = \lambda \hat{q}.$$

Note that $\hat{q}$ is an imaginary object as it squares to -1:

$$\begin{aligned}
\hat{q}^2 &= [0, \ \hat{\mathbf{v}}][0, \ \hat{\mathbf{v}}] \\
&= [-\hat{\mathbf{v}} \cdot \hat{\mathbf{v}}, \ \hat{\mathbf{v}} \times \hat{\mathbf{v}}] \\
&= [-1, \ \mathbf{0}] \\
&= -1
\end{aligned}$$

which is not too surprising, bearing in mind Hamilton's original invention!

11.10 Additive Form of a Quaternion

We now come to the idea of splitting a quaternion into its constituent parts: a real quaternion and a pure quaternion. Again, intuition suggests that we can write a quaternion as

$$\begin{aligned}
q &= [s, \ \mathbf{v}] \\
&= [s, \ \mathbf{0}] + [0, \ \mathbf{v}]
\end{aligned}$$

and we can test this by forming the algebraic product of two quaternions represented in this way:

$$\begin{aligned}
q_a &= [s_a, \ \mathbf{0}] + [0, \ \mathbf{a}] \\
q_b &= [s_b, \ \mathbf{0}] + [0, \ \mathbf{b}] \\
q_a q_b &= \big([s_a, \ \mathbf{0}] + [0, \ \mathbf{a}]\big)\big([s_b, \ \mathbf{0}] + [0, \ \mathbf{b}]\big) \\
&= [s_a, \ \mathbf{0}][s_b, \ \mathbf{0}] + [s_a, \ \mathbf{0}][0, \ \mathbf{b}] + [0, \ \mathbf{a}][s_b, \ \mathbf{0}] + [0, \ \mathbf{a}][0, \ \mathbf{b}] \\
&= [s_a s_b, \ \mathbf{0}] + [0, \ s_a \mathbf{b}] + [0, \ s_b \mathbf{a}] + [-\mathbf{a} \cdot \mathbf{b}, \ \mathbf{a} \times \mathbf{b}] \\
&= [s_a s_b - \mathbf{a} \cdot \mathbf{b}, \ s_a \mathbf{b} + s_b \mathbf{a} + \mathbf{a} \times \mathbf{b}]
\end{aligned}$$

which is correct, and confirms that the additive form works.

11.11 Binary Form of a Quaternion

Having shown that the additive form of a quaternion works, and discovered the unit quaternion, we can join the two objects together as follows:

$$q = [s, \ \mathbf{v}]$$

$$= [s, \ \mathbf{0}] + [0, \ \mathbf{v}]$$
$$= [s, \ \mathbf{0}] + \lambda[0, \ \hat{\mathbf{v}}]$$
$$= s + \lambda\hat{q}.$$

Just to recap, s is a scalar, λ is the length of the vector term, and $\hat{q}$ is the unit quaternion $[0, \ \hat{\mathbf{v}}]$.

Look how similar this notation is to a complex number:

$$z = a + bi$$
$$q = s + \lambda\hat{q}$$

where a, b, s, λ are scalars, i is the unit imaginary and $\hat{q}$ is the unit quaternion.

11.12   The Complex Conjugate of a Quaternion

We have already discovered that the conjugate of a complex number $z = a + bi$ is given by

$$z^* = a - bi$$

and is very useful in computing the inverse of z. The *quaternion conjugate* plays a similar role in computing the inverse of a quaternion. Therefore, given

$$q = [s, \ \mathbf{v}]$$

the quaternion conjugate is defined as

$$q^* = [s, \ -\mathbf{v}].$$

For example:

$$q = [2, \ 3\mathbf{i} - 4\mathbf{j} + 5\mathbf{k}]$$
$$q^* = [2, \ -3\mathbf{i} + 4\mathbf{j} - 5\mathbf{k}]$$

If we compute the product qq^* we obtain

$$qq^* = [s, \ \mathbf{v}][s, \ -\mathbf{v}]$$
$$= \left[s^2 - \mathbf{v} \cdot (-\mathbf{v}), \ -s\mathbf{v} + s\mathbf{v} + \mathbf{v} \times (-\mathbf{v})\right]$$
$$= \left[s^2 + \mathbf{v} \cdot \mathbf{v}, \ \mathbf{0}\right]$$
$$= \left[s^2 + v^2, \ \mathbf{0}\right].$$

Let's show that $qq^* = q^*q$:

$$q^*q = [s, \ -\mathbf{v}][s, \ \mathbf{v}]$$
$$= \left[s^2 - (-\mathbf{v}) \cdot \mathbf{v}, \ s\mathbf{v} - s\mathbf{v} + (-\mathbf{v}) \times \mathbf{v}\right]$$
$$= \left[s^2 + \mathbf{v} \cdot \mathbf{v}, \ \mathbf{0}\right]$$
$$= \left[s^2 + v^2, \ \mathbf{0}\right]$$
$$= qq^*.$$

Now let's show that $(q_a q_b)^* = q_b^* q_a^*$.

$$q_a = [s_a, \ \mathbf{a}]$$
$$q_b = [s_b, \ \mathbf{b}]$$
$$q_a q_b = [s_a, \ \mathbf{a}][s_b, \ \mathbf{b}]$$
$$= [s_a s_b - \mathbf{a} \cdot \mathbf{b}, \ s_a \mathbf{b} + s_b \mathbf{a} + \mathbf{a} \times \mathbf{b}]$$
$$(q_a q_b)^* = [s_a s_b - \mathbf{a} \cdot \mathbf{b}, \ -s_a \mathbf{b} - s_b \mathbf{a} - \mathbf{a} \times \mathbf{b}]. \tag{11.10}$$

Next, we compute $q_b^* q_a^*$

$$q_a^* = [s_a, \ -\mathbf{a}]$$
$$q_b^* = [s_b, \ -\mathbf{b}]$$
$$q_b^* q_a^* = [s_b, \ -\mathbf{b}][s_a, \ -\mathbf{a}]$$
$$= [s_a s_b - \mathbf{a} \cdot \mathbf{b}, \ -s_a \mathbf{b} - s_b \mathbf{a} - \mathbf{a} \times \mathbf{b}]. \tag{11.11}$$

And as (11.10) equals (11.11), $(q_a q_b)^* = q_b^* q_a^*$.

11.13 Norm of a Quaternion

The *norm* of a complex number $z = a + bi$ is defined as:

$$|z| = \sqrt{a^2 + b^2}$$

which allows us to write

$$zz^* = |z|^2.$$

Similarly, the norm of a quaternion q is defined as:

$$q = [s, \ \mathbf{v}]$$
$$= [s, \ \lambda \hat{\mathbf{v}}]$$
$$|q| = \sqrt{s^2 + \lambda^2}$$

where $\lambda = \|\mathbf{v}\|$ which allows us to write

$$qq^* = |q|^2.$$

For example:

$$
\begin{aligned}
q &= [1, \ 4\mathbf{i} + 4\mathbf{j} - 4\mathbf{k}] \\
|q| &= \sqrt{1^2 + 4^2 + 4^2 + (-4)^2} \\
&= \sqrt{49} \\
&= 7.
\end{aligned}
$$

11.14   Normalised Quaternion

A quaternion with a unit norm is called a *normalised quaternion*. For example, the quaternion $q = [s, \ \mathbf{v}]$ is *normalised* by dividing it by $|q|$:

$$q' = \frac{q}{\sqrt{s^2 + \lambda^2}}.$$

We must be careful not to confuse the unit quaternion with a unit-norm quaternion. The unit quaternion is $[0, \ \hat{\mathbf{v}}]$ with a unit-vector part, whereas a unit-norm quaternion is normalised such that $s^2 + \lambda^2 = 1$.

I will be careful to distinguish between these two terms as many authors—including myself—use the term unit quaternion to describe a quaternion with a unit norm. For example:

$$q = [1, \ 4\mathbf{i} + 4\mathbf{j} - 4\mathbf{k}]$$

has a norm of 7, and q is normalised by dividing by 7:

$$q' = \tfrac{1}{7}[1, \ 4\mathbf{i} + 4\mathbf{j} - 4\mathbf{k}].$$

The type of unit-norm quaternion we will be using takes the form:

$$q = \left[\cos\left(\tfrac{\theta}{2}\right), \ \sin\left(\tfrac{\theta}{2}\right)\hat{\mathbf{v}}\right]$$

because $\cos^2\theta + \sin^2\theta = 1$.

11.15   Quaternion Products

Having shown that ordered pairs can represent a quaternion and its various manifestations, let's summarise the products we will eventually encounter. To start, we have the product of two normal quaternions:

$$q_a q_b = [s_a, \ \mathbf{a}][s_b, \ \mathbf{b}]$$
$$= [s_a s_b - \mathbf{a} \cdot \mathbf{b}, \ s_a \mathbf{b} + s_b \mathbf{a} + \mathbf{a} \times \mathbf{b}].$$

11.15.1   Product of Pure Quaternions

Given two pure quaternions:

$$q_a = [0, \ \mathbf{a}]$$
$$q_b = [0, \ \mathbf{b}]$$

their product is

$$q_a q_b = [0, \ \mathbf{a}][0, \ \mathbf{b}]$$
$$= [-\mathbf{a} \cdot \mathbf{b}, \ \mathbf{a} \times \mathbf{b}].$$

11.15.2   Product of Unit-Norm Quaternions

Given two unit-norm quaternions:

$$q_a = [s_a, \ \mathbf{a}]$$
$$q_b = [s_b, \ \mathbf{b}]$$

where $|q_a| = |q_b| = 1$. Their product is another unit-norm quaternion, which is proved as follows.

We assume $q_c = [s_c, \ \mathbf{c}]$ and show that $|q_c| = s_c^2 + c^2 = 1$ where

$$[s_c, \ \mathbf{c}] = [s_a, \ \mathbf{a}][s_b, \ \mathbf{b}]$$
$$= [s_a s_b - \mathbf{a} \cdot \mathbf{b}, \ s_a \mathbf{b} + s_b \mathbf{a} + \mathbf{a} \times \mathbf{b}].$$

Let's assume the angle between $\mathbf{a}$ and $\mathbf{b}$ is θ, which permits us to write:

$$s_c = s_a s_b - ab \cos \theta$$
$$\mathbf{c} = s_a b \hat{\mathbf{b}} + s_b a \hat{\mathbf{a}} + ab \sin \theta \left(\hat{\mathbf{a}} \times \hat{\mathbf{b}} \right).$$

Therefore,

$$s_c^2 = (s_a s_b - ab \cos \theta)(s_a s_b - ab \cos \theta)$$
$$= s_a^2 s_b^2 - 2 s_a s_b ab \cos \theta + a^2 b^2 \cos^2 \theta.$$

Fig. 11.2 Geometry for $s_a b\hat{\mathbf{b}} + s_b a\hat{\mathbf{a}} + ab \sin\theta\,(\hat{\mathbf{a}} \times \hat{\mathbf{b}})$

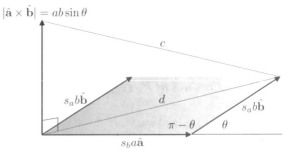

Figure 11.2 shows the geometry representing **c**.

$$d^2 = s_b^2 a^2 + s_a^2 b^2 - 2s_a s_b ab \cos(\pi - \theta)$$
$$= s_b^2 a^2 + s_a^2 b^2 + 2s_a s_b ab \cos\theta$$
$$c^2 = d^2 + a^2 b^2 \sin^2\theta$$
$$= s_b^2 a^2 + s_a^2 b^2 + 2s_a s_b ab \cos\theta + a^2 b^2 \sin^2\theta$$
$$s_c^2 + c^2 = s_a^2 s_b^2 - 2s_a s_b ab \cos\theta + a^2 b^2 \cos^2\theta + s_b^2 a^2 + s_a^2 b^2 + 2s_a s_b ab \cos\theta + a^2 b^2 \sin^2\theta$$
$$= s_a^2 s_b^2 + a^2 b^2 + s_b^2 a^2 + s_a^2 b^2$$
$$= s_a^2\left(s_b^2 + b^2\right) + a^2\left(s_b^2 + b^2\right)$$
$$= s_a^2 + a^2$$
$$= 1.$$

Therefore, the product of two unit-norm quaternions is another unit-norm quaternion. Consequently, multiplying a quaternion by a unit-norm quaternion, does not change its norm:

$$q_a = [s_a, \ \mathbf{a}]$$
$$|q_a| = 1$$
$$q_b = [s_b, \ \mathbf{b}]$$
$$|q_a q_b| = |q_b|.$$

11.15.3   Square of a Quaternion

The square of a quaternion is given by:

$$\mathbf{v} = x\mathbf{i} + y\mathbf{j} + z\mathbf{k}$$
$$q = [s, \ \mathbf{v}]$$
$$q^2 = [s, \ \mathbf{v}][s, \ \mathbf{v}]$$

$$= \left[s^2 - \mathbf{v} \cdot \mathbf{v}, \ 2s\mathbf{v} + \mathbf{v} \times \mathbf{v}\right]$$
$$= \left[s^2 - \mathbf{v} \cdot \mathbf{v}, \ 2s\mathbf{v}\right]$$
$$= \left[s^2 - x^2 - y^2 - z^2, \ 2s(x\mathbf{i} + y\mathbf{j} + z\mathbf{k})\right].$$

For example:

$$q = [7, \ 2\mathbf{i} + 3\mathbf{j} + 4\mathbf{k}]$$
$$q^2 = \left[7^2 - 2^2 - 3^2 - 4^2, \ 14(2\mathbf{i} + 3\mathbf{j} + 4\mathbf{k})\right]$$
$$= [20, \ 28\mathbf{i} + 42\mathbf{j} + 56\mathbf{k}].$$

The square of a pure quaternion is

$$\mathbf{v} = x\mathbf{i} + y\mathbf{j} + z\mathbf{k}$$
$$q = [0, \ \mathbf{v}]$$
$$q^2 = [0, \ \mathbf{v}][0, \ \mathbf{v}]$$
$$= [0 - \mathbf{v} \cdot \mathbf{v}, \ \mathbf{v} \times \mathbf{v}]$$
$$= [0 - \mathbf{v} \cdot \mathbf{v}, \ \mathbf{0}]$$
$$= \left[-\left(x^2 + y^2 + z^2\right), \ \mathbf{0}\right]$$

which makes the square of a pure, unit-norm quaternion equal to -1, and was one of the results, to which some 19th-century mathematicians objected.

11.15.4 Norm of the Quaternion Product

In proving that the product of two unit-norm quaternions is another unit-norm quaternion we saw that

$$q_a = [s_a, \ \mathbf{a}]$$
$$q_b = [s_b, \ \mathbf{b}]$$
$$q_c = q_a q_b$$
$$|q_c|^2 = s_a^2 \left(s_b^2 + b^2\right) + a^2 \left(s_b^2 + b^2\right)$$
$$= \left(s_a^2 + a^2\right)\left(s_b^2 + b^2\right)$$

which, if we ignore the constraint of unit-norm quaternions, shows that the norm of a quaternion product equals the product of the individual norms:

$$|q_a q_b|^2 = |q_a|^2 |q_b|^2$$
$$|q_a q_b| = |q_a||q_b|.$$

11.16   Inverse Quaternion

An important feature of quaternion algebra is the ability to divide two quaternions q_b/q_a, as long as q_a does not vanish.

By definition, the inverse q^{-1} of q satisfies

$$qq^{-1} = [1, \ \mathbf{0}] = 1. \tag{11.12}$$

To isolate q^{-1}, we multiply (11.12) by q^*

$$q^*qq^{-1} = q^*$$
$$|q|^2 q^{-1} = q^* \tag{11.13}$$

and from (11.13) we can write

$$q^{-1} = \frac{q^*}{|q|^2}.$$

If q is a unit-norm quaternion, then

$$q^{-1} = q^*$$

which is useful in the context of rotations.

Furthermore, as

$$(q_a q_b)^* = q_b^* q_a^*$$

then

$$(q_a q_b)^{-1} = q_b^{-1} q_a^{-1}.$$

Note that $qq^{-1} = q^{-1}q$:

$$qq^{-1} = \frac{qq^*}{|q|^2} = 1$$
$$q^{-1}q = \frac{q^*q}{|q|^2} = 1.$$

Thus, we represent the quotient q_b/q_a as

$$q_c = \frac{q_b}{q_a}$$
$$= q_b q_a^{-1}$$
$$= \frac{q_b q_a^*}{|q_a|^2}.$$

For completeness let's evaluate the inverse of q where

$$q = \left[1, \ \tfrac{1}{\sqrt{3}}\mathbf{i} + \tfrac{1}{\sqrt{3}}\mathbf{j} + \tfrac{1}{\sqrt{3}}\mathbf{k}\right]$$

$$q^* = \left[1, \ -\tfrac{1}{\sqrt{3}}\mathbf{i} - \tfrac{1}{\sqrt{3}}\mathbf{j} - \tfrac{1}{\sqrt{3}}\mathbf{k}\right]$$

$$|q|^2 = 1 + \tfrac{1}{3} + \tfrac{1}{3} + \tfrac{1}{3} = 2$$

$$q^{-1} = \frac{q^*}{|q|^2} = \tfrac{1}{2}\left[1, \ -\tfrac{1}{\sqrt{3}}\mathbf{i} - \tfrac{1}{\sqrt{3}}\mathbf{j} - \tfrac{1}{\sqrt{3}}\mathbf{k}\right].$$

It should be clear that $q^{-1}q = 1$:

$$q^{-1}q = \tfrac{1}{2}\left[1, \ -\tfrac{1}{\sqrt{3}}\mathbf{i} - \tfrac{1}{\sqrt{3}}\mathbf{j} - \tfrac{1}{\sqrt{3}}\mathbf{k}\right]\left[1, \ \tfrac{1}{\sqrt{3}}\mathbf{i} + \tfrac{1}{\sqrt{3}}\mathbf{j} + \tfrac{1}{\sqrt{3}}\mathbf{k}\right]$$

$$= \tfrac{1}{2}\left[1 + \tfrac{1}{3} + \tfrac{1}{3} + \tfrac{1}{3}, \ \mathbf{0}\right]$$

$$= 1.$$

11.17 Matrices

Matrices provide another way to express a quaternion product. For convenience, let's repeat (11.8) again and show it in matrix form:

$$
\begin{aligned}
[s_a, \ \mathbf{a}][s_b, \ \mathbf{b}] = [s_a s_b &- x_a x_b - y_a y_b - z_a z_b, \\
&+ s_a\,(x_b\mathbf{i} + y_b\mathbf{j} + z_b\mathbf{k}) + s_b(x_a\mathbf{i} + y_a\mathbf{j} + z_a\mathbf{k}) \\
&+ (y_a z_b - y_b z_a)\mathbf{i} + (z_a x_b - z_b x_a)\mathbf{j} + (x_a y_b - x_b y_a)\mathbf{k}]
\end{aligned}
$$

$$
= \begin{bmatrix}
s_a & -x_a & -y_a & -z_a \\
x_a & s_a & -z_a & y_a \\
y_a & z_a & s_a & -x_a \\
z_a & -y_a & x_a & s_a
\end{bmatrix}
\begin{bmatrix}
s_b \\ x_b \\ y_b \\ z_b
\end{bmatrix}.
\tag{11.14}
$$

Let's recompute the product $q_a q_b$ using the above matrix:

$$q_a = [1, \ 2\mathbf{i} + 3\mathbf{j} + 4\mathbf{k}]$$

$$q_b = [2, \ 3\mathbf{i} + 4\mathbf{j} + 5\mathbf{k}]$$

$$
q_a q_b = \begin{bmatrix}
1 & -2 & -3 & -4 \\
2 & 1 & -4 & 3 \\
3 & 4 & 1 & -2 \\
4 & -3 & 2 & 1
\end{bmatrix}
\begin{bmatrix}
2 \\ 3 \\ 4 \\ 5
\end{bmatrix}
$$

$$= \begin{bmatrix} -36 \\ 6 \\ 12 \\ 12 \end{bmatrix}$$

$$= [-36, \ 6\mathbf{i} + 12\mathbf{j} + 12\mathbf{k}].$$

11.17.1   Orthogonal Matrix

We can demonstrate that the unit-norm quaternion matrix is orthogonal by showing that the product with its transpose equals the identity matrix. As we are dealing with matrices, $\mathbf{Q}$ will represent the matrix for q:

$$q = [s, \ x\mathbf{i} + y\mathbf{j} + z\mathbf{k}]$$
$$\text{where} \quad 1 = s^2 + x^2 + y^2 + z^2$$

$$\mathbf{Q} = \begin{bmatrix} s & -x & -y & -z \\ x & s & -z & y \\ y & z & s & -x \\ z & -y & x & s \end{bmatrix}$$

$$\mathbf{Q}^\mathrm{T} = \begin{bmatrix} s & x & y & z \\ -x & s & z & -y \\ -y & -z & s & x \\ -z & y & -x & s \end{bmatrix}$$

$$\mathbf{Q}\mathbf{Q}^\mathrm{T} = \begin{bmatrix} s & -x & -y & -z \\ x & s & -z & y \\ y & z & s & -x \\ z & -y & x & s \end{bmatrix} \begin{bmatrix} s & x & y & z \\ -x & s & z & -y \\ -y & -z & s & x \\ -z & y & -x & s \end{bmatrix}$$

$$= \begin{bmatrix} 1 & 0 & 0 & 0 \\ 0 & 1 & 0 & 0 \\ 0 & 0 & 1 & 0 \\ 0 & 0 & 0 & 1 \end{bmatrix}$$

For this to occur, $\mathbf{Q}^\mathrm{T} = \mathbf{Q}^{-1}$.

11.18   Quaternion Algebra

Ordered pairs provide a simple notation for representing quaternions, and allow us to represent the real unit 1 as $[1, \ \mathbf{0}]$, and the imaginaries $i, \ j, \ k$ as $[0, \ \mathbf{i}]$, $[0, \ \mathbf{j}]$, $[0, \ \mathbf{k}]$ respectively. A quaternion then becomes a linear combination of these elements with associated real coefficients. Under such conditions, the elements form the *basis* for an algebra over the field of reals.

Furthermore, because quaternion algebra supports division, and obeys the normal axioms of algebra, except that multiplication is non-commutative, it is called a *division algebra*. The German mathematician Ferdinand Georg Frobenius (1849–1917) proved that only three such real associative division algebras exist: real numbers, complex numbers and quaternions [8].

The *Cayley numbers* $\mathbb{O}$, constitute a real division algebra, but the Cayley numbers are 8-dimensional and are not associative, i.e. $a(bc) \neq (ab)c$ for all $a, b, c \in \mathbb{O}$.

11.19 Summary

Quaternions are very similar to complex numbers, apart from the fact that they have three imaginary terms, rather than one. Consequently, they inherit some of the properties associated with complex numbers, such as norm, complex conjugate, unit norm and inverse. They can also be added, subtracted, multiplied and divided. However, unlike complex numbers, they anticommute when multiplied.

11.19.1 Summary of Definitions

Quaternion

$$q_a = [s_a, \ \mathbf{a}] = [s_a, \ x_a\mathbf{i} + y_a\mathbf{j} + z_a\mathbf{k}]$$
$$q_b = [s_b, \ \mathbf{b}] = [s_b, \ x_b\mathbf{i} + y_b\mathbf{j} + z_b\mathbf{k}] \, .$$

Adding and subtracting

$$q_a \pm q_b = [s_a \pm s_b, \ \mathbf{a} \pm \mathbf{b}].$$

Product

$$q_a q_b = [s_a, \ \mathbf{a}][s_b, \ \mathbf{b}]$$
$$= [s_a s_b - \mathbf{a} \cdot \mathbf{b}, \ s_a \mathbf{b} + s_b \mathbf{a} + \mathbf{a} \times \mathbf{b}]$$
$$= \begin{bmatrix} s_a & -x_a & -y_a & -z_a \\ x_a & s_a & -z_a & y_a \\ y_a & z_a & s_a & -x_a \\ z_a & -y_a & x_a & s_a \end{bmatrix} \begin{bmatrix} s_b \\ x_b \\ y_b \\ z_b \end{bmatrix} .$$

Square

$$\mathbf{v} = x\mathbf{i} + y\mathbf{j} + z\mathbf{k}$$
$$q^2 = [s, \ \mathbf{v}][s, \ \mathbf{v}]$$

$$= \left[s^2 - x^2 - y^2 - z^2, \ 2s(x\mathbf{i} + y\mathbf{j} + z\mathbf{k})\right].$$

Pure

$$\mathbf{v} = x\mathbf{i} + y\mathbf{j} + z\mathbf{k}$$
$$q^2 = [0, \ \mathbf{v}][0, \ \mathbf{v}]$$
$$= \left[-(x^2 + y^2 + z^2), \ \mathbf{0}\right].$$

Norm

$$\mathbf{v} = \lambda \hat{\mathbf{v}}$$
$$q = [s, \ \lambda \hat{\mathbf{v}}]$$
$$|q| = \sqrt{s^2 + \lambda^2}.$$

Unit norm

$$|q| = \sqrt{s^2 + \lambda^2} = 1.$$

Conjugate

$$q^* = [s, \ -\mathbf{v}]$$
$$(q_a q_b)^* = q_b^* q_a^*.$$

Inverse

$$q^{-1} = \frac{q^*}{|q|^2}$$
$$(q_a q_b)^{-1} = q_b^{-1} q_a^{-1}.$$

11.20   Worked Examples

Here are some further worked examples that employ the ideas described above. In some cases, a test is included to confirm the result.

11.20.1   Adding and Subtracting Quaternions

Add and subtract the following quaternions:

$$q_a = [2, \ -2\mathbf{i} + 3\mathbf{j} - 4\mathbf{k}]$$
$$q_b = [1, \ -2\mathbf{i} + 5\mathbf{j} - 6\mathbf{k}]$$

$$q_a + q_b = [3, \; -4\mathbf{i} + 8\mathbf{j} - 10\mathbf{k}]$$
$$q_a - q_b = [1, \; 0\mathbf{i} - 2\mathbf{j} + 2\mathbf{k}].$$

11.20.2 Norm of a Quaternion

Find the norm of the following quaternions:

$$q_a = [2, \; -2\mathbf{i} + 3\mathbf{j} - 4\mathbf{k}]$$
$$q_b = [1, \; -2\mathbf{i} + 5\mathbf{j} - 6\mathbf{k}]$$
$$|q_a| = \sqrt{2^2 + (-2)^2 + 3^2 + (-4)^2} = \sqrt{33}$$
$$|q_b| = \sqrt{1^2 + (-2)^2 + 5^2 + (-6)^2} = \sqrt{66}.$$

11.20.3 Unit-norm Quaternions

Convert these quaternions to their unit-norm form:

$$q_a = [2, \; -2\mathbf{i} + 3\mathbf{j} - 4\mathbf{k}]$$
$$q_b = [1, \; -2\mathbf{i} + 5\mathbf{j} - 6\mathbf{k}]$$
$$|q_a| = \sqrt{33}$$
$$|q_b| = \sqrt{66}$$
$$q'_a = \tfrac{1}{\sqrt{33}}[2, \; -2\mathbf{i} + 3\mathbf{j} - 4\mathbf{k}]$$
$$q'_b = \tfrac{1}{\sqrt{66}}[1, \; -2\mathbf{i} + 5\mathbf{j} - 6\mathbf{k}].$$

11.20.4 Quaternion Product

Compute the product and reverse product of the following quaternions.

$$q_a = [2, \; -2\mathbf{i} + 3\mathbf{j} - 4\mathbf{k}]$$
$$q_b = [1, \; -2\mathbf{i} + 5\mathbf{j} - 6\mathbf{k}]$$
$$q_a q_b = [2, \; -2\mathbf{i} + 3\mathbf{j} - 4\mathbf{k}][1, \; -2\mathbf{i} + 5\mathbf{j} - 6\mathbf{k}]$$
$$= [2 \times 1 - ((-2) \times (-2) + 3 \times 5 + (-4) \times (-6)),$$
$$+ 2(-2\mathbf{i} + 5\mathbf{j} - 6\mathbf{k}) + 1(-2\mathbf{i} + 3\mathbf{j} - 4\mathbf{k})$$
$$+ (3 \times (-6) - (-4) \times 5)\mathbf{i} - ((-2) \times (-6) - (-4) \times (-2))\mathbf{j} + ((-2) \times 5 - 3 \times (-2))\mathbf{k}]$$
$$= [-41, \; -6\mathbf{i} + 13\mathbf{j} - 16\mathbf{k} + 2\mathbf{i} - 4\mathbf{j} - 4\mathbf{k}]$$
$$= [-41, \; -4\mathbf{i} + 9\mathbf{j} - 20\mathbf{k}].$$

$$q_b q_a = [1, \; -2\mathbf{i} + 5\mathbf{j} - 6\mathbf{k}][2 - 2\mathbf{i} + 3\mathbf{j} - 4\mathbf{k}]$$
$$= [1 \times 2 - ((-2) \times (-2) + 5 \times 3 + (-6) \times (-4)),$$
$$+ \, 1(-2\mathbf{i} + 3\mathbf{j} - 4\mathbf{k}) + 2(-2\mathbf{i} + 5\mathbf{j} - 6\mathbf{k})$$
$$+ \, (5 \times (-4) - (-6) \times 3)\mathbf{i} - ((-2) \times (-4) - (-6) \times (-2))\mathbf{j} + ((-2) \times 3 - 5 \times (-2))\mathbf{k}]$$
$$= [-41, \; -6\mathbf{i} + 13\mathbf{j} - 16\mathbf{k} - 2\mathbf{i} + 4\mathbf{j} + 4\mathbf{k}]$$
$$= [-41, \; -8\mathbf{i} + 17\mathbf{j} - 12\mathbf{k}].$$

Note: The only thing that has changed in this computation is the sign of the cross-product axial vector.

11.20.5   Square of a Quaternion

Compute the square of this quaternion:

$$q = [2, \; -2\mathbf{i} + 3\mathbf{j} - 4\mathbf{k}]$$
$$q^2 = [2, \; -2\mathbf{i} + 3\mathbf{j} - 4\mathbf{k}][2, \; -2\mathbf{i} + 3\mathbf{j} - 4\mathbf{k}]$$
$$= [2 \times 2 - ((-2) \times (-2) + 3 \times 3 + (-4) \times (-4)),$$
$$+ \, 2 \times 2(-2\mathbf{i} + 3\mathbf{j} - 4\mathbf{k})]$$
$$= [-25, \; -8\mathbf{i} + 12\mathbf{j} - 16\mathbf{k}].$$

11.20.6   Inverse of a Quaternion

Compute the inverse of this quaternion:

$$q = [2, \; -2\mathbf{i} + 3\mathbf{j} - 4\mathbf{k}]$$
$$q^* = [2, \; 2\mathbf{i} - 3\mathbf{j} + 4\mathbf{k}]$$
$$|q|^2 = 2^2 + (-2)^2 + 3^2 + (-4)^2 = 33$$
$$q^{-1} = \tfrac{1}{33}[2, \; 2\mathbf{i} - 3\mathbf{j} + 4\mathbf{k}].$$

References

1. Hamilton WR (1844) On quaternions: or a new system of imaginaries in algebra. Phil Mag 3rd ser. 25
2. Hamilton WR (1853) Lectures on quaternions. Hodges & Smith, Dublin
3. Hamilton WR (1899–1901) Elements of quaternions, 2nd edn. Longmans, Green & Co., London (Jolly, C.J. (ed.) 2 vols.)

4. Tait PG (1867) An elementary treatise on quaternions. Cambridge University Press, Cambridge
5. Gauss CF (1819) Mutation des Raumes In: Carl Friedrich Gauss Werke, Achter Band, pp 357–361, König. Gesell. Wissen. Göttingen, 1900
6. Wilson EB (1901) Vector analysis. Yale University Press, New Haven
7. Feynman RP. Symmetry and physical laws. In: Feynman lectures in physics, vol 1
8. Altmann SL (2005) Rotations. Quaternions and Double Groups, Dover, New York

Chapter 12
Quaternions in Space

12.1 Introduction

In this chapter we show how quaternions are used to rotate vectors about an arbitrary axis. We begin by reviewing some of the history associated with quaternions, and the development of octonions.

We then examine various quaternion products to discover their rotational properties. This begins with two orthogonal quaternions, and moves towards the general case of using qpq^{-1} where q is a unit-norm quaternion, and p is a pure quaternion.

A technique shows how to express a quaternion product as a matrix.

We continue to represent a quaternion as an ordered pair, with italic, lower-case letters to represent quaternions, and bold lower-case letters to represent vectors.

12.2 Some History

Hamilton invented quaternions in October 1843, and by December of the same year, his friend, Irish mathematician John Thomas Graves (1806–1870), had invented *octaves*, which would eventually be called *octonions*. Arthur Cayley had also been intrigued by Hamilton's quaternions, and independently invented octonions in 1845. Octonions eventually became known as *Cayley numbers* rather than *octaves*, simply because Graves did not publish his results until 1848—three years after Cayley!

Just as quaternions can be defined in terms of ordered pairs of complex numbers, the octaves, or octonions, can be defined as ordered pairs of quaternions.

12.3 Quaternion Products

A quaternion q is the union of a scalar s and a vector $\mathbf{v}$:

$$q = [s, \mathbf{v}], \quad s \in \mathbb{R}, \quad \mathbf{v} \in \mathbb{R}^3.$$

© Springer-Verlag London Ltd., part of Springer Nature 2022
J. Vince, *Mathematics for Computer Graphics*, Undergraduate Topics in Computer Science, https://doi.org/10.1007/978-1-4471-7520-9_12

If we express $\mathbf{v}$ in terms of its components, we have

$$q = [s, \ x\mathbf{i} + y\mathbf{j} + z\mathbf{k}], \quad s, x, y, z \in \mathbb{R}.$$

When two such quaternions are multiplied together, we obtain a third quaternion:

$$q_a = [s_a, \ \mathbf{v}_a]$$
$$q_b = [s_b, \ \mathbf{v}_b]$$
$$q_a q_b = [s_a, \ \mathbf{v}_a][s_b, \ \mathbf{v}_b]$$
$$= [s_a s_b - \mathbf{v}_a \cdot \mathbf{v}_b, \ s_a \mathbf{v}_b + s_b \mathbf{v}_a + \mathbf{v}_a \times \mathbf{v}_b].$$

Naturally, if s_a or s_b are zero, as in the case of a pure quaternion, the product is simplified. Therefore, in future I will omit any zero terms, to simplify the algebra.

Hamilton had hoped that a quaternion could be used like a complex rotor, where

$$\mathbf{R}_\theta = \cos\theta + i\sin\theta$$

rotates a complex number by θ. Could a unit-norm quaternion q be used to rotate a vector stored as a pure quaternion p? Well yes, but only as a special case. To understand this, let's construct the product of a unit-norm quaternion q and a pure quaternion p. The unit-norm quaternion q is defined as

$$q = [s, \ \lambda\hat{\mathbf{v}}], \quad s, \lambda \in \mathbb{R}, \quad \hat{\mathbf{v}} \in \mathbb{R}^3, \qquad (12.1)$$
$$\|\hat{\mathbf{v}}\| = 1$$
$$s^2 + \lambda^2 = 1$$

and the pure quaternion p stores the vector $\mathbf{p}$ to be rotated:

$$p = [0, \ \mathbf{p}], \quad \mathbf{p} \in \mathbb{R}^3.$$

Let's compute the product $p' = qp$ and examine the vector part of p' to see if $\mathbf{p}$ is rotated:

$$q = [s, \ \lambda\hat{\mathbf{v}}]$$
$$p = [0, \ \mathbf{p}]$$
$$p' = qp$$
$$= [s, \ \lambda\hat{\mathbf{v}}][0, \ \mathbf{p}]$$
$$= [-\lambda\hat{\mathbf{v}} \cdot \mathbf{p}, \ s\mathbf{p} + \lambda\hat{\mathbf{v}} \times \mathbf{p}]. \qquad (12.2)$$

We can see from (12.2) that the result is a general quaternion with a scalar and a vector component.

12.3.1 Special Case

The 'special case' referred to above is that $\hat{\mathbf{v}}$ must be perpendicular to $\mathbf{p}$, which makes the dot product term $-\lambda\hat{\mathbf{v}} \cdot \mathbf{p}$ in (12.2) vanish, and we are left with the pure quaternion:

$$p' = [0, \ s\mathbf{p} + \lambda\hat{\mathbf{v}} \times \mathbf{p}]. \tag{12.3}$$

Figure 12.1 illustrates this scenario, where $\mathbf{p}$ is perpendicular to $\hat{\mathbf{v}}$, and $\hat{\mathbf{v}} \times \mathbf{p}$ is perpendicular to the plane containing $\mathbf{p}$ and $\hat{\mathbf{v}}$.

Now as $\hat{\mathbf{v}}$ is a unit vector, $\|\mathbf{p}\| = \|\hat{\mathbf{v}} \times \mathbf{p}\|$, which means that we have two orthogonal vectors, i.e. $\mathbf{p}$ and $\hat{\mathbf{v}} \times \mathbf{p}$, with the same length. Therefore, to rotate $\mathbf{p}$ about $\hat{\mathbf{v}}$, all that we have to do is make $s = \cos\theta$ and $\lambda = \sin\theta$ in (12.3):

$$p' = [0, \ \mathbf{p}']$$
$$= [0, \ \cos\theta\mathbf{p} + \sin\theta\hat{\mathbf{v}} \times \mathbf{p}].$$

For example, to rotate a vector about the z-axis, q's vector $\hat{\mathbf{v}}$ must be aligned with the z-axis as shown in Fig. 12.2. If we make the angle of rotation $\theta = 45°$ then

$$q = [s, \ \lambda\hat{\mathbf{v}}]$$
$$= [\cos\theta, \ \sin\theta\mathbf{k}]$$

Fig. 12.1 Three orthogonal vectors $\mathbf{p}$, $\hat{\mathbf{v}}$ and $\hat{\mathbf{v}} \times \mathbf{p}$

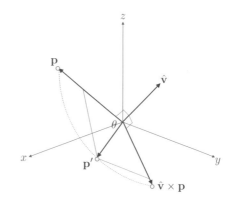

Fig. 12.2 The vector $2\mathbf{i}$ is rotated $45°$ by the quaternion $q = \left[\frac{\sqrt{2}}{2}, \ \frac{\sqrt{2}}{2}\mathbf{k}\right]$

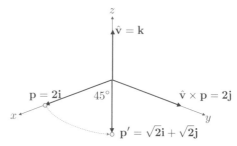

$$= \left[\tfrac{\sqrt{2}}{2}, \ \tfrac{\sqrt{2}}{2}\mathbf{k} \right]$$
$$= \tfrac{\sqrt{2}}{2}[1, \ \mathbf{k}]$$

and if the vector to be rotated is $\mathbf{p} = 2\mathbf{i}$, then

$$p = [0, \ \mathbf{p}]$$
$$= [0, \ 2\mathbf{i}]$$
$$= 2[0, \ \mathbf{i}].$$

There are now four product combinations worth exploring: qp, pq, $q^{-1}p$ and pq^{-1}. It's not worth considering qp^{-1} and $p^{-1}q$ as p^{-1} simply reverses the direction of $\mathbf{p}$. Let's start with qp:

$$q = \tfrac{\sqrt{2}}{2}[1, \ \mathbf{k}]$$
$$p = 2[0, \ \mathbf{i}]$$
$$p' = qp$$
$$= \sqrt{2}[1, \ \mathbf{k}][0, \ \mathbf{i}]$$
$$= \sqrt{2}[0, \ \mathbf{i} + \mathbf{j}]$$

and $\mathbf{p}$ has been rotated $45°$ to $\mathbf{p}' = \sqrt{2}\mathbf{i} + \sqrt{2}\mathbf{j}$.

Next, pq:

$$p = 2[0, \ \mathbf{i}]$$
$$q = \tfrac{\sqrt{2}}{2}[1, \ \mathbf{k}]$$
$$p' = pq$$
$$= \sqrt{2}[0, \ \mathbf{i}][1, \ \mathbf{k}]$$
$$= \sqrt{2}[0, \ \mathbf{i} + \mathbf{i} \times \mathbf{k}]$$
$$= \sqrt{2}[0, \ \mathbf{i} - \mathbf{j}]$$

and $\mathbf{p}$ has been rotated $-45°$ to $\mathbf{p}' = \sqrt{2}\mathbf{i} - \sqrt{2}\mathbf{j}$.

Next, $q^{-1}p$, and as q is a unit-norm quaternion, $q^{-1} = q^*$:

$$q = \tfrac{\sqrt{2}}{2}[1, \ \mathbf{k}]$$
$$q^{-1} = \tfrac{\sqrt{2}}{2}[1, \ -\mathbf{k}]$$
$$p = 2[0, \ \mathbf{i}]$$
$$p' = q^{-1}p$$

$$= \sqrt{2}[1, \ -\mathbf{k}][0, \ \mathbf{i}]$$
$$= \sqrt{2}[0, \ \mathbf{i} - \mathbf{k} \times \mathbf{i}]$$
$$= \sqrt{2}[0, \ \mathbf{i} - \mathbf{j}]$$

and $\mathbf{p}$ has been rotated $-45°$ to $\mathbf{p}' = \sqrt{2}\mathbf{i} - \sqrt{2}\mathbf{j}$.
Finally, pq^{-1}:

$$p = 2[0, \ \mathbf{i}]$$
$$q = \tfrac{\sqrt{2}}{2}[1, \ \mathbf{k}]$$
$$q^{-1} = \tfrac{\sqrt{2}}{2}[1, \ -\mathbf{k}]$$
$$p' = pq^{-1}$$
$$= \sqrt{2}[0, \ \mathbf{i}][1, \ -\mathbf{k}]$$
$$= \sqrt{2}[0, \ \mathbf{i} - \mathbf{i} \times \mathbf{k}]$$
$$= \sqrt{2}[0, \ \mathbf{i} + \mathbf{j}]$$

and $\mathbf{p}$ has been rotated $45°$ to $\mathbf{p}' = \sqrt{2}\mathbf{i} + \sqrt{2}\mathbf{j}$. Thus, for orthogonal quaternions, θ is the angle of rotation, then

$$qp = pq^{-1}$$
$$pq = q^{-1}p.$$

Before moving on, let's see what happens to the product qp when $\theta = 180°$:

$$q = [\cos\theta, \ \sin\theta\mathbf{k}]$$
$$= [-1, \ \mathbf{0}]$$
$$p = 2[0, \ \mathbf{i}]$$
$$p' = qp$$
$$= 2[-1, \ \mathbf{0}][0, \ \mathbf{i}]$$
$$= 2[0, \ -\mathbf{i} + \mathbf{0} \times \mathbf{i}]$$
$$= [0, \ -2\mathbf{i}]$$

and $\mathbf{p}$ has been rotated $180°$ to $\mathbf{p}' = -2\mathbf{i}$.

Note that in all the above products, the vector has not been scaled during the rotation. This is because q is a unit-norm quaternion.

Now let's see what happens if we change the angle between $\hat{\mathbf{v}}$ and $\mathbf{p}$. Let's reduce the angle to $45°$ and retain q's unit vector, as shown in Fig. 12.3, such that $\hat{\mathbf{v}}$ is directed along the z-axis, and $\mathbf{p} = \mathbf{i} + \mathbf{k}$. Therefore,

$$\hat{\mathbf{v}} = \mathbf{k}$$

Fig. 12.3 Rotating the
vector $\mathbf{p} = \mathbf{i} + \mathbf{k}$ by the
quaternion
$q = [\cos\theta, \ \sin\theta\hat{\mathbf{v}}]$

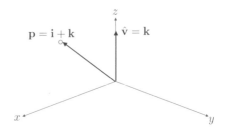

$$q = [\cos\theta, \ \sin\theta\hat{\mathbf{v}}]$$
$$p = [0, \ \mathbf{p}].$$

This time we must include the dot product term $-\sin\theta\hat{\mathbf{v}} \cdot \mathbf{p}$, as it is no longer zero:

$$q = [\cos\theta, \ \sin\theta\hat{\mathbf{v}}]$$
$$p = [0, \ \mathbf{p}]$$
$$p' = qp$$
$$= [\cos\theta, \ \sin\theta\hat{\mathbf{v}}][0, \ \mathbf{p}]$$
$$= [-\sin\theta\hat{\mathbf{v}} \cdot \mathbf{p}, \ \cos\theta\mathbf{p} + \sin\theta\hat{\mathbf{v}} \times \mathbf{p}]. \qquad (12.4)$$

Substituting $\hat{\mathbf{v}}$, $\mathbf{p}$ and $\theta = 45°$ in (12.4), we have

$$\hat{\mathbf{v}} = \mathbf{k}$$
$$\mathbf{p} = \mathbf{i} + \mathbf{k}$$
$$p' = \left[-\tfrac{\sqrt{2}}{2}\mathbf{k} \cdot (\mathbf{i} + \mathbf{k}), \ \tfrac{\sqrt{2}}{2}(\mathbf{i} + \mathbf{k}) + \tfrac{\sqrt{2}}{2}\mathbf{k} \times (\mathbf{i} + \mathbf{k}) \right]$$
$$= \left[-\tfrac{\sqrt{2}}{2}, \ \tfrac{\sqrt{2}}{2}\mathbf{i} + \tfrac{\sqrt{2}}{2}\mathbf{k} + \tfrac{\sqrt{2}}{2}\mathbf{j} \right]$$
$$= \tfrac{\sqrt{2}}{2}[-1, \ \mathbf{i} + \mathbf{j} + \mathbf{k}] \qquad (12.5)$$

which, unfortunately, is no longer a pure quaternion. Multiplying the vector by a non-orthogonal quaternion has converted some of the vector information into the quaternion's scalar component.

12.3.2   General Case

Not to worry. Could it be that an inverse quaternion reverses the operation? Let's see what happens if we post-multiply qp by q^{-1}.

Given

$$q = [\cos \theta, \ \sin \theta \mathbf{k}]$$

then

$$q^{-1} = [\cos \theta, \ -\sin \theta \mathbf{k}]$$
$$= \left[\tfrac{\sqrt{2}}{2}, \ -\tfrac{\sqrt{2}}{2}\mathbf{k} \right]$$
$$= \tfrac{\sqrt{2}}{2}[1, \ -\mathbf{k}].$$

Therefore, post-multiplying (12.5) by q^{-1} we have:

$$qp = \tfrac{\sqrt{2}}{2}[-1, \ \mathbf{i}+\mathbf{j}+\mathbf{k}]$$
$$q^{-1} = \tfrac{\sqrt{2}}{2}[1, \ -\mathbf{k}]$$
$$qpq^{-1} = \tfrac{\sqrt{2}}{2}[-1, \ \mathbf{i}+\mathbf{j}+\mathbf{k}]\tfrac{\sqrt{2}}{2}[1, \ -\mathbf{k}]$$
$$= \tfrac{1}{2}[-1, \ \mathbf{i}+\mathbf{j}+\mathbf{k}][1, \ -\mathbf{k}]$$
$$= \tfrac{1}{2}[-1 + 1, \ \mathbf{k}+\mathbf{i}+\mathbf{j}+\mathbf{k} + (\mathbf{i}+\mathbf{j}+\mathbf{k}) \times -\mathbf{k})]$$
$$= \tfrac{1}{2}[0, \ \mathbf{i}+\mathbf{j}+2\mathbf{k} - \mathbf{i}+\mathbf{j}]$$
$$= [0, \ \mathbf{j}+\mathbf{k}]. \qquad (12.6)$$

Equation (12.6) is a pure quaternion, with a norm of $\sqrt{2}$, which is the same as $\mathbf{p}$. However, the vector has been rotated 90° rather than 45°, twice the desired angle, as shown in Fig. 12.4.

If this 'sandwiching' of the vector in the form of a pure quaternion by q and q^{-1} is correct, it suggests that increasing θ to 90° should rotate $\mathbf{p} = \mathbf{i}+\mathbf{k}$ by 180° to $-\mathbf{i}+\mathbf{k}$. Let's try this.

Let $\theta = 90°$, therefore,

$$q = [\cos 90°, \ \sin 90°\mathbf{k}]$$
$$= [0, \ \mathbf{k}]$$
$$p = [0, \ \mathbf{i}+\mathbf{k}]$$
$$qp = [0, \ \mathbf{k}][0, \ \mathbf{i}+\mathbf{k}]$$

Fig. 12.4 The vector $\mathbf{i}+\mathbf{k}$ is rotated 90° to $\mathbf{j}+\mathbf{k}$

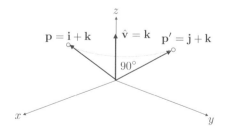

$$= [-1, \ \mathbf{k} \times (\mathbf{i} + \mathbf{k})]$$
$$= [-1, \ \mathbf{j}].$$

Next, we post-multiply qp by q^{-1}:

$$q^{-1} = [0, \ -\mathbf{k}]$$
$$qpq^{-1} = [-1, \ \mathbf{j}][0, \ -\mathbf{k}]$$
$$= [0, \ \mathbf{k} + (\mathbf{j} \times -\mathbf{k})]$$
$$= [0, \ -\mathbf{i} + \mathbf{k}]$$

which confirms our prediction and suggests that qpq^{-1} works.

12.3.3   Double Angle

Now let's show how this double angle arises. We begin by defining a unit-norm quaternion q:

$$q = [s, \ \lambda \hat{\mathbf{v}}]$$

where $s^2 + \lambda^2 = 1$. The vector $\mathbf{p}$ to be rotated is encoded as a pure quaternion:

$$p = [0, \ \mathbf{p}]$$

and the inverse quaternion q^{-1} is

$$q^{-1} = [s, \ -\lambda \hat{\mathbf{v}}].$$

Therefore, the product qpq^{-1} is

$$
\begin{aligned}
qpq^{-1} &= \left[s, \ \lambda \hat{\mathbf{v}} \right] [0, \ \mathbf{p}][s, \ -\lambda \hat{\mathbf{v}}] \\
&= \left[-\lambda \hat{\mathbf{v}} \cdot \mathbf{p}, \ s\mathbf{p} + \lambda \hat{\mathbf{v}} \times \mathbf{p} \right] [s, \ -\lambda \hat{\mathbf{v}}] \\
&= \left[-\lambda s \hat{\mathbf{v}} \cdot \mathbf{p} + \lambda s \mathbf{p} \cdot \hat{\mathbf{v}} + \lambda^2 (\hat{\mathbf{v}} \times \mathbf{p}) \cdot \hat{\mathbf{v}}, \right. \\
&\quad + \lambda^2 (\hat{\mathbf{v}} \cdot \mathbf{p}) \hat{\mathbf{v}} + s^2 \mathbf{p} + \lambda s \hat{\mathbf{v}} \times \mathbf{p} \\
&\quad \left. - \lambda s \mathbf{p} \times \hat{\mathbf{v}} - \lambda^2 (\hat{\mathbf{v}} \times \mathbf{p}) \times \hat{\mathbf{v}} \right] \\
&= \left[\lambda^2 (\hat{\mathbf{v}} \times \mathbf{p}) \cdot \hat{\mathbf{v}}, \ \lambda^2 (\hat{\mathbf{v}} \cdot \mathbf{p}) \hat{\mathbf{v}} + s^2 \mathbf{p} + 2\lambda s \hat{\mathbf{v}} \times \mathbf{p} - \lambda^2 (\hat{\mathbf{v}} \times \mathbf{p}) \times \hat{\mathbf{v}} \right].
\end{aligned}
$$

Note that

$$(\hat{\mathbf{v}} \times \mathbf{p}) \cdot \hat{\mathbf{v}} = 0$$

and

$$(\hat{\mathbf{v}} \times \mathbf{p}) \times \hat{\mathbf{v}} = (\hat{\mathbf{v}} \cdot \hat{\mathbf{v}})\mathbf{p} - (\mathbf{p} \cdot \hat{\mathbf{v}})\hat{\mathbf{v}} = \mathbf{p} - (\mathbf{p} \cdot \hat{\mathbf{v}})\hat{\mathbf{v}}.$$

Therefore,

$$qpq^{-1} = \left[0,\ \lambda^2 \left(\hat{\mathbf{v}} \cdot \mathbf{p}\right) \hat{\mathbf{v}} + s^2\mathbf{p} + 2\lambda s\hat{\mathbf{v}} \times \mathbf{p} - \lambda^2\mathbf{p} + \lambda^2 \left(\mathbf{p} \cdot \hat{\mathbf{v}}\right) \hat{\mathbf{v}}\right]$$
$$= \left[0,\ 2\lambda^2 \left(\hat{\mathbf{v}} \cdot \mathbf{p}\right) \hat{\mathbf{v}} + \left(s^2 - \lambda^2\right)\mathbf{p} + 2\lambda s\hat{\mathbf{v}} \times \mathbf{p}\right]. \qquad (12.7)$$

Clearly, (12.7) is a pure quaternion as the scalar component is zero. However, it is not obvious where the angle doubling comes from. But look what happens when we make $s = \cos\theta$ and $\lambda = \sin\theta$:

$$qpq^{-1} = \left[0,\ 2\sin^2\theta \left(\hat{\mathbf{v}} \cdot \mathbf{p}\right) \hat{\mathbf{v}} + \left(\cos^2\theta - \sin^2\theta\right)\mathbf{p} + 2\sin\theta\cos\theta\hat{\mathbf{v}} \times \mathbf{p}\right]$$
$$= \left[0,\ \left(1 - \cos(2\theta)\right)\left(\hat{\mathbf{v}} \cdot \mathbf{p}\right) \hat{\mathbf{v}} + \cos(2\theta)\mathbf{p} + \sin(2\theta)\hat{\mathbf{v}} \times \mathbf{p}\right].$$

The double-angle trigonometric terms emerge! Now, if we want this product to actually rotate the vector by θ, then we must build this in from the outset by halving θ in q:

$$q = \left[\cos\left(\tfrac{\theta}{2}\right),\ \sin\left(\tfrac{\theta}{2}\right)\hat{\mathbf{v}}\right] \qquad (12.8)$$

which makes

$$qpq^{-1} = \left[0,\ \left(1 - \cos\theta\right)\left(\hat{\mathbf{v}} \cdot \mathbf{p}\right)\hat{\mathbf{v}} + \cos\theta\mathbf{p} + \sin\theta\hat{\mathbf{v}} \times \mathbf{p}\right]. \qquad (12.9)$$

The product qpq^{-1} was discovered by Hamilton who failed to publish the result. Cayley, also discovered the product and published the result in 1845 [1]. However, Altmann notes that 'in Cayley's collected papers he concedes priority to Hamilton.' [2], which was a nice gesture. However, the person who had recognised the importance of the half-angle parameters in (12.8) before Hamilton and Cayley was Rodrigues—who published a solution that was not seen by Hamilton, but apparently, was seen by Cayley.

Let's test (12.9) using the previous example where we rotated a vector $\mathbf{p} = \mathbf{i} + \mathbf{k}$, $\theta = 90°$ about the quaternion's vector $\hat{\mathbf{v}} = \mathbf{k}$.

$$qpq^{-1} = \left[0,\ \left(1 - \cos\theta\right)\left(\hat{\mathbf{v}} \cdot \mathbf{p}\right)\hat{\mathbf{v}} + \cos\theta\mathbf{p} + \sin\theta\hat{\mathbf{v}} \times \mathbf{p}\right]$$
$$= \left[0,\ \left(\hat{\mathbf{v}} \cdot \mathbf{p}\right)\hat{\mathbf{v}} + \hat{\mathbf{v}} \times \mathbf{p}\right]$$
$$= \left[0,\ \left(\mathbf{k} \cdot \left(\mathbf{i} + \mathbf{k}\right)\right)\mathbf{k} + \mathbf{j}\right]$$
$$= \left[0,\ \mathbf{j} + \mathbf{k}\right]$$

which agrees with (12.6). Thus, when a unit-norm quaternion takes the form

$$q = \left[\cos\left(\tfrac{\theta}{2}\right),\ \sin\left(\tfrac{\theta}{2}\right)\hat{\mathbf{v}}\right]$$

and a pure quaternion storing a vector to be rotated takes the form

$$p = \left[0,\ \mathbf{p}\right]$$

the pure quaternion

$$p' = qpq^{-1}$$

stores the rotated vector $\mathbf{p}'$. Let's show why this product preserves the magnitude of the rotated vector.

$$|p'| = |qp||q^{-1}|$$
$$= |q||p||q^{-1}|$$
$$= |q|^2|p|$$

and if q is a unit-norm quaternion, $|q| = 1$, then $|p'| = |p|$.

You may be wondering what happens if the product is reversed to $q^{-1}pq$? A guess would suggest that the rotation sequence is reversed, but let's see what an algebraic analysis confirms.

$$q^{-1}pq = [s, \ -\lambda\hat{\mathbf{v}}][0, \ \mathbf{p}][s, \ \lambda\hat{\mathbf{v}}]$$
$$= [\lambda\hat{\mathbf{v}} \cdot \mathbf{p}, \ s\mathbf{p} - \lambda\hat{\mathbf{v}} \times \mathbf{p}][s, \ \lambda\hat{\mathbf{v}}]$$
$$= \big[\lambda s\hat{\mathbf{v}} \cdot \mathbf{p} - \lambda s\mathbf{p} \cdot \hat{\mathbf{v}},$$
$$\lambda^2\hat{\mathbf{v}} \times \mathbf{p} \cdot \hat{\mathbf{v}} + \lambda^2\hat{\mathbf{v}} \cdot \mathbf{p}\hat{\mathbf{v}} + s^2\mathbf{p} - \lambda s\hat{\mathbf{v}} \times \mathbf{p} + \lambda s\mathbf{p} \times \hat{\mathbf{v}} - \lambda^2\hat{\mathbf{v}} \times \mathbf{p} \times \hat{\mathbf{v}}\big]$$
$$= \big[\lambda^2(\hat{\mathbf{v}} \times \mathbf{p}) \cdot \hat{\mathbf{v}}, \ \lambda^2(\hat{\mathbf{v}} \cdot \mathbf{p})\hat{\mathbf{v}} + s^2\mathbf{p} - 2\lambda s\hat{\mathbf{v}} \times \mathbf{p} - \lambda^2(\hat{\mathbf{v}} \times \mathbf{p}) \times \hat{\mathbf{v}}\big].$$

Once again

$$(\hat{\mathbf{v}} \times \mathbf{p}) \cdot \hat{\mathbf{v}} = 0$$

and

$$(\hat{\mathbf{v}} \times \mathbf{p}) \times \hat{\mathbf{v}} = \mathbf{p} - (\mathbf{p} \cdot \hat{\mathbf{v}})\hat{\mathbf{v}}.$$

Therefore,

$$q^{-1}pq = \big[0, \ \lambda^2(\hat{\mathbf{v}} \cdot \mathbf{p})\hat{\mathbf{v}} + s^2\mathbf{p} - 2\lambda s\hat{\mathbf{v}} \times \mathbf{p} - \lambda^2\mathbf{p} + \lambda^2(\mathbf{p} \cdot \hat{\mathbf{v}})\hat{\mathbf{v}}\big]$$
$$= \big[0, \ 2\lambda^2(\hat{\mathbf{v}} \cdot \mathbf{p})\hat{\mathbf{v}} + (s^2 - \lambda^2)\mathbf{p} - 2\lambda s\hat{\mathbf{v}} \times \mathbf{p}\big].$$

Again, let's make $s = \cos\theta$ and $\lambda = \sin\theta$:

$$q^{-1}pq = \big[0, \ (1 - \cos(2\theta))(\hat{\mathbf{v}} \cdot \mathbf{p})\hat{\mathbf{v}} + \cos(2\theta)\mathbf{p} - \sin(2\theta)\hat{\mathbf{v}} \times \mathbf{p}\big]$$

and the only thing that has changed from qpq^{-1} is the sign of the cross-product term, which reverses the direction of its vector. However, we must remember to compensate for the angle-doubling by halving θ:

$$q^{-1}pq = \big[0, \ (1 - \cos\theta)(\hat{\mathbf{v}} \cdot \mathbf{p})\hat{\mathbf{v}} + \cos\theta\mathbf{p} - \sin\theta\hat{\mathbf{v}} \times \mathbf{p}\big]. \qquad (12.10)$$

Let's see what happens when we employ (12.10) to rotate $\mathbf{p} = \mathbf{i} + \mathbf{k}$, $90°$ about the quaternion's vector $\hat{\mathbf{v}} = \mathbf{k}$:

Fig. 12.5 The point
$P(0, 1, 1)$ is rotated $90°$ to
$P'(1, 1, 0)$ about the y-axis

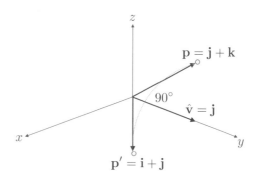

$$q^{-1}pq = [0, \ (\mathbf{k} \cdot (\mathbf{i} + \mathbf{k})\mathbf{k}) - \mathbf{j}]$$
$$= [0, \ -\mathbf{j} + \mathbf{k}]$$

which has rotated $\mathbf{p}$ clockwise $90°$ about the quaternion's vector. Therefore, the rotor qpq^{-1} rotates a vector counter-clockwise, and $q^{-1}pq$ rotates a vector clockwise:

$$qpq^{-1} = \left[0, \ (1 - \cos\theta)(\hat{\mathbf{v}} \cdot \mathbf{p})\hat{\mathbf{v}} + \cos\theta\mathbf{p} + \sin\theta\hat{\mathbf{v}} \times \mathbf{p}\right]$$
$$q^{-1}pq = \left[0, \ (1 - \cos\theta)(\hat{\mathbf{v}} \cdot \mathbf{p})\hat{\mathbf{v}} + \cos\theta\mathbf{p} - \sin\theta\hat{\mathbf{v}} \times \mathbf{p}\right].$$

Let's compute another example. Consider the point $P(0, 1, 1)$ in Fig. 12.5 which is to be rotated $90°$ about the y-axis. We can see that the rotated point P' has the coordinates $(1, 1, 0)$ which we will confirm algebraically. The point P is represented by its position vector $\mathbf{p}$ in the pure quaternion

$$p = [0, \ \mathbf{p}].$$

The axis of rotation is $\hat{\mathbf{v}} = \mathbf{j}$, and the vector to be rotated is $\mathbf{p} = \mathbf{j} + \mathbf{k}$. Therefore,

$$qpq^{-1} = \left[0, \ (1 - \cos\theta)(\hat{\mathbf{v}} \cdot \mathbf{p})\hat{\mathbf{v}} + \cos\theta\mathbf{p} + \sin\theta\hat{\mathbf{v}} \times \mathbf{p}\right]$$
$$= [0, \ \mathbf{j} \cdot (\mathbf{j} + \mathbf{k})\,\mathbf{j} + \mathbf{j} \times (\mathbf{j} + \mathbf{k})]$$
$$= [0, \ \mathbf{i} + \mathbf{j}]$$

and confirms that P is indeed rotated to $(1, 1, 0)$.

Now let's explore how this product is represented in matrix form.

12.4 Quaternions in Matrix Form

Having discovered a vector equation to represent qpq^{-1}, let's continue and convert it into a matrix. We will explore two methods: the first is a simple vectorial method which translates the vector equation representing qpq^{-1} directly into matrix form. The second method uses matrix algebra to develop a rather cunning solution.

12.4.1   Vector Method

For the vector method it is convenient to describe the unit-norm quaternion as

$$q = [s, \ \mathbf{v}]$$
$$= [s, \ x\mathbf{i} + y\mathbf{j} + z\mathbf{k}]$$

where

$$s^2 + \|\mathbf{v}\|^2 = 1$$

and the pure quaternion as

$$p = [0, \ \mathbf{p}]$$
$$= [0, \ x_p\mathbf{i} + y_p\mathbf{j} + z_p\mathbf{k}].$$

A simple way to compute qpq^{-1} is to use (12.9) and substitute $\|\mathbf{v}\|$ for λ:

$$qpq^{-1} = \left[0, \ 2\lambda^2 \left(\hat{\mathbf{v}} \cdot \mathbf{p}\right)\hat{\mathbf{v}} + \left(s^2 - \lambda^2\right)\mathbf{p} + 2\lambda s\hat{\mathbf{v}} \times \mathbf{p}\right]$$
$$= \left[0, \ 2\|\mathbf{v}\|^2 \left(\hat{\mathbf{v}} \cdot \mathbf{p}\right)\hat{\mathbf{v}} + \left(s^2 - \|\mathbf{v}\|^2\right)\mathbf{p} + 2\|\mathbf{v}\|s\hat{\mathbf{v}} \times \mathbf{p}\right].$$

Next, we substitute $\mathbf{v}$ for $\|\mathbf{v}\|\hat{\mathbf{v}}$:

$$qpq^{-1} = \left[0, \ 2\left(\mathbf{v} \cdot \mathbf{p}\right)\mathbf{v} + \left(s^2 - \|\mathbf{v}\|^2\right)\mathbf{p} + 2s\mathbf{v} \times \mathbf{p}\right].$$

Finally, as we are working with unit-norm quaternions to prevent scaling

$$s^2 + \|\mathbf{v}\|^2 = 1$$

and

$$s^2 - \|\mathbf{v}\|^2 = 2s^2 - 1$$

therefore,

$$qpq^{-1} = \left[0, \ 2(\mathbf{v} \cdot \mathbf{p})\mathbf{v} + \left(2s^2 - 1\right)\mathbf{p} + 2s\mathbf{v} \times \mathbf{p}\right].$$

If we let $p' = qpq^{-1}$, which is a pure quaternion, we have

$$p' = qpq^{-1}$$
$$= [0, \ \mathbf{p}']$$
$$= \left[0, \ 2(\mathbf{v} \cdot \mathbf{p})\mathbf{v} + \left(2s^2 - 1\right)\mathbf{p} + 2s\mathbf{v} \times \mathbf{p}\right]$$
$$\mathbf{p}' = 2(\mathbf{v} \cdot \mathbf{p})\mathbf{v} + \left(2s^2 - 1\right)\mathbf{p} + 2s\mathbf{v} \times \mathbf{p}.$$

We are only interested in the rotated vector $\mathbf{p}'$ comprising the three terms $2(\mathbf{v} \cdot \mathbf{p})\mathbf{v}$, $(2s^2 - 1)\mathbf{p}$ and $2s\mathbf{v} \times \mathbf{p}$, which can be represented by three individual matrices and summed together.

$$2(\mathbf{v} \cdot \mathbf{p})\mathbf{v} = 2(xx_p + yy_p + zz_p)(x\mathbf{i} + y\mathbf{j} + z\mathbf{k})$$

$$= \begin{bmatrix} 2x^2 & 2xy & 2xz \\ 2xy & 2y^2 & 2yz \\ 2xz & 2yz & 2z^2 \end{bmatrix} \begin{bmatrix} x_p \\ y_p \\ z_p \end{bmatrix}$$

$$(2s^2 - 1)\mathbf{p} = (2s^2 - 1)x_p\mathbf{i} + (2s^2 - 1)y_p\mathbf{j} + (2s^2 - 1)z_p\mathbf{k}$$

$$= \begin{bmatrix} 2s^2 - 1 & 0 & 0 \\ 0 & 2s^2 - 1 & 0 \\ 0 & 0 & 2s^2 - 1 \end{bmatrix} \begin{bmatrix} x_p \\ y_p \\ z_p \end{bmatrix}$$

$$2s\mathbf{v} \times \mathbf{p} = 2s\left((yz_p - zy_p)\mathbf{i} + (zx_p - xz_p)\mathbf{j} + (xy_p - yx_p)\mathbf{k}\right)$$

$$= \begin{bmatrix} 0 & -2sz & 2sy \\ 2sz & 0 & -2sx \\ -2sy & 2sx & 0 \end{bmatrix} \begin{bmatrix} x_p \\ y_p \\ z_p \end{bmatrix}.$$

Adding these matrices together:

$$\mathbf{p}' = \begin{bmatrix} 2(s^2 + x^2) - 1 & 2(xy - sz) & 2(xz + sy) \\ 2(xy + sz) & 2(s^2 + y^2) - 1 & 2(yz - sx) \\ 2(xz - sy) & 2(yz + sx) & 2(s^2 + z^2) - 1 \end{bmatrix} \begin{bmatrix} x_p \\ y_p \\ z_p \end{bmatrix} \qquad (12.11)$$

or

$$\mathbf{p}' = \begin{bmatrix} 1 - 2(y^2 + z^2) & 2(xy - sz) & 2(xz + sy) \\ 2(xy + sz) & 1 - 2(x^2 + z^2) & 2(yz - sx) \\ 2(xz - sy) & 2(yz + sx) & 1 - 2(x^2 + y^2) \end{bmatrix} \begin{bmatrix} x_p \\ y_p \\ z_p \end{bmatrix} \qquad (12.12)$$

where

$$[0, \ \mathbf{p}'] = qpq^{-1}.$$

Now let's reverse the product. To compute the vector part of $q^{-1}pq$ all that we have to do is reverse the sign of $2s\mathbf{v} \times \mathbf{p}$:

$$\mathbf{p}' = \begin{bmatrix} 2(s^2 + x^2) - 1 & 2(xy + sz) & 2(xz - sy) \\ 2(xy - sz) & 2(s^2 + y^2) - 1 & 2(yz + sx) \\ 2(xz + sy) & 2(yz - sx) & 2(s^2 + z^2) - 1 \end{bmatrix} \begin{bmatrix} x_p \\ y_p \\ z_p \end{bmatrix} \qquad (12.13)$$

or

$$\mathbf{p}' = \begin{bmatrix} 1 - 2(y^2 + z^2) & 2(xy + sz) & 2(xz - sy) \\ 2(xy - sz) & 1 - 2(x^2 + z^2) & 2(yz + sx) \\ 2(xz + sy) & 2(yz - sx) & 1 - 2(x^2 + y^2) \end{bmatrix} \begin{bmatrix} x_p \\ y_p \\ z_p \end{bmatrix} \qquad (12.14)$$

where

$$[0,\ \mathbf{p}'] = q^{-1}pq.$$

Observe that (12.13) is the transpose of (12.11), and (12.14) is the transpose of (12.12).

12.4.2 Geometric Verification

Let's illustrate the action of (12.11) by rotating the point $(0, 1, 1)$, $90°$ about the y-axis, as shown in Fig. 12.6. The quaternion takes the form

$$q = \left[\cos\left(\tfrac{\theta}{2}\right),\ \sin\left(\tfrac{\theta}{2}\right)\hat{\mathbf{v}}\right]$$

which means that $\theta = 90°$ and $\hat{\mathbf{v}} = \mathbf{j}$, therefore,

$$q = \left[\cos 45°,\ \sin 45°\hat{\mathbf{j}}\right].$$

Consequently,

$$s = \tfrac{\sqrt{2}}{2},\quad x = 0,\quad y = \tfrac{\sqrt{2}}{2},\quad z = 0.$$

Substituting these values in (12.11) gives

$$\mathbf{p}' = \begin{bmatrix} 2\left(s^2 + x^2\right) - 1 & 2(xy - sz) & 2(xz + sy) \\ 2(xy + sz) & 2\left(s^2 + y^2\right) - 1 & 2(yz - sx) \\ 2(xz - sy) & 2(yz + sx) & 2\left(s^2 + z^2\right) - 1 \end{bmatrix} \begin{bmatrix} x_p \\ y_p \\ z_p \end{bmatrix}$$

$$\begin{bmatrix} 1 \\ 1 \\ 0 \end{bmatrix} = \begin{bmatrix} 0 & 0 & 1 \\ 0 & 1 & 0 \\ -1 & 0 & 0 \end{bmatrix} \begin{bmatrix} 0 \\ 1 \\ 1 \end{bmatrix}$$

where $(0,\ 1,\ 1)$ is rotated to $(1,\ 1,\ 0)$, which is correct.

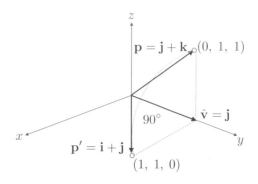

Fig. 12.6 The point $P(0, 1, 1)$ is rotated $90°$ to $P'(1, 1, 0)$ about the y-axis

So now we have a transform that rotates a point about an arbitrary axis intersecting the origin without the problems of gimbal lock associated with Euler transforms.

Before moving on, let's evaluate one more example. Let's perform a 180° rotation about a vector $\mathbf{v} = \mathbf{i} + \mathbf{k}$. To begin with, we will deliberately forget to convert the $\mathbf{v}$ into a unit vector, just to see what happens to the final matrix. The quaternion takes the form

$$q = \left[\cos\left(\tfrac{\theta}{2}\right), \ \sin\left(\tfrac{\theta}{2}\right)\hat{\mathbf{v}}\right]$$

but we will use $\mathbf{v}$ as specified. Therefore, with $\theta = 180°$

$$s = 0, \quad x = 1, \quad y = 0, \quad z = 1.$$

Substituting these values in (12.11) gives

$$\mathbf{p}' = \begin{bmatrix} 2\left(s^2 + x^2\right) - 1 & 2(xy - sz) & 2(xz + sy) \\ 2(xy + sz) & 2\left(s^2 + y^2\right) - 1 & 2(yz - sx) \\ 2(xz - sy) & 2(yz + sx) & 2\left(s^2 + z^2\right) - 1 \end{bmatrix} \begin{bmatrix} x_p \\ y_p \\ z_p \end{bmatrix}$$

$$= \begin{bmatrix} 1 & 0 & 2 \\ 0 & -1 & 0 \\ 2 & 0 & 1 \end{bmatrix} \begin{bmatrix} 1 \\ 0 \\ 0 \end{bmatrix}$$

which looks nothing like a rotation matrix, and reminds us how important it is to have a unit vector to represent the axis. Let's repeat these calculations normalising the vector to $\hat{\mathbf{v}} = \frac{\sqrt{2}}{2}\mathbf{i} + \frac{\sqrt{2}}{2}\mathbf{k}$:

$$s = 0, \quad x = \tfrac{\sqrt{2}}{2}, \quad y = 0, \quad z = \tfrac{\sqrt{2}}{2}.$$

Substituting these values in (12.11) gives

$$\mathbf{p}' = \begin{bmatrix} 2\left(s^2 + x^2\right) - 1 & 2(xy - sz) & 2(xz + sy) \\ 2(xy + sz) & 2\left(s^2 + y^2\right) - 1 & 2(yz - sx) \\ 2(xz - sy) & 2(yz + sx) & 2\left(s^2 + z^2\right) - 1 \end{bmatrix} \begin{bmatrix} x_p \\ y_p \\ z_p \end{bmatrix}$$

$$\begin{bmatrix} 0 \\ 0 \\ 1 \end{bmatrix} = \begin{bmatrix} 0 & 0 & 1 \\ 0 & -1 & 0 \\ 1 & 0 & 0 \end{bmatrix} \begin{bmatrix} 1 \\ 0 \\ 0 \end{bmatrix}$$

which not only looks like a rotation matrix, but has a determinant of 1 and rotates the point $(1, \ 0, \ 0)$ to $(0, \ 0, \ 1)$ as shown in Fig. 12.7.

Fig. 12.7 The point
$P(1,\ 0,\ 0)$ is rotated $180°$ to
$P'(0,\ 0,\ 1)$ about $\hat{\mathbf{v}}$

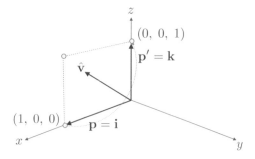

12.5   Multiple Rotations

Say a vector or frame of reference is subjected to two rotations specified by q_1 followed by q_2. There is a temptation to convert both quaternions to their respective matrix and multiply the matrices together. However, this not the most efficient way of combining the rotations. It is best to accumulate the rotations as quaternions and then convert to matrix notation, if required.

To illustrate this, consider the pure quaternion p subjected to the first quaternion q_1:

$$q_1 p q_1^{-1}$$

followed by a second quaternion q_2

$$q_2 \left(q_1 p q_1^{-1} \right) q_2^{-1}$$

which can be expressed as

$$(q_2 q_1)\, p\, (q_2 q_1)^{-1}.$$

Extra quaternions can be added accordingly. Let's illustrate this with two examples. To keep things simple, the first quaternion q_1 rotates $30°$ about the y-axis:

$$q_1 = \left[\cos 15°,\ \sin 15°\mathbf{j}\right].$$

The second quaternion q_2 rotates $60°$ also about the y-axis:

$$q_2 = \left[\cos 30°,\ \sin 30°\mathbf{j}\right].$$

Together, the two quaternions rotate $90°$ about the y-axis. To accumulate these rotations, we multiply them together:

$$
\begin{aligned}
q_1 q_2 &= \left[\cos 15°,\ \sin 15°\mathbf{j}\right]\left[\cos 30°,\ \sin 30°\mathbf{j}\right] \\
&= \left[\cos 15° \cos 30° - \sin 15° \sin 30°,\ \cos 15° \sin 30°\mathbf{j} + \cos 30° \sin 15°\mathbf{j}\right] \\
&= \tfrac{\sqrt{2}}{2}[1,\ \mathbf{j}]
\end{aligned}
$$

which is a quaternion that rotates 90° about the y-axis. Using the matrix (12.11) we have

$$
\mathbf{p}' = \begin{bmatrix} 2\left(s^2 + x^2\right) - 1 & 2\left(xy - sz\right) & 2\left(xz + sy\right) \\ 2\left(xy + sz\right) & 2\left(s^2 + y^2\right) - 1 & 2\left(yz - sx\right) \\ 2\left(xz - sy\right) & 2\left(yz + sx\right) & 2\left(s^2 + z^2\right) - 1 \end{bmatrix} \begin{bmatrix} x_p \\ y_p \\ z_p \end{bmatrix}
$$

$$
= \begin{bmatrix} 0 & 0 & 1 \\ 0 & 1 & 0 \\ -1 & 0 & 0 \end{bmatrix} \begin{bmatrix} x_p \\ y_p \\ z_p \end{bmatrix}
$$

which rotates points about the y-axis by 90°.

For a second example, let's just evaluate the quaternions. The first quaternion q_1 rotates 90° about the x-axis, and q_2 rotates 90° about the y-axis:

$$
q_1 = \tfrac{\sqrt{2}}{2}\,[1,\ \mathbf{i}]
$$
$$
q_2 = \tfrac{\sqrt{2}}{2}\,[1,\ \mathbf{j}]
$$
$$
p = [0,\ \mathbf{i} + \mathbf{j}].
$$

Therefore,

$$
\begin{aligned}
q_2 q_1 &= \tfrac{\sqrt{2}}{2}\,[1,\ \mathbf{i}]\,\tfrac{\sqrt{2}}{2}\,[1,\ \mathbf{j}] \\
&= \tfrac{1}{2}\,[1,\ \mathbf{i} + \mathbf{j} - \mathbf{k}] \\
(q_2 q_1)^{-1} &= \tfrac{1}{2}\,[1,\ -\mathbf{i} - \mathbf{j} + \mathbf{k}] \\
(q_2 q_1)\, p &= \tfrac{1}{2}\,[1,\ \mathbf{i} + \mathbf{j} - \mathbf{k}]\,[0,\ \mathbf{i} + \mathbf{j}] \\
&= \tfrac{1}{2}\,[-2,\ (\mathbf{i} + \mathbf{j}) + \mathbf{i} - \mathbf{j}] \\
&= [-1,\ \mathbf{i}] \\
(q_2 q_1)\, p\,(q_2 q_1)^{-1} &= \tfrac{1}{2}\,[-1,\ \mathbf{i}]\,[1,\ -\mathbf{i} - \mathbf{j} + \mathbf{k}] \\
&= \tfrac{1}{2}\,[-1 + 1,\ \mathbf{i} + \mathbf{j} - \mathbf{k} + \mathbf{i} - \mathbf{j} - \mathbf{k}] \\
&= [0,\ \mathbf{i} - \mathbf{k}].
\end{aligned}
$$

Thus the point $(1,\ 1,\ 0)$ is rotated to $(1,\ 0,\ -1)$, which is correct.

12.6 Rotating About an Off-Set Axis

Now that we have a matrix to represent a quaternion rotor, we can employ it to resolve problems such as rotating a point about an off-set axis using the same techniques associated with normal rotation transforms. We use the following notation to rotate a point about a fixed axis parallel with the y-axis:

$$\begin{bmatrix} x' \\ y' \\ z' \\ 1 \end{bmatrix} = \mathbf{T}_{(t_x,\,0,\,t_z)} \mathbf{R}_{\beta,\,y} \mathbf{T}_{(-t_x,\,0,\,-t_z)} \begin{bmatrix} x \\ y \\ z \\ 1 \end{bmatrix}$$

Therefore, by substituting the matrix qpq^{-1} for $\mathbf{R}_{\beta,\,y}$ we have:

$$\begin{bmatrix} x' \\ y' \\ z' \\ 1 \end{bmatrix} = \mathbf{T}_{(t_x,\,0,\,t_z)} \left(qpq^{-1} \right) \mathbf{T}_{(-t_x,\,0,\,-t_z)} \begin{bmatrix} x \\ y \\ z \\ 1 \end{bmatrix}.$$

Let's test this by rotating our unit cube 90° about the axis intersecting vertices 4 and 6 as shown in Fig. 12.8. The unit-norm quaternion to achieve this is

$$q = \left[\cos 45°, \ \sin 45° \mathbf{j} \right]$$

with the pure quaternion

$$p = [0, \ \mathbf{p}].$$

Consequently,

$$s = \tfrac{\sqrt{2}}{2}, \quad x = 0, \quad y = \tfrac{\sqrt{2}}{2}, \quad z = 0$$

and using (12.11) in a homogeneous form we have

$$\mathbf{p}' = \begin{bmatrix} 2\left(s^2 + x^2\right) - 1 & 2(xy - sz) & 2(xz + sy) & 0 \\ 2(xy + sz) & 2\left(s^2 + y^2\right) - 1 & 2(yz - sx) & 0 \\ 2(xz - sy) & 2(yz + sx) & 2\left(s^2 + z^2\right) - 1 & 0 \\ 0 & 0 & 0 & 1 \end{bmatrix} \begin{bmatrix} x_p \\ y_p \\ z_p \\ 1 \end{bmatrix}$$

$$= \begin{bmatrix} 0 & 0 & 1 & 0 \\ 0 & 1 & 0 & 0 \\ -1 & 0 & 0 & 0 \\ 0 & 0 & 0 & 1 \end{bmatrix} \begin{bmatrix} x_p \\ y_p \\ z_p \\ 1 \end{bmatrix}.$$

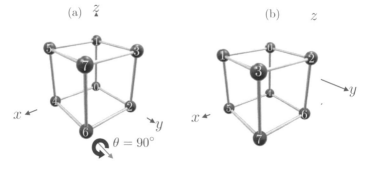

(a)   (b)

$\theta = 90°$

Fig. 12.8 The cube is rotated 90° about the axis intersecting vertices 4 and 6

The other two matrices are

$$T_{(-t_x, 0, 0)} = \begin{bmatrix} 1 & 0 & 0 & -1 \\ 0 & 1 & 0 & 0 \\ 0 & 0 & 1 & 0 \\ 0 & 0 & 0 & 1 \end{bmatrix}$$

$$T_{(t_x, 0, 0)} = \begin{bmatrix} 1 & 0 & 0 & 1 \\ 0 & 1 & 0 & 0 \\ 0 & 0 & 1 & 0 \\ 0 & 0 & 0 & 1 \end{bmatrix}.$$

Multiplying these three matrices together creates:

$$\mathbf{p}'T_{(-t_x, 0, 0)} = \begin{bmatrix} 0 & 0 & 1 & 0 \\ 0 & 1 & 0 & 0 \\ -1 & 0 & 0 & 0 \\ 0 & 0 & 0 & 1 \end{bmatrix} \begin{bmatrix} 1 & 0 & 0 & -1 \\ 0 & 1 & 0 & 0 \\ 0 & 0 & 1 & 0 \\ 0 & 0 & 0 & 1 \end{bmatrix} = \begin{bmatrix} 0 & 0 & 1 & 0 \\ 0 & 1 & 0 & 0 \\ -1 & 0 & 0 & 1 \\ 0 & 0 & 0 & 1 \end{bmatrix}$$

$$T_{(t_x, 0, 0)}\mathbf{p}'T_{(-t_x, 0, 0)} = \begin{bmatrix} 1 & 0 & 0 & 1 \\ 0 & 1 & 0 & 0 \\ 0 & 0 & 1 & 0 \\ 0 & 0 & 0 & 1 \end{bmatrix} \begin{bmatrix} 0 & 0 & 1 & 0 \\ 0 & 1 & 0 & 0 \\ -1 & 0 & 0 & 1 \\ 0 & 0 & 0 & 1 \end{bmatrix} = \begin{bmatrix} 0 & 0 & 1 & 1 \\ 0 & 1 & 0 & 0 \\ -1 & 0 & 0 & 1 \\ 0 & 0 & 0 & 1 \end{bmatrix}$$

$$T_{(t_x, 0, 0)}\mathbf{p}'T_{(-t_x, 0, 0)} = \begin{bmatrix} 0 & 0 & 1 & 1 \\ 0 & 1 & 0 & 0 \\ -1 & 0 & 0 & 1 \\ 0 & 0 & 0 & 1 \end{bmatrix}. \tag{12.15}$$

Although not mathematically correct, the following statement shows the matrix (12.15) and the array of coordinates representing a unit cube, followed by the rotated cube's coordinates.

$$\begin{bmatrix} 0 & 0 & 1 & 1 \\ 0 & 1 & 0 & 0 \\ -1 & 0 & 0 & 1 \\ 0 & 0 & 0 & 1 \end{bmatrix} \begin{bmatrix} 0 & 0 & 0 & 0 & 1 & 1 & 1 & 1 \\ 0 & 0 & 1 & 1 & 0 & 0 & 1 & 1 \\ 0 & 1 & 0 & 1 & 0 & 1 & 0 & 1 \\ 1 & 1 & 1 & 1 & 1 & 1 & 1 & 1 \end{bmatrix} = \begin{bmatrix} 1 & 2 & 1 & 2 & 1 & 2 & 1 & 2 \\ 0 & 0 & 1 & 1 & 0 & 0 & 1 & 1 \\ 1 & 1 & 1 & 1 & 0 & 0 & 0 & 0 \\ 1 & 1 & 1 & 1 & 1 & 1 & 1 & 1 \end{bmatrix}.$$

These coordinates are confirmed by Fig. 12.8.

12.7 Converting a Rotation Matrix to a Quaternion

Very often one has a 3D rotation matrix which would be nice to see as a quaternion. So let's see how this can be realised. The matrix transform equivalent to qpq^{-1} is

$$qpq^{-1} = \begin{bmatrix} 2\left(s^2 + x^2\right) - 1 & 2(xy - sz) & 2(xz + sy) \\ 2(xy + sz) & 2\left(s^2 + y^2\right) - 1 & 2(yz - sx) \\ 2(xz - sy) & 2(yz + sx) & 2\left(s^2 + z^2\right) - 1 \end{bmatrix} \begin{bmatrix} x_p \\ y_p \\ z_p \end{bmatrix} \quad (12.16)$$

$$= \begin{bmatrix} a_{11} & a_{12} & a_{13} \\ a_{21} & a_{22} & a_{23} \\ a_{31} & a_{32} & a_{33} \end{bmatrix} \begin{bmatrix} x_p \\ y_p \\ z_p \end{bmatrix}. \quad (12.17)$$

Inspection of (12.16) and (12.17) shows that by combining various elements we can isolate the terms of a quaternion s, x, y, z. For example, by adding the diagonal terms of (12.17): $a_{11} + a_{22} + a_{33}$, we obtain

$$a_{11} + a_{22} + a_{33} = \left[2\left(s^2 + x^2\right) - 1\right] + \left[2\left(s^2 + y^2\right) - 1\right] + \left[2\left(s^2 + z^2\right) - 1\right]$$
$$= 6s^2 + 2\left(x^2 + y^2 + z^2\right) - 3$$
$$= 4s^2 - 1$$

therefore,

$$s = \tfrac{1}{2}\sqrt{1 + a_{11} + a_{22} + a_{33}}.$$

To isolate x, y, z we employ:

$$x = \frac{1}{4s}\left(a_{32} - a_{23}\right)$$

$$y = \frac{1}{4s}\left(a_{13} - a_{31}\right)$$

$$z = \frac{1}{4s}\left(a_{21} - a_{12}\right).$$

12.8 Summary

This chapter has shown how a quaternion is used to rotate a vector about a quaternion's vector. It would have been useful if this could have been achieved by the simple product qp, like complex numbers. But as we saw, this only works when the quaternion is orthogonal to the vector. The product qpq^{-1}—discovered by Hamilton and Cayley—works for all orientations between a quaternion and a vector. It is also relatively easy to compute. We also saw that the product can be represented as a matrix, which can be integrated with other matrices.

Perhaps one of the most interesting features of quaternions that has emerged in this chapter, is that their imaginary qualities are not required in any calculations, because they are embedded within the algebra.

The reverse product $q^{-1}pq$ reverses the angle of rotation, and is equivalent to changing the sign of the rotation angle in qpq^{-1}. Consequently, it can be used to rotate a frame of reference in the same direction as qpq^{-1}.

12.8.1 Summary of Definitions

Rotating a vector by a quaternion

$$q = [s, \ \mathbf{v}]$$
$$s^2 + \|\mathbf{v}\|^2 = 1$$
$$p = [0, \ \mathbf{p}]$$
$$qpq^{-1} = \left[0, \ 2(\mathbf{v} \cdot \mathbf{p})\mathbf{v} + \left(2s^2 - 1\right)\mathbf{p} + 2s\mathbf{v} \times \mathbf{p}\right].$$

$$q = \left[\cos\left(\tfrac{\theta}{2}\right), \ \sin\left(\tfrac{\theta}{2}\right)\hat{\mathbf{v}}\right]$$
$$p = [0, \ \mathbf{p}]$$
$$qpq^{-1} = \left[0, \ (1 - \cos\theta)(\hat{\mathbf{v}} \cdot \mathbf{p})\hat{\mathbf{v}} + \cos\theta\mathbf{p} + \sin\theta\hat{\mathbf{v}} \times \mathbf{p}\right].$$

Matrix for rotating a vector by a quaternion

$$\mathbf{p}' = \begin{bmatrix} 1 - 2\left(y^2 + z^2\right) & 2(xy - sz) & 2(xz + sy) \\ 2(xy + sz) & 1 - 2\left(x^2 + z^2\right) & 2(yz - sx) \\ 2(xz - sy) & 2(yz + sx) & 1 - 2\left(x^2 + y^2\right) \end{bmatrix} \begin{bmatrix} x_p \\ y_p \\ z_p \end{bmatrix}.$$

12.9 Worked Examples

Here are some further worked examples that employ the ideas described above.

12.9.1 Special Case Quaternion

Use qp to rotate $p = [0, \ \mathbf{j}]$ $90°$ about the x-axis.
For this to work q must be orthogonal to p:

$$q = [\cos\theta, \ \sin\theta\mathbf{i}]$$
$$= [0, \ \mathbf{i}]$$

and

$$p' = qp$$
$$= [0, \ \mathbf{i}][0, \ \mathbf{j}]$$
$$= [0, \ \mathbf{k}].$$

12.9.2   *Rotating a Vector Using a Quaternion*

Use qpq^{-1} to rotate $p = [0, \; \mathbf{j}]$ $90°$ about the x-axis.
For this to work:

$$q = \left[\cos\left(\tfrac{\theta}{2}\right), \; \sin\left(\tfrac{\theta}{2}\right)\mathbf{i}\right]$$
$$= \left[\tfrac{\sqrt{2}}{2}, \; \tfrac{\sqrt{2}}{2}\mathbf{i}\right]$$
$$= \tfrac{\sqrt{2}}{2}[1, \; \mathbf{i}]$$

and

$$p' = qpq^{-1}$$
$$= \tfrac{\sqrt{2}}{2}\tfrac{\sqrt{2}}{2}[1, \; \mathbf{i}]\,[0, \; \mathbf{j}]\,[1, \; -\mathbf{i}]$$
$$= \tfrac{1}{2}\,[0, \; \mathbf{j}+\mathbf{k}]\,[1, \; -\mathbf{i}]$$
$$= \tfrac{1}{2}\,[0, \; (\mathbf{j}+\mathbf{k})-\mathbf{j}+\mathbf{k}]$$
$$= [0, \; \mathbf{k}]\,.$$

12.9.3   *Evaluate qpq^{-1}*

Evaluate qpq^{-1} for $p = [0, \; \mathbf{p}]$ and $q = \left[\cos\left(\tfrac{\theta}{2}\right), \; \sin\left(\tfrac{\theta}{2}\right)\mathbf{v}\right]$, where $\theta = 360°$.

$$q = [-1, \; \mathbf{0}]$$
$$qpq^{-1} = [-1, \; \mathbf{0}]\,[0, \; \mathbf{p}]\,[-1, \; \mathbf{0}]$$
$$= [0, \; -\mathbf{p}]\,[-1, \; \mathbf{0}]$$
$$= [0, \; \mathbf{p}]$$

which confirms that the vector remains unmoved, as expected.

12.9.4   *Evaluate qpq^{-1} Using a Matrix*

Compute the matrix for $q = \left[\tfrac{1}{2}, \; \tfrac{\sqrt{3}}{2}\mathbf{k}\right]$.
   From q:

$$s = \tfrac{1}{2}, \quad x = 0, \quad y = 0, \quad z = \tfrac{\sqrt{3}}{2}$$

$$\mathbf{p}' = \begin{bmatrix} 2\left(s^2 + x^2\right) - 1 & 2\left(xy - sz\right) & 2\left(xz + sy\right) \\ 2\left(xy + sz\right) & 2\left(s^2 + y^2\right) - 1 & 2\left(yz - sx\right) \\ 2\left(xz - sy\right) & 2\left(yz + sx\right) & 2\left(s^2 + z^2\right) - 1 \end{bmatrix} \begin{bmatrix} x_p \\ y_p \\ z_p \end{bmatrix}$$

$$= \begin{bmatrix} -\frac{1}{2} & -\frac{\sqrt{3}}{2} & 0 \\ \frac{\sqrt{3}}{2} & -\frac{1}{2} & 0 \\ 0 & 0 & 1 \end{bmatrix} \begin{bmatrix} x_p \\ y_p \\ z_p \end{bmatrix}.$$

If we plug in the point $(1, \ 0, \ 0)$, it is rotated about the z-axis by $120°$:

$$\begin{bmatrix} -\frac{1}{2} \\ \frac{\sqrt{3}}{2} \\ 1 \end{bmatrix} = \begin{bmatrix} -\frac{1}{2} & -\frac{\sqrt{3}}{2} & 0 \\ \frac{\sqrt{3}}{2} & -\frac{1}{2} & 0 \\ 0 & 0 & 1 \end{bmatrix} \begin{bmatrix} 1 \\ 0 \\ 0 \end{bmatrix}.$$

References

1. Cayley A (1848) The collected mathematical papers, vol I, p 586, note 20
2. Altmann SL (1986) Rotations, quaternions and double groups, p 16. Dover Publications
3. Vince JA (2017) Mathematics for computer graphics, 5th edn. Springer

Chapter 13
Interpolation

13.1 Introduction

This chapter covers linear and non-linear interpolation of scalars, and includes trigonometric and cubic polynomials. It also includes the interpolation of vectors and quaternions.

13.2 Background

Interpolation is not a branch of mathematics but rather a collection of techniques the reader will find useful when solving computer graphic problems. Basically, an *interpolant* is a strategy for selecting a number between two limits. For example, if the limits are 2 and 4, a parameter t can be used to select the sequence 2.0, 2.2, 2.4, 2.6, 2.8, 3.0, 3.2, 3.4, 3.6, 3.8, and 4. These numbers could then be used to translate, scale, rotate an object, move a virtual camera, or change the position, colour or brightness of a virtual light source.

To implement the above interpolant for different limits we require a general algorithm, which is one of the first exercises of this chapter. We also need to explore ways of controlling the spacing between the interpolated values. In animation, for example, we often need to move an object very slowly and gradually increase its speed. Conversely, we may want to bring an object to a halt, making its speed less and less. The interpolant function includes a parameter within its algorithm, which permits any interpolated value to be created at will. The parameter can depend upon time, or operate over a distance in space.

© Springer-Verlag London Ltd., part of Springer Nature 2022
J. Vince, *Mathematics for Computer Graphics*, Undergraduate Topics in Computer Science, https://doi.org/10.1007/978-1-4471-7520-9_13

13.3   Linear Interpolation

A *linear interpolant* generates equal spacing between the interpolated values for equal changes in the interpolating parameter. In the above example the increment 0.2 is calculated by subtracting the first number from the second and dividing the result by 10, i.e. $(4 - 2)/10 = 0.2$. Although this works, it is not in a very flexible form, so let's express the problem differently.

Given two numbers n_1 and n_2, which represent the start and final values of the interpolant, we require an interpolated value controlled by a parameter t that varies between 0 and 1. When $t = 0$, the result is n_1, and when $t = 1$, the result is n_2. A solution to this problem is given by

$$n = n_1 + t(n_2 - n_1)$$

for when $n_1 = 2$, $n_2 = 4$ and $t = 0.5$:

$$n = 2 + \tfrac{1}{2}(4 - 2) = 3$$

which is a halfway point. Furthermore, when $t = 0, n = n_1$, and when $t = 1, n = n_2$, which confirms that we have a sound interpolant. However, it can be expressed differently:

$$n = n_1(1 - t) + n_2 t \tag{13.1}$$

which shows what is really going on, and forms the basis for further development. Figure 13.1 shows the graphs of $n = 1 - t$ and $n = t$ over the range $0 \leq t \leq 1$.

With reference to (13.1), we see that as t changes from 0 to 1, the $(1 - t)$ term varies from 1 to 0. This attenuates the value of n_1 to zero over the range of t, while the t term scales n_2 from zero to its actual value. Figure 13.2 illustrates these two actions with $n_1 = 1$ and $n_2 = 5$.

Observe that the terms $(1 - t)$ and t sum to unity—this is not a coincidence. This type of interpolant ensures that if it takes a quarter of n_1, it balances it with three-quarters of n_2, and vice versa. Obviously we could design an interpolant that takes arbitrary portions of n_1 and n_2, but would lead to arbitrary results.

Fig. 13.1 The graphs of $n = 1 - t$ and $n = t$ over the range $0 \leq t \leq 1$

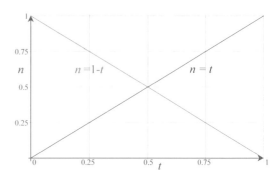

Fig. 13.2 The green line shows the result of linearly interpolating between 1 and 5

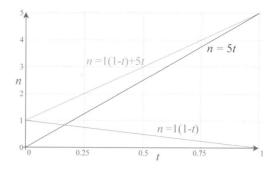

Fig. 13.3 Interpolating between the points $(1,\ 1)$ and $(4,\ 5)$

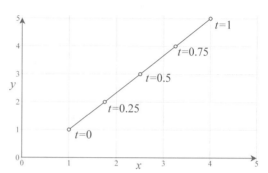

Although this interpolant is extremely simple, it is widely used in computer graphics software. Just to put it into context, consider the task of moving an object between two locations $(x_1,\ y_1,\ z_1)$ and $(x_2,\ y_2,\ z_2)$. The interpolated position is given by

$$\left.\begin{array}{l} x = x_1(1-t) + x_2 t \\ y = y_1(1-t) + y_2 t \\ z = x_1(1-t) + z_2 t \end{array}\right\} \quad 0 \le t \le 1.$$

The parameter t could be generated from two frame values within an animation. What is assured by this interpolant, is that equal steps in t result in equal steps in x, y, and z. Figure 13.3 illustrates this linear spacing with a 2D example where we interpolate between the points $(1,\ 1)$ and $(4,\ 5)$. Note the equal spacing between the intermediate interpolated points.

We can write (13.1) in matrix form as follows:

$$n = [(1-t)\ t]\begin{bmatrix} n_1 \\ n_2 \end{bmatrix}$$

or as

$$n = [t\ 1]\begin{bmatrix} -1 & 1 \\ 1 & 0 \end{bmatrix}\begin{bmatrix} n_1 \\ n_2 \end{bmatrix}.$$

The reader can confirm that this generates identical results to the algebraic form.

13.4   Non-Linear Interpolation

A linear interpolant ensures that equal steps in the parameter t give rise to equal steps in the interpolated values; but it is often required that equal steps in t give rise to unequal steps in the interpolated values. We can achieve this using a variety of mathematical techniques. For example, we could use trigonometric functions or polynomials. To begin with, let's look at a trigonometric solution.

13.4.1   Trigonometric Interpolation

In Chap. 4 we noted that $\sin^2 t + \cos^2 t = 1$, which satisfies one of the requirements of an interpolant: the terms must sum to 1. If t varies between 0 and $\pi/2$, $\cos^2 t$ varies between 1 and 0, and $\sin^2 t$ varies between 0 and 1, which can be used to modify the two interpolated values n_1 and n_2 as follows:

$$n = n_1 \cos^2 t + n_2 \sin^2 t, \qquad 0 \le t \le \tfrac{\pi}{2}. \qquad (13.2)$$

The interpolation curves are shown in Fig. 13.4.

If $n_1 = 1$ and $n_2 = 3$ in (13.2), we obtain the curves shown in Fig. 13.5. If we apply this interpolant to two 2D points in space: (1, 1) and (4, 5), we obtain a

Fig. 13.4 The curves for $n = \cos^2 t$ and $n = \sin^2 t$

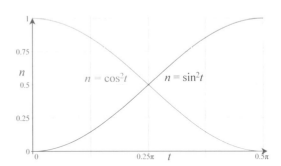

Fig. 13.5 Interpolating between 1 and 3 using a trigonometric interpolant

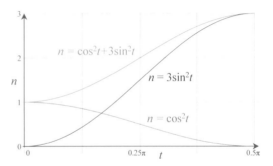

Fig. 13.6 Interpolating
between two points (1, 1)
and (4, 5)

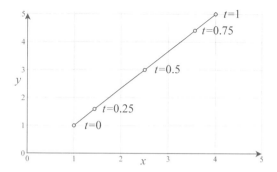

straight-line interpolation, but the distribution of points is non-linear, as shown in
Fig. 13.6. In other words, equal steps in t give rise to unequal distances in space.

The main problem with this approach is that it is impossible to change the nature
of the curve—it is a *sinusoid*, and its slope is determined by the interpolated values.
One way of gaining control over the interpolated curve is to use a *polynomial*, which
is the subject of the next section.

13.4.2 Cubic Interpolation

To begin with, let's develop a cubic blending function that will be similar to the
previous sinusoidal one. This can then be extended to provide extra flexibility. A
cubic polynomial will form the basis of the interpolant:

$$v_1 = at^3 + bt^2 + ct + d$$

and the final interpolant will be of the form

$$n = [v_1 \quad v_2] \begin{bmatrix} n_1 \\ n_2 \end{bmatrix}.$$

The task is to find the values of the constants associated with the polynomials v_1 and
v_2. The requirements are:

1. The cubic function v_2 must grow from 0 to 1 for $0 \le t \le 1$.
2. The slope at a point t must equal the slope at the point $(1 - t)$. This ensures
   slope continuity over the range of the function.
3. The value v_2 at any point t must also produce $(1 - v_2)$ at $(1 - t)$. This ensures
   curve continuity.
   - To satisfy the first requirement:

$$v_2 = at^3 + bt^2 + ct + d$$

and when $t = 0$, $v_2 = 0$ and $d = 0$. Similarly, when $t = 1$, $v_2 = a + b + c$.
- We now need some calculus, which is described in a later chapter. To satisfy the second requirement, differentiate v_2 to obtain the slope:

$$\frac{dv_2}{dt} = 3at^2 + 2bt + c = 3a(1-t)^2 + 2b(1-t) + c$$

and equating constants we discover $c = 0$ and $0 = 3a + 2b$.
- To satisfy the third requirement:

$$at^3 + bt^2 = 1 - [a(1-t)^3 + b(1-t)^2]$$

where we discover $1 = a + b$. But $0 = 3a + 2b$, therefore $a = 2$ and $b = 3$.
Therefore,

$$v_2 = -2t^3 + 3t^2. \tag{13.3}$$

To find the curve's mirror curve, which starts at 1 and collapses to 0 as t moves from 0 to 1, we subtract (13.3) from 1:

$$v_1 = 2t^3 - 3t^2 + 1.$$

Therefore, the two polynomials are

$$v_1 = 2t^3 - 3t^2 + 1 \tag{13.4}$$
$$v_2 = -2t^3 + 3t^2 \tag{13.5}$$

and are shown in Fig. 13.7. They are used as interpolants as follows:

$$n = v_1 n_1 + v_2 n_2$$

or in matrix form:

$$n = [2t^3 - 3t^2 + 1 \quad -2t^3 + 3t^2] \begin{bmatrix} n_1 \\ n_2 \end{bmatrix}$$

Fig. 13.7 Two cubic polynomials

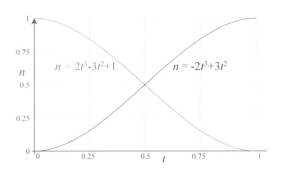

Fig. 13.8 Interpolating between 1 and 3 using a cubic interpolant

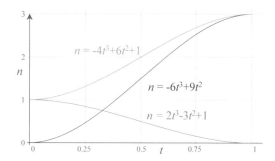

$n = -4t^3 + 6t^2 + 1$

$n = -6t^3 + 9t^2$

$n = 2t^3 - 3t^2 + 1$

Fig. 13.9 A cubic interpolant between points (1, 1) and (8, 3)

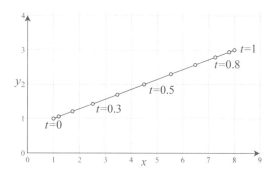

$t=1$
$t=0.8$
$t=0.5$
$t=0.3$
$t=0$

$$n = [t^3 \quad t^2 \quad t \quad 1] \begin{bmatrix} 2 & -2 \\ -3 & 3 \\ 0 & 0 \\ 1 & 0 \end{bmatrix} \begin{bmatrix} n_1 \\ n_2 \end{bmatrix}. \tag{13.6}$$

If we let $n_1 = 1$ and $n_2 = 3$ we obtain the curves shown in Fig. 13.8. And if we apply the interpolant to the points (1, 1) and (8, 3) we obtain the line shown in Fig. 13.9. This interpolant can be used to blend any pair of numbers together.

Now let's examine the scenario where we interpolate between two points P_1 and P_2, and have to arrange that the interpolated curve is tangential with a vector at each point. Such *tangent vectors* forces the curve into a desired shape, as shown in Fig. 13.11. Unfortunately, calculus is required to compute the slope of the cubic polynomial, which is covered in a later chapter.

As this interpolant can be applied to 2D and 3D points, P_1 and P_2 are represented by their position vectors $\mathbf{P}_1$ and $\mathbf{P}_2$, which are unpacked for each Cartesian component.

We now have two position vectors $\mathbf{P}_1$ and $\mathbf{P}_2$ and their respective tangent vectors $\mathbf{s}_1$ and $\mathbf{s}_2$. The requirement is to modulate the interpolating curve in Fig. 13.8 with two further cubic curves. One that blends out the tangent vector $\mathbf{s}_1$ associated with P_1, and the other that blends in the tangent vector $\mathbf{s}_2$ associated with P_2. Let's begin with a cubic polynomial to blend $\mathbf{s}_1$ to zero:

$$v_{out} = at^3 + bt^2 + ct + d.$$

v_{out} must equal zero when $t = 0$ and $t = 1$, otherwise it will disturb the start and end values. Therefore $d = 0$, and

$$a + b + c = 0.$$

The rate of change of v_{out} relative to t (i.e. dv_{out}/dt) must equal 1 when $t = 0$, so it can be used to multiply s_1. When $t = 1$, dv_{out}/dt must equal 0 to attenuate any trace of s_1:

$$\frac{dv_{out}}{dt} = 3at^2 + 2bt + c$$

but $dv_{out}/dt = 1$ when $t = 0$, and $dv_{out}/dt = 0$ when $t = 1$. Therefore, $c = 1$, and

$$3a + 2b + 1 = 0.$$

Using (13.6) implies that $b = -2$ and $a = 1$. Therefore, the polynomial v_{out} has the form

$$v_{out} = t^3 - 2t^2 + t. \tag{13.7}$$

Using a similar argument, one can prove that the function to blend in s_2 equals

$$v_{in} = t^3 - t^2. \tag{13.8}$$

Graphs of (13.4), (13.5), (13.7) and (13.8) are shown in Fig. 13.10.
    The complete interpolating function looks like

$$n = [2t^3 - 3t^2 + 1 \quad -2t^3 + 3t^2 \quad t^3 - 2t^2 + t \quad t^3 - t^2] \begin{bmatrix} P_1 \\ P_2 \\ s_1 \\ s_2 \end{bmatrix}$$

and unpacking the constants and polynomial terms we obtain

Fig. 13.10 The four
Hermite interpolating curves

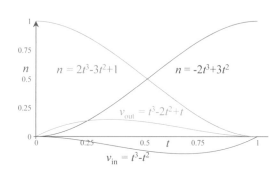

$$n = [t^3 \ t^2 \ t^1 \ 1] \begin{bmatrix} 2 & -2 & 1 & 1 \\ -3 & 3 & -2 & -1 \\ 0 & 0 & 1 & 0 \\ 1 & 0 & 0 & 0 \end{bmatrix} \begin{bmatrix} \mathbf{P}_1 \\ \mathbf{P}_2 \\ \mathbf{s}_1 \\ \mathbf{s}_2 \end{bmatrix}.$$

This type of interpolation is called *Hermite interpolation*, after the French mathematician Charles Hermite (1822–1901). Hermite also proved in 1873 that e is transcendental.

Now let's illustrate Hermite interpolation with a 2D example. It is also very easy to implement the same technique in 3D. Figure 13.11 shows how two points $(0, \ 0)$ and $(1, \ 1)$ are to be connected by a cubic curve that responds to the initial and final tangent vectors. At the start point $(0, \ 0)$ the tangent vector is $[-5 \ \ 0]^T$, and at the final point $(1, \ 1)$ the tangent vector is $[0 \ \ -5]^T$. The x and y interpolants are

$$x = [t^3 \ t^2 \ t^1 \ 1] \begin{bmatrix} 2 & -2 & 1 & 1 \\ -3 & 3 & -2 & -1 \\ 0 & 0 & 1 & 0 \\ 1 & 0 & 0 & 0 \end{bmatrix} \begin{bmatrix} 0 \\ 1 \\ -5 \\ 0 \end{bmatrix}$$

$$y = [t^3 \ t^2 \ t^1 \ 1] \begin{bmatrix} 2 & -2 & 1 & 1 \\ -3 & 3 & -2 & -1 \\ 0 & 0 & 1 & 0 \\ 1 & 0 & 0 & 0 \end{bmatrix} \begin{bmatrix} 0 \\ 1 \\ 0 \\ -5 \end{bmatrix}$$

which become

$$x = [t^3 \ t^2 \ t^1 \ 1] \begin{bmatrix} -7 \\ 13 \\ -5 \\ 0 \end{bmatrix} = -7t^3 + 13t^2 - 5t$$

$$y = [t^3 \ t^2 \ t^1 \ 1] \begin{bmatrix} -7 \\ 8 \\ 0 \\ 0 \end{bmatrix} = -7t^3 + 8t^2.$$

Fig. 13.11 A Hermite curve between the points $(0, \ 0)$ and $(1, \ 1)$ with tangent vectors $[-5 \ 0]^T$ and $[0 \ \ -5]^T$ not drawn to scale

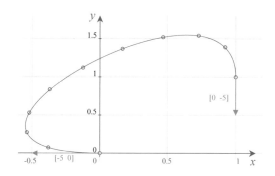

When these polynomials are plotted over the range $0 \le t \le 1$ we obtain the curve shown in Fig. 13.11

We have now reached a point where we are starting to discover how parametric polynomials can be used to generate space curves, which is the subject of the next chapter. So, to conclude this chapter on interpolants, we will take a look at interpolating vectors.

13.5   Interpolating Vectors

So far we have been interpolating between a pair of numbers. Now the question arises: can we use the same interpolants for vectors? We can if we interpolate both the magnitude and direction of a vector. However, if we linearly interpolate only the x- and y-components of two vectors, the in-between vectors would neither respect their orientation nor their magnitude. But if we defined two 2D vectors as $l_1,\ \theta_1$ and $l_2,\ \theta_2$, where l is the magnitude and θ the rotated angle, then a linearly interpolated vector is given by

$$l = l_1(1 - t) + l_2 t$$
$$\theta = \theta_1(1 - t) + \theta_2 t$$

and the x- and y-components of the interpolated vector are:

$$l_x = l \cos \theta$$
$$l_y = l \sin \theta.$$

Figure 13.12 shows the trace of interpolating between vector 2, $45°$ and vector 3, $135°$. The half-way point, when $t = 0.5$, generates the vector 2.5, $90°$. The same technique can be used with 3D vectors using the equivalent polar notation.

We can interpolate between x- y- and z-coordinates if we respect the magnitude and orientation of the encoded vectors using the following technique. Figure 13.13

Fig. 13.12   The trace of interpolating between vectors 2, $45°$ and 3, $135°$

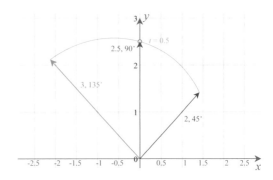

Fig. 13.13 Vector **v** is
derived from part a of of $\mathbf{v}_1$
and part b of $\mathbf{v}_2$

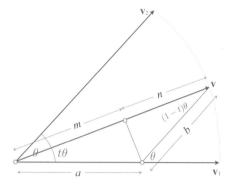

shows two unit vectors $\mathbf{v}_1$ and $\mathbf{v}_2$ separated by an angle θ. The interpolated vector **v**
is defined as a proportion of $\mathbf{v}_1$ and a proportion of $\mathbf{v}_2$:

$$\mathbf{v} = a\mathbf{v}_1 + b\mathbf{v}_2.$$

Let's define the values of a and b such that they are a function of the separating angle
θ. Vector **v** is $t\theta$ from $\mathbf{v}_1$ and $(1 - t)\theta$ from $\mathbf{v}_2$, and it is evident from Fig. 13.13 that
using the sine rule

$$\frac{a}{\sin[(1 - t)\theta]} = \frac{b}{\sin(t\theta)} \tag{13.9}$$

and furthermore:

$$m = a\cos(t\theta)$$
$$n = b\cos[(1 - t)\theta]$$

where

$$m + n = 1. \tag{13.10}$$

From (13.9)

$$b = \frac{a\sin(t\theta)}{\sin[(1 - t)\theta]}$$

and from (13.10) we get

$$a\cos(t\theta) + \frac{a\sin(t\theta)\cos[(1 - t)\theta]}{\sin[(1 - t)\theta]} = 1.$$

Solving for a we find

$$a = \frac{\sin[(1 - t)\theta]}{\sin\theta}$$

$$b = \frac{\sin(t\theta)}{\sin\theta}.$$

Therefore, the final interpolant is

$$\mathbf{v} = \frac{\sin[(1-t)\theta]}{\sin\theta}\mathbf{v}_1 + \frac{\sin(t\theta)}{\sin\theta}\mathbf{v}_2. \qquad (13.11)$$

To see how (13.11) operates, let's consider a simple exercise of interpolating between two unit vectors $[1 \ \ 0]^T$ and $[-1/\sqrt{2} \ \ 1/\sqrt{2}]^T$. The angle between the vectors θ is $135°$. Equation (13.11) is used to interpolate the x- and the y-components individually:

$$v_x = \frac{\sin[(1-t)135°]}{\sin 135°} \times (1) + \frac{\sin(t\,135°)}{\sin 135°} \times \left(-\frac{1}{\sqrt{2}}\right)$$

$$v_y = \frac{\sin[(1-t)135°]}{\sin 135°} \times (0) + \frac{\sin(t\,135°)}{\sin 135°} \times \frac{1}{\sqrt{2}}.$$

Figure 13.14 shows the interpolating curves and Fig. 13.15 shows a trace of the interpolated vectors.

Two observations to note with (13.11):

- The angle θ is the angle between the two vectors, which, if not known, can be computed using the dot product.
- Secondly, the range of θ is given by $0 \le \theta \le 180°$, but when $\theta = 180°$ the denominator collapses to zero.

So far, we have only considered unit vectors. Now let's see how the interpolant reacts to vectors of different magnitudes. As a test, we can input the following vectors to (13.11):

$$\mathbf{v}_1 = [2 \ \ 0]^T, \quad \text{and} \quad \mathbf{v}_2 = [0 \ \ 1]^T.$$

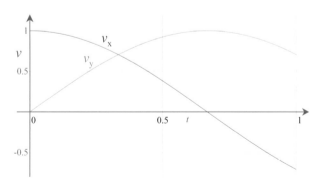

Fig. 13.14 Curves of v_x and v_y using (13.11)

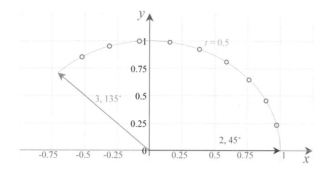

Fig. 13.15 A trace of the interpolated vectors $[1 \quad 0]^T$ and $[-1/\sqrt{2} \quad 1/\sqrt{2}]^T$

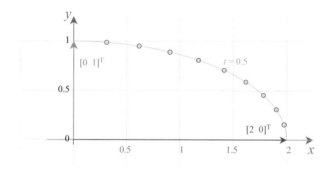

Fig. 13.16 Interpolating between the vectors $[2 \quad 0]^T$ and $[0 \quad 1]^T$

The separating angle $\theta = 90°$, and the result is shown in Fig. 13.16. Note how the initial length of $\mathbf{v}_1$ reduces from 2 to 1 over 90°. It is left to the reader to examine other combinations of vectors. There is one more application for this interpolant, and that is with quaternions.

13.6 Interpolating Quaternions

It just so happens that the interpolant used for vectors also works with quaternions. Which means that, given two quaternions $\mathbf{q}_1$ and $\mathbf{q}_2$, the interpolated quaternion $\mathbf{q}$ is given by

$$\mathbf{q} = \frac{\sin[(1-t)\theta]}{\sin\theta}\mathbf{q}_1 + \frac{\sin(t\theta)}{\sin\theta}\mathbf{q}_2. \tag{13.12}$$

The interpolant is applied individually to the four terms of the quaternion.

When interpolating vectors, θ is the angle between the two vectors. If this is not known, it can be derived using the dot product formula:

$$\cos \theta = \frac{\mathbf{v}_1 \cdot \mathbf{v}_2}{\|\mathbf{v}_1\|\|\mathbf{v}_2\|}$$
$$= \frac{x_1 x_2 + y_1 y_2 + z_1 z_2}{\|\mathbf{v}_1\|\|\mathbf{v}_2\|}.$$

Similarly, when interpolating quaternions, θ is computed by taking the 4-D dot product of the two quaternions:

$$\cos \theta = \frac{\mathbf{q}_1 \cdot \mathbf{q}_2}{|\mathbf{q}_1||\mathbf{q}_2|}$$
$$= \frac{s_1 s_2 + x_1 x_2 + y_1 y_2 + z_1 z_2}{|\mathbf{q}_1||\mathbf{q}_2|}.$$

If we are using unit quaternions

$$\cos \theta = s_1 s_2 + x_1 x_2 + y_1 y_2 + z_1 z_2. \tag{13.13}$$

We are now in a position to demonstrate how to interpolate between a pair of quaternions. For example, say we have two quaternions $\mathbf{q}_1$ and $\mathbf{q}_2$ that rotate $0°$ and $90°$ about the z-axis respectively:

$$\mathbf{q}_1 = \left[\cos\left(\tfrac{0°}{2}\right), \ \sin\left(\tfrac{0°}{2}\right)(0\mathbf{i} + 0\mathbf{j} + 1\mathbf{k})\right]$$
$$\mathbf{q}_1 = \left[\cos\left(\tfrac{90°}{2}\right), \ \sin\left(\tfrac{90°}{2}\right)(0\mathbf{i} + 0\mathbf{j} + 1\mathbf{k})\right]$$

which become

$$\mathbf{q}_1 = [1, \ 0\mathbf{i} + 0\mathbf{j} + 0\mathbf{k}]$$
$$\mathbf{q}_2 \approx [0.7071, \ 0\mathbf{i} + 0\mathbf{j} + 0.7071\mathbf{k}].$$

Any interpolated quaternion is found by the application of (13.12). But first, we need to find the value of θ using (13.13):

$$\cos \theta \approx 0.7071$$
$$\theta = 45°.$$

Now when $t = 0.5$, the interpolated quaternion is given by

$$\mathbf{q} \approx \frac{\sin(45°/2)}{\sin 45°}[1, \ 0\mathbf{i} + 0\mathbf{j} + 0\mathbf{k}] + \frac{\sin(45°/2)}{\sin 45°}[0.7071, \ 0\mathbf{i} + 0\mathbf{j} + 0.7071\mathbf{k}]$$
$$\approx 0.541196[1, \ 0\mathbf{i} + 0\mathbf{j} + 0\mathbf{k}] + 0.541196[0.7071, \ 0\mathbf{i} + 0\mathbf{j} + 0.7071\mathbf{k}]$$
$$\approx [0.541196, \ 0\mathbf{i} + 0\mathbf{j} + 0\mathbf{k}] + [0.382683, \ 0\mathbf{i} + 0\mathbf{j} + 0.382683\mathbf{k}]$$
$$\approx [0.923879, \ 0\mathbf{i} + 0\mathbf{j} + 0.382683\mathbf{k}].$$

Although it is not obvious, this interpolated quaternion is also a unit quaternion, as the square root of the sum of the squares is 1. It should rotate a point about the z-axis, halfway between $0°$ and $90°$, i.e. $45°$. We can test that this works with a simple example.

Take the point $(1, 0, 0)$ and subject it to the standard quaternion operation:

$$\mathbf{P}' = \mathbf{q}\mathbf{P}\mathbf{q}^{-1}.$$

To keep the arithmetic work to a minimum, we substitute $a = 0.923879$ and $b = 0.382683$. Therefore,

$$\mathbf{q} = [a, \; 0\mathbf{i} + 0\mathbf{j} + b\mathbf{k}]$$
$$\mathbf{q}^{-1} = [a, \; -0\mathbf{i} - 0\mathbf{j} - b\mathbf{k}]$$
$$\mathbf{P}' = [a, \; 0\mathbf{i} + 0\mathbf{j} + b\mathbf{k}][0, \; 1\mathbf{i} + 0\mathbf{j} + 0\mathbf{k}][a, \; -0\mathbf{i} - 0\mathbf{j} - b\mathbf{k}]$$
$$= [0, \; a\mathbf{i} + b\mathbf{j} + 0\mathbf{k}][a, \; -0\mathbf{i} - 0\mathbf{j} - b\mathbf{k}]$$
$$= [0, \; (a^2 - b^2)\mathbf{i} + 2ab\mathbf{j} + 0\mathbf{k}]$$
$$\mathbf{P}' \approx [0, \; 0.7071\mathbf{i} + 0.7071\mathbf{j} + 0\mathbf{k}].$$

Therefore, $(1, 0, 0)$ is rotated to $(0.7071, 0.7071, 0)$, which is correct!

13.7 Summary

This chapter has covered some very interesting, yet simple ideas about changing one number into another. In the following chapter we will develop these ideas and see how we design algebraic solutions to curves and surfaces.

Chapter 14
Curves and Patches

14.1   Introduction

In this chapter we investigate the foundations of curves and surface patches. This is a very large and complex subject and it will be impossible to delve too deeply. However, we can explore many of the ideas that are essential to understanding the mathematics behind 2D and 3D curves and how they are developed to produce surface patches. Once you have understood these ideas you will be able to read more advanced texts and develop a wider knowledge of the subject.

14.2   Background

Two people, working for competing French car manufacturers, are associated with what are now called *Bézier curves*: the French physicist and mathematician Paul de Casteljau (1930–), who worked for Citröen, and the French engineer Pierre Bézier (1910–1999), who worked for Rénault. De Casteljau's work was slightly ahead of Bézier's, but because of Citröen's policy of secrecy it was never published, so Bézier's name has since been associated with the theory of polynomial curves and surfaces. Casteljau started his research work in 1959, but his reports were only discovered in 1975, by which time Bézier had become known for his special curves and surfaces.

In the previous chapter we saw how polynomials are used as interpolants and blending functions. We will now see how these form the basis of parametric curves and patches. To begin with, let's start with the humble circle.

© Springer-Verlag London Ltd., part of Springer Nature 2022
J. Vince, *Mathematics for Computer Graphics*, Undergraduate Topics
in Computer Science, https://doi.org/10.1007/978-1-4471-7520-9_14

Fig. 14.1 The circle is
drawn by tracing out a series
of points on the
circumference

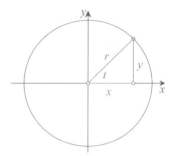

14.3   The Circle

The circle has a very simple equation:

$$x^2 + y^2 = r^2$$

where r is the radius and (x, y) is a point on the circumference. Although this equation has its uses, it is not very convenient for drawing the curve. What we really want are two functions that generate the coordinates of any point on the circumference in terms of some parameter t. Figure 14.1 shows a scenario where the x- and y-coordinates are given by

$$\left.\begin{array}{l} x = r \cos t \\ y = r \sin t \end{array}\right\} \qquad 0 \le t \le 2\pi.$$

By varying the parameter t over the range 0 to 2π, we trace out the curve of the circumference. In fact, by selecting a suitable range of t we can isolate any portion of the circle's circumference.

14.4   The Ellipse

The equation for an ellipse is

$$\frac{x^2}{r_{maj}^2} + \frac{y^2}{r_{min}^2} = 1$$

and its parametric form is

$$\left.\begin{array}{l} x = r_{maj} \cos t \\ y = r_{min} \sin t \end{array}\right\} \qquad 0 \le t \le 2\pi$$

Fig. 14.2 An ellipse
showing the major and minor
radii

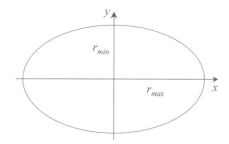

where r_{maj} and r_{min} are the major and minor radii respectively, and (x, y) is a
point on the circumference, as shown in Fig. 14.2. We now examine a very useful
parametric curve called a Bézier curve.

14.5   Bézier Curves

14.5.1   Bernstein Polynomials

Bézier curves employ *Bernstein polynomials* which were described by the Russian
mathematician Sergei Bernstein (1880–1968) in 1912. They are expressed as follows:

$$B_i^n(t) = \binom{n}{i} t^i (1 - t)^{n-i} \tag{14.1}$$

where $\binom{n}{i}$ is shorthand for the number of selections of i different items from n
distinguishable items when the order of selection is ignored, and equals

$$\binom{n}{i} = \frac{n!}{(n - i)! i!} \tag{14.2}$$

where, for example, 3! (factorial 3) is shorthand for $3 \times 2 \times 1$. When (14.2) is evaluated for different values of i and n, we discover the pattern of numbers shown in Table
14.1. This pattern of numbers is known as *Pascal's triangle*. In western countries
they are named after a 17th century French mathematician, even though they had
been described in China as early as 1303 in *Precious Mirror of the Four Elements* by
the Chinese mathematician Chu Shih-chieh. The pattern represents the coefficients
found in binomial expansions. For example, the expansion of $(x + a)^n$ for different
values of n is

Table 14.1 Pascal's triangle

n	i						
	0	1	2	3	4	5	6
0	1						
1	1	1					
2	1	2	1				
3	1	3	3	1			
4	1	4	6	4	1		
5	1	5	10	10	5	1	
6	1	6	15	20	15	6	1

Table 14.2 Expansion of the terms t and $(1 - t)$

n	i				
	0	1	2	3	4
1	t	$(1 - t)$			
2	t^2	$t(1 - t)$	$(1 - t)^2$		
3	t^3	$t^2(1 - t)$	$t(1 - t)^2$	$(1 - t)^3$	
4	t^4	$t^3(1 - t)$	$t^2(1 - t)^2$	$t(1 - t)^3$	$(1 - t)^4$

$$(x + a)^0 = 1$$
$$(x + a)^1 = 1x + 1a$$
$$(x + a)^2 = 1x^2 + 2ax + 1a^2$$
$$(x + a)^3 = 1x^3 + 3ax^2 + 3a^2x + 1a^3$$
$$(x + a)^4 = 1x^4 + 4ax^3 + 6a^2x^2 + 4a^3x + 1a^4$$

which reveal Pascal's triangle as coefficients of the polynomial terms. Thus the $\binom{n}{i}$ term in (14.1) is nothing more than a generator for Pascal's triangle. The powers of t and $(1 - t)$ in (14.1) appear as shown in Table 14.2 for different values of n and i. When the two sets of results are combined we get the complete Bernstein polynomial terms shown in Table 14.3. One very important property of these terms is that they sum to unity, which is an important feature of any interpolant.

The sum of $(1 - t)$ and t is 1, therefore,

$$[(1 - t) + t]^n = 1 \tag{14.3}$$

which is why we can use the binomial expansion of $(1 - t)$ and t as interpolants. For example, when $n = 2$ we obtain the quadratic form:

Table 14.3 The Bernstein polynomial terms

n	i				
	0	1	2	3	4
1	$1t$	$1(1-t)$			
2	$1t^2$	$2t(1-t)$	$1(1-t)^2$		
3	$1t^3$	$3t^2(1-t)$	$3t(1-t)^2$	$1(1-t)^3$	
4	$1t^4$	$4t^3(1-t)$	$6t^2(1-t)^2$	$4t(1-t)^3$	$1(1-t)^4$

Fig. 14.3 Graphs of the quadratic Bernstein polynomials

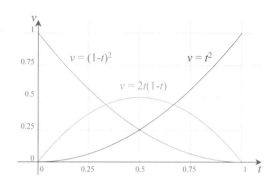

$$(1-t)^2 + 2t(1-t) + t^2 = 1. \tag{14.4}$$

Figure 14.3 shows the graphs of the three polynomial terms of (14.4). The $(1-t)^2$ graph starts at 1 and decays to zero, whereas the t^2 graph starts at zero and rises to 1. The $2t(1-t)$ graph starts at zero reaches a maximum of 0.5 and returns to zero. Thus the central polynomial term has no influence at the end conditions, where $t = 0$ and $t = 1$. We can use these three terms to interpolate between a pair of values as follows:

$$v = v_1(1-t)^2 + 2t(1-t) + v_2 t^2.$$

If $v_1 = 1$ and $v_2 = 3$ we obtain the curve shown in Fig. 14.4. However, there is nothing preventing us from multiplying the middle term $2t(1-t)$ by any arbitrary number v_c:

$$v = v_1(1-t)^2 + v_c 2t(1-t) + v_2 t^2. \tag{14.5}$$

For example, if $v_c = 3$, we obtain the graph shown in Fig. 14.5, which is totally different to the curve in Fig. 14.4. As Bézier observed, the value of v_c provides an excellent mechanism for determining the rate of change between two values. Figure 14.6 shows a variety of graphs for different values of v_c. A very interesting effect occurs when the value of v_c is set midway between v_1 and v_2. For example, when $v_1 = 1$, $v_2 = 3$ and $v_c = 2$, we obtain linear interpolation between v_1 and v_2, as shown in Fig. 14.5.

Fig. 14.4 Bernstein interpolation between the values 1 and 3

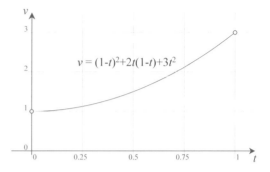

Fig. 14.5 Bernstein interpolation between the values 1 and 3 with $v_c = 3$

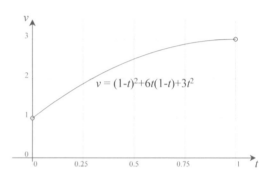

Fig. 14.6 Bernstein interpolation between the values 1 for different values of v_c

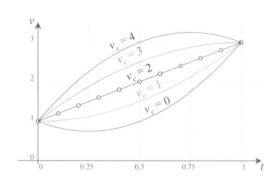

14.5.2　*Quadratic Bézier Curves*

Quadratic Bézier curves are formed by using Bernstein polynomials to interpolate between the x-, y- and z-coordinates associated with the start- and end-points forming the curve. For example, we can draw a 2D quadratic Bézier curve between $(1,\ 1)$ and $(4,\ 3)$ using the following equations:

Fig. 14.7 Quadratic Bézier curve between (1, 1) and (4, 3), with (3, 4) as the control vertex

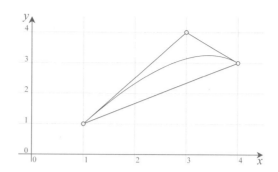

$$x = 1(1 - t)^2 + x_c 2t(1 - t) + 4t^2 \tag{14.6}$$

$$y = 1(1 - t)^2 + y_c 2t(1 - t) + 3t^2. \tag{14.7}$$

But what should be the values of (x_c, y_c)? Well, this is entirely up to us; the position of this *control vertex* determines how the curve moves between (1, 1) and (4, 3).

A Bézier curve possesses interpolating and approximating qualities: the interpolating feature ensures that the curve passes through the end points, while the approximating feature shows how the curve passes close to the control point. To illustrate this, if we make $x_c = 3$ and $y_c = 4$ we obtain the curve shown in Fig. 14.7, which shows how the curve intersects the end-points, but misses the control point. It also highlights two important features of Bézier curves: the *convex hull* property, and the end slopes of the curve.

The convex hull property implies that the curve is always contained within the polygon connecting the start, end and control points. In this case the curve is inside the triangle formed by the vertices (1, 1), (3, 4) and (4, 3). The slope of the curve at (1, 1) is equal to the slope of the line connecting the start point to the control point (3, 4), and the slope of the curve at (4, 3) is equal to the slope of the line connecting the control point (3, 4) to the end point (4, 3). Naturally, these two qualities of Bézier curves can be proved mathematically.

14.5.3 Cubic Bernstein Polynomials

Before moving on, there are two further points to note:
- No restrictions are placed upon the position of (x_c, y_c)—it can be anywhere.
- Simply including z-coordinates for the start, end and control vertices creates 3D curves.

One of the drawbacks with quadratic curves is that they are perhaps, too simple. If we want to construct a complex curve with several peaks and valleys, we would have to join together a large number of such curves. A *cubic curve*, on the other hand, naturally supports one peak and one valley, which simplifies the construction of more complex curves.

When $n = 3$ in (14.3) we obtain the following terms:

$$[(1 - t) + t]^3 = (1 - t)^3 + 3t(1 - t)^2 + 3t^2(1 - t) + t^3$$

which can be used as a cubic interpolant, as

$$v = v_1(1 - t)^3 + v_{c1}3t(1 - t)^2 + v_{c2}3t^2(1 - t) + v_2t^3.$$

Once more, the terms sum to unity, and the convex hull and slope properties also hold. Figure 14.8 shows the graphs of the four polynomial terms.

This time we have two control values v_{c1} and v_{c2}. These are set to any value, independent of the values chosen for v_1 and v_2. To illustrate this, let's consider an example of blending between values 1 and 3, with v_{c1} and v_{c2} set to 2.5 and -2.5 respectively. The blending curve is shown in Fig. 14.9.

The next step is to associate the blending polynomials with x- and y-coordinates:

$$x = x_1(1 - t)^3 + x_{c1}3t(1 - t)^2 + x_{c2}3t^2(1 - t) + x_2t^3 \qquad (14.8)$$
$$y = y_1(1 - t)^3 + y_{c1}3t(1 - t)^2 + y_{c2}3t^2(1 - t) + y_2t^3. \qquad (14.9)$$

Fig. 14.8 The cubic Bernstein polynomial curves

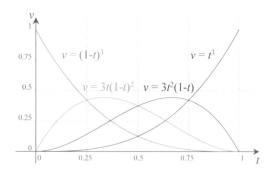

Fig. 14.9 The cubic Bernstein polynomial through the values 1, 2.5, −2.5, 3

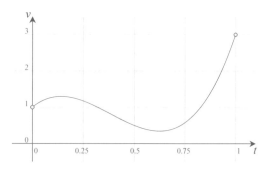

Fig. 14.10 A cubic Bézier
curve

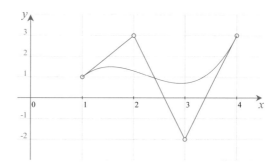

Evaluating (14.8) and (14.9) with the following points:

$$(x_1,\ y_1) = (1,\ 1),\quad (x_2,\ y_2) = (4,\ 3)$$
$$(x_{c1},\ y_{c1}) = (2,\ 3),\quad (x_{c2},\ y_{c2}) = (3,\ -2)$$

we obtain the cubic Bézier curve shown in Fig. 14.10, which also shows the guidelines
between the end and control points.

Just to show how consistent Bernstein polynomials are, let's set the values to

$$(x_1,\ y_1) = (1,\ 1),\qquad (x_2,\ y_2) = (4,\ 3)$$
$$(x_{c1},\ y_{c1}) = (2,\ 1.666),\quad (x_{c2},\ y_{c2}) = (3,\ 2.333)$$

where $(x_{c1},\ y_{c1})$ and $(x_{c2},\ y_{c2})$ are points one-third and two-thirds respectively,
between the start and final values. As we found in the quadratic case, where the
single control point was halfway between the start and end values, we obtain linear
interpolation as shown in Fig. 14.11.

As mathematicians are interested in expressing a formula succinctly, there is
an elegant way of abbreviating Bernstein polynomials. Equations (14.6) and (14.7)
describe the three polynomial terms for generating a quadratic Bézier curve and
(14.8) and (14.9) describe the four polynomial terms for generating a cubic Bézier

Fig. 14.11 A cubic Bézier
line

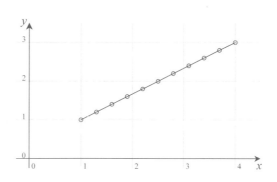

curve. To begin with, quadratic equations are called *second-degree equations*, and cubics are called *third-degree equations*. In the original Bernstein formulation:

$$B_i^n(t) = \binom{n}{i} t^i (1-t)^{n-i}$$

n represents the degree of the polynomial, and i, which has values between 0 and n, creates the individual polynomial terms. These terms are then used to multiply the coordinates of the end and control points. If these points are stored as a vector $\mathbf{P}$, the position vector $\mathbf{p}(t)$ for a point on the curve is written:

$$\mathbf{p}(t) = \sum_{i=0}^{n} \binom{n}{i} t^i (1-t)^{n-i} \mathbf{P}_i, \qquad 0 \leq i \leq n \tag{14.10}$$

or

$$\mathbf{p}(t) = \sum_{i=0}^{n} B_i^n(t) \mathbf{P}_i, \qquad 0 \leq i \leq n. \tag{14.11}$$

For example, a point $\mathbf{p}(t)$ on a quadratic curve is represented by

$$\mathbf{p}(t) = 1t^0(1-t)^2 \mathbf{P}_0 + 2t^1(1-t)^1 \mathbf{P}_1 + 1t^2(1-t)^0 \mathbf{P}_2.$$

You will discover (14.10) and (14.11) used in more advanced texts to describe Bézier curves. Although they initially appear intimidating, you should now find them relatively easy to understand.

14.6   A Recursive Bézier Formula

Note that (14.10) explicitly describes the polynomial terms needed to construct the blending terms. With the use of *recursive functions* (a recursive function is a function that calls itself), it is possible to arrive at another formulation that leads towards an understanding of *B-splines*. To begin, we need to express $\binom{n}{i}$ in terms of lower terms, and because the coefficients of any row in Pascal's triangle are the sum of the two coefficients immediately above, we can write

$$\binom{n}{i} = \binom{n-1}{i} + \binom{n-1}{i-1}.$$

Therefore, we can write:

$$B_i^n(t) = \binom{n-1}{i} t^i (1-t)^{n-i} + \binom{n-1}{i-1} t^i (1-t)^{n-i}$$

$$B_i^n(t) = (1-t) B_i^{n-1}(t) + t B_{i-1}^{n-1}(t).$$

As with all recursive functions, some condition must terminate the process; in this case, it is when the degree is zero. Consequently, $B_0^0(t) = 1$ and $B_j^n(t) = 0$ for $j < 0$.

14.7 Bézier Curves Using Matrices

As we have already seen, matrices provide a very compact notation for algebraic formulae. So let's see how Bernstein polynomials lend themselves to this form of notation. Recall (14.4) which defines the three terms associated with a quadratic Bernstein polynomial. These are expanded to

$$1 - 2t + t^2, \quad 2t - 2t^2, \quad t^2$$

and written as the product:

$$\begin{bmatrix} t^2 & t & 1 \end{bmatrix} \begin{bmatrix} 1 & -2 & 1 \\ -2 & 2 & 0 \\ 1 & 0 & 0 \end{bmatrix}.$$

This means that (14.5) can be written:

$$v = \begin{bmatrix} t^2 & t & 1 \end{bmatrix} \begin{bmatrix} 1 & -2 & 1 \\ -2 & 2 & 0 \\ 1 & 0 & 0 \end{bmatrix} \begin{bmatrix} v_1 \\ v_c \\ v_2 \end{bmatrix}$$

or

$$\mathbf{p}(t) = \begin{bmatrix} t^2 & t & 1 \end{bmatrix} \begin{bmatrix} 1 & -2 & 1 \\ -2 & 2 & 0 \\ 1 & 0 & 0 \end{bmatrix} \begin{bmatrix} \mathbf{P}_1 \\ \mathbf{P}_c \\ \mathbf{P}_2 \end{bmatrix}$$

where $\mathbf{p}(t)$ points to any point on the curve, and $\mathbf{P}_1$, $\mathbf{P}_c$ and $\mathbf{P}_2$ point to the start, control and end points respectively.

A similar development is used for a cubic Bézier curve, which has the following matrix formulation:

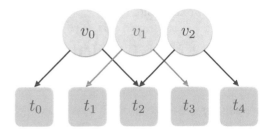

$$\mathbf{p}(t) = \begin{bmatrix} t^3 & t^2 & t & 1 \end{bmatrix} \begin{bmatrix} -1 & 3 & -3 & 1 \\ 3 & -6 & 3 & 0 \\ -3 & 3 & 0 & 0 \\ 1 & 0 & 0 & 0 \end{bmatrix} \begin{bmatrix} \mathbf{P}_1 \\ \mathbf{P}_{c1} \\ \mathbf{P}_{c2} \\ \mathbf{P}_2 \end{bmatrix}.$$

There is no doubt that Bézier curves are very useful, and they find their way into all sorts of applications. But, perhaps their one weakness is that whenever an end or control vertex is repositioned, the entire curve is modified. So let's examine another type of curve that prevents this from happening—B-splines. But before we consider this form, let's revisit linear interpolation between multiple values.

14.7.1   Linear Interpolation

To interpolate linearly between two numbers v_0 and v_1, we use the following interpolant:

$$v(t) = v_0(1-t) + v_1 t, \qquad 0 \le t \le 1.$$

But say we have to interpolate continuously between three values on a linear basis, i.e. v_0, v_1, v_2, with the possibility of extending the technique to any number of values. One solution is to use a sequence of parameter values t_1, t_2, t_3 that are associated with the given values of v, as shown in Fig. 14.12. For the sake of symmetry:

> v_0 is associated with the parameter range t_0 to t_2,
> v_1 is associated with the parameter range t_1 to t_3,
> v_2 is associated with the parameter range t_2 to t_4.

This sequence of parameters is called a *knot vector*. The only assumption we make about the knot vector is that $t_0 \le t_1 \le t_2 \le$, etc.

Now let's invent a linear blending function $B_i^1(t)$ whose subscript i is used to reference values in the knot vector. We want to use the blending function to compute the influence of the three values on any interpolated value $v(t)$ as follows:

$$v(t) = B_0^1(t)v_0 + B_1^1(t)v_1 + B_2^1(t)v_2. \tag{14.12}$$

It's obvious from this arrangement that v_0 will influence $v(t)$ only when t is between t_0 and t_2. Similarly, v_1 and v_2 will influence $v(t)$ only when t is between t_1 and t_3, and t_2 and t_4 respectively.

To understand the action of the blending function let's concentrate upon one particular value $B_1^1(t)$. When t is less than t_1 or greater than t_3, the function $B_1^1(t)$ must be zero. When $t_1 \leq t \leq t_3$, the function must return a value reflecting the proportion of v_1 that influences $v(t)$. During the span $t_1 \leq t \leq t_2$, v_1 has to be blended in, and during the span $t_1 \leq t \leq t_3$, v_1 has to be blended out. The blending in is effected by the ratio

$$\left(\frac{t - t_1}{t_2 - t_1} \right)$$

and the blending out is effected by the ratio

$$\left(\frac{t_3 - t}{t_3 - t_2} \right).$$

Thus $B_1^1(t)$ has to incorporate both ratios, but it must ensure that they only become active during the appropriate range of t. Let's remind ourselves of this requirement by subscripting the ratios accordingly:

$$B_1^1(t) = \left(\frac{t - t_1}{t_2 - t_1} \right)_{1,2} + \left(\frac{t_3 - t}{t_3 - t_2} \right)_{2,3}.$$

We can now write the other two blending terms $B_0^1(t)$ and $B_2^1(t)$ as

$$B_0^1(t) = \left(\frac{t - t_0}{t_1 - t_0} \right)_{0,1} + \left(\frac{t_2 - t}{t_2 - t_1} \right)_{1,2}$$

$$B_2^1(t) = \left(\frac{t - t_2}{t_3 - t_2} \right)_{2,3} + \left(\frac{t_4 - t}{t_4 - t_3} \right)_{3,4}.$$

You should be able to see a pattern linking the variables with their subscripts, and the possibility of writing a general linear blending term $B_i^1(t)$ as

$$B_i^1(t) = \left(\frac{t - t_i}{t_{i+1} - t_i} \right)_{i,i+1} + \left(\frac{t_{i+2} - t}{t_{i+2} - t_{i+1}} \right)_{i+1,i+2}.$$

This enables us to write (14.12) in a general form as

$$v(t) = \sum_{i=0}^{2} B_i^1(t) v_i.$$

But there is still a problem concerning the values associated with the knot vector. Fortunately, there is an easy solution. One simple approach is to keep the differences between t_1, t_2 and t_3 whole numbers, e.g. 0, 1 and 2. But what about the end conditions t_0 and t_4? To understand the resolution of this problem let's examine the action of the three terms over the range of the parameter t. The three terms are

$$\left[\left(\frac{t - t_0}{t_1 - t_0} \right)_{0,1} + \left(\frac{t_2 - t}{t_2 - t_1} \right)_{1,2} \right] v_0 \tag{14.13}$$

$$\left[\left(\frac{t - t_1}{t_2 - t_1} \right)_{1,2} + \left(\frac{t_3 - t}{t_3 - t_2} \right)_{2,3} \right] v_1 \tag{14.14}$$

$$\left[\left(\frac{t - t_2}{t_3 - t_2} \right)_{2,3} + \left(\frac{t_4 - t}{t_4 - t_3} \right)_{3,4} \right] v_2 \tag{14.15}$$

and I propose to initialise the knot vector as follows:

t_0	t_1	t_2	t_3	t_4
0	0	1	2	2

- Remember that the subscripts of the ratios are the subscripts of t, not the values of t.
- Over the range $t_0 \leq t \leq t_1$, i.e. 0 to 0. Only the first ratio in (14.13) is active and returns $\frac{0}{0}$. The algorithm must detect this condition and take no action.
- Over the range $t_1 \leq t \leq t_2$. i.e. 0 to 1. The first ratio of (14.13) is active again, and over the range of t blends out v_0. The first ratio of (14.14) is also active, and over the range of t blends in v_1.
- Over the range $t_2 \leq t \leq t_3$. i.e. 1 to 2. The second ratio of (14.14) is active, and over the range of t blends out v_1. The first ratio of (14.15) is also active, and over the range of t blends in v_2.
- Finally, over the range $t_3 \leq t \leq t_4$. i.e. 2 to 2. The second ratio of (14.15) is active and returns $\frac{0}{0}$. The algorithm must detect this condition and take no action.

This process results in a linear interpolation between v_0, v_1 and v_2. If (14.13)–(14.15) are applied to coordinate values, the result is two straight lines. This seems like a lot of work just to draw two lines, but the beauty of the technique is that it will work with any number of points, and can be developed for quadratic and higher order interpolants.

The New Zealand mathematician Alexander Aitken (1895–1967), developed the following recursive interpolant:

$$\mathbf{p}_i^r(t) = \left(\frac{t_{i+r} - t}{t_{i+r} - t_i} \right) \mathbf{p}_i^{r-1}(t) + \left(\frac{t - t_i}{t_{i+r} - t_i} \right) \mathbf{p}_{i+1}^{r-1}(t) \quad \begin{cases} r = 1, \dots, n \\ i = 0, \dots, n - r \end{cases}$$

which interpolates between a series of points using repeated linear interpolation.

14.8   B-Splines

B-splines, like Bézier curves, use polynomials to generate a curve segment. But, unlike Bézier curves, B-splines employ a series of control points that determine the curve's local geometry. This feature ensures that only a small portion of the curve is changed when a control point is moved.

There are two types of B-splines: *rational* and *non-rational* splines, which divide into two further categories: *uniform* and *non-uniform*. Rational B-splines are formed from the ratio of two polynomials such as

$$x(t) = \frac{X(t)}{W(t)}, \quad y(t) = \frac{Y(t)}{W(t)}, \quad z(t) = \frac{Z(t)}{W(t)}.$$

Although this appears to introduce an unnecessary complication, the division by a second polynomial brings certain advantages:

- They describe perfect circles, ellipses, parabolas and hyperbolas, whereas non-rational curves can only approximate these curves.
- They are invariant of their control points when subjected to rotation, scaling, translation and perspective transformations, whereas non-rational curves lose this geometric integrity.
- They allow weights to be used at the control points to push and pull the curve.

An explanation of uniform and non-uniform types is best left until you understand the idea of splines. So, without knowing the meaning of uniform, let's begin with uniform B-splines.

14.8.1   Uniform B-Splines

A B-spline is constructed from a string of curve segments whose geometry is determined by a group of local control points. These curves are known as *piecewise polynomials*. A curve segment does not have to pass through a control point, although this may be desirable at the two end points.

Cubic B-splines are very common, as they provide a geometry that is one step away from simple quadratics, and possess continuity characteristics that make the joins between the segments invisible. In order to understand their construction consider the scenario in Fig. 14.13. Here we see a group of $(m + 1)$ control points $\mathbf{P}_0, \mathbf{P}_1,$ $\mathbf{P}_2, \ldots, \mathbf{P}_m$ which determine the shape of a cubic curve constructed from a series of curve segments $\mathbf{S}_0, \mathbf{S}_1, \mathbf{S}_2, \ldots, \mathbf{S}_{m-3}$.

As the curve is cubic, curve segment $\mathbf{S}_i$ is influenced by $\mathbf{P}_i, \mathbf{P}_{i+1}, \mathbf{P}_{i+2}, \mathbf{P}_{i+3}$, and curve segment $\mathbf{S}_{i+1}$ is influenced by $\mathbf{P}_{i+1}, \mathbf{P}_{i+2}, \mathbf{P}_{i+3}, \mathbf{P}_{i+4}$. And as there are $(m + 1)$ control points, there are $(m - 2)$ curve segments.

A single segment $\mathbf{S}_i(t)$ of a B-spline curve is defined by

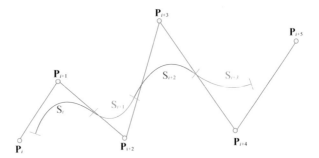

Fig. 14.13 The construction of a uniform non-rational B-spline curve

$$\mathbf{S}_i(t) = \sum_{r=0}^{3} \mathbf{P}_{i+r} B_r(t), \qquad 0 \leq t \leq 1$$

where

$$B_0(t) = \tfrac{1}{6}\left(-t^3 + 3t^2 - 3t + 1\right) = \tfrac{1}{6}(1 - t)^3 \qquad (14.16)$$
$$B_1(t) = \tfrac{1}{6}\left(3t^3 - 6t^2 + 4\right) \qquad (14.17)$$
$$B_2(t) = \tfrac{1}{6}\left(-3t^3 + 3t^2 + 3t + 1\right) \qquad (14.18)$$
$$B_3(t) = \tfrac{1}{6}t^3. \qquad (14.19)$$

These are the B-spline *basis functions* and are shown in Fig. 14.14.

Although it is not apparent, these four curve segments are part of one curve. The basis function $B_3(t)$ starts at zero and rises to 0.1666 at $t = 1$. It is taken over by $B_2(t)$ at $t = 0$, which rises to 0.666 at $t = 1$. The next segment is $B_1(t)$ and takes over at $t = 0$ and falls to 0.1666 at $t = 1$. Finally, $B_0(t)$ takes over at 0.1666 and falls to zero at $t = 1$. Equations (14.16)–(14.19) are represented in matrix form by

Fig. 14.14 The B-spline basis functions

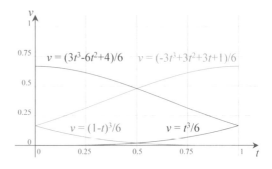

Fig. 14.15 Four curve segments forming a B-spline curve

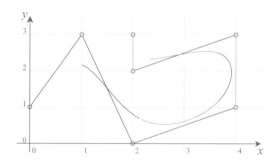

$$\mathbf{Q}_1(t) = \tfrac{1}{6} \begin{bmatrix} t^3 & t^2 & t & 1 \end{bmatrix} \begin{bmatrix} -1 & 3 & -3 & 1 \\ 3 & -6 & 3 & 0 \\ -3 & 0 & 3 & 0 \\ 1 & 4 & 1 & 0 \end{bmatrix} \begin{bmatrix} \mathbf{P}_i \\ \mathbf{P}_{i+1} \\ \mathbf{P}_{i+2} \\ \mathbf{P}_{i+3} \end{bmatrix}. \tag{14.20}$$

Let's now illustrate how (14.20) works. We first identify the control points $\mathbf{P}_i$, $\mathbf{P}_{i+1}, \mathbf{P}_{i+2}$, etc. Let these be $(0, 1), (1, 3), (2, 0), (4, 1), (4, 3), (2, 2)$ and $(2, 3)$. They can be seen in Fig. 14.15 connected together by straight lines. If we take the first four control points: $(0, 1), (1, 3), (2, 0), (4, 1)$, and subject the x- and y-coordinates to the matrix in (14.20) over the range $0 \leq t \leq 1$ we obtain the first B-spline curve segment shown in Fig. 14.15. If we move along one control point and take the next group of control points $(1, 3), (2, 0), (4, 1), (4, 3)$, we obtain the second B-spline curve segment. This is repeated a further two times.

Figure 14.15 shows the four curve segments, and it is obvious that even though there are four discrete segments, they join together perfectly. This is no accident. The slopes at the end points of the basis curves are designed to match the slopes of their neighbours and ultimately keep the geometric curve continuous.

14.8.2 *Continuity*

In order to explain continuity, it is necessary to employ differentiation. Therefore, you may wish to read the chapter on calculus before continuing.

Constructing curves from several segments can only succeed if the slope of the abutting curves match. As we are dealing with curves whose slopes are changing everywhere, it will be necessary to ensure that even the rate of change of slopes is matched at the join. This aspect of curve design is called *geometric continuity* and is determined by the continuity properties of the basis function. Let's explore such features.

The *first level* of curve continuity C^0, ensures that the physical end of one basis curve corresponds with the following, e.g. $\mathbf{S}_i(1) = \mathbf{S}_{i+1}(0)$. We know that this occurs from the basis graphs shown in Fig. 14.14. The *second level* of curve continuity C^1,

Table 14.4 Continuity properties of cubic B-splines

t			t			t		
C^0	0	1	C^1	0	1	C^2	0	1
$B_3(t)$	0	1/6	$B_3'(t)$	0	0.5	$B_3''(t)$	0	1
$B_2(t)$	1/6	2/3	$B_2'(t)$	0.5	0	$B_2''(t)$	1	-2
$B_1(t)$	2/3	1/6	$B_1'(t)$	0	-0.5	$B_1''(t)$	-2	1
$B_0(t)$	1/6	0	$B_0'(t)$	-0.5	0	$B_0''(t)$	1	0

ensures that the slope at the end of one basis curve matches that of the following curve. This is confirmed by differentiating the basis functions (14.16)–(14.19):

$$B_0'(t) = \tfrac{1}{6}\left(-3t^2 + 6t - 3\right) \tag{14.21}$$

$$B_1'(t) = \tfrac{1}{6}\left(9t^2 - 12t\right) \tag{14.22}$$

$$B_2'(t) = \tfrac{1}{6}\left(-9t^2 + 6t + 3\right) \tag{14.23}$$

$$B_3'(t) = \tfrac{1}{6}3t^2. \tag{14.24}$$

Evaluating (14.21)–(14.24) for $t = 0$ and $t = 1$, we discover the slopes 0.5, 0, −0.5, 0 for the joins between B_3, B_2, B_1, B_0. The *third level* of curve continuity C^2, ensures that the rate of change of slope at the end of one basis curve matches that of the following curve. This is confirmed by differentiating (14.21)–(14.24):

$$B_0''(t) = -t + 1 \tag{14.25}$$

$$B_1''(t) = 3t - 2 \tag{14.26}$$

$$B_2''(t) = -3t + 1 \tag{14.27}$$

$$B_3''(t) = t. \tag{14.28}$$

Evaluating (14.25)–(14.28) for $t = 0$ and $t = 1$, we discover the values 1, 2, 1, 0 for the joins between B_3, B_2, B_1, B_0. These combined continuity results are tabulated in Table 14.4.

14.8.3 Non-uniform B-Splines

Uniform B-splines are constructed from curve segments where the parameter spacing is at equal intervals. *Non-uniform B-splines*, with the support of a knot vector, provide extra shape control and the possibility of drawing periodic shapes. Unfortunately an explanation of the underlying mathematics would take us beyond the introductory nature of this text, and readers are advised to seek out other books dealing in such matters.

14.8.4 Non-uniform Rational B-Splines

Non-uniform rational B-splines (NURBS) combine the advantages of non-uniform B-splines and rational polynomials: they support periodic shapes such as circles, and they accurately describe curves associated with the conic sections. They also play a very important role in describing geometry used in the modeling of computer animation characters.

NURBS surfaces also have a patch formulation and play a very important role in surface modelling in computer animation and CAD. However, tempting though it is to give a description of NURBS surfaces here, they have been omitted because their inclusion would unbalance the introductory nature of this text.

14.9 Surface Patches

14.9.1 Planar Surface Patch

The simplest form of surface geometry consists of a patchwork of polygons or triangles, where three or more vertices provide the basis for describing the associated planar surface. For example, given four vertices P_{00}, P_{10}, P_{01}, P_{11} as shown in Fig. 14.16, a point P_{uv} can be defined as follows. To begin with, a point along the edge $P_{00} - P_{10}$ is defined as

$$P_{u1} = (1 - u)P_{00} + uP_{10}$$

and a point along the edge $P_{01} - P_{11}$ is defined as

$$P_{u2} = (1 - u)P_{01} + uP_{11}.$$

Therefore, any point P_{uv} is defined as

$$\begin{aligned} P_{uv} &= (1 - v)P_{u1} + vP_{u2} \\ &= (1 - v)[(1 - u)P_{00} + uP_{10}] + v[(1 - u)P_{01} + uP_{11}] \\ &= (1 - u)(1 - v)P_{00} + u(1 - v)P_{10} + v(1 - u)P_{01} + uvP_{11} \end{aligned}$$

and is written in matrix form as

$$P_{uv} = [1 - u \quad u] \begin{bmatrix} P_{00} & P_{01} \\ P_{10} & P_{11} \end{bmatrix} \begin{bmatrix} 1 - v \\ v \end{bmatrix}$$

which expands to

$$P_{uv} = [u \quad 1] \begin{bmatrix} -1 & 1 \\ 1 & 0 \end{bmatrix} \begin{bmatrix} P_{00} & P_{01} \\ P_{10} & P_{11} \end{bmatrix} \begin{bmatrix} -1 & 1 \\ 1 & 0 \end{bmatrix} \begin{bmatrix} v \\ 1 \end{bmatrix}.$$

Fig. 14.16 A flat patch
defined by u and v
parameters

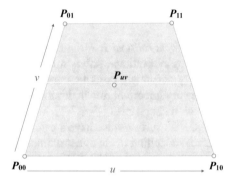

Let's illustrate this with an example. Given the following four points: $P_{00} = (0, 0, 0)$, $P_{10} = (0, 0, 4)$, $P_{01} = (2, 2, 1)$, $P_{11} = (2, 2, 3)$, we can write the coordinates of any point on the patch as

$$x_{uv} = [u \quad 1] \begin{bmatrix} -1 & 1 \\ 1 & 0 \end{bmatrix} \begin{bmatrix} 0 & 2 \\ 0 & 2 \end{bmatrix} \begin{bmatrix} -1 & 1 \\ 1 & 0 \end{bmatrix} \begin{bmatrix} v \\ 1 \end{bmatrix}$$

$$y_{uv} = [u \quad 1] \begin{bmatrix} -1 & 1 \\ 1 & 0 \end{bmatrix} \begin{bmatrix} 0 & 2 \\ 0 & 2 \end{bmatrix} \begin{bmatrix} -1 & 1 \\ 1 & 0 \end{bmatrix} \begin{bmatrix} v \\ 1 \end{bmatrix}$$

$$z_{uv} = [u \quad 1] \begin{bmatrix} -1 & 1 \\ 1 & 0 \end{bmatrix} \begin{bmatrix} 0 & 1 \\ 4 & 3 \end{bmatrix} \begin{bmatrix} -1 & 1 \\ 1 & 0 \end{bmatrix} \begin{bmatrix} v \\ 1 \end{bmatrix}$$

$$x_{uv} = 2v \tag{14.29}$$
$$y_{uv} = 2v \tag{14.30}$$
$$z_{uv} = u(4 - 2v) + v. \tag{14.31}$$

By substituting values of u and v in (14.29)–(14.31) between the range $0 \leq (u, v) \leq 1$, we obtain the coordinates of any point on the surface of the patch.

If we now introduce the ideas of Bézier control points into a surface patch definition, we provide a very powerful way of creating smooth 3D surface patches.

14.9.2 Quadratic Bézier Surface Patch

Bézier proposed a matrix of nine control points to determine the geometry of a quadratic patch, as shown in Fig. 14.17. Any point on the patch is defined by

Fig. 14.17 A quadratic Bézier surface patch

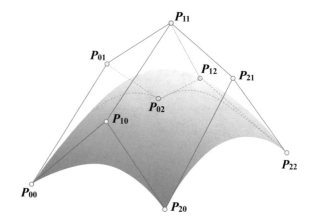

$$P_{uv} = \begin{bmatrix} u^2 & u & 1 \end{bmatrix} \begin{bmatrix} 1 & -2 & 1 \\ -2 & 2 & 0 \\ 1 & 0 & 0 \end{bmatrix} \begin{bmatrix} P_{00} & P_{01} & P_{02} \\ P_{10} & P_{11} & P_{12} \\ P_{20} & P_{21} & P_{22} \end{bmatrix} \begin{bmatrix} 1 & -2 & 1 \\ -2 & 2 & 0 \\ 1 & 0 & 0 \end{bmatrix} \begin{bmatrix} v^2 \\ v \\ 1 \end{bmatrix}.$$

The individual x-, y- and z-coordinates are obtained by substituting the x-, y- and z-values for the central P matrix.

Let's illustrate the process with an example. Given the following points:

$$P_{00} = (0,\ 0,\ 0),\quad P_{01} = (1,\ 1,\ 0),\quad P_{02} = (2,\ 0,\ 0)$$
$$P_{10} = (0,\ 1,\ 1),\quad P_{11} = (1,\ 2,\ 1),\quad P_{12} = (2,\ 1,\ 1)$$
$$P_{20} = (0,\ 0,\ 2),\quad P_{21} = (1,\ 1,\ 2),\quad P_{22} = (2,\ 0,\ 2)$$

we can write

$$x_{uv} = \begin{bmatrix} u^2 & u & 1 \end{bmatrix} \begin{bmatrix} 1 & -2 & 1 \\ -2 & 2 & 0 \\ 1 & 0 & 0 \end{bmatrix} \begin{bmatrix} 0 & 1 & 2 \\ 0 & 1 & 2 \\ 0 & 1 & 2 \end{bmatrix} \begin{bmatrix} 1 & -2 & 1 \\ -2 & 2 & 0 \\ 1 & 0 & 0 \end{bmatrix} \begin{bmatrix} v^2 \\ v \\ 1 \end{bmatrix}$$

$$x_{uv} = \begin{bmatrix} u^2 & u & 1 \end{bmatrix} \begin{bmatrix} 0 & 0 & 0 \\ 0 & 0 & 0 \\ 0 & 2 & 0 \end{bmatrix} \begin{bmatrix} v^2 \\ v \\ 1 \end{bmatrix}$$

$$x_{uv} = 2v$$

$$y_{uv} = \begin{bmatrix} u^2 & u & 1 \end{bmatrix} \begin{bmatrix} 1 & -2 & 1 \\ -2 & 2 & 0 \\ 1 & 0 & 0 \end{bmatrix} \begin{bmatrix} 0 & 1 & 0 \\ 1 & 2 & 1 \\ 0 & 1 & 0 \end{bmatrix} \begin{bmatrix} 1 & -2 & 1 \\ -2 & 2 & 0 \\ 1 & 0 & 0 \end{bmatrix} \begin{bmatrix} v^2 \\ v \\ 1 \end{bmatrix}$$

$$y_{uv} = \begin{bmatrix} u^2 & u & 1 \end{bmatrix} \begin{bmatrix} 0 & 0 & -2 \\ 0 & 0 & 2 \\ -2 & 2 & 0 \end{bmatrix} \begin{bmatrix} v^2 \\ v \\ 1 \end{bmatrix}$$

$$y_{uv} = 2\left(u + v - u^2 - v^2\right)$$

Table 14.5 The x-, y-, z-coordinates for different values of u and v

u	v		
	0	0.5	1
0	(0, 0, 0)	(1, 0.5, 0)	(2, 0, 0)
0.5	(0, 0.5, 1)	(1, 0.5, 1)	(2, 0.5, 1)
1	(0, 0, 2)	(1, 0.5, 2)	(2, 0, 2)

$$z_{uv} = \begin{bmatrix} u^2 & u & 1 \end{bmatrix} \begin{bmatrix} 1 & -2 & 1 \\ -2 & 2 & 0 \\ 1 & 0 & 0 \end{bmatrix} \begin{bmatrix} 0 & 0 & 0 \\ 1 & 1 & 1 \\ 2 & 2 & 2 \end{bmatrix} \begin{bmatrix} 1 & -2 & 1 \\ -2 & 2 & 0 \\ 1 & 0 & 0 \end{bmatrix} \begin{bmatrix} v^2 \\ v \\ 1 \end{bmatrix}$$

$$z_{uv} = \begin{bmatrix} u^2 & u & 1 \end{bmatrix} \begin{bmatrix} 0 & 0 & 0 \\ 0 & 0 & 2 \\ 0 & 0 & 0 \end{bmatrix} \begin{bmatrix} v^2 \\ v \\ 1 \end{bmatrix}$$

$$z_{uv} = 2u.$$

Therefore, any point on the surface patch has coordinates

$$x_{uv} = 2v, \quad y_{uv} = 2\left(u + v - u^2 - v^2\right), \quad z_{uv} = 2u.$$

Table 14.5 shows the coordinate values for different values of u and v. In this example, the y-coordinates provide the surface curvature, which could be enhanced by modifying the y-coordinates of the control points.

14.9.3   Cubic Bézier Surface Patch

As we saw earlier in this chapter, cubic Bézier curves require two end-points, and two central control points. In the surface patch formulation a 4×4 matrix is required as follows:

$$P_{uv} = \begin{bmatrix} u^3 & u^2 & u & 1 \end{bmatrix} \begin{bmatrix} -1 & 3 & -3 & 1 \\ 3 & -6 & 3 & 0 \\ -3 & 3 & 0 & 0 \\ 1 & 0 & 0 & 0 \end{bmatrix} \begin{bmatrix} P_{00} & P_{01} & P_{02} & P_{03} \\ P_{10} & P_{11} & P_{12} & P_{13} \\ P_{20} & P_{21} & P_{22} & P_{23} \\ P_{30} & P_{31} & P_{32} & P_{33} \end{bmatrix}$$
$$\begin{bmatrix} -1 & 3 & -3 & 1 \\ 3 & -6 & 3 & 0 \\ -3 & 3 & 0 & 0 \\ 1 & 0 & 0 & 0 \end{bmatrix} \begin{bmatrix} v^3 \\ v^2 \\ v \\ 1 \end{bmatrix}$$

which is illustrated with an example.
   Given the points:

$$P_{00} = (0, 0, 0), \quad P_{01} = (1, 1, 0), \quad P_{02} = (2, 1, 0), \quad P_{03} = (3, 0, 0)$$
$$P_{10} = (0, 1, 1), \quad P_{11} = (1, 2, 1), \quad P_{12} = (2, 2, 1), \quad P_{13} = (3, 1, 1)$$
$$P_{20} = (0, 1, 2), \quad P_{21} = (1, 2, 2), \quad P_{22} = (2, 2, 2), \quad P_{23} = (3, 1, 2)$$
$$P_{30} = (0, 0, 3), \quad P_{31} = (1, 1, 3), \quad P_{32} = (2, 1, 3), \quad P_{33} = (3, 0, 3)$$

we can write the following matrix equations:

$$x_{uv} = \begin{bmatrix} u^3 & u^2 & u & 1 \end{bmatrix} \begin{bmatrix} -1 & 3 & -3 & 1 \\ 3 & -6 & 3 & 0 \\ -3 & 3 & 0 & 0 \\ 1 & 0 & 0 & 0 \end{bmatrix} \begin{bmatrix} 0 & 1 & 2 & 3 \\ 0 & 1 & 2 & 3 \\ 0 & 1 & 2 & 3 \\ 0 & 1 & 2 & 3 \end{bmatrix}$$
$$\begin{bmatrix} -1 & 3 & -3 & 1 \\ 3 & -6 & 3 & 0 \\ -3 & 3 & 0 & 0 \\ 1 & 0 & 0 & 0 \end{bmatrix} \begin{bmatrix} v^3 \\ v^2 \\ v \\ 1 \end{bmatrix}$$

$$x_{uv} = \begin{bmatrix} u^3 & u^2 & u & 1 \end{bmatrix} \begin{bmatrix} 0 & 0 & 0 & 0 \\ 0 & 0 & 0 & 0 \\ 0 & 0 & 0 & 0 \\ 0 & 0 & 3 & 0 \end{bmatrix} \begin{bmatrix} v^3 \\ v^2 \\ v \\ 1 \end{bmatrix}$$

$$x_{uv} = 3v$$

$$y_{uv} = \begin{bmatrix} u^3 & u^2 & u & 1 \end{bmatrix} \begin{bmatrix} -1 & 3 & -3 & 1 \\ 3 & -6 & 3 & 0 \\ -3 & 3 & 0 & 0 \\ 1 & 0 & 0 & 0 \end{bmatrix} \begin{bmatrix} 0 & 1 & 1 & 0 \\ 1 & 2 & 2 & 1 \\ 1 & 2 & 2 & 1 \\ 0 & 1 & 1 & 0 \end{bmatrix}$$
$$\begin{bmatrix} -1 & 3 & -3 & 1 \\ 3 & -6 & 3 & 0 \\ -3 & 3 & 0 & 0 \\ 1 & 0 & 0 & 0 \end{bmatrix} \begin{bmatrix} v^3 \\ v^2 \\ v \\ 1 \end{bmatrix}$$

$$y_{uv} = \begin{bmatrix} u^3 & u^2 & u & 1 \end{bmatrix} \begin{bmatrix} 0 & 0 & 0 & 0 \\ 0 & 0 & 0 & -3 \\ 0 & 0 & 0 & 3 \\ 0 & -3 & 3 & 0 \end{bmatrix} \begin{bmatrix} v^3 \\ v^2 \\ v \\ 1 \end{bmatrix}$$

$$y_{uv} = 3 \left(u + v - u^2 - v^2 \right)$$

$$z_{uv} = \begin{bmatrix} u^3 & u^2 & u & 1 \end{bmatrix} \begin{bmatrix} -1 & 3 & -3 & 1 \\ 3 & -6 & 3 & 0 \\ -3 & 3 & 0 & 0 \\ 1 & 0 & 0 & 0 \end{bmatrix} \begin{bmatrix} 0 & 0 & 0 & 0 \\ 1 & 1 & 1 & 1 \\ 2 & 2 & 2 & 2 \\ 3 & 3 & 3 & 3 \end{bmatrix}$$

Table 14.6 The x-, y-, z-coordinates for different values of u and v

u	v		
	0	0.5	1
0	(0, 0, 0)	(1.5, 0.75, 0)	(3, 0, 0)
0.5	(0, 0.75, 1.5)	(1.5, 1.5, 1.5)	(3, 0.75, 1.5)
1	(0, 0, 3)	(1.5, 0.75, 3)	(3, 0, 3)

$$z_{uv} = \begin{bmatrix} u^3 & u^2 & u & 1 \end{bmatrix} \begin{bmatrix} -1 & 3 & -3 & 1 \\ 3 & -6 & 3 & 0 \\ -3 & 3 & 0 & 0 \\ 1 & 0 & 0 & 0 \end{bmatrix} \begin{bmatrix} v^3 \\ v^2 \\ v \\ 1 \end{bmatrix}$$

$$z_{uv} = \begin{bmatrix} u^3 & u^2 & u & 1 \end{bmatrix} \begin{bmatrix} 0 & 0 & 0 & 0 \\ 0 & 0 & 0 & 0 \\ 0 & 0 & 0 & 3 \\ 0 & 0 & 0 & 0 \end{bmatrix} \begin{bmatrix} v^3 \\ v^2 \\ v \\ 1 \end{bmatrix}$$

$$z_{uv} = 3u.$$

Therefore, any point on the surface patch has coordinates

$$x_{uv} = 3v, \quad y_{uv} = 3\left(u + v - u^2 - v^2\right), \quad z_{uv} = 3u.$$

Table 14.6 shows the coordinate values for different values of u and v. In this example, the y-coordinates provide the surface curvature, which could be enhanced by modifying the y-coordinates of the control points.

Complex 3D surfaces are readily modeled using Bézier patches. One simply creates a mesh of patches such that their control points are shared at the joins. Surface continuity is controlled using the same mechanism for curves. But where the slopes of trailing and starting control edges apply for curves, the corresponding slopes of control tiles apply for patches.

14.10   Summary

This subject has been the most challenging one to describe. On the one hand, the subject is vital to every aspect of computer graphics, and on the other, the reader is required to wrestle with cubic polynomials and a little calculus. However, I do hope that I have managed to communicate some essential concepts behind curves and surfaces, and that you will be tempted to implement some of the mathematics.

Chapter 15
Analytic Geometry

15.1 Introduction

This chapter explores some basic elements of geometry and analytic geometry that are frequently encountered in computer graphics. For completeness, I have included a short review of important elements of Euclidean geometry with which you should be familiar. Perhaps the most important topics that you should try to understand concern the definitions of straight lines in space, 3D planes, and how points of intersection are computed. Another useful topic is the role of parameters in describing lines and line segments, and their intersection.

15.2 Background

In the third century BCE, Euclid laid the foundations of geometry that have been taught in schools for centuries. In the 19th century, mathematicians such as Bernhard Riemann (1809–1900) and Nicolai Lobachevsky transformed this Euclidean geometry with ideas such as curved space, and spaces with higher dimensions. Although none of these developments affect computer graphics, they do place Euclid's theorems in a specific context: a set of axioms that apply to flat surfaces. We have probably all been taught that parallel lines never meet, and that the internal angles of a triangle sum to 180°, but these are only true in specific situations. As soon as the surface or space becomes curved, such rules break down. So let's review some rules and observations that apply to shapes drawn on a flat surface.

15.2.1 Angles

By definition, 360° or 2π [radians] measure one revolution. You should be familiar with both units of measurement, and how to convert from one to the other. Figure 15.1

© Springer-Verlag London Ltd., part of Springer Nature 2022
J. Vince, *Mathematics for Computer Graphics*, Undergraduate Topics
in Computer Science, https://doi.org/10.1007/978-1-4471-7520-9_15

Fig. 15.1 Examples of
adjacent, supplementary,
opposite and complementary
angles

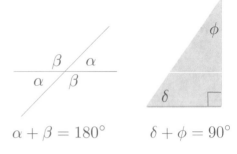

$$\alpha + \beta = 180° \qquad \delta + \phi = 90°$$

Fig. 15.2 The first intercept
theorem

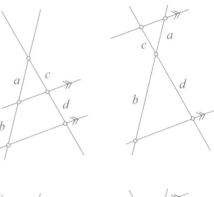

Fig. 15.3 The second
intercept theorem

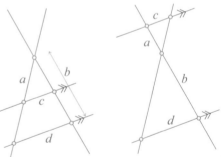

shows examples of *adjacent/supplementary* angles (sum to 180°), *opposite* angles
(equal), and *complementary* angles (sum to 90°).

15.2.2 Intercept Theorems

The *Intercept Theorems* are attributed to the Greek philosopher and mathematician Thales of Miletus (c.624–c.546 BC) and involve intersecting and parallel lines.
Figures 15.2 and 15.3 show two scenarios that give rise to the following observations:

- First intercept theorem:

$$\frac{a+b}{a} = \frac{c+d}{c}, \quad \frac{b}{a} = \frac{d}{c}.$$

- Second intercept theorem:

$$\frac{a}{b} = \frac{c}{d}.$$

15.2.3 Golden Section

The *golden section* is widely used in art and architecture to represent an 'ideal' ratio for the height and width of an object. Its origins stem from the interaction between a circle and triangle and give rise to the following relationship:

$$b = \frac{a}{2}\left(\sqrt{5} - 1\right) \approx 0.618a.$$

The rectangle in Fig. 15.4 has proportions:

$$\text{height} = 0.618 \times \text{width}.$$

15.2.4 Triangles

The rules associated with *interior* and *exterior* angles of a triangle are very useful in solving all sorts of geometric problems. Figure 15.5 shows two diagrams identifying interior and exterior angles. We can see that the sum of the interior angles is 180°, and that the exterior angles of a triangle are equal to the sum of the opposite angles:

$$\alpha + \beta + \theta = 180°$$
$$\alpha' = \theta + \beta$$
$$\beta' = \alpha + \theta$$
$$\theta' = \alpha + \beta.$$

Fig. 15.4 A rectangle with a height to width ratio equal to the golden section

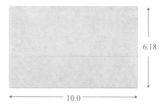

6.18

10.0

Fig. 15.5 Relationship
between interior and exterior
angles

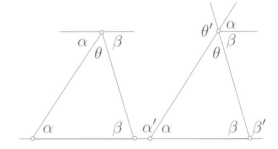

15.2.5   Centre of Gravity of a Triangle

A *median* is a straight line joining a vertex of a triangle to the mid-point of the
opposite side. When all three medians are drawn, they intersect at a common point,
which is also the triangle's *centre of gravity*. The centre of gravity divides all the
medians in the ratio 2 : 1. Figure 15.6 illustrates this arrangement.

15.2.6   Isosceles Triangle

Figure 15.7 shows an *isosceles* triangle, which has two equal sides of length l and
equal base angles α. The triangle's altitude and area are

$$h = \sqrt{l^2 - \left(\frac{c}{2}\right)^2}, \qquad A = \tfrac{1}{2}ch.$$

Fig. 15.6 The three medians
of a triangle intersect at its
centre of gravity

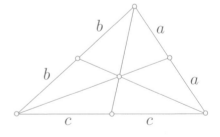

Fig. 15.7 An isosceles
triangle

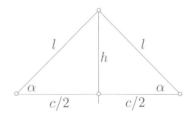

15.2.7 Equilateral Triangle

An *equilateral* triangle has three equal sides of length l and equal angles of $60°$. The triangle's altitude and area are

$$h = \tfrac{\sqrt{3}}{2}l, \qquad A = \tfrac{\sqrt{3}}{4}l^2.$$

15.2.8 Right Triangle

Figure 15.8 shows a right triangle with its obligatory right angle. The triangle's altitude and area are

$$h = \frac{ab}{c}, \qquad A = \tfrac{1}{2}ab.$$

15.2.9 Theorem of Thales

Figure 15.9 illustrates the theorem of Thales, which states that the right angle of a right triangle lies on the circumcircle over the hypotenuse.

15.2.10 Theorem of Pythagoras

Although this theorem is named after Pythagoras there is substantial evidence to show that it was known by the Babylonians a millennium earlier. However, Pythagoras is

Fig. 15.8 A right triangle

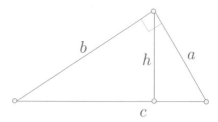

Fig. 15.9 The theorem of Thales

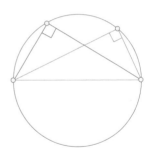

Fig. 15.10 The theorem of
Pythagoras states that
$a^2 = b^2 + c^2$

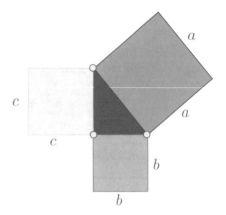

credited with its proof. Figure 15.10 illustrates the well-known relationship:

$$a^2 = b^2 + c^2$$

from which one can show that

$$\sin^2 \theta + \cos^2 \theta = 1.$$

15.2.11 Quadrilateral

Quadrilaterals have four sides and include the square, rectangle, trapezoid, parallel-ogram and rhombus, whose interior angles sum to 360°. As the square and rectangle are familiar shapes, we will only consider the other three.

15.2.12 Trapezoid

Figure 15.11 shows a *trapezoid* which has one pair of parallel sides h apart. The mid-line m and area are given by

Fig. 15.11 A trapezoid with
one pair of parallel sides

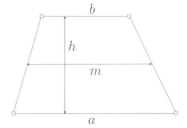

Fig. 15.12 A parallelogram
formed by two pairs of
parallel lines

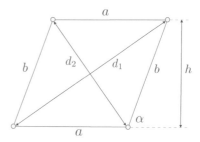

$$m = \tfrac{1}{2}(a + b)$$
$$A = mh.$$

15.2.13   Parallelogram

Figure 15.12 shows a *parallelogram*, which is formed from two pairs of intersecting parallel lines, so it has equal opposite sides and equal opposite angles. The altitude, diagonal lengths and area are given by

$$h = b \sin \alpha$$
$$d_{1,2} = \sqrt{a^2 + b^2 \pm 2a\sqrt{b^2 - h^2}}$$
$$A = ah.$$

15.2.14   Rhombus

Figure 15.13 shows a *rhombus*, which is a parallelogram with four sides of equal length a. The area is given by

Fig. 15.13 A rhombus is a
parallelogram with four
equal sides

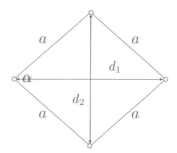

Fig. 15.14 Part of a regular gon showing the inner and outer radii and the edge length

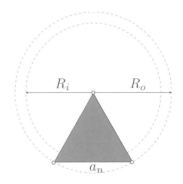

$$A = a^2 \sin \alpha = \tfrac{1}{2} d_1 d_2.$$

15.2.15   Regular Polygon

Figure 15.14 shows part of a regular n-gon with outer radius R_o, inner radius R_i and edge length a_n. Table 15.1 shows the relationship between the area, a_n, R_i and R_o for different polygons.

15.2.16   Circle

The circumference C and area A of a circle are given by

$$C = \pi d = 2\pi r$$

Table 15.1   The area A_n, edge length a_n, inner radius R_i, and outer radius R_o for different polygons

n	$a_n = 2R_i \tan(180°/n)$	$R_i = R_o \cos(180°/n)$	$R_o^2 = R_i^2 + \tfrac{1}{4}a_n^2$
n	$A_n = \tfrac{n}{4}a_n^2 \cot(180°/n)$	$A_n = \tfrac{n}{2}R_o^2 \sin(360°/n)$	$A_n = nR_i^2 \tan(180°/n)$
5	$a_5 = 2R_i\sqrt{5 - 2\sqrt{5}}$	$R_i = \tfrac{R_o}{4}(\sqrt{5}+1)$	$R_o = R_i(\sqrt{5}-1)$
5	$A_5 = \tfrac{a_5^2}{4}\sqrt{25 + 10\sqrt{5}}$	$A_5 = \tfrac{5}{8}R_o^2\sqrt{10 + 2\sqrt{5}}$	$A_5 = 5R_i^2\sqrt{5 - 2\sqrt{5}}$
6	$a_6 = \tfrac{2}{3}R_i\sqrt{3}$	$R_i = \tfrac{R_o}{2}\sqrt{3}$	$R_o = \tfrac{2}{3}R_i\sqrt{3}$
6	$A_6 = \tfrac{3}{2}a_6^2\sqrt{3}$	$A_6 = \tfrac{3}{2}R_o^2\sqrt{3}$	$A_6 = 2R_i^2\sqrt{3}$
8	$a_8 = 2R_i(\sqrt{2}-1)$	$R_i = \tfrac{R_o}{2}\sqrt{2 + \sqrt{2}}$	$R_o = R_i\sqrt{4 - 2\sqrt{2}}$
8	$A_8 = 2a_8^2\left(\sqrt{2}+1\right)$	$A_8 = 2R_o^2\sqrt{2}$	$A_8 = 8R_i^2\left(\sqrt{2}-1\right)$
10	$a_{10} = \tfrac{2}{5}R_i\sqrt{25 - 10\sqrt{5}}$	$R_i = \tfrac{R_o}{4}\sqrt{10 + 2\sqrt{5}}$	$R_o = \tfrac{R_i}{5}\sqrt{50 - 10\sqrt{5}}$
10	$A_{10} = \tfrac{5}{2}a_{10}^2\sqrt{5 + 2\sqrt{5}}$	$A_{10} = \tfrac{5}{4}R_o^2\sqrt{10 - 2\sqrt{5}}$	$A_{10} = 2R_i^2\sqrt{25 - 10\sqrt{5}}$

Fig. 15.15 An annulus formed from two concentric circles

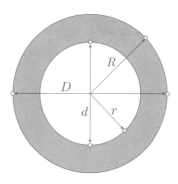

Fig. 15.16 A sector of a circle defined by the angle α

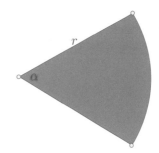

$$A = \pi r^2 = \tfrac{1}{4}\pi d^2$$

where the diameter $d = 2r$.

An *annulus* is the area between two concentric circles as shown in Fig. 15.15, and its area A is given by

$$A = \pi \left(R^2 - r^2 \right) = \tfrac{1}{4}\pi \left(D^2 - d^2 \right)$$

where $D = 2R$ and $d = 2r$.

Figure 15.16 shows a sector of a circle, whose area is given by

$$A = \frac{\alpha^\circ}{360^\circ} \pi r^2.$$

Figure 15.17 shows a *segment* of a circle, whose area is given by

$$A = \tfrac{1}{2}r^2(\alpha - \sin \alpha), \quad \text{where } \alpha \text{ is in radians.}$$

The area of an *ellipse* with major and minor radii a and b is

$$A = \pi ab.$$

Fig. 15.17 A segment of a
circle defined by the angle α

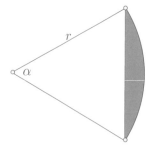

15.3   2D Analytic Geometry

In this section we briefly examine familiar descriptions of geometric elements and
ways of computing intersections.

15.3.1   Equation of a Straight Line

The well-known equation of a line is

$$y = mx + c$$

where m is the slope and c the intersection with the y-axis, as shown in Fig. 15.18.
This is called the *normal form*.

Given two points $(x_1, \; y_1)$ and $(x_2, \; y_2)$ we can state that for any other point $(x, \; y)$

$$\frac{y - y_1}{x - x_1} = \frac{y_2 - y_1}{x_2 - x_1}$$

Fig. 15.18 The normal form
of the straight line is
$y = mx + c$

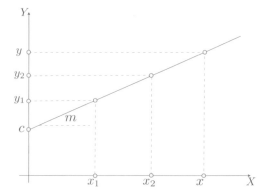

which yields

$$y = (x - x_1)\frac{y_2 - y_1}{x_2 - x_1} + y_1.$$

Although these equations have their uses, the more general form is much more convenient:

$$ax + by + c = 0.$$

As we shall see, this equation possesses some interesting qualities.

15.3.2   The Hessian Normal Form

Figure 15.19 shows a line whose orientation is controlled by a normal unit vector $\mathbf{n} = [a \ \ b]^T$. If $P(x, \ y)$ is any point on the line, then $\mathbf{p}$ is a position vector where $\mathbf{p} = [x \ \ y]^T$ and d is the perpendicular distance from the origin to the line. Therefore,

$$\frac{d}{\|\mathbf{p}\|} = \cos\alpha$$

and

$$d = \|\mathbf{p}\| \cos\alpha.$$

But the dot product $\mathbf{n} \cdot \mathbf{p}$ is given by

$$\mathbf{n} \cdot \mathbf{p} = \|\mathbf{n}\|\|\mathbf{p}\| \cos\alpha = ax + by$$

which implies that

$$ax + by = d\|\mathbf{n}\|$$

Fig. 15.19 The orientation of a line can be controlled by a normal vector **n** and a distance d

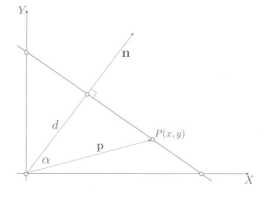

and because $\|\mathbf{n}\| = 1$ we can write

$$ax + by - d = 0$$

where (x, y) is a point on the line, a and b are the components of a unit vector normal to the line, and d is the perpendicular distance from the origin to the line. The distance d is positive when the normal vector points away from the origin, otherwise it is negative. For example, let's find the equation of a line whose normal vector is $[3 \quad 4]^T$ and the perpendicular distance from the origin to the line is 1.

We begin by normalising the normal vector to its unit form. Therefore, if $\mathbf{n} = [3 \quad 4]^T$, $\|\mathbf{n}\| = \sqrt{3^2 + 4^2} = 5$ The equation of the line is

$$\tfrac{3}{5}x + \tfrac{4}{5}y - 1 = 0.$$

Similarly, let's find the Hessian normal form of $y = 2x + 1$.
Rearranging the equation we get

$$2x - y = -1$$

which gives a negative distance. If we want the normal vector to point away from the origin we multiply by -1:

$$-2x + y - 1 = 0.$$

Normalise the normal vector to a unit form

$$\text{i.e.} \quad \sqrt{(-2)^2 + 1^2} = \sqrt{5}$$

$$-\tfrac{2}{\sqrt{5}}x + \tfrac{1}{\sqrt{5}}y - \tfrac{1}{\sqrt{5}} = 0.$$

Therefore, the perpendicular distance from the origin to the line, and the unit normal vector are respectively

$$\tfrac{1}{\sqrt{5}} \quad \text{and} \quad \left[\tfrac{-2}{\sqrt{5}} \quad \tfrac{1}{\sqrt{5}}\right]^T.$$

As the Hessian normal form involves a unit normal vector, we can incorporate the vector's direction cosines within the equation:

$$x \cos \alpha + y \sin \alpha - d = 0$$

where α is the angle between the normal vector and the x-axis.

Fig. 15.20 The Hessian
normal form of the line
equation partitions space into
two zones

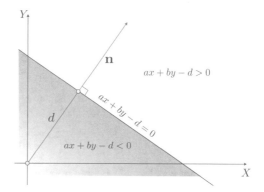

15.3.3 Space Partitioning

The Hessian normal form provides a very useful way of partitioning space into two
zones: the partition that includes the normal vector, and the opposite partition. This
is illustrated in Fig. 15.20.

Given the equation

$$ax + by - d = 0$$

a point (x, y) on the line satisfies the equation. But if the point (x, y) is in the
partition in the direction of the normal vector, it creates the inequality

$$ax + by - d > 0.$$

Conversely, if (x, y) is in the partition opposite to the direction of the normal vector
creates the inequality

$$ax + by - d < 0.$$

This space-partitioning feature of the Hessian normal form is useful in clipping lines
against polygonal windows.

15.3.4 The Hessian Normal Form from Two Points

Given two points (x_1, y_1) and (x_2, y_2) we compute the values of a, b and d for the
Hessian normal form as follows.

The vector joining the two points is $\mathbf{v} = [\Delta_x \quad \Delta_y]^\mathsf{T}$ where

$$\Delta_x = x_2 - x_1$$
$$\Delta_y = y_2 - y_1$$

$$\|\mathbf{v}\| = \sqrt{\Delta_x^2 + \Delta_y^2}$$

The unit vector normal to $\mathbf{v}$ is $\mathbf{n} = [-\Delta_y' \quad \Delta_x']^T$, where

$$\Delta_x' = \frac{\Delta_x}{\|\mathbf{v}\|}$$

$$\Delta_y' = \frac{\Delta_y}{\|\mathbf{v}\|}$$

Therefore, let $\mathbf{p} = [x \quad y]^T$ be any point on the line, and using the Hessian Normal Form, we can write:

$$\mathbf{n} \cdot \mathbf{p} = -\Delta_y' x + \Delta_x' y = -\Delta_y' x_1 + \Delta_x' y_1$$

and

$$- \Delta_y' x + \Delta_x' y + (\Delta_y' x_1 - \Delta_x' y_1) = 0 \qquad (15.1)$$

For example, given the following points: $(x_1, \ y_1) = (0, \ 1)$ and $(x_2, \ y_2) = (1, \ 0)$; then $\Delta_x' = 1/\sqrt{2}$ and $\Delta_y' = -1/\sqrt{2}$. Therefore, using (15.1)

$$\frac{x}{\sqrt{2}} + \frac{y}{\sqrt{2}} + \left(0 \times \tfrac{-1}{\sqrt{2}} - 1 \times \tfrac{1}{\sqrt{2}}\right) = 0$$

$$\frac{x}{\sqrt{2}} + \frac{y}{\sqrt{2}} - \frac{1}{\sqrt{2}} = 0.$$

15.4   Intersection Points

15.4.1   Intersecting Straight Lines

Given two line equations of the form

$$a_1 x + b_1 y + d_1 = 0$$
$$a_2 x + b_2 y + d_2 = 0$$

the intersection point $(x_i, \ y_i)$ is given by

$$x_i = \frac{b_1 d_2 - b_2 d_1}{a_1 b_2 - a_2 b_1}$$

$$y_i = \frac{d_1 a_2 - d_2 a_1}{a_1 b_2 - a_2 b_1}$$

or using determinants:

$$x_i = \frac{\begin{vmatrix} b_1 & d_1 \\ b_2 & d_2 \end{vmatrix}}{\begin{vmatrix} a_1 & b_1 \\ a_2 & b_2 \end{vmatrix}}$$

$$y_i = \frac{\begin{vmatrix} d_1 & a_1 \\ d_2 & a_2 \end{vmatrix}}{\begin{vmatrix} a_1 & b_1 \\ a_2 & b_2 \end{vmatrix}}.$$

If the denominator is zero, the equations are linearly dependent, indicating that there is no intersection.

15.4.2 Intersecting Line Segments

We are often concerned with line segments in computer graphics as they represent the edges of shapes and objects. So let's investigate how to compute the intersection of two 2D-line segments. Figure 15.21 shows two line segments defined by their end points P_1, P_2, P_3, P_4 and respective position vectors $\mathbf{p}_1$, $\mathbf{p}_2$, $\mathbf{p}_3$ and $\mathbf{p}_4$. We can write the following vector equations to identify the point of intersection:

$$\mathbf{p}_i = \mathbf{p}_1 + t(\mathbf{p}_2 - \mathbf{p}_1) \qquad (15.2)$$
$$\mathbf{p}_i = \mathbf{p}_3 + s(\mathbf{p}_4 - \mathbf{p}_3) \qquad (15.3)$$

where parameters s and t vary between 0 and 1. For the point of intersection, we can write

Fig. 15.21 Two line segments with their associated position vectors

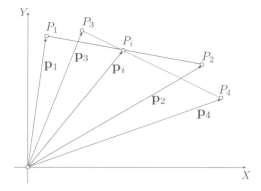

$$\mathbf{p}_1 + t(\mathbf{p}_2 - \mathbf{p}_1) = \mathbf{p}_3 + s(\mathbf{p}_4 - \mathbf{p}_3).$$

Therefore, the parameters s and t are given by

$$s = \frac{(\mathbf{p}_1 - \mathbf{p}_3) + t(\mathbf{p}_2 - \mathbf{p}_1)}{\mathbf{p}_4 - \mathbf{p}_3} \tag{15.4}$$

$$t = \frac{(\mathbf{p}_3 - \mathbf{p}_1) + s(\mathbf{p}_4 - \mathbf{p}_3)}{\mathbf{p}_2 - \mathbf{p}_1}. \tag{15.5}$$

From (15.5) we can write

$$t = \frac{(x_3 - x_1) + s(x_4 - x_3)}{x_2 - x_1}$$

$$t = \frac{(y_3 - y_1) + s(y_4 - y_3)}{y_2 - y_1}$$

which yields

$$s = \frac{x_1(y_3 - y_2) + x_2(y_3 - y_1) + x_3(y_2 - y_1)}{(x_2 - x_1)(y_4 - y_3) - (x_4 - x_3)(y_2 - y_1)} \tag{15.6}$$

similarly,

$$t = \frac{x_1(y_4 - y_3) + x_3(y_1 - y_4) + x_4(y_3 - y_1)}{(x_4 - x_3)(y_2 - y_1) - (x_2 - x_1)(y_4 - y_3)}. \tag{15.7}$$

Let's test (15.6) and (15.7) with two examples to illustrate how the equations are used in practice. The first example demonstrates an intersection condition, and the second demonstrates a touching condition.

Figure 15.22a shows two line segments intersecting, with an obvious intersection point of $(1.5,\ 0)$. The coordinates of the line segments are

$$(x_1,\ y_1) = (1,\ 0), \qquad (x_2,\ y_2) = (2,\ 0)$$
$$(x_3,\ y_3) = (1.5,\ -1), \qquad (x_4,\ y_4) = (1.5,\ 1)$$

therefore,

$$t = \frac{1(1 - (-1)) + 1.5(0 - 1) + 1.5(-1 - 0)}{(0 - 0)(1.5 - 1.5) - (2 - 1)(1 - (-1))} = 0.5$$

and

$$s = \frac{1(-1 - 0) + 2(0 - (-1)) + 1.5(0 - 0)}{(1 - (-1))(2 - 1) - (1.5 - 1.5)(0 - 0)} = 0.5.$$

Substituting s and t in (15.2) and (15.3) we get $(x_i,\ y_i) = (1.5,\ 0)$ as predicted.

Figure 15.22b shows two line segments touching at $(1.5,\ 0)$. The coordinates of the line segments are

Fig. 15.22 a Shows two line
segments intersecting **b**
Shows two line segments
touching

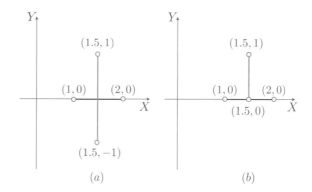

(a)   (b)

$$(x_1, \; y_1) = (1, \; 0), \qquad (x_2, \; y_2) = (2, \; 0)$$
$$(x_3, \; y_3) = (1.5, \; 0), \qquad (x_4, \; y_4) = (1.5, \; 1)$$

therefore,

$$t = \frac{1(1 - 0) + 1.5(0 - 1) + 1.5(0 - 0)}{(0 - 0)(1.5 - 1.5) - (2 - 1)(1 - 0)} = 0.5$$

and

$$s = \frac{1(0 - 0) + 2(0 - 0) + 1.5(0 - 0)}{(1 - 0)(2 - 1) - (1.5 - 1.5)(0 - 0)} = 0.$$

The zero value of s confirms that the lines touch, rather than intersect, and $t = 0.5$
confirms that the touching takes place halfway along the line segment.

15.5 Point Inside a Triangle

We often require to test whether a point is inside, outside or touching a triangle. Let's
examine two ways of performing this operation. The first is related to finding the
area of a triangle.

15.5.1 Area of a Triangle

Let's declare a triangle formed by the anticlockwise points $P_1(x_1, \; y_1)$, $P_2(x_2, \; y_2)$
and $P_3(x_3, \; y_3)$ as shown in Fig. 15.23. The area of the triangle is given by:

$$A = (x_2 - x_1)(y_3 - y_1) - \tfrac{1}{2}(x_2 - x_1)(y_2 - y_1) - \tfrac{1}{2}(x_2 - x_3)(y_3 - y_2) - \tfrac{1}{2}(x_3 - x_1)(y_3 - y_1)$$

which simplifies to

Fig. 15.23 The area of the
triangle is computed by
subtracting the smaller
triangles from the
rectangular area

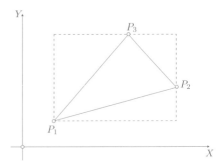

$$A = \tfrac{1}{2}[x_1(y_2 - y_3) + x_2(y_3 - y_1) + x_3(y_1 - y_2)]$$

and this can be further simplified to

$$A = \tfrac{1}{2} \begin{vmatrix} x_1 & y_1 & 1 \\ x_2 & y_2 & 1 \\ x_3 & y_3 & 1 \end{vmatrix}.$$

Figure 15.24 shows two triangles with opposing vertex sequences. If we calculate
the area of the top triangle with anticlockwise vertices, we obtain

$$A = \tfrac{1}{2}[1(2 - 4) + 3(4 - 2) + 2(2 - 2)] = 2$$

whereas the area of the bottom triangle with clockwise vertices is

$$A = \tfrac{1}{2}[1(2 - 0) + 3(0 - 2) + 2(2 - 2)] = -2$$

which shows that the technique is sensitive to vertex direction. We can exploit this
sensitivity to test if a point is inside or outside a triangle.

Fig. 15.24 The top triangle
has anticlockwise vertices,
and the bottom triangle
clockwise vertices

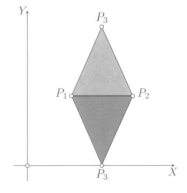

Fig. 15.25 If the point P_t is inside the triangle, it is always to the left as the boundary is traversed in an anticlockwise direction

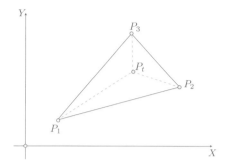

Consider the scenario shown in Fig. 15.25, where the point P_t is inside the triangle (P_1, P_2, P_3).

- If the area of triangle (P_1, P_2, P_t) is positive, P_t must be to the left of the line (P_1, P_2).
- If the area of triangle (P_2, P_3, P_t) is positive, P_t must be to the left of the line (P_2, P_3).
- If the area of triangle (P_3, P_1, P_t) is positive, P_t must be to the left of the line (P_3, P_1).

If all the above tests are positive, P_t is inside the triangle. Furthermore, if one area is zero and the other areas are positive, the point is on the boundary, and if two areas are zero and the other positive, the point is on a vertex.

Let's now investigate how the Hessian normal form provides a similar function.

15.5.2   *Hessian Normal Form*

We can determine whether a point is inside, touching or outside a triangle by representing the triangle's edges in the Hessian normal form, and testing in which partition the point is located. If we arrange that the normal vectors are pointing towards the inside of the triangle, any point inside the triangle will create a positive result when tested against the edge equation. In the following calculations there is no need to ensure that the normal vector is a unit vector, therefore (15.1) can be written:

$$-\Delta_y x + \Delta_x y + (\Delta_y x_1 - \Delta_x y_1) = 0$$

To illustrate this, consider the scenario shown in Fig. 15.26 where a triangle is formed by the points $(1, 1), (3, 1)$ and $(2, 3)$. With reference to (15.1) we compute the three line equations:

Fig. 15.26 The triangle is
represented by three line
equations expressed in the
Hessian normal form. Any
point inside the triangle is
found by evaluating their
equations

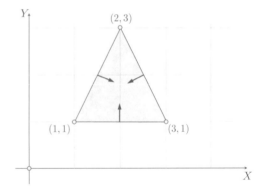

The line between $(1,\ 1)$ and $(3,\ 1)$:

$$\Delta_x = 2$$
$$\Delta_y = 0$$
$$-0 \times x + 2 \times y - 2 \times 1 = 0$$
$$2y - 2 = 0.$$

The line between $(3,\ 1)$ and $(2,\ 3)$:

$$\Delta_x = -1$$
$$\Delta_y = 2$$
$$-2 \times x - 1 \times y + (2 \times 3 + 1 \times 1) = 0$$
$$-2x - y + 7 = 0.$$

The line between $(2,\ 3)$ and $(1,\ 1)$:

$$\Delta_x = -1$$
$$\Delta_y = -2$$
$$2 \times x - 1 \times y + (-2 \times 2 - 1 \times 3) = 0$$
$$2x - y - 1 = 0.$$

Thus the three line equations for the triangle are

$$2y - 2 = 0$$
$$-2x - y + 7 = 0$$
$$2x - y - 1 = 0.$$

We are only interested in the signs of the equations:

$$2y - 2 \tag{15.8}$$
$$- 2x - y + 7 \tag{15.9}$$
$$2x - y - 1 \tag{15.10}$$

which can be tested for any arbitrary point (x, y). If they are all positive, the point is inside the triangle. If one expression is negative, the point is outside. If one expression is zero, the point is on an edge, and if two expressions are zero, the point is on a vertex.

Just as a quick test, consider the point $(2, 2)$. The three expressions (15.8) to (15.10) are positive, which confirms that the point is inside the triangle. The point $(3, 3)$ is obviously outside the triangle, which is confirmed by two positive results and one negative. Finally, the point $(2, 3)$, which is a vertex, creates one positive result and two zero results.

15.6 Intersection of a Circle with a Straight Line

The equation of a circle has already been given in the previous chapter, so we will now consider how to compute its intersection with a straight line. We begin by testing the equation of a circle with the normal form of the line equation:

$$x^2 + y^2 = r^2 \quad \text{and} \quad y = mx + c.$$

By substituting the line equation in the circle's equation we discover the two intersection points:

$$x_{1,2} = \frac{-mc \pm \sqrt{r^2(1 + m^2) - c^2}}{1 + m^2} \tag{15.11}$$
$$y_{1,2} = \frac{c \pm m\sqrt{r^2(1 + m^2) - c^2}}{1 + m^2}. \tag{15.12}$$

Let's test this result with the scenario shown in Fig. 15.27. Using the normal form of the line equation we have

$$y = x + 1, \quad m = 1, \quad \text{and} \quad c = 1.$$

Substituting these values in (15.11) and (15.12) yields

$$x_{1,2} = -1, \ 0, \quad y_{1,2} = 0, \ 1.$$

The actual points of intersection are $(-1, 0)$ and $(0, 1)$.

Fig. 15.27 The intersection of a circle with a line

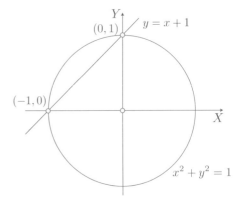

Testing the equation of the circle with the general equation of the line $ax + by + c = 0$ yields intersections given by

$$x_{1,2} = \frac{-ac \pm b\sqrt{r^2(a^2 + b^2) - c^2}}{a^2 + b^2} \qquad (15.13)$$

$$y_{1,2} = \frac{-bc \pm a\sqrt{r^2(a^2 + b^2) - c^2}}{a^2 + b^2}. \qquad (15.14)$$

The general form of the line equation $y = x + 1$ is

$$x - y + 1 = 0 \quad \text{where} \quad a = 1, \ b = -1 \quad \text{and} \quad c = 1.$$

Substituting these values in (15.13) and (15.14) yields

$$x_{1,2} = -1, \ 0, \quad \text{and} \quad y_{1,2} = 0, \ 1$$

which gives the same intersection points found above.

Finally, using the Hessian normal form of the line $ax + by - d = 0$ yields intersections given by

$$x_{1,2} = ad \pm b\sqrt{r^2 - d^2} \qquad (15.15)$$

$$y_{1,2} = bd \pm a\sqrt{r^2 - d^2}. \qquad (15.16)$$

The Hessian normal form of the line equation $x - y + 1 = 0$ is

$$-0.707x + 0.707y - 0.707 \approx 0$$

where $a \approx -0.707$, $b \approx 0.707$ and $d \approx 0.707$. Substituting these values in (15.15) and (15.16) yields

$$x_{1,2} = -1, \ 0 \quad \text{and} \quad y_{1,2} = 0, \ 1$$

which gives the same intersection points found above. One can readily see the computational benefits of using the Hessian normal form over the other forms of equations.

15.7 3D Geometry

3D straight lines are best described using vector notation, and readers are urged to develop strong skills in these techniques if they wish to solve problems in 3D geometry. Let's begin this short survey of 3D analytic geometry by describing the equation of a straight line.

15.7.1 Equation of a Straight Line

We start by using a vector **b** to define the orientation of the line, and a point a in space through which the line passes. This scenario is shown in Fig. 15.28. Given another point P on the line we can define a vector $t\mathbf{b}$ between a and P, where t is a scalar. The position vector **p** for P is given by

$$\mathbf{p} = \mathbf{a} + t\mathbf{b}$$

from which we can obtain the coordinates of the point P:

$$x_p = x_a + tx_b$$
$$y_p = y_a + ty_b$$
$$z_p = z_a + tz_b.$$

Fig. 15.28 The line equation is based upon the point a and the vector **b**

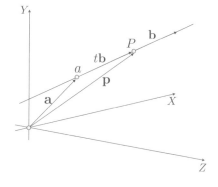

For example, if $\mathbf{b} = [1 \ \ 2 \ \ 3]^{\mathrm{T}}$ and $a = (2, \ 3, \ 4)$, then by setting $t = 1$ we can identify a second point on the line:

$$x_p = 2 + 1 = 3$$
$$y_p = 3 + 2 = 5$$
$$z_p = 4 + 3 = 7.$$

In fact, by using different values of t we can slide up and down the line with ease.

If we have two points P_1 and P_2, such as the vertices of an edge, we can represent the line equation using the above vector technique:

$$\mathbf{p} = \mathbf{p}_1 + t(\mathbf{p}_2 - \mathbf{p}_1)$$

where $\mathbf{p}_1$ and $\mathbf{p}_2$ are position vectors to their respective points. Once more, we can write the coordinates of any point P as follows:

$$x_p = x_1 + t(x_2 - x_1)$$
$$y_p = y_1 + t(y_2 - y_1)$$
$$z_p = z_1 + t(z_2 - z_1).$$

15.7.2   Intersecting Two Straight Lines

Given two straight lines we can test for a point of intersection, but must be prepared for three results:

- a real intersection point
- no intersection point
- an infinite number of intersections (identical lines).

If the line equations are of the form

$$\mathbf{p} = \mathbf{a}_1 + r\mathbf{b}_1$$
$$\mathbf{p} = \mathbf{a}_2 + r\mathbf{b}_2$$

for an intersection we can write

$$\mathbf{a}_1 + r\mathbf{b}_1 = \mathbf{a}_2 + s\mathbf{b}_2$$

which yields

$$x_{a1} + rx_{b1} = x_{a2} + sx_{b2} \qquad (15.17)$$
$$y_{a1} + ry_{b1} = y_{a2} + sy_{b2} \qquad (15.18)$$

$$z_{a1} + r z_{b1} = z_{a2} + s z_{b2}. \tag{15.19}$$

We now have three equations in two unknowns, and any value of r and s must hold for all three equations. We begin by selecting two equations that are linearly independent (i.e. one equation is not a scalar multiple of the other) and solve for r and s, which must then satisfy the third equation. If this final substitution fails, then there is no intersection. If all three equations are linearly dependent, they describe two parallel lines, which can never intersect.

To check for linear dependency we rearrange (15.17)–(15.19) as follows:

$$r x_{b1} - s x_{b2} = x_{a2} - x_{a1} \tag{15.20}$$

$$r y_{b1} - s y_{b2} = y_{a2} - y_{a1} \tag{15.21}$$

$$r z_{b1} - s z_{b2} = z_{a2} - z_{a1}. \tag{15.22}$$

If the determinant Δ of any pair of these equations is zero, then they are dependent. For example, (15.20) and (15.21) form the determinant

$$\Delta = \begin{vmatrix} x_{b1} & -x_{b2} \\ y_{b1} & -y_{b2} \end{vmatrix}$$

which, if zero, implies that the two equations can not yield a solution. As it is impossible to predict which pair of equations from (15.20) to (15.22) will be independent, let's express two independent equations as follows:

$$r a_{11} - s a_{12} = b_1$$
$$r a_{21} - s a_{22} = b_2$$

which yields

$$r = \frac{a_{22} b_1 - a_{12} b_2}{\Delta}$$

$$s = \frac{a_{21} b_1 - a_{11} b_2}{\Delta}$$

where

$$\Delta = \begin{vmatrix} a_{11} & a_{12} \\ a_{21} & a_{22} \end{vmatrix}.$$

Solving for r and s we obtain

$$r = \frac{y_{b2}(x_{a2} - x_{a1}) - x_{b2}(y_{a2} - y_{a1})}{x_{b1} y_{b2} - y_{b1} x_{b2}} \tag{15.23}$$

$$s = \frac{y_{b1}(x_{a2} - x_{a1}) - x_{b1}(y_{a2} - y_{a1})}{x_{b1} y_{b2} - y_{b1} x_{b2}}. \tag{15.24}$$

As a quick test, consider the intersection of the lines encoded by the following vectors:

$$\mathbf{a}_1 = \begin{bmatrix} 0 \\ 1 \\ 0 \end{bmatrix}, \quad \mathbf{b}_1 = \begin{bmatrix} 3 \\ 3 \\ 3 \end{bmatrix}, \quad \mathbf{a}_2 = \begin{bmatrix} 0 \\ 0.5 \\ 0 \end{bmatrix}, \quad \mathbf{b}_2 = \begin{bmatrix} 2 \\ 3 \\ 2 \end{bmatrix}.$$

Substituting the x- and y-components in (15.23) and (15.24) we discover

$$r = \tfrac{1}{3} \quad \text{and} \quad s = \tfrac{1}{2}$$

but for these to be consistent, they must satisfy the z-component of the original equation:

$$r z_{b1} - s z_{b2} = z_{a2} - z_{a1}$$

$$\tfrac{1}{3} \times 3 - \tfrac{1}{2} \times 2 = 0$$

which is correct. Therefore, the point of intersection is given by either

$$\mathbf{p}_i = \mathbf{a}_1 + r\mathbf{b}_1, \quad \text{or}$$
$$\mathbf{p}_i = \mathbf{a}_2 + s\mathbf{b}_2.$$

Let's try both, just to prove the point:

$$x_i = 0 + \tfrac{1}{3}3 = 1, \quad x_i = 0 + \tfrac{1}{2}2 = 1$$
$$y_i = 1 + \tfrac{1}{3}3 = 2, \quad y_i = \tfrac{1}{2} + \tfrac{1}{2}3 = 2$$
$$z_i = 0 + \tfrac{1}{3}3 = 1, \quad z_i = 0 + \tfrac{1}{2}2 = 1.$$

Therefore, the point of intersection point is $(1, 2, 1)$.

Now let's take two lines that don't intersect, and also exhibit some linear dependency:

$$\mathbf{a}_1 = \begin{bmatrix} 0 \\ 1 \\ 0 \end{bmatrix}, \quad \mathbf{b}_1 = \begin{bmatrix} 2 \\ 2 \\ 0 \end{bmatrix}, \quad \mathbf{a}_2 = \begin{bmatrix} 0 \\ 2 \\ 0 \end{bmatrix}, \quad \mathbf{b}_2 = \begin{bmatrix} 2 \\ 2 \\ 1 \end{bmatrix}.$$

Taking the x- and y-components we discover that the determinant Δ is zero, which has identified the linear dependency. Taking the y- and z-components the determinant is non-zero, which permits us to compute r and s using

$$r = \frac{z_{b2}(y_{a2} - y_{a1}) - y_{b2}(z_{a2} - z_{a1})}{y_{b1}z_{b2} - z_{b1}y_{b2}}$$

$$s = \frac{z_{b1}(y_{a2} - y_{a1}) - y_{b1}(z_{a2} - z_{a1})}{y_{b1}z_{b2} - z_{b1}y_{b2}}$$

$$r = \frac{1(2-1) - 2(0-0)}{2 \times 1 - 0 \times 2} = \frac{1}{2}$$

$$s = \frac{0(2-1) - 2(0-0)}{2 \times 1 - 0 \times 2} = 0.$$

But these values of r and s must also apply to the x-components:

$$r x_{b1} - s x_{b2} = x_{a2} - x_{a1}$$

$$\tfrac{1}{2} \times 2 - 0 \times 2 \neq 0$$

which they clearly do not, therefore the lines do not intersect.

Now let's proceed with the equation of a plane, and then look at how to compute the intersection of a line with a plane using a similar technique.

15.8 Equation of a Plane

We now consider four ways of representing a plane equation: the Cartesian form, general form, parametric form and a plane from three points.

15.8.1 Cartesian Form of the Plane Equation

One popular method of representing a plane equation is the Cartesian form, which employs a vector normal to the plane's surface and a point on the plane. The equation is derived as follows.

Let $\mathbf{n}$ be a nonzero vector normal to the plane and $P_0(x_0,\ y_0,\ z_0)$ a point on the plane. $P(x,\ y,\ z)$ is any other point on the plane. Figure 15.29 illustrates the scenario.

The normal vector is defined as

Fig. 15.29 The vector $\mathbf{n}$ is normal to the plane, which contains a point P_0. P is any other point on the plane

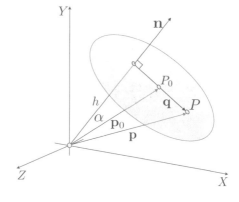

$$\mathbf{n} = a\mathbf{i} + b\mathbf{j} + c\mathbf{k}$$

and the position vectors for P_0 and P are

$$\mathbf{p}_0 = x_0\mathbf{i} + y_0\mathbf{j} + z_0\mathbf{k}$$
$$\mathbf{p} = x\mathbf{i} + y\mathbf{j} + z\mathbf{k}$$

respectively. From Fig. 15.29 we observe that

$$\mathbf{q} = \mathbf{p} - \mathbf{p}_0$$

and as $\mathbf{n}$ is orthogonal to $\mathbf{q}$

$$\mathbf{n} \cdot \mathbf{q} = 0$$

therefore,

$$\mathbf{n} \cdot (\mathbf{p} - \mathbf{p}_0) = 0$$

which expands into

$$\mathbf{n} \cdot \mathbf{p} = \mathbf{n} \cdot \mathbf{p}_0. \tag{15.25}$$

Writing (15.25) in its Cartesian form we obtain

$$ax + by + cz = ax_0 + by_0 + cz_0$$

but $ax_0 + by_0 + cz_0$ is a scalar quantity associated with the plane and can be replaced by d . Therefore,

$$ax + by + cz = d \tag{15.26}$$

which is the Cartesian form of the plane equation.

The value of d has the following geometric interpretation.
In Fig. 15.29 the perpendicular distance from the origin to the plane is

$$h = \|\mathbf{p}_0\| \cos \alpha$$

therefore,

$$\mathbf{n} \cdot \mathbf{p}_0 = \|\mathbf{n}\| \|\mathbf{p}_0\| \cos \alpha = h \|\mathbf{n}\|$$

therefore, the plane equation is also expressed as

$$ax + by + cz = h \|\mathbf{n}\|. \tag{15.27}$$

Dividing (15.27) by $\|\mathbf{n}\|$ we obtain

$$\frac{a}{\|\mathbf{n}\|}x + \frac{b}{\|\mathbf{n}\|}y + \frac{c}{\|\mathbf{n}\|}z = h$$

Fig. 15.30 A plane
represented by the normal
vector **n** and a point
$P_0(0, 1, 0)$

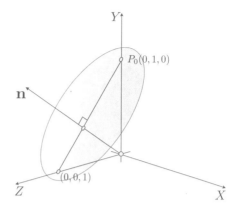

where
$$\|\mathbf{n}\| = \sqrt{a^2 + b^2 + c^2}.$$

This means that when a unit normal vector is used, h is the perpendicular distance
from the origin to the plane. Let's investigate this equation with an example.

Figure 15.30 shows a plane represented by the normal vector $\mathbf{n} = \mathbf{j} + \mathbf{k}$ and a
point on the plane $P_0(0, 1, 0)$. Using (15.26) we have

$$0x + 1y + 1z = 0 \times 0 + 1 \times 1 + 1 \times 0 = 1$$

therefore, the plane equation is
$$y + z = 1.$$

If we normalise the equation to create a unit normal vector, we have

$$\frac{y}{\sqrt{2}} + \frac{z}{\sqrt{2}} = \frac{1}{\sqrt{2}}$$

where the perpendicular distance from the origin to the plane is $1/\sqrt{2}$.

15.8.2 General Form of the Plane Equation

The general form of the equation of a plane is expressed as

$$Ax + By + Cz + D = 0$$

which means that the Cartesian form is translated into the general form by making

Fig. 15.31 A plane is
defined by the vectors **a** and
b and the point $T(x_t, \ y_t, \ z_t)$

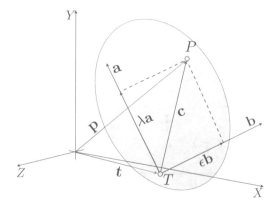

$$A = a, \quad B = b, \quad C = c, \quad D = -d.$$

15.8.3   *Parametric Form of the Plane Equation*

Another method of representing a plane is to employ two vectors and a point that
lie on the plane. Figure 15.31 illustrates a scenario where vectors **a** and **b**, and the
point $T(x_t, \ y_t, \ z_t)$ lie on a plane. We now identify any other point on the plane
$P(x_p, \ y_p, \ z_p)$ with its associated position vector **p**. The point T also has its associated position vector **t**.

Using vector addition we can write

$$\mathbf{c} = \lambda \mathbf{a} + \epsilon \mathbf{b}$$

where λ and ϵ are two scalars such that **c** locates the point P . We can now write

$$\mathbf{p} = \mathbf{t} + \mathbf{c} \tag{15.28}$$

therefore,

$$x_p = x_t + \lambda x_a + \epsilon x_b$$
$$y_p = y_t + \lambda y_a + \epsilon y_b$$
$$z_p = z_t + \lambda z_a + \epsilon z_b$$

which means that the coordinates of any point on the plane are formed from the
coordinates of the known point on the plane, and a linear mixture of the components
of the two vectors. Let's illustrate this vector approach with an example.

Fig. 15.32 The plane is
defined by the vectors **a** and
b and the point $T(1, 1, 1)$

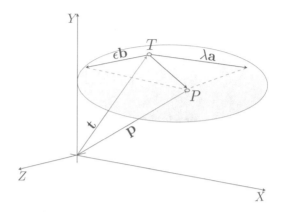

Figure 15.32 shows a plane containing the vectors $\mathbf{a} = \mathbf{i}$ and $\mathbf{b} = \mathbf{k}$, and the point $T(1, 1, 1)$ with its position vector $\mathbf{t} = \mathbf{i} + \mathbf{j} + \mathbf{k}$. By inspection, the plane is parallel with the xz-plane and intersects the y-axis at $y = 1$.

From (15.28) we can write

$$\mathbf{p} = \mathbf{t} + \lambda \mathbf{a} + \epsilon \mathbf{b}$$

where λ and ϵ are arbitrary scalars.

For example, if $\lambda = 2$ and $\epsilon = 1$:

$$x_p = 1 + 2 \times 1 + 1 \times 0 = 3$$
$$y_p = 1 + 2 \times 0 + 1 \times 0 = 1$$
$$z_p = 1 + 2 \times 0 + 1 \times 1 = 2.$$

Therefore, the point $(3, 1, 2)$ is on the plane.

15.8.4 Converting from the Parametric to the General Form

It is possible to convert from the parametric form to the general form of the plane equation using the following formulae:

$$\lambda = \frac{(\mathbf{a} \cdot \mathbf{b})(\mathbf{b} \cdot \mathbf{t}) - (\mathbf{a} \cdot \mathbf{t})\|\mathbf{b}\|^2}{\|\mathbf{a}\|^2 \|\mathbf{b}\|^2 - (\mathbf{a} \cdot \mathbf{b})^2}$$

$$\epsilon = \frac{(\mathbf{a} \cdot \mathbf{b})(\mathbf{a} \cdot \mathbf{t}) - (\mathbf{b} \cdot \mathbf{t})\|\mathbf{a}\|^2}{\|\mathbf{a}\|^2 \|\mathbf{b}\|^2 - (\mathbf{a} \cdot \mathbf{b})^2}.$$

The resulting point $P(x_p, y_p, z_p)$ is perpendicular to the origin.

If vectors **a** and **b** are unit vectors, λ and ϵ become

$$\lambda = \frac{(\mathbf{a} \cdot \mathbf{b})(\mathbf{b} \cdot \mathbf{t}) - \mathbf{a} \cdot \mathbf{t}}{1 - (\mathbf{a} \cdot \mathbf{b})^2} \tag{15.29}$$

$$\epsilon = \frac{(\mathbf{a} \cdot \mathbf{b})(\mathbf{a} \cdot \mathbf{t}) - \mathbf{b} \cdot \mathbf{t}}{1 - (\mathbf{a} \cdot \mathbf{b})^2}. \tag{15.30}$$

P's position vector **p** is also the plane's normal vector. Therefore,

$$x_p = x_t + \lambda x_a + \epsilon x_b$$
$$y_p = y_t + \lambda y_a + \epsilon y_b$$
$$z_p = z_t + \lambda z_a + \epsilon z_b.$$

The normal vector is

$$\mathbf{p} = x_p \mathbf{i} + y_p \mathbf{j} + z_p \mathbf{k}$$

and because $\|\mathbf{p}\|$ is the perpendicular distance from the plane to the origin we can state

$$\frac{x_p}{\|\mathbf{p}\|}x + \frac{y_p}{\|\mathbf{p}\|}y + \frac{z_p}{\|\mathbf{p}\|}z = \|\mathbf{p}\|$$

or in the general form of the plane equation:

$$Ax + By + Cz + D = 0$$

where

$$A = \frac{x_p}{\|\mathbf{p}\|}, \quad B = \frac{y_p}{\|\mathbf{p}\|}, \quad C = \frac{z_p}{\|\mathbf{p}\|}, \quad D = -\|\mathbf{p}\|.$$

As an example, Fig. 15.33 shows a plane inclined 45° to the y- and z-axis and parallel with the x-axis. The vectors for the parametric equation are

Fig. 15.33 The vectors **a** and **b** are parallel to the plane and the point $(0, 0, 1)$ is on the plane

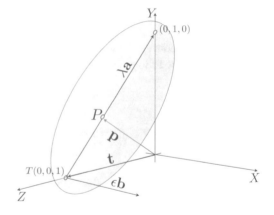

$$a = j - k$$
$$b = i$$
$$t = k.$$

Substituting these components in (15.29) and (15.30) we have

$$\lambda = \frac{(0)(0) - (-1) \times 1}{2 \times 1 - (0)} = 0.5$$

$$\epsilon = \frac{(0)(-1) - (0) \times 2}{2 \times 1 - (0)} = 0.$$

Therefore,

$$x_p = 0 + 0.5 \times 0 + 0 \times 1 = 0$$
$$y_p = 0 + 0.5 \times 1 + 0 \times 0 = 0.5$$
$$z_p = 1 + 0.5 \times (-1) + 0 \times 0 = 0.5.$$

The point $(0,\ 0.5,\ 0.5)$ has position vector $\mathbf{p}$, where

$$\|\mathbf{p}\| = \sqrt{0^2 + 0.5^2 + 0.5^2} = \tfrac{\sqrt{2}}{2}$$

the plane equation is

$$0x + \frac{0.5}{\sqrt{2}/2} y + \frac{0.5}{\sqrt{2}/2} z - \sqrt{2}/2 = 0$$

which simplifies to

$$y + z - 1 = 0.$$

15.8.5 Plane Equation from Three Points

Very often in computer graphics problems we need to find the plane equation from three known points. To begin with, the three points must be distinct and not lie on a line. Figure 15.34 shows three points R, S and T, from which we create two vectors $\mathbf{u} = \overrightarrow{RS}$ and $\mathbf{v} = \overrightarrow{RT}$. The vector product $\mathbf{u} \times \mathbf{v}$ then provides a vector normal to the plane containing the original points. We now take another point $P(x,\ y,\ z)$ and form a vector $\mathbf{w} = \overrightarrow{RP}$. The scalar product $\mathbf{w} \cdot (\mathbf{u} \times \mathbf{v}) = 0$ if P is in the plane

Fig. 15.34 The vectors used to determine a plane equation from three points R, S and T

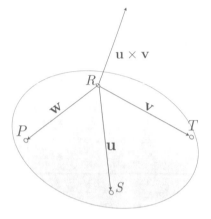

containing the original points. This condition can be expressed as a determinant and converted into the general equation of a plane. The three points are assumed to be in an anticlockwise sequence viewed from the direction of the surface normal.

We begin with

$$\mathbf{u} \times \mathbf{v} = \begin{vmatrix} \mathbf{i} & \mathbf{j} & \mathbf{k} \\ x_u & y_u & z_u \\ x_v & y_v & z_v \end{vmatrix}.$$

As $\mathbf{w}$ is perpendicular to $\mathbf{u} \times \mathbf{v}$

$$\mathbf{w} \cdot (\mathbf{u} \times \mathbf{v}) = \begin{vmatrix} x_w & y_w & z_w \\ x_u & y_u & z_u \\ x_v & y_v & z_v \end{vmatrix} = 0.$$

Expanding the determinant we obtain

$$x_w \begin{vmatrix} y_u & z_u \\ y_v & z_v \end{vmatrix} + y_w \begin{vmatrix} z_u & x_u \\ z_v & x_v \end{vmatrix} + z_w \begin{vmatrix} x_u & y_u \\ x_v & y_v \end{vmatrix} = 0$$

which becomes

$$(x - x_R) \begin{vmatrix} y_S - y_R & z_S - z_R \\ y_T - y_R & z_T - z_R \end{vmatrix} + (y - y_R) \begin{vmatrix} z_S - z_R & x_S - x_R \\ z_T - z_R & x_T - x_R \end{vmatrix}$$
$$+ (z - z_R) \begin{vmatrix} x_S - x_R & y_S - y_R \\ x_T - x_R & y_T - y_R \end{vmatrix} = 0.$$

This can be arranged in the form $ax + by + cz + d = 0$ where

$$a = \begin{vmatrix} y_S - y_R & z_S - z_R \\ y_T - y_R & z_T - z_R \end{vmatrix}, \quad b = \begin{vmatrix} z_S - z_R & x_S - x_R \\ z_T - z_R & x_T - x_R \end{vmatrix}$$

$$c = \begin{vmatrix} x_S - x_R & y_S - y_R \\ x_T - x_R & y_T - y_R \end{vmatrix}, \quad d = -(ax_R + by_R + cz_R)$$

or

$$a = \begin{vmatrix} 1 & y_R & z_R \\ 1 & y_S & z_S \\ 1 & y_T & z_T \end{vmatrix}, \quad b = \begin{vmatrix} x_R & 1 & z_R \\ x_S & 1 & z_S \\ x_T & 1 & z_T \end{vmatrix}, \quad c = \begin{vmatrix} x_R & y_R & 1 \\ x_S & y_S & 1 \\ x_T & y_T & 1 \end{vmatrix}$$

$$d = -(ax_R + by_R + cz_R).$$

As an example, consider the three points $R(0, 0, 1)$, $S(1, 0, 0)$, $T(0, 1, 0)$. Therefore,

$$a = \begin{vmatrix} 1 & 0 & 1 \\ 1 & 0 & 0 \\ 1 & 1 & 0 \end{vmatrix} = 1, \quad b = \begin{vmatrix} 0 & 1 & 1 \\ 1 & 1 & 0 \\ 0 & 1 & 0 \end{vmatrix} = 1, \quad c = \begin{vmatrix} 0 & 0 & 1 \\ 1 & 0 & 1 \\ 0 & 1 & 1 \end{vmatrix} = 1$$

$$d = -(1 \times 0 + 1 \times 0 + 1 \times 1) = -1$$

and the plane equation is

$$x + y + z - 1 = 0.$$

15.9 Intersecting Planes

When two non-parallel planes intersect, they form a straight line at the intersection, which is parallel to both planes. This line can be represented as a vector, whose direction is revealed by the vector product of the planes' surface normals. However, we require a point on this line to establish a unique vector equation; a useful point is chosen as P_0, whose position vector $\mathbf{p}_0$ is perpendicular to the line.

Fig. 15.35 Two intersecting planes create a line of intersection

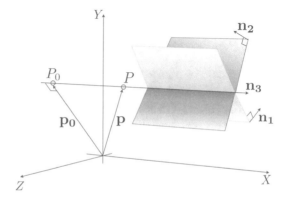

Figure 15.35 shows two planes with normal vectors n_1 and n_2 intersecting to create a line represented by n_3, whilst $P_0(x_0, y_0, z_0)$ is a particular point on n_3 and $P(x, y, z)$ is any point on the line.

We start the analysis by defining the surface normals:

$$n_1 = a_1i + b_1j + c_1k$$
$$n_2 = a_2i + b_2j + c_2k$$

next we define p and p_0:

$$p = xi + yj + zk$$
$$p_0 = x_0i + y_0j + z_0k.$$

Now we state the plane equations in vector form:

$$n_1 \cdot p + d_1 = 0$$
$$n_2 \cdot p + d_2 = 0.$$

The geometric significance of the scalars d_1 and d_2 has already been described above. Let's now define the line of intersection as

$$p = p_0 + \lambda n_3$$

where λ is a scalar.

As the line of intersection must be orthogonal to n_1 and n_2:

$$n_3 = a_3i + b_3j + c_3k = n_1 \times n_2.$$

Now we introduce P_0 as this must satisfy both plane equations, therefore,

$$n_1 \cdot p_0 = -d_1 \qquad\qquad (15.31)$$
$$n_2 \cdot p_0 = -d_2 \qquad\qquad (15.32)$$

and as P_0 is such that p_0 is orthogonal to n_3

$$n_3 \cdot p_0 = 0. \qquad\qquad (15.33)$$

Equations (15.31)–(15.33) form three simultaneous equations, which reveal the point P_0. These are represented in matrix form as

$$\begin{bmatrix} -d_1 \\ -d_2 \\ 0 \end{bmatrix} = \begin{bmatrix} a_1 & b_1 & c_1 \\ a_2 & b_2 & c_2 \\ a_3 & b_3 & c_3 \end{bmatrix} \begin{bmatrix} x_0 \\ y_0 \\ z_0 \end{bmatrix}$$

or

$$\begin{bmatrix} d_1 \\ d_2 \\ 0 \end{bmatrix} = - \begin{bmatrix} a_1 & b_1 & c_1 \\ a_2 & b_2 & c_2 \\ a_3 & b_3 & c_3 \end{bmatrix} \begin{bmatrix} x_0 \\ y_0 \\ z_0 \end{bmatrix}$$

therefore,

$$\frac{x_0}{\begin{vmatrix} d_1 & b_1 & c_1 \\ d_2 & b_2 & c_2 \\ 0 & b_3 & c_3 \end{vmatrix}} = \frac{y_0}{\begin{vmatrix} a_1 & d_1 & c_1 \\ a_2 & d_2 & c_2 \\ a_3 & 0 & c_3 \end{vmatrix}} = \frac{z_0}{\begin{vmatrix} a_1 & b_1 & d_1 \\ a_2 & b_2 & d_2 \\ a_3 & b_3 & 0 \end{vmatrix}} = \frac{-1}{DET}$$

which enables us to state

$$x_0 = \frac{d_2 \begin{vmatrix} b_1 & c_1 \\ b_3 & c_3 \end{vmatrix} - d_1 \begin{vmatrix} b_2 & c_2 \\ b_3 & c_3 \end{vmatrix}}{DET}$$

$$y_0 = \frac{d_2 \begin{vmatrix} a_3 & c_3 \\ a_1 & c_1 \end{vmatrix} - d_1 \begin{vmatrix} a_3 & c_3 \\ a_2 & c_2 \end{vmatrix}}{DET}$$

$$z_0 = \frac{d_2 \begin{vmatrix} a_1 & b_1 \\ a_3 & b_3 \end{vmatrix} - d_1 \begin{vmatrix} a_2 & b_2 \\ a_3 & b_3 \end{vmatrix}}{DET}$$

where

$$DET = \begin{vmatrix} a_1 & b_1 & c_1 \\ a_2 & b_2 & c_2 \\ a_3 & b_3 & c_3 \end{vmatrix}.$$

The line of intersection is then given by

$$\mathbf{p} = \mathbf{p}_0 + \lambda \mathbf{n}_3.$$

If $DET = 0$ the line and plane are parallel.

To illustrate this, let the two intersecting planes be the xy-plane and the yz-plane, which means that the line of intersection will be the y-axis, as shown in Fig. 15.36.

The plane equations are $z = 0$ and $x = 0$, therefore,

$$\mathbf{n}_1 = \mathbf{k}$$
$$\mathbf{n}_2 = \mathbf{i}$$

and $d_1 = d_2 = 0$.

We now compute $\mathbf{n}_3$, DET, x_0, y_0, z_0:

Fig. 15.36 Two intersecting planes creating a line of intersection coincident with the y-axis

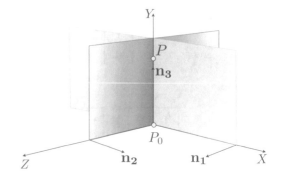

$$\mathbf{n}_3 = \begin{vmatrix} \mathbf{i} & \mathbf{j} & \mathbf{k} \\ 0 & 0 & 1 \\ 1 & 0 & 0 \end{vmatrix} = \mathbf{j}$$

$$DET = \begin{vmatrix} 0 & 0 & 1 \\ 1 & 0 & 0 \\ 0 & 1 & 0 \end{vmatrix} = 1$$

$$x_0 = \frac{0\begin{vmatrix} 0 & 1 \\ 1 & 0 \end{vmatrix} - 0\begin{vmatrix} 0 & 0 \\ 1 & 0 \end{vmatrix}}{1} = 0$$

$$y_0 = \frac{0\begin{vmatrix} 0 & 0 \\ 0 & 1 \end{vmatrix} - 0\begin{vmatrix} 0 & 0 \\ 1 & 0 \end{vmatrix}}{1} = 0$$

$$z_0 = \frac{0\begin{vmatrix} 0 & 0 \\ 0 & 1 \end{vmatrix} - 0\begin{vmatrix} 1 & 0 \\ 0 & 1 \end{vmatrix}}{1} = 0.$$

Therefore, the line equation is $\mathbf{p} = \lambda\mathbf{n}_3$, where $\mathbf{n}_3 = \mathbf{j}$, which is the y-axis.

15.9.1   Intersection of Three Planes

Three mutually intersecting planes will intersect at a point as shown in Fig. 15.37, and we can find this point by using a similar strategy to the one used in two intersecting planes by creating three simultaneous plane equations using determinants.

Figure 15.37 shows the common point $P(x, y, z)$. The three planes can be defined by the following equations:

$$a_1x + b_1y + c_1z + d_1 = 0$$
$$a_2x + b_1y + c_2z + d_2 = 0$$

Fig. 15.37 Three mutually
intersecting planes

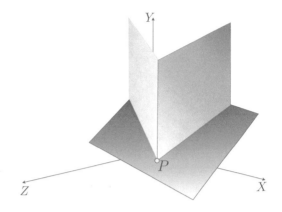

$$a_3x + b_1y + c_3z + d_3 = 0$$

which means that they can be rewritten as

$$\begin{bmatrix} -d_1 \\ -d_2 \\ -d_3 \end{bmatrix} = \begin{bmatrix} a_1 \ b_1 \ c_1 \\ a_2 \ b_2 \ c_2 \\ a_3 \ b_3 \ c_3 \end{bmatrix} \begin{bmatrix} x \\ y \\ z \end{bmatrix}$$

or

$$\begin{bmatrix} d_1 \\ d_2 \\ d_3 \end{bmatrix} = - \begin{bmatrix} a_1 \ b_1 \ c_1 \\ a_2 \ b_2 \ c_2 \\ a_3 \ b_3 \ c_3 \end{bmatrix} \begin{bmatrix} x \\ y \\ z \end{bmatrix}$$

or in determinant form

$$\frac{x}{\begin{vmatrix} d_1 \ b_1 \ c_1 \\ d_2 \ b_2 \ c_2 \\ d_3 \ b_3 \ c_3 \end{vmatrix}} = \frac{y}{\begin{vmatrix} a_1 \ d_1 \ c_1 \\ a_2 \ d_2 \ c_2 \\ a_3 \ d_3 \ c_3 \end{vmatrix}} = \frac{z}{\begin{vmatrix} a_1 \ b_1 \ d_1 \\ a_2 \ b_2 \ d_2 \\ a_3 \ b_3 \ d_3 \end{vmatrix}} = \frac{-1}{DET}$$

where

$$DET = \begin{vmatrix} a_1 \ b_1 \ c_1 \\ a_2 \ b_2 \ c_2 \\ a_3 \ b_3 \ c_3 \end{vmatrix}.$$

Therefore, we can state that

$$x = - \frac{\begin{vmatrix} d_1 \ b_1 \ c_1 \\ d_2 \ b_2 \ c_2 \\ d_3 \ b_3 \ c_3 \end{vmatrix}}{DET}$$

Fig. 15.38 Three planes
intersecting at a point P

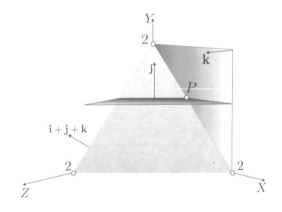

$$y = -\frac{\begin{vmatrix} a_1 & d_1 & c_1 \\ a_2 & d_2 & c_2 \\ a_3 & d_3 & c_3 \end{vmatrix}}{DET}$$

$$z = -\frac{\begin{vmatrix} a_1 & b_1 & d_1 \\ a_2 & b_2 & d_2 \\ a_3 & b_3 & d_3 \end{vmatrix}}{DET}.$$

If $DET = 0$, two of the planes at least, are parallel. Let's test these equations with a simple example.

The planes shown in Fig. 15.38 have the following equations:

$$x + y + z - 2 = 0$$
$$z = 0$$
$$y - 1 = 0$$

therefore,

$$DET = \begin{vmatrix} 1 & 1 & 1 \\ 0 & 0 & 1 \\ 0 & 1 & 0 \end{vmatrix} = -1$$

$$x = -\frac{\begin{vmatrix} -2 & 1 & 1 \\ 0 & 0 & 1 \\ -1 & 1 & 0 \end{vmatrix}}{-1} = 1$$

$$y = -\frac{\begin{vmatrix} 1 & -2 & 1 \\ 0 & 0 & 1 \\ 0 & -1 & 0 \end{vmatrix}}{-1} = 1$$

$$z = -\frac{\begin{vmatrix} 1 & 1 & -2 \\ 0 & 0 & 0 \\ 0 & 1 & -1 \end{vmatrix}}{-1} = 0$$

which means that the intersection point is $(1, 1, 0)$, which is correct.

15.9.2 Angle Between Two Planes

Calculating the angle between two planes is relatively easy and can be found by taking the dot product of the planes' normals. Figure 15.39 shows two planes with α representing the angle between the two surface normals $\mathbf{n}_1$ and $\mathbf{n}_2$.

Let the plane equations be

$$ax_1 + by_1 + cz_1 + d_1 = 0$$
$$ax_2 + by_2 + cz_2 + d_2 = 0$$

therefore, their surface normals are

$$\mathbf{n}_1 = a_1\mathbf{i} + b_1\mathbf{j} + c_1\mathbf{k}$$
$$\mathbf{n}_2 = a_2\mathbf{i} + b_2\mathbf{j} + c_2\mathbf{k}.$$

Taking the dot product of $\mathbf{n}_1$ and $\mathbf{n}_2$:

$$\mathbf{n}_1 \cdot \mathbf{n}_2 = \|\mathbf{n}_1\| \, \|\mathbf{n}_2\| \cos \alpha$$

Fig. 15.39 The angle between two planes is the angle between their surface normals

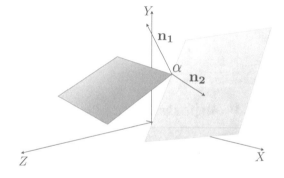

Fig. 15.40 α is the angle between two planes

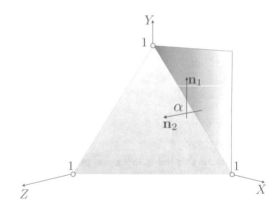

and

$$\alpha = \cos^{-1}\left(\frac{\mathbf{n}_1 \cdot \mathbf{n}_2}{\|\mathbf{n}_1\| \ \|\mathbf{n}_2\|}\right).$$

For example, Fig. 15.40 shows two planes with normal vectors $\mathbf{n}_1$ and $\mathbf{n}_2$. The plane equations are

$$x + y + z - 1 = 0$$
$$z = 0$$

therefore,

$$\mathbf{n}_1 = \mathbf{i} + \mathbf{j} + \mathbf{k}$$
$$\mathbf{n}_2 = \mathbf{k}$$

therefore,

$$\|\mathbf{n}_1\| = \sqrt{3}$$
$$\|\mathbf{n}_2\| = 1$$

and

$$\alpha = \cos^{-1}\left(\tfrac{1}{\sqrt{3}}\right) \approx 54.74°.$$

15.9.3   Angle Between a Line and a Plane

The angle between a line and a plane is calculated using a similar technique used for calculating the angle between two planes. If the line equation employs a direction

Fig. 15.41 α is the angle between the plane's surface normal and the line's direction vector

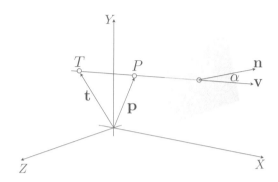

vector, the angle is determined by taking the dot product of this vector and between the plane's normal. Figure 15.41 shows such a scenario where $\mathbf{n}$ is the plane's surface normal and $\mathbf{v}$ is the line's direction vector.

Let the plane equation be

$$ax + by + cz + d = 0$$

then its surface normal is

$$\mathbf{n} = a\mathbf{i} + b\mathbf{j} + c\mathbf{k}.$$

Let the line's direction vector be $\mathbf{v}$ and $T(x_t\ y_t,\ z_t)$ is a point on the line, then any point on the line is given by the position vector $\mathbf{p}$:

$$\mathbf{p} = \mathbf{t} + \lambda\mathbf{v}$$

therefore, we can write

$$\mathbf{n} \cdot \mathbf{v} = \|\mathbf{n}\|\ \|\mathbf{v}\| \cos\alpha$$

$$\alpha = \cos^{-1}\left(\frac{\mathbf{n} \cdot \mathbf{v}}{\|\mathbf{n}\|\ \|\mathbf{v}\|}\right).$$

When the line is parallel to the plane $\mathbf{n} \cdot \mathbf{v} = 0$.

Consider the scenario illustrated in Fig. 15.42 where the plane equation is

$$x + y + z - 1 = 0$$

therefore, the surface normal is given by $\mathbf{n}$:

$$\mathbf{n} = \mathbf{i} + \mathbf{j} + \mathbf{k}$$

Fig. 15.42 The required
angle is between **a** and **b**

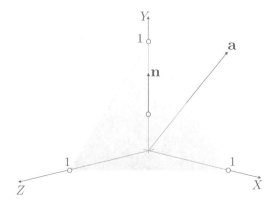

and the line's direction vector is **a**:

$$\mathbf{a} = \mathbf{i} + \mathbf{j}$$

therefore,

$$\|\mathbf{n}\| = \sqrt{3}$$
$$\|\mathbf{a}\| = \sqrt{2}$$

and

$$\alpha = \cos^{-1}\left(\tfrac{2}{\sqrt{6}}\right) \approx 35.26°.$$

15.9.4   Intersection of a Line with a Plane

Given a line and a plane, they will either intersect, or not, if they are parallel. Either
way, both conditions can be found using some simple vector analysis, as shown in
Fig. 15.43. The objective is to identify a point P that is on the line and the plane.
    Let the plane equation be

$$ax + by + cz + d = 0$$

where

$$\mathbf{n} = a\mathbf{i} + b\mathbf{j} + c\mathbf{k}.$$

P is a point on the plane with position vector

$$\mathbf{p} = x\mathbf{i} + y\mathbf{j} + z\mathbf{k}$$

Fig. 15.43 The vectors
required to determine
whether a line and plane
intersect

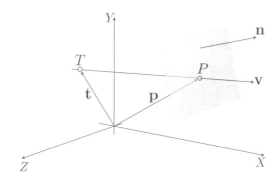

therefore,

$$\mathbf{n} \cdot \mathbf{p} + d = 0.$$

Let the line equation be

$$\mathbf{p} = \mathbf{t} + \lambda \mathbf{v}$$

where

$$\mathbf{t} = x_t \mathbf{i} + y_t \mathbf{j} + z_t \mathbf{k}$$

and

$$\mathbf{v} = x_v \mathbf{i} + y_v \mathbf{j} + z_v \mathbf{k}$$

therefore, the line and plane will intersect for some λ such that

$$\mathbf{n} \cdot (\mathbf{t} + \lambda \mathbf{v}) + d = \mathbf{n} \cdot \mathbf{t} + \lambda \mathbf{n} \cdot \mathbf{v} + d = 0.$$

Therefore,

$$\lambda = \frac{-(\mathbf{n} \cdot \mathbf{t} + d)}{\mathbf{n} \cdot \mathbf{v}}$$

for the intersection point. The position vector for P is $\mathbf{p} = \mathbf{t} + \lambda \mathbf{v}$.
If $\mathbf{n} \cdot \mathbf{v} = 0$ the line and plane are parallel.
Let's test this result with the scenario shown in Fig. 15.44.
Given the plane

$$x + y + z - 1 = 0$$
$$\mathbf{n} = \mathbf{i} + \mathbf{j} + \mathbf{k}$$

and the line

$$\mathbf{p} = \mathbf{t} + \lambda \mathbf{v}$$

Fig. 15.44 *P* identifies the intersection point of the line and the plane

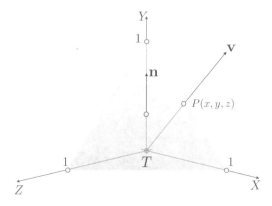

where

$$\mathbf{t} = 0$$
$$\mathbf{v} = \mathbf{i} + \mathbf{j}$$

then

$$\lambda = \frac{-(1 \times 0 + 1 \times 0 + 1 \times 0 - 1)}{1 \times 1 + 1 \times 1 + 1 \times 0} = 0.5$$

and the point of intersection is $P(0.5, \ 0.5, \ 0)$.

15.10   Summary

Mixing vectors with geometry is a powerful analytical tool, and helps us solve many problems associated with computer graphics, encountered in rendering, modelling, collision detection and physically-based animation. Unfortunately, there has not been space to investigate every topic, but hopefully, what has been covered, will enable the reader solve other problems with greater confidence.

Chapter 16
Barycentric Coordinates

16.1 Introduction

Cartesian coordinates are a fundamental concept in mathematics and are central to computer graphics. Such rectangular coordinates are just offsets relative to some origin. Other coordinate systems also exist such as polar, spherical and cylindrical coordinates, and they too, require an origin. Barycentric coordinates, on the other hand, locate points relative to existing points, rather than to an origin and are known as *local coordinates*.

16.2 Background

The German mathematician August Möbius is credited with their discovery. '*barus*' is the Greek entomological root for '*heavy*', and barycentric coordinates were originally used for identifying the centre of mass of shapes and objects. It is interesting to note that the prefixes '*bari*', '*bary*' and '*baro*' have also influenced other words such as baritone, baryon (heavy atomic particle) and barometer.

Although barycentric coordinates are used in geometry, computer graphics, relativity and global time systems, they do not appear to be a major topic in a typical math syllabus. Nevertheless, they are important and I would like to describe what they are and how they can be used in computer graphics.

The idea behind barycentric coordinates can be approached from different directions, and I have chosen mass points and linear interpolation. But before we begin this analysis, it will be useful to investigate a rather elegant theorem known as Ceva's Theorem, which we will invoke later in this chapter.

© Springer-Verlag London Ltd., part of Springer Nature 2022
J. Vince, *Mathematics for Computer Graphics*, Undergraduate Topics in Computer Science, https://doi.org/10.1007/978-1-4471-7520-9_16

16.3 Ceva's Theorem

The Italian mathematician Giovanni Ceva (1647–1734) is credited with a theorem associated with the concurrency of lines in a triangle. It states that: In a triangle $\triangle ABC$, the lines AA', BB' and CC', where A', B' and C' are points on the opposite sides facing vertices A, B and C respectively, are concurrent (intersect at a common point) if, and only if

$$\frac{AC'}{C'B} \cdot \frac{BA'}{A'C} \cdot \frac{CB'}{B'A} = 1.$$

Figure 16.1 shows such a scenario.

There are various ways of proving this theorem, and Alfred Posamentier provides one [1]; but perhaps the simplest proof is as follows.

Figure 16.2 shows triangle $\triangle ABC$ with line AA' extended to R and BB' extended to S, where line SR is parallel to line AB. The resulting geometry creates a number of similar triangles:

$$\triangle ABA' \ : \ \triangle RCA' \quad \Rightarrow \quad \frac{A'C}{BA'} = \frac{CR}{AB} \tag{16.1}$$

Fig. 16.1 The geometry associated with Ceva's Theorem

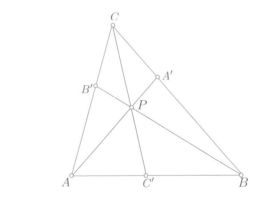

Fig. 16.2 The geometry for proving Ceva's Theorem

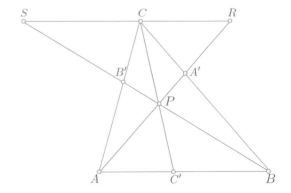

$$\Delta ABB' \ : \ \Delta CSB' \quad \Rightarrow \quad \frac{B'A}{CB'} = \frac{AB}{SC} \tag{16.2}$$

$$\Delta BPC' \ : \ \Delta CSP \quad \Rightarrow \quad \frac{C'B}{SC} = \frac{C'P}{PC} \tag{16.3}$$

$$\Delta AC'P \ : \ \Delta RCP \quad \Rightarrow \quad \frac{AC'}{CR} = \frac{C'P}{PC}. \tag{16.4}$$

From (16.3) and (16.4) we get

$$\frac{C'B}{SC} = \frac{AC'}{CR}$$

which can be rewritten as

$$\frac{C'B}{AC'} = \frac{SC}{CR}. \tag{16.5}$$

The product of (16.1), (16.2) and (16.5) is

$$\frac{A'C}{BA'} \cdot \frac{B'A}{CB'} \cdot \frac{C'B}{AC'} = \frac{CR}{AB} \cdot \frac{AB}{SC} \cdot \frac{SC}{CR} = 1. \tag{16.6}$$

Rearranging the terms of (16.6) we get

$$\frac{AC'}{C'B} \cdot \frac{BA'}{A'C} \cdot \frac{CB'}{B'A} = 1$$

which is rather an elegant relationship.

16.4 Ratios and Proportion

Central to barycentric coordinates are ratios and proportion, so let's begin by revising some fundamental formulae used in calculating ratios.

Imagine the problem of dividing £100 between two people in the ratio 2 : 3. The solution lies in the fact that the money is divided into 5 parts $(2 + 3)$, where 2 parts go to one person and 3 parts to the other person. In this case, one person receives £40 and the other £60. At a formal level, we can describe this as follows.

A scalar A can be divided into the ratio $r : s$ using the following expressions:

$$\frac{r}{r+s}A \quad \text{and} \quad \frac{s}{r+s}A.$$

Note that

$$\frac{r}{r+s} + \frac{s}{r+s} = 1$$

and

$$1 - \frac{r}{r+s} = \frac{s}{r+s}.$$

Furthermore, the above formulae can be extended to incorporate any number of ratio divisions. For example, A can be divided into the ratio $r : s : t$ by the following:

$$\frac{r}{r+s+t}A, \quad \frac{s}{r+s+t}A \quad \text{and} \quad \frac{t}{r+s+t}A$$

similarly,

$$\frac{r}{r+s+t} + \frac{s}{r+s+t} + \frac{t}{r+s+t} = 1.$$

These expressions are very important as they show the emergence of barycentric coordinates. For the moment though, just remember their structure and we will investigate some ideas associated with balancing weights.

16.5 Mass Points

We begin by calculating the centre of mass—the centroid—of two masses. Consider the scenario shown in Fig. 16.3 where two masses m_A and m_B are placed at the ends of a massless rod.

If $m_A = m_B$ a state of equilibrium is achieved by placing the fulcrum mid-way between the masses. If the fulcrum is moved towards m_A, mass m_B will have a turning advantage and the rod rotates clockwise.

To calculate a state of equilibrium for a general system of masses, consider the geometry illustrated in Fig. 16.4, where two masses m_A and m_B are positioned x_A

Fig. 16.3 Two masses fixed at the ends of a massless rod

Fig. 16.4 The geometry used for equating turning moments

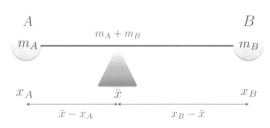

and x_B at A and B respectively. When the system is in balance we can replace the two masses by a single mass $m_A + m_B$ at the centroid denoted by $\bar{x}$ (pronounced 'x bar').

A balance condition arises when the LHS turning moment equals the RHS turning moment. The turning moment being the product of a mass by its offset from the fulcrum. Equating turning moments, equilibrium is reached when

$$m_B(x_B - \bar{x}) = m_A(\bar{x} - x_A)$$
$$m_B x_B - m_B \bar{x} = m_A \bar{x} - m_A x_A$$
$$(m_A + m_B)\bar{x} = m_A x_A + m_B x_B$$

$$\bar{x} = \frac{m_A x_A + m_B x_B}{m_A + m_B} = \frac{m_A}{m_A + m_B} x_A + \frac{m_B}{m_A + m_B} x_B. \qquad (16.7)$$

For example, if $m_A = 6$ and $m_B = 12$, and positioned at $x_A = 0$ and $x_B = 12$ respectively, the centroid is located at

$$\bar{x} = \tfrac{6}{18} \times 0 + \tfrac{12}{18} \times 12 = 8.$$

Thus we can replace the two masses by a single mass of 18 located at $\bar{x} = 8$.

Note that the terms in (16.7) $m_A/(m_A + m_B)$ and $m_B/(m_A + m_B)$ sum to 1 and are identical to those used above for calculating ratios. They are also called the *barycentric coordinates* of $\bar{x}$ relative to the points A and B.

Using the general form of (16.7) any number of masses can be analysed using

$$\bar{x} = \frac{\sum\limits_{i=1}^{n} m_i x_i}{\sum\limits_{i=1}^{n} m_i}$$

where m_i is a mass located at x_i. Furthermore, we can compute the y-component of the centroid $\bar{y}$ using

$$\bar{y} = \frac{\sum\limits_{i=1}^{n} m_i y_i}{\sum\limits_{i=1}^{n} m_i}$$

and in 3D the z-component of the centroid $\bar{z}$ is

$$\bar{z} = \frac{\sum\limits_{i=1}^{n} m_i z_i}{\sum\limits_{i=1}^{n} m_i}.$$

Fig. 16.5 The geometry used for equating turning moments

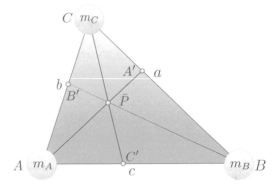

To recap, (16.7) states that

$$\bar{x} = \frac{m_A}{m_A + m_B} x_A + \frac{m_B}{m_A + m_B} x_B$$

therefore, we can write

$$\bar{y} = \frac{m_A}{m_A + m_B} y_A + \frac{m_B}{m_A + m_B} y_B$$

which allows us to state

$$\bar{\mathbf{P}} = \frac{m_A}{m_A + m_B} \mathbf{A} + \frac{m_B}{m_A + m_B} \mathbf{B}$$

where $\mathbf{A}$ and $\mathbf{B}$ are the position vectors for the mass locations A and B respectively, and $\bar{\mathbf{P}}$ is the position vector for the centroid $\bar{P}$.

If we extend the number of masses to three: m_A, m_B and m_C, which are organised as a triangle, then we can write

$$\bar{\mathbf{P}} = \frac{m_A}{m_A + m_B + m_C} \mathbf{A} + \frac{m_B}{m_A + m_B + m_C} \mathbf{B} + \frac{m_C}{m_A + m_B + m_C} \mathbf{C}. \qquad (16.8)$$

The three multipliers of $\mathbf{A}$, $\mathbf{B}$ and $\mathbf{C}$ are the barycentric coordinates of $\bar{P}$ relative to the points A, B and C. Note that the number of coordinates is not associated with the number of spatial dimensions, but the number of reference points.

Now consider the scenario shown in Fig. 16.5. If $m_A = m_B = m_C$ then we can determine the location of A', B' and C' as follows:

1. We begin by placing a fulcrum under A mid-way along BC as shown in Fig. 16.6.
   The triangle will balance because $m_B = m_C$ and A' is $\frac{1}{2}a$ from C and $\frac{1}{2}a$ from B.

2. Now we place the fulcrum under B mid-way along CA as shown in Fig. 16.7. Once more the triangle will balance, because $m_C = m_A$ and B' is $\frac{1}{2}b$ from C and $\frac{1}{2}b$ from A.
3. Finally, we do the same for C and AB. Figure 16.8 shows the final scenario.

Ceva's Theorem confirms that the medians AA', BB' and CC' are concurrent at $\bar{P}$ because

Fig. 16.6 Balancing the triangle along AA'

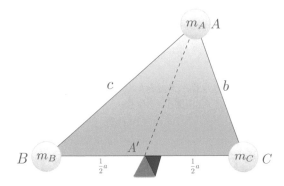

Fig. 16.7 Balancing the triangle along BB'

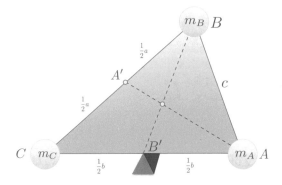

Fig. 16.8 $\bar{P}$ is the centroid of the triangle

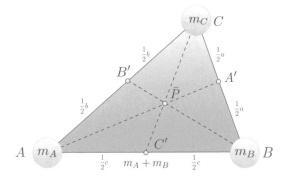

$$\frac{AC'}{C'B} \cdot \frac{BA'}{A'C} \cdot \frac{CB'}{B'A} = \frac{\frac{1}{2}c}{\frac{1}{2}c} \cdot \frac{\frac{1}{2}a}{\frac{1}{2}a} \cdot \frac{\frac{1}{2}b}{\frac{1}{2}b} = 1.$$

Arbitrarily, we select the median $C'C$. At C' we have an effective mass of $m_A + m_B$ and m_C at C. For a balance condition

$$(m_A + m_B) \times C'\bar{P} = m_C \times \bar{P}C$$

and as the masses are equal, $C'\bar{P}$ must be $\frac{1}{3}$ along the median $C'C$.
    If we use (16.8) we obtain

$$\bar{P} = \tfrac{1}{3}\mathbf{A} + \tfrac{1}{3}\mathbf{B} + \tfrac{1}{3}\mathbf{C}$$

which locates the coordinates of the centroid correctly.
    Now let's consider another example where $m_A = 1$, $m_B = 2$ and $m_C = 3$, as shown in Fig. 16.9. For a balance condition A' must be $\frac{3}{5}a$ from B and $\frac{2}{5}a$ from C. Equally, B' must be $\frac{1}{4}b$ from C and $\frac{3}{4}b$ from A. Similarly, C' must be $\frac{2}{3}c$ from A and $\frac{1}{3}c$ from B.
    Ceva's Theorem confirms that the lines AA', BB' and CC' are concurrent at $\bar{P}$ because

$$\frac{AC'}{C'B} \cdot \frac{BA'}{A'C} \cdot \frac{CB'}{B'A} - \frac{\frac{2}{3}c}{\frac{1}{3}c} \cdot \frac{\frac{3}{5}a}{\frac{2}{5}a} \cdot \frac{\frac{1}{4}b}{\frac{3}{4}b} = 1.$$

Arbitrarily select $C'C$. At C' we have an effective mass of 3 $(1+2)$ and 3 at C, which means that for a balance condition $\bar{P}$ is mid-way along $C'C$. Similarly, $\bar{P}$ is $\frac{1}{6}$ along $A'A$ and $\frac{1}{3}$ along $B'B$.
    Once more, using (16.8) in this scenario we obtain

$$\bar{P} = \tfrac{1}{6}\mathbf{A} + \tfrac{1}{3}\mathbf{B} + \tfrac{1}{2}\mathbf{C}.$$

Note that the multipliers of $\mathbf{A}$, $\mathbf{B}$ and $\mathbf{C}$ are identical to the proportions of $\bar{P}$ along $A'A$, $B'B$ and $C'C$. Let's prove why this is so.

Fig. 16.9 How the masses determine the positions of A', B' and C'

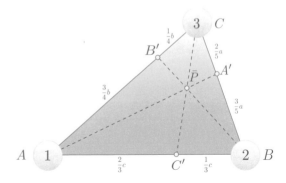

Fig. 16.10 How the masses determine the positions of A', B' and C'

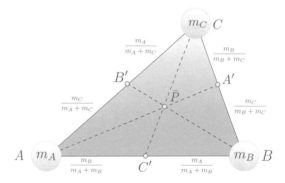

Figure 16.10 shows three masses with the triangle's sides divided into their various proportions to derive $\bar{P}$.

On the line $A'A$ we have m_A at A and effectively $m_B + m_C$ at A', which means that $\bar{P}$ divides $A'A$ in the ratio $m_A/(m_A + m_B + m_C) : (m_B + m_C)/(m_A + m_B + m_C)$.

On the line $B'B$ we have m_B at B and effectively $m_A + m_C$ at B', which means that $\bar{P}$ divides $B'B$ in the ratio $m_B/(m_A + m_B + m_C) : (m_A + m_C)/(m_A + m_B + m_C)$.

Similarly, on the line $C'C$ we have m_C at C and effectively $m_A + m_B$ at C', which means that $\bar{P}$ divides $C'C$ in the ratio $m_C/(m_A + m_B + m_C) : (m_A + m_B)/(m_A + m_B + m_C)$.

To summarise, given three masses m_A, m_B and m_C located at A, B and C respectively, the centroid $\bar{P}$ is given by

$$\bar{\mathbf{P}} = \frac{m_A}{m_A + m_B + m_C}\mathbf{A} + \frac{m_B}{m_A + m_B + m_C}\mathbf{B} + \frac{m_C}{m_A + m_B + m_C}\mathbf{C}. \qquad (16.9)$$

If we accept that m_A, m_B and m_C can have any value, including zero, then the barycentric coordinates of $\bar{P}$ will be affected by these values. For example, if $m_B = m_C = 0$ and $m_A = 1$, then $\bar{P}$ will be located at A with barycentric coordinates $(1, 0, 0)$. Similarly, if $m_A = m_C = 0$ and $m_B = 1$, then $\bar{P}$ will be located at B with barycentric coordinates $(0, 1, 0)$. And if $m_A = m_B = 0$ and $m_C = 1$, then $\bar{P}$ will be located at C with barycentric coordinates $(0, 0, 1)$.

Now let's examine a 3D example as illustrated in Fig. 16.11. The figure shows three masses 4, 8 and 12 and their equivalent mass 24 located at $(\bar{x}, \bar{y}, \bar{z})$.

The magnitude and coordinates of three masses are shown in Table 16.1, together with the barycentric coordinate t_i. The column headed t_i expresses the masses as fractions of the total mass: i.e.

$$t_i = \frac{m_i}{m_1 + m_2 + m_3}$$

and we see that the centroid is located at $(5, 5, 3)$.

Having discovered barycentric coordinates in weight balancing, let's see how they emerge in linear interpolation.

Fig. 16.11 Three masses
can be represented by a
single mass located at the
centroid

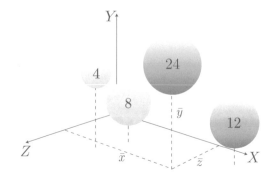

Table 16.1 The magnitude and coordinates of three masses

m_i	t_i	x_i	y_i	z_i	$t_i x_i$	$t_i y_i$	$t_i z_i$
12	$\frac{1}{2}$	8	6	2	4	3	1
8	$\frac{1}{3}$	2	3	3	$\frac{2}{3}$	1	1
4	$\frac{1}{6}$	2	6	6	$\frac{1}{3}$	1	1
					$\bar{x} = 5$	$\bar{y} = 5$	$\bar{z} = 3$

16.6   Linear Interpolation

Suppose that we wish to find a value mid-way between two scalars A and B. We
could proceed as follows:

$$V = A + \tfrac{1}{2}(B - A)$$
$$= \tfrac{1}{2}A + \tfrac{1}{2}B$$

which seems rather obvious. Similarly, to find a value one-third between A and B,
we can write:

$$V = A + \tfrac{1}{3}(B - A)$$
$$= \tfrac{2}{3}A + \tfrac{1}{3}B.$$

Generalising, to find some fraction t between A and B we can write

$$V = (1 - t)A + tB. \qquad (16.10)$$

For example, to find a value $\frac{3}{4}$ between 10 and 18 we have

$$V = \left(1 - \tfrac{3}{4}\right) \times 10 + \tfrac{3}{4} \times 18 = 16.$$

Although this is a trivial formula, it is very useful when interpolating between two numerical values. Let us explore (16.10) in greater detail.

To begin with, it is worth noting that the multipliers of A and B sum to 1:

$$(1 - t) + t = 1.$$

Rather than using $(1 - t)$ as a multiplier, it is convenient to make a substitution such as $s = 1 - t$, and we have

$$V = sA + tB$$

where

$$s = 1 - t$$

and

$$s + t = 1.$$

Equation (16.10) is called a *linear interpolant* as it linearly interpolates between A and B using the parameter t. It is also known as a *lerp*. The terms s and t are the barycentric coordinates of V as they determine the value of V relative to A and B.

Now let's see what happens when we substitute coordinates for scalars. We start with 2D coordinates $A(x_A, y_A)$ and $B(x_B, y_B)$, and position vectors $\mathbf{A}$, $\mathbf{B}$ and $\mathbf{C}$ and the following linear interpolant

$$\mathbf{V} = s\mathbf{A} + t\mathbf{B}$$

where

$$s = 1 - t$$

and

$$s + t = 1$$

then

$$x_V = sx_A + tx_B$$
$$y_V = sy_A + ty_B.$$

Figure 16.12 illustrates what happens when t varies between 0 and 1.

The point V slides along the line connecting A and B. When $t = 0$, V is coincident with A, and when $t = 1$, V is coincident with B. You should not be surprised that the same technique works in 3D.

Now let's extend the number of vertices to three in the form of a triangle as shown in Fig. 16.13. This time we will use r, s and t to control the interpolation. We would start as follows:

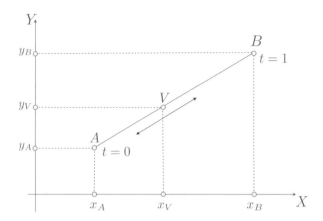

Fig. 16.12   The position of V slides between A and B as t varies between 0 and 1

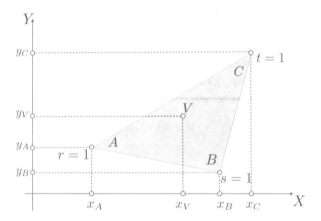

Fig. 16.13   The position of V moves between A, B and C depending on the value r, s and t

$$V = rA + sB + tC$$

where A, B and C are the position vectors for A, B and C respectively, and V is the position vector for the point V.

Let

$$r = 1 - s - t$$

and

$$r + s + t = 1.$$

Once more, we begin with 2D coordinates $A(x_A,\ y_A)$, $B(x_B,\ y_B)$ and $C(x_C,\ y_C)$ where

$$x_V = rx_A + sx_B + tx_C$$
$$y_V = ry_A + sy_B + ty_C.$$

When

$$r = 1, \quad V \text{ is coincident with } A$$
$$s = 1, \quad V \text{ is coincident with } B$$
$$t = 1, \quad V \text{ is coincident with } C.$$

Similarly, when

$$r = 0, \quad V \text{ is located on the edge } BC$$
$$s = 0, \quad V \text{ is located on the edge } CA$$
$$t = 0, \quad V \text{ is located on the edge } AB.$$

For all other values of r, s and t, where $r + s + t = 1$ and $0 \leq r, s, t \leq 1$, V is inside triangle $\triangle ABC$, otherwise it is outside the triangle.

The triple (r, s, t) are barycentric coordinates and locate points relative to A, B and C, rather than an origin. For example, the barycentric coordinates of A, B and C are $(1, 0, 0)$, $(0, 1, 0)$ and $(0, 0, 1)$ respectively.

All of the above formulae work equally well in three dimensions, so let's investigate how barycentric coordinates can locate points inside a 3D triangle. However, before we start, let's clarify what we mean by *inside* a triangle. Fortunately, barycentric coordinates can distinguish points within the triangle's three sides; points coincident with the sides; and points outside the triangle's boundary. The range and value of the barycentric coordinates provide the mechanism for detecting these three conditions.

As an example, Fig. 16.14 illustrates a scenario with the points $P_1(x_1, y_1, z_1)$, $P_2(x_2, y_2, z_2)$ and $P_3(x_3, y_3, z_3)$. Using barycentric coordinates we can state that any point $P_0(x_0, y_0, z_0)$ inside or on the edge of triangle $\triangle P_1 P_2 P_3$ is defined by

Fig. 16.14 A 3D triangle

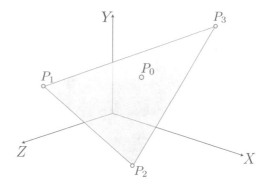

$$x_0 = rx_1 + sx_2 + tx_3$$
$$y_0 = ry_1 + sy_2 + ty_3$$
$$z_0 = rz_1 + sx_2 + tz_3$$

where $r + s + t = 1$ and $0 \le r, s, t, \le 1$.

If the triangle's vertices are $P_1(0,\ 2,\ 0)$, $P_2(0,\ 0,\ 4)$ and $P_3(3,\ 1,\ 2)$ then we can choose different values of r, s and t to locate P_0 inside the triangle. However, I would also like to confirm that P_0 lies on the plane containing the three points. To do this we require the plane equation for the three points, which can be derived as follows.

Given $P_1(x_1,\ y_1,\ z_1)$, $P_2(x_2,\ y_2,\ z_2)$ and $P_3(x_3,\ y_3,\ z_3)$, and the target plane equation $ax + by + cz + d = 0$, then

$$a = \begin{vmatrix} 1 & y_1 & z_1 \\ 1 & y_2 & z_2 \\ 1 & y_3 & z_3 \end{vmatrix}$$

$$b = \begin{vmatrix} x_1 & 1 & z_1 \\ x_2 & 1 & z_2 \\ x_3 & 1 & z_3 \end{vmatrix}$$

$$c = \begin{vmatrix} x_1 & y_1 & 1 \\ x_2 & y_2 & 1 \\ x_3 & z_2 & 1 \end{vmatrix}$$

$$d = -(ax_1 + by_1 + cz_1)$$

thus

$$a = \begin{vmatrix} 1 & 2 & 0 \\ 1 & 0 & 4 \\ 1 & 1 & 2 \end{vmatrix} = 0$$

$$b = \begin{vmatrix} 0 & 1 & 0 \\ 0 & 1 & 4 \\ 3 & 1 & 2 \end{vmatrix} = 12$$

$$c = \begin{vmatrix} 0 & 2 & 1 \\ 0 & 0 & 1 \\ 3 & 1 & 1 \end{vmatrix} = 6$$

$$d = -24$$

therefore, the plane equation is

$$12y + 6z = 24. \tag{16.11}$$

Table 16.2 The barycentric coordinates of P_0

r	s	t	x_0	y_0	z_0	$12y_0 + 6z_0$
1	0	0	0	2	0	24
0	1	0	0	0	4	24
0	0	1	3	1	2	24
$\frac{1}{4}$	$\frac{1}{4}$	$\frac{1}{2}$	$1\frac{1}{2}$	1	2	24
0	$\frac{1}{2}$	$\frac{1}{2}$	$1\frac{1}{2}$	$\frac{1}{2}$	3	24
$\frac{1}{2}$	$\frac{1}{2}$	0	0	1	2	24
$\frac{1}{3}$	$\frac{1}{3}$	$\frac{1}{3}$	1	1	2	24

If we substitute a point $(x_0,\ y_0,\ z_0)$ in the LHS of (16.11) and obtain a value of 24, then the point is on the plane.

Table 16.2 shows various values of r, s and t, and the corresponding position of P_0. The table also confirms that P_0 is always on the plane containing the three points.

Now we are in a position to test whether a point is inside, on the boundary or outside a 3D triangle.

We begin by writing the three simultaneous equations defining P_0 in matrix form

$$\begin{bmatrix} x_0 \\ y_0 \\ z_0 \end{bmatrix} = \begin{bmatrix} x_1 & x_2 & x_3 \\ y_1 & y_2 & y_3 \\ z_1 & z_2 & z_3 \end{bmatrix} \begin{bmatrix} r \\ s \\ t \end{bmatrix}$$

therefore,

$$\frac{r}{\begin{vmatrix} x_0 & x_2 & x_3 \\ y_0 & y_2 & y_3 \\ z_0 & z_2 & z_3 \end{vmatrix}} = \frac{s}{\begin{vmatrix} x_1 & x_0 & x_3 \\ y_1 & y_0 & y_3 \\ z_1 & z_0 & z_3 \end{vmatrix}} = \frac{t}{\begin{vmatrix} x_1 & x_2 & x_0 \\ y_1 & y_2 & y_0 \\ z_1 & z_2 & z_0 \end{vmatrix}} = \frac{1}{\begin{vmatrix} x_1 & x_2 & x_3 \\ y_1 & y_2 & y_3 \\ z_1 & z_2 & z_3 \end{vmatrix}}$$

and

$$r = \frac{\begin{vmatrix} x_0 & x_2 & x_3 \\ y_0 & y_2 & y_3 \\ z_0 & z_2 & z_3 \end{vmatrix}}{DET}$$

$$s = \frac{\begin{vmatrix} x_1 & x_0 & x_3 \\ y_1 & y_0 & y_3 \\ z_1 & z_0 & z_3 \end{vmatrix}}{DET}$$

$$t = \frac{\begin{vmatrix} x_1 & x_2 & x_0 \\ y_1 & y_2 & y_0 \\ z_1 & z_2 & z_0 \end{vmatrix}}{DET}$$

$$DET = \begin{vmatrix} x_1 & x_2 & x_3 \\ y_1 & y_2 & y_3 \\ z_1 & z_2 & z_3 \end{vmatrix}.$$

Using the three points $P_1(0,\ 2,\ 0)$, $P_2(0,\ 0,\ 4)$, $P_3(3,\ 1,\ 2)$ and arbitrary positions of P_0, the values of r, s and t identify whether P_0 is inside or outside triangle $\triangle ABC$. For example, the point $P_0(0,\ 2,\ 0)$ is a vertex and is classified as being on the boundary. To confirm this we calculate r, s and t, and show that $r + s + t = 1$:

$$DET = \begin{vmatrix} 0 & 0 & 3 \\ 2 & 0 & 1 \\ 0 & 4 & 2 \end{vmatrix} = 24$$

$$r = \frac{\begin{vmatrix} 0 & 0 & 3 \\ 2 & 0 & 1 \\ 0 & 4 & 2 \end{vmatrix}}{24} = 1$$

$$s = \frac{\begin{vmatrix} 0 & 0 & 3 \\ 2 & 2 & 1 \\ 0 & 0 & 2 \end{vmatrix}}{24} = 0$$

$$t = \frac{\begin{vmatrix} 0 & 0 & 0 \\ 2 & 0 & 2 \\ 0 & 4 & 0 \end{vmatrix}}{24} = 0$$

therefore $r + s + t = 1$, but both s and t are zero which confirms that the point $(0,\ 2,\ 0)$ is on the boundary. In fact, as both coordinates are zero it confirms that the point is located on a vertex.

Now let's deliberately choose a point outside the triangle. For example, $P_0(4,\ 0,\ 3)$ is outside the triangle, which is confirmed by the corresponding values of r, s and t:

$$r = \frac{\begin{vmatrix} 4 & 0 & 3 \\ 0 & 0 & 1 \\ 3 & 4 & 2 \end{vmatrix}}{24} = -\frac{2}{3}$$

$$s = \frac{\begin{vmatrix} 0 & 4 & 3 \\ 2 & 0 & 1 \\ 0 & 3 & 2 \end{vmatrix}}{24} = \frac{3}{4}$$

$$t = \frac{\begin{vmatrix} 0 & 0 & 4 \\ 2 & 0 & 0 \\ 0 & 4 & 3 \end{vmatrix}}{24} = \frac{4}{3}$$

therefore,

$$r + s + t = -\tfrac{2}{3} + \tfrac{3}{4} + \tfrac{4}{3} = 1\tfrac{5}{12}$$

which confirms that the point $(4,\ 0,\ 3)$ is outside the triangle. Note that $r < 0$ and $t > 1$, which individually confirm that the point is outside the triangle's boundary.

16.7 Convex Hull Property

We have already shown that it is possible to determine whether a point is inside or outside a triangle. But remember that triangles are always convex. So can we test whether a point is inside or outside any polygon? Well the answer is no, unless the polygon is convex. The reason for this can be understood by considering the concave polygon shown in Fig. 16.15.

Let the barycentric coordinates for a point P_0 be

$$\mathbf{P}_0 = r\mathbf{A} + s\mathbf{B} + t\mathbf{C} + u\mathbf{D}$$

where $r + s + t + u = 1$. When $t = 0$, P_0 can exist anywhere inside triangle $\triangle ABD$. Thus, if any vertex creates a concavity, it will be ignored by barycentric coordinates.

16.8 Areas

Barycentric coordinates are also known as *areal coordinates* due to their area dividing properties. For example, in Fig. 16.16 the areas of the three internal triangles are in proportion to the barycentric coordinates of the point P.

To prove this, let P have barycentric coordinates

$$\mathbf{P} = r\mathbf{A} + s\mathbf{B} + t\mathbf{C}$$

Fig. 16.15 A concave polygon

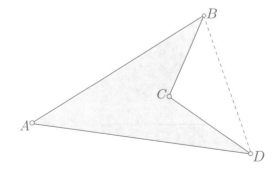

Fig. 16.16 The areas of the internal triangles are directly proportional to the barycentric coordinates of P

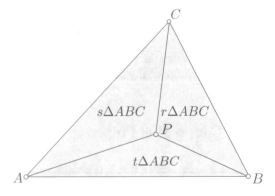

where

$$r + s + t = 1, \quad \text{and} \quad 0 \le (r, \ s, \ t) \le 1.$$

If we use the notation area($\triangle ABC$) to represent the area of the triangle formed from the vertices A, B and C then area($\triangle ABC$) is the sum of the areas of the smaller triangles:

$$\text{area}(\triangle ABC) = \text{area}(\triangle ABP) + \text{area}(\triangle BCP) + \text{area}(\triangle CAP).$$

But the area of any 2D triangle $\triangle P_1 P_2 P_3$ is

$$\text{area}(\triangle P_1 P_2 P_3) = \tfrac{1}{2} \begin{vmatrix} x_1 & y_1 & 1 \\ x_2 & y_2 & 1 \\ x_3 & y_3 & 1 \end{vmatrix}$$

therefore,

$$\text{area}(\triangle ABP) = \tfrac{1}{2} \begin{vmatrix} x_A & y_A & 1 \\ x_B & y_B & 1 \\ x_P & y_P & 1 \end{vmatrix}$$

but

$$x_P = r x_A + s x_B + t x_C$$

and

$$y_P = r y_A + s y_B + t y_C$$

therefore,

$$\text{area}(\varDelta ABP) = \tfrac{1}{2}\begin{vmatrix} x_A & y_A & 1 \\ x_B & y_B & 1 \\ rx_A + sx_B + tx_C & ry_A + sy_B + ty_C & 1 \end{vmatrix}$$

which expands to

$$\text{area}(\varDelta ABP) = \tfrac{1}{2}\left[\begin{array}{l} x_A y_B + rx_B y_A + sx_B y_B + tx_B y_C + rx_A y_A + sx_B y_A + tx_C y_A \\ -rx_A y_A - sx_A y_B - tx_A y_C - x_B y_A - rx_A y_B - sx_B y_B - tx_C y_B \end{array}\right]$$

$$= \tfrac{1}{2}\left[\begin{array}{l} x_A y_B - x_B y_A + r(x_B y_A - x_A y_B) + s(x_B y_A - x_A y_B) \\ +t(x_B y_C - x_C y_B) + t(x_C y_A - x_A y_C) \end{array}\right]$$

$$= \tfrac{1}{2}\left[\begin{array}{l} x_A y_B - x_B y_A + (1-t)(x_B y_A - x_A y_B) + t(x_B y_C - x_C y_B) \\ +t(x_C y_A - x_A y_C) \end{array}\right]$$

$$= \tfrac{1}{2}[-tx_B y_A + tx_A y_B + tx_B y_C - tx_C y_B + tx_C y_A - tx_A y_C]$$

and simplifies to

$$\text{area}(\varDelta ABP) = \tfrac{1}{2}t\begin{vmatrix} x_A & y_A & 1 \\ x_B & y_B & 1 \\ x_C & y_C & 1 \end{vmatrix} = t \times \text{area}(\varDelta ABC)$$

therefore,

$$t = \frac{\text{area}(\varDelta ABP)}{\text{area}(\varDelta ABC)}$$

similarly,

$$\text{area}(\varDelta BCP) = \tfrac{1}{2}r\begin{vmatrix} x_A & y_A & 1 \\ x_B & y_B & 1 \\ x_C & y_C & 1 \end{vmatrix} = r \times \text{area}(\varDelta ABC)$$

$$r = \frac{\text{area}(\varDelta BCP)}{\text{area}(\varDelta ABC)}$$

and

$$\text{area}(\varDelta CAP) = \tfrac{1}{2}s\begin{vmatrix} x_A & y_A & 1 \\ x_B & y_B & 1 \\ x_C & y_C & 1 \end{vmatrix} = s \times \text{area}(\varDelta ABC)$$

$$s = \frac{\text{area}(\varDelta CAP)}{\text{area}(\varDelta ABC)}.$$

Thus, we see that the areas of the internal triangles are directly proportional to the barycentric coordinates of P.

This is quite a useful relationship and can be used to resolve various geometric problems. For example, let's use it to find the radius and centre of the inscribed circle for a triangle. We could approach this problem using classical Euclidean geome-

Fig. 16.17 The inscribed
circle in triangle $\triangle ABC$

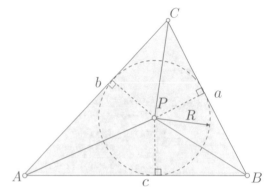

try, but barycentric coordinates provide a powerful analytical tool for resolving the problem very quickly. Consider triangle $\triangle ABC$ with sides a, b and c as shown in Fig. 16.17. The point P is the centre of the inscribed circle with radius R. From our knowledge of barycentric coordinates we know that

$$\mathbf{P} = r\mathbf{A} + s\mathbf{B} + t\mathbf{C}$$

where

$$r + s + t = 1. \tag{16.12}$$

We also know that the area properties of barycentric coordinates permit us to state

$$\text{area}(\triangle BCP) = r \times \text{area}(\triangle ABC) = \tfrac{1}{2}aR$$
$$\text{area}(\triangle CAP) = s \times \text{area}(\triangle ABC) = \tfrac{1}{2}bR$$
$$\text{area}(\triangle ABP) = t \times \text{area}(\triangle ABC) = \tfrac{1}{2}cR$$

therefore,

$$r = \frac{aR}{2 \times \text{area}(\triangle ABC)}, \quad s = \frac{bR}{2 \times \text{area}(\triangle ABC)}, \quad t = \frac{cR}{2 \times \text{area}(\triangle ABC)}$$

substituting r, s and t in (16.12) we get

$$\frac{R}{2 \times \text{area}(\triangle ABC)}(a + b + c) = 1$$

and

$$R = \frac{2 \times \text{area}(\triangle ABC)}{a + b + c}.$$

Substituting R in the definitions of r, s and t we obtain

Fig. 16.18 The inscribed
circle for a triangle

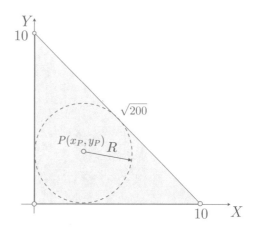

$$r = \frac{a}{a+b+c} \quad s = \frac{b}{a+b+c} \quad t = \frac{c}{a+b+c}$$

and

$$x_P = r x_A + s x_B + t x_C$$
$$y_P = r y_A + s y_B + t y_C.$$

To test this solution, consider the right-angled triangle in Fig. 16.18, where $a = \sqrt{200}$, $b = 10$, $c = 10$ and area($\triangle ABC$) $= 50$. Therefore

$$R = \frac{2 \times 50}{10 + 10 + \sqrt{200}} \approx 2.929$$

and

$$r = \frac{\sqrt{200}}{34.1421} \approx 0.4142, \quad s = \frac{10}{34.1421} \approx 0.2929, \quad t = \frac{10}{34.1421} \approx 0.2929$$

therefore,

$$x_P = 0.4142 \times 0 + 0.2929 \times 10 + 0.2929 \times 0 \approx 2.929$$
$$y_P = 0.4142 \times 0 + 0.2929 \times 0 + 0.2929 \times 0 \approx 2.929.$$

Therefore, the inscribed circle has a radius of 2.929 and a centre with coordinates (2.929, 2.929).

Let's explore another example where we determine the barycentric coordinates of a point using virtual mass points.

Fig. 16.19 Triangle $\triangle ABC$
with sides divided in the
ratio $1 : 2$.

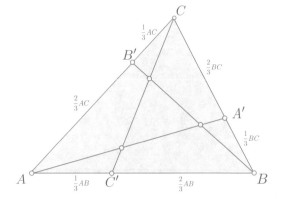

Fig. 16.20 The masses
assigned to A, B and C to
determine D

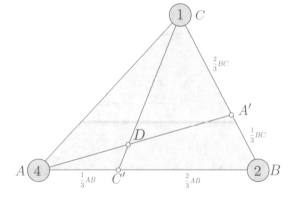

Figure 16.19 shows triangle $\triangle ABC$ where A', B' and C' divide BC, CA and AB respectively, in the ratio $1 : 2$. The objective is to find the barycentric coordinates of D, E and F, and the area of triangle $\triangle DEF$ as a proportion of triangle $\triangle ABC$.

We can approach the problem using mass points. For example, if we assume D is the centroid, all we have to do is determine the mass points that create this situation. Then the barycentric coordinates of D are given by (16.8). We proceed as follows.

The point D is on the intersection of lines CC' and AA'. Therefore, we begin by placing a mass of 1 at C. Then, for line BC to balance at A' a mass of 2 must be placed at B. Similarly, for line AB to balance at C' a mass of 4 must be placed at A. This configuration is shown in Fig. 16.20.

The total mass is $7 = (1 + 2 + 4)$, therefore,

$$D = \tfrac{4}{7}A + \tfrac{2}{7}B + \tfrac{1}{7}C.$$

The point E is on the intersection of lines BB' and AA'. Therefore, we begin by placing a mass of 1 at A. Then, for line CA to balance at B' a mass of 2 must be placed at C. Similarly, for line BC to balance at A' a mass of 4 must be placed at B.

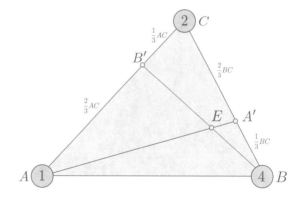

Fig. 16.21 The masses assigned to A, B and C to determine E

This configuration is shown in Fig. 16.21. The total mass is still 7, therefore,

$$E = \tfrac{1}{7}A + \tfrac{4}{7}B + \tfrac{2}{7}C.$$

From the symmetry of the triangle we can state that

$$F = \tfrac{2}{7}A + \tfrac{1}{7}B + \tfrac{4}{7}C.$$

Thus we can locate the points and using the vector equations

$$\mathbf{D} = \tfrac{4}{7}\mathbf{A} + \tfrac{2}{7}\mathbf{B} + \tfrac{1}{7}\mathbf{C}$$
$$\mathbf{E} = \tfrac{1}{7}\mathbf{A} + \tfrac{4}{7}\mathbf{B} + \tfrac{2}{7}\mathbf{C}$$
$$\mathbf{F} = \tfrac{2}{7}\mathbf{A} + \tfrac{1}{7}\mathbf{B} + \tfrac{4}{7}\mathbf{C}.$$

The important feature of these equations is that the barycentric coordinates of D, E and F are independent of $\mathbf{A}$, $\mathbf{B}$ and $\mathbf{C}$ they arise from the ratio used to divide the triangle's sides.

Although it was not the original intention, we can quickly explore what the barycentric coordinates of D, E and F would be if the triangle's sides had been $1 : 3$ instead of $1 : 2$. Without repeating all of the above steps, we would proceed as follows.

The point D is on the intersection of lines CC' and AA'. Therefore, we begin by placing a mass of 1 at C. Then, for line BC to balance at A' a mass of 3 must be placed at B. Similarly, for line AB to balance at C' a mass of 9 must be placed at A. This configuration is shown in Fig. 16.22. The total mass is $13 = (1 + 3 + 9)$, therefore,

$$\mathbf{D} = \tfrac{9}{13}\mathbf{A} + \tfrac{3}{13}\mathbf{B} + \tfrac{1}{13}\mathbf{C}$$
$$\mathbf{E} = \tfrac{1}{13}\mathbf{A} + \tfrac{9}{13}\mathbf{B} + \tfrac{3}{13}\mathbf{C}$$
$$\mathbf{F} = \tfrac{3}{13}\mathbf{A} + \tfrac{1}{13}\mathbf{B} + \tfrac{9}{13}\mathbf{C}.$$

Fig. 16.22 The masses
assigned to A, B and C to
determine D

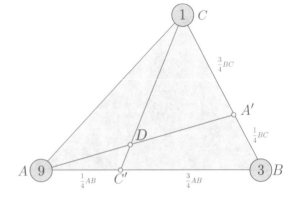

We could even develop the general equations for a ratio $1 : n$. It is left to the reader
to show that

$$D = \frac{n^2}{n^2+n+1}A + \frac{n}{n^2+n+1}B + \frac{1}{n^2+n+1}C$$

$$E = \frac{1}{n^2+n+1}A + \frac{n^2}{n^2+n+1}B + \frac{n}{n^2+n+1}C$$

$$F = \frac{n}{n^2+n+1}A + \frac{1}{n^2+n+1}B + \frac{n^2}{n^2+n+1}C.$$

As a quick test for the above equations, let $n = 1$, which make D, E and F
concurrent at the triangle's centroid:

$$D = \tfrac{1}{3}A + \tfrac{1}{3}B + \tfrac{1}{3}C$$
$$E = \tfrac{1}{3}A + \tfrac{1}{3}B + \tfrac{1}{3}C$$
$$F = \tfrac{1}{3}A + \tfrac{1}{3}B + \tfrac{1}{3}C$$

which is rather reassuring!

Now let's return to the final part of the problem and determine the area of triangle
$\triangle DEF$ in terms of $\triangle ABC$. The strategy is to split triangle $\triangle ABC$ into four triangles:
$\triangle BCF$, $\triangle CAD$, $\triangle ABE$ and $\triangle DEF$ as shown in Fig. 16.23.

Therefore,

$$\text{area}(\triangle ABC) = \text{area}(\triangle BCF) + \text{area}(\triangle CAD) + \text{area}(\triangle ABE) + \text{area}(\triangle DEF)$$

and

$$1 = \frac{\text{area}(\triangle BCF)}{\text{area}(\triangle ABC)} + \frac{\text{area}(\triangle CAD)}{\text{area}(\triangle ABC)} + \frac{\text{area}(\triangle ABE)}{\text{area}(\triangle ABC)} + \frac{\text{area}(\triangle DEF)}{\text{area}(\triangle ABC)} \qquad (16.13)$$

Fig. 16.23 Triangle $\triangle ABC$
divides into four triangles
$\triangle ABE$, $\triangle BCF$, $\triangle CAD$
and $\triangle DEF$

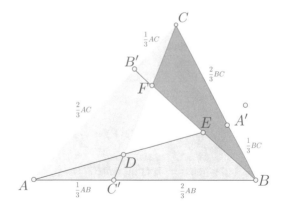

But we have just discovered that the barycentric coordinates are intimately connected with the ratios of triangles. For example, if F has barycentric coordinates
(r_F, s_F, t_F) relative to the points A, B and C respectively, then

$$r_F = \frac{\text{area}(\triangle BCF)}{\text{area}(\triangle ABC)}.$$

And if D has barycentric coordinates (r_D, s_D, t_D) relative to the points A, B and C
respectively, then

$$s_D = \frac{\text{area}(\triangle CAD)}{\text{area}(\triangle ABC)}.$$

Similarly, if E has barycentric coordinates (r_E, s_E, t_E) relative to the points A, B
and C respectively, then

$$t_E = \frac{\text{area}(\triangle ABE)}{\text{area}(\triangle ABC)}.$$

Substituting r_F, s_E and t_D in (16.13) we obtain

$$1 = r_F + s_D + t_E + \frac{\text{area}(\triangle DEF)}{\text{area}(\triangle ABC)}.$$

From (16.12) we see that

$$r_F = \tfrac{2}{7}, \quad s_D = \tfrac{2}{7}, \quad t_E = \tfrac{2}{7}$$

therefore,

$$1 = \tfrac{6}{7} + \frac{\text{area}(\triangle DEF)}{\text{area}(\triangle ABC)}$$

and

$$\text{area}(\Delta DEF) = \tfrac{1}{7}\text{area}(\Delta ABC)$$

which is rather neat!

Before we leave this example, let's state a general expression for the area(ΔDEF) for a triangle whose sides are divided in the ratio $1 : n$. Once again, I'll leave it to the reader to prove that

$$\text{area}(\Delta DEF) = \frac{n^2 - 2n + 1}{n^2 + n + 1} \times \text{area}(\Delta ABC).$$

Note that when $n = 1$, area(ΔDEF) = 0, which is correct.

[Hint: The corresponding values of r_F, s_D and t_E are $n/(n^2 + n + 1)$.]

16.9 Volumes

We have now seen that barycentric coordinates can be used to locate a scalar within a 1D domain, a point within a 2D area, so it seems logical that the description should extend to 3D volumes, which is the case.

To demonstrate this, consider the tetrahedron shown in Fig. 16.24. The volume of a tetrahedron is give by

$$V = \tfrac{1}{6} \begin{vmatrix} x_1 & y_1 & z_1 \\ x_2 & y_2 & z_2 \\ x_3 & y_3 & z_3 \end{vmatrix}$$

where $[x_1 \ \ y_1 \ \ z_1]^\mathrm{T}$, $[x_2 \ \ y_2 \ \ z_2]^\mathrm{T}$ and $[x_3 \ \ y_3 \ \ z_3]^\mathrm{T}$ are the three vectors extending from the fourth vertex to the other three vertices. However, if we locate the fourth vertex at the origin, $(x_1, \ y_1, \ z_1)$, $(x_2, \ y_2, \ z_2)$ and $(x_3, \ y_3, \ z_3)$ become the coordinates of the three vertices.

Fig. 16.24 A tetrahedron

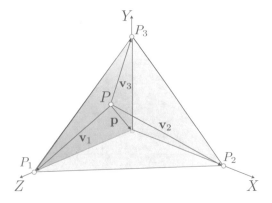

Let's locate a point $P(x_P, y_P, z_P)$ inside the tetrahedron with the following barycentric definition

$$\mathbf{P} = r\mathbf{P}_1 + s\mathbf{P}_2 + t\mathbf{P}_3 + u\mathbf{P}_0 \tag{16.14}$$

where $\mathbf{P}$, $\mathbf{P}_1$, $\mathbf{P}_2$, $\mathbf{P}_3$ and $\mathbf{P}_0$ are the position vectors for P, P_1, P_2, P_3 and P_0 respectively. The fourth barycentric term $u\mathbf{P}_0$ can be omitted as P_0 has coordinates $(0, 0, 0)$.

Therefore, we can state that the volume of the tetrahedron formed by the three vectors $\mathbf{P}$, $\mathbf{P}_2$ and $\mathbf{P}_3$ is given by

$$V = \frac{1}{6} \begin{vmatrix} x_P & y_P & z_P \\ x_2 & y_2 & z_2 \\ x_3 & y_3 & z_3 \end{vmatrix}. \tag{16.15}$$

Substituting (16.14) in (16.15) we obtain

$$V = \frac{1}{6} \begin{vmatrix} rx_1 + sx_2 + tx_3 & ry_1 + sy_2 + ty_3 & rz_1 + sz_2 + tz_3 \\ x_2 & y_2 & z_2 \\ x_3 & y_3 & z_3 \end{vmatrix} \tag{16.16}$$

which expands to

$$V = \frac{1}{6} \begin{bmatrix} y_2z_3(rx_1 + sx_2 + tx_3) + x_2y_3(rz_1 + sz_2 + tz_3) + x_3z_2(ry_1 + sy_2 + ty_3) \\ -y_3z_2(rx_1 + sx_2 + tx_3) - x_3y_2(rz_1 + sz_2 + tz_3) - x_2z_3(ry_1 + sy_2 + ty_3) \end{bmatrix}$$

$$= \frac{1}{6} \begin{bmatrix} r(x_1y_2z_3 + x_2y_3z_1 + x_3y_1z_2 - x_1y_3z_2 - x_3y_2z_1 - x_2y_1z_3) \\ +s(x_2y_2z_3 + x_2y_3z_2 + x_3y_1z_2 - x_2y_3z_2 - x_3y_1z_2 - x_2y_2z_3) \\ +t(x_3y_2z_3 + x_2y_3z_3 + x_3y_3z_2 - x_3y_3z_2 - x_3y_2z_3 - x_2y_3z_3) \end{bmatrix}$$

and simplifies to

$$V = \frac{1}{6}r \begin{vmatrix} x_1 & y_1 & z_1 \\ x_2 & y_2 & z_2 \\ x_3 & y_3 & z_3 \end{vmatrix}.$$

This states that the volume of the smaller tetrahedron is r times the volume of the larger tetrahedron V_T, where r is the barycentric coordinate modifying the vertex not included in the volume. By a similar process we can develop volumes for the other tetrahedra:

$$V(P, P_2, P_4, P_3) = rV_T$$
$$V(P, P_1, P_3, P_4) = sV_T$$
$$V(P, P_1, P_2, P_4) = tV_T$$
$$V(P, P_1, P_2, P_3) = uV_T$$

where $r + s + t + u = 1$. Similarly, the barycentric coordinates of a point inside the volume sum to unity.

Let's test the above statements with an example. Given $P_1(0, 0, 1)$, $P_2(1, 0, 0)$, $P_3(0, 1, 0)$ and $P\left(\frac{1}{3}, \frac{1}{3}, \frac{1}{3}\right)$ which is located inside the tetrahedron, the volume of the tetrahedron V_T is

$$V_T = \frac{1}{6}\begin{vmatrix} 0 & 0 & 1 \\ 1 & 0 & 0 \\ 0 & 1 & 0 \end{vmatrix} = \frac{1}{6}$$

$$r = \frac{V(P, P_2, P_4, P_3)}{V_T} = \frac{6}{6}\begin{vmatrix} \frac{2}{3} & -\frac{1}{3} & -\frac{1}{3} \\ -\frac{1}{3} & -\frac{1}{3} & -\frac{1}{3} \\ -\frac{1}{3} & \frac{2}{3} & -\frac{1}{3} \end{vmatrix} = \frac{1}{3}$$

$$s = \frac{V(P, P_1, P_3, P_4)}{V_T} = \frac{6}{6}\begin{vmatrix} -\frac{1}{3} & -\frac{1}{3} & \frac{2}{3} \\ -\frac{1}{3} & \frac{2}{3} & -\frac{1}{3} \\ -\frac{1}{3} & -\frac{1}{3} & -\frac{1}{3} \end{vmatrix} = \frac{1}{3}$$

$$t = \frac{V(P, P_1, P_2, P_4)}{V_T} = \frac{6}{6}\begin{vmatrix} -\frac{1}{3} & -\frac{1}{3} & \frac{2}{3} \\ \frac{2}{3} & -\frac{1}{3} & -\frac{1}{3} \\ -\frac{1}{3} & -\frac{1}{3} & -\frac{1}{3} \end{vmatrix} = \frac{1}{3}$$

$$u = \frac{V(P, P_1, P_2, P_3)}{V_T} = \frac{6}{6}\begin{vmatrix} -\frac{1}{3} & -\frac{1}{3} & \frac{2}{3} \\ \frac{2}{3} & -\frac{1}{3} & -\frac{1}{3} \\ -\frac{1}{3} & \frac{2}{3} & -\frac{1}{3} \end{vmatrix} = 0.$$

The barycentric coordinates (r, s, t, u) confirm that the point is located at the centre of triangle $\Delta P_1 P_2 P_3$. Note that the above determinants will create a negative volume if the vector sequences are reversed.

16.10   Bézier Curves and Patches

In Chap. 14 we examined Bézier curves and surface patches which are based on Bernstein polynomials:

$$\mathbf{B}_i^n(t) = \binom{n}{i} t^i (1 - t)^{n-i}.$$

We discovered that these polynomials create the quadratic terms

$$(1 - t)^2, \quad 2t(1 - t), \quad t^2$$

and the cubic terms

$$(1 - t)^3, \quad 3t(1 - t)^2, \quad 3t^2(1 - t), \quad t^3$$

which are used as scalars to multiply sequences of control points to create a parametric curve. Furthermore, these terms sum to unity, therefore they are also another form of barycentric coordinates. The only difference between these terms and the others described above is that they are controlled by a common parameter t. Another property of Bézier curves and patches is that they are constrained within the convex hull formed by the control points, which is also a property of barycentric coordinates.

16.11   Summary

Barycentric coordinates provide another way to locate points in space, which permit them to be used for ratios and proportion, areas, volumes, and centres of gravity.

Reference

1. Posamentier A (2008) Advanced euclidean geometry. Blackwell

Chapter 17
Geometric Algebra

17.1 Introduction

This can only be a brief introduction to geometric algebra as the subject really demands an entire book. Those readers who wish to pursue the subject further should consult the author's books [1, 2].

17.2 Background

Although geometric algebra introduces some new ideas, the subject should not be regarded as difficult. If you have read and understood the previous chapters, you should be familiar with vectors, vector products, transforms, and the idea that the product of two transforms is sensitive to the transform sequence. For example, in general, scaling an object after it has been translated, is not the same as translating an object after it has been scaled. Similarly, given two vectors **r** and **s** their vector product **r** × **s** creates a third vector **t**, using the right-hand rule, perpendicular to the plane containing **r** and **s**. However, just by reversing the vectors to **s** × **r**, creates a similar vector but in the opposite direction −**t**.

We regard vectors as *directed* lines or *oriented* lines, but if they exist, why shouldn't oriented planes and oriented volumes exist? Well, the answer to this question is that they do, which is what geometric algebra is about. Unfortunately, when vectors were invented, geometric algebra was overlooked, and it has taken a further century for it to emerge through the work of William Kingdon Clifford and the theoretical physicist David Hestenes (1933–). So let's continue and discover an exciting new algebra that will, in time, be embraced by the computer graphics community.

J. Vince, *Mathematics for Computer Graphics*, Undergraduate Topics in Computer Science, https://doi.org/10.1007/978-1-4471-7520-9_17

17.3 Symmetric and Antisymmetric Functions

It is possible to classify functions into two categories: *symmetric* (*even*) and *anti-symmetric* (*odd*) functions. For example, given two symmetric functions $f(x)$ and $f(x, y)$:

$$f(-x) = f(x)$$

and

$$f(y, x) = f(x, y)$$

an example being $\cos x$ where $\cos(-x) = \cos x$. Figure 17.1 illustrates how the cosine function is reflected about the origin. However, if the functions are anti-symmetric:

$$f(-x) = -f(x)$$

and

$$f(y, x) = -f(x, y)$$

an example being $\sin x$ where $\sin(-x) = -\sin x$. Figure 17.2 illustrates how the sine function is reflected about the origin.

Fig. 17.1 The graph of the symmetric cosine function

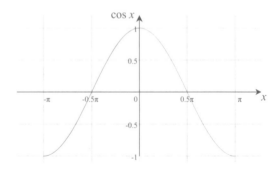

Fig. 17.2 The graph of the antisymmetric sine function

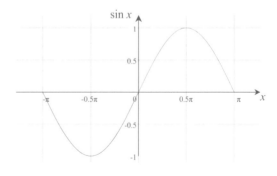

The reason why we have covered symmetric and antisymmetric functions is that they play an important role in geometric algebra. Now let's continue with this introduction and explore some important trigonometric foundations.

17.4 Trigonometric Foundations

Figure 17.3 shows two line segments a and b with coordinates (a_1, a_2), (b_1, b_2) respectively. The lines are separated by an angle θ, and we will compute the expressions $ab \cos \theta$ and $ab \sin \theta$, as these play an important role in geometric algebra.

Using the trigonometric identities

$$\sin(\theta + \phi) = \sin \theta \cos \phi + \cos \theta \sin \phi \qquad (17.1)$$
$$\cos(\theta + \phi) = \cos \theta \cos \phi - \sin \theta \sin \phi \qquad (17.2)$$

and the following observations

$$\cos \phi = \frac{a_1}{a}, \quad \sin \phi = \frac{a_2}{a}, \quad \cos(\theta + \phi) = \frac{b_1}{b}, \quad \sin(\theta + \phi) = \frac{b_2}{b}$$

we can rewrite (17.1) and (17.2) as

$$\frac{b_2}{b} = \frac{a_1}{a} \sin \theta + \frac{a_2}{a} \cos \theta \qquad (17.3)$$
$$\frac{b_1}{b} = \frac{a_1}{a} \cos \theta - \frac{a_2}{a} \sin \theta. \qquad (17.4)$$

To isolate $\cos \theta$ we multiply (17.3) by a_2 and (17.4) by a_1:

Fig. 17.3 Two line segments a and b separated by $+\theta$

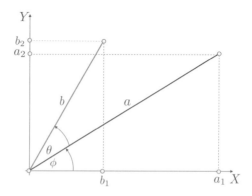

$$\frac{a_2 b_2}{b} = \frac{a_1 a_2}{a} \sin\theta + \frac{a_2^2}{a} \cos\theta \tag{17.5}$$

$$\frac{a_1 b_1}{b} = \frac{a_1^2}{a} \cos\theta - \frac{a_1 a_2}{a} \sin\theta. \tag{17.6}$$

Adding (17.5) and (17.6) we obtain

$$\frac{a_1 b_1 + a_2 b_2}{b} = \frac{a_1^2 + a_2^2}{a} \cos\theta = a\cos\theta$$

therefore,

$$ab\cos\theta = a_1 b_1 + a_2 b_2.$$

To isolate $\sin\theta$ we multiply (17.3) by a_1 and (17.4) by a_2

$$\frac{a_1 b_2}{b} = \frac{a_1^2}{a} \sin\theta + \frac{a_1 a_2}{a} \cos\theta \tag{17.7}$$

$$\frac{a_2 b_1}{b} = \frac{a_1 a_2}{a} \cos\theta - \frac{a_2^2}{a} \sin\theta \tag{17.8}$$

Subtracting (17.8) from (17.7) we obtain

$$\frac{a_1 b_2 - a_2 b_1}{b} = \frac{a_1^2 + a_2^2}{a} \sin\theta = a\sin\theta$$

therefore,

$$ab\sin\theta = a_1 b_2 - a_2 b_1.$$

If we form the product of b's projection on a with a, we get $ab\cos\theta$ which we have shown equals $a_1 b_1 + a_2 b_2$. Similarly, if we form the product $ab\sin\theta$ we compute the area of the parallelogram formed by sweeping a along b, which equals $a_1 b_2 - a_2 b_1$. What is noteworthy, is that the product $ab\cos\theta$ is independent of the sign of the angle θ, whereas the product $ab\sin\theta$ is sensitive to the sign of θ. Consequently, if we construct the lines a and b such that b is rotated $-\theta$ relative to a as shown in Fig. 17.4, $ab\cos\theta = a_1 b_1 + a_2 b_2$, but $ab\sin\theta = -(a_1 b_2 - a_2 b_1)$. The antisymmetric nature of the sine function reverses the sign of the area.

Having shown that area is a signed quantity just by using trigonometric identities, let's explore how vector algebra responds to this idea.

Fig. 17.4 Two line segments
a and b separated by $-\theta$

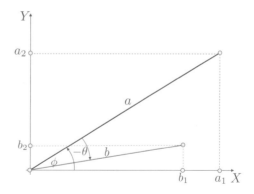

17.5 Vectorial Foundations

When we form the algebraic product of two 2D vectors **a** and **b**:

$$\mathbf{a} = a_1\mathbf{i} + a_2\mathbf{j}$$
$$\mathbf{b} = b_1\mathbf{i} + b_2\mathbf{j}$$

we obtain

$$\mathbf{ab} = a_1b_1\mathbf{i}^2 + a_2b_2\mathbf{j}^2 + a_1b_2\mathbf{ij} + a_2b_1\mathbf{ji} \qquad (17.9)$$

and it is clear that $a_1b_1\mathbf{i}^2 + a_2b_2\mathbf{j}^2$ has something to do with $ab\cos\theta$, and $a_1b_2\mathbf{ij} + a_2b_1\mathbf{ji}$ has something to do with $ab\sin\theta$. The product **ab** creates the terms $\mathbf{i}^2$, $\mathbf{j}^2$, $\mathbf{ij}$ and $\mathbf{ji}$, which are resolved as follows.

17.6 Inner and Outer Products

I like to believe that mathematics is a game—a game where we make the rules. Some rules might take us nowhere; others might take us so far in a particular direction and then restrict any further development; whilst other rules might open up a fantastic landscape that would have remained hidden had we not stumbled upon them. There are no 'wrong' or 'right' rules—there are just rules where some work better than others. Fortunately, the rules behind geometric algebra have been tested for over a hundred years, so we know they work. But these rules were not hiding somewhere waiting to be discovered, they arose due to the collective intellectual endeavour of many mathematicians over several decades.

Let's begin with the products **ij** and **ji** in (17.9) and assume that they anticommute: $\mathbf{ji} = -\mathbf{ij}$. Therefore,

$$\mathbf{ab} = a_1b_1\mathbf{i}^2 + a_2b_2\mathbf{j}^2 + (a_1b_2 - a_2b_1)\mathbf{ij} \qquad (17.10)$$

and if we reverse the product to **ba** we obtain

$$\mathbf{ba} = a_1 b_1 \mathbf{i}^2 + a_2 b_2 \mathbf{j}^2 - (a_1 b_2 - a_2 b_1)\mathbf{ij}. \qquad (17.11)$$

From (17.10) and (17.11) we see that the product of two vectors contains a symmetric component

$$a_1 b_1 \mathbf{i}^2 + a_2 b_2 \mathbf{j}^2$$

and an antisymmetric component

$$(a_1 b_2 - a_2 b_1)\mathbf{ij}.$$

It is interesting to observe that the symmetric component has $0°$ between its vector pairs ($\mathbf{i}^2$ or $\mathbf{j}^2$), whereas the antisymmetric component has $90°$ between its vector pairs ($\mathbf{i}$ and $\mathbf{j}$). Therefore, the sine and cosine functions play a natural role in our rules. What we are looking for are two functions that, when given our vectors **a** and **b**, one function returns the symmetric component and the other returns the antisymmetric component. We call these the *inner* and *outer* functions respectively.

It should be clear that if the inner function includes the cosine of the angle between the two vectors it will reject the antisymmetric component and return the symmetric element. Similarly, if the outer function includes the sine of the angle between the vectors, the symmetric component is rejected, and returns the antisymmetric element.

If we declare the inner function as the *inner product*:

$$\mathbf{a} \cdot \mathbf{b} = \|\mathbf{a}\| \|\mathbf{b}\| \cos \theta \qquad (17.12)$$

then

$$\begin{aligned}
\mathbf{a} \cdot \mathbf{b} &= (a_1 \mathbf{i} + a_2 \mathbf{j}) \cdot (b_1 \mathbf{i} + b_2 \mathbf{j}) \\
&= a_1 b_1 \mathbf{i} \cdot \mathbf{i} + a_1 b_2 \mathbf{i} \cdot \mathbf{j} + a_2 b_1 \mathbf{j} \cdot \mathbf{i} + a_2 b_2 \mathbf{j} \cdot \mathbf{j} \\
&= a_1 b_1 + a_2 b_2
\end{aligned}$$

which is perfect!

Next, we declare the outer function as the *outer product* using the wedge '∧' symbol; which is why it is also called the *wedge product*:

$$\mathbf{a} \wedge \mathbf{b} = \|\mathbf{a}\| \|\mathbf{b}\| \sin \theta \ \mathbf{i} \wedge \mathbf{j}. \qquad (17.13)$$

Note that product includes a strange $\mathbf{i} \wedge \mathbf{j}$ term. This is included as we just can't ignore the **ij** term in the antisymmetric component:

$$\mathbf{a} \wedge \mathbf{b} = (a_1\mathbf{i} + a_2\mathbf{j}) \wedge (b_1\mathbf{i} + b_2\mathbf{j})$$
$$= a_1b_1\mathbf{i} \wedge \mathbf{i} + a_1b_2\mathbf{i} \wedge \mathbf{j} + a_2b_1\mathbf{j} \wedge \mathbf{i} + a_2b_2\mathbf{j} \wedge \mathbf{j}$$
$$= (a_1b_2 - a_2b_1)\mathbf{i} \wedge \mathbf{j}$$

which enables us to write

$$\mathbf{ab} = \mathbf{a} \cdot \mathbf{b} + \mathbf{a} \wedge \mathbf{b} \qquad\qquad (17.14)$$
$$\mathbf{ab} = \|\mathbf{a}\| \|\mathbf{b}\| \cos\theta + \|\mathbf{a}\| \|\mathbf{b}\| \sin\theta \; \mathbf{i} \wedge \mathbf{j}. \qquad\qquad (17.15)$$

17.7 The Geometric Product in 2D

Clifford named the sum of the two products the *geometric product*, which means that (17.14) reads: The geometric product $\mathbf{ab}$ is the sum of the inner product 'a dot b' and the outer product 'a wedge b'. Remember that all this assumes that $\mathbf{ji} = -\mathbf{ij}$ which seems a reasonable assumption.

Given the definition of the geometric product, let's evaluate $\mathbf{i}^2$

$$\mathbf{ii} = \mathbf{i} \cdot \mathbf{i} + \mathbf{i} \wedge \mathbf{i}.$$

Using the definition for the inner product (17.12) we have

$$\mathbf{i} \cdot \mathbf{i} = 1 \times 1 \times \cos 0° = 1$$

whereas, using the definition of the outer product (17.13) we have

$$\mathbf{i} \wedge \mathbf{i} = 1 \times 1 \times \sin 0° \; \mathbf{i} \wedge \mathbf{i} = 0.$$

Thus $\mathbf{i}^2 = 1$ and $\mathbf{j}^2 = 1$, and $\mathbf{aa} = \|\mathbf{a}\|^2$:

$$\mathbf{aa} = \mathbf{a} \cdot \mathbf{a} + \mathbf{a} \wedge \mathbf{a}$$
$$= \|\mathbf{a}\| \|\mathbf{a}\| \cos 0° + \|\mathbf{a}\| \|\mathbf{a}\| \sin 0° \mathbf{i} \wedge \mathbf{j}$$
$$\mathbf{aa} = \|\mathbf{a}\|^2.$$

Now let's evaluate $\mathbf{ij}$:
$$\mathbf{ij} = \mathbf{i} \cdot \mathbf{j} + \mathbf{i} \wedge \mathbf{j}.$$

Using the definition for the inner product (17.12) we have

$$\mathbf{i} \cdot \mathbf{j} = 1 \times 1 \times \cos 90° = 0$$

whereas using the definition of the outer product (17.13) we have

Fig. 17.5 An anticlockwise
and clockwise bivector

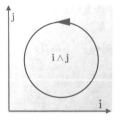 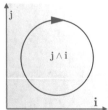

$$\mathbf{i} \wedge \mathbf{j} = 1 \times 1 \times \sin 90° \ \mathbf{i} \wedge \mathbf{j} = \mathbf{i} \wedge \mathbf{j}.$$

Thus $\mathbf{ij} = \mathbf{i} \wedge \mathbf{j}$. But what is $\mathbf{i} \wedge \mathbf{j}$? Well, it is a new object and is called a '*bivector*' and defines the orientation of the plane containing $\mathbf{i}$ and $\mathbf{j}$.

As the order of the vectors is from $\mathbf{i}$ to $\mathbf{j}$, the angle is $+90°$ and $\sin(+90)° = 1$. Whereas, if the order is from $\mathbf{j}$ to $\mathbf{i}$ the angle is $-90°$ and $\sin(-90°) = -1$. Consequently,

$$\mathbf{ji} = \mathbf{j} \cdot \mathbf{i} + \mathbf{j} \wedge \mathbf{i}$$
$$= 0 + 1 \times 1 \times \sin(-90°)\mathbf{i} \wedge \mathbf{j}$$
$$\mathbf{ji} = -\mathbf{i} \wedge \mathbf{j}.$$

Thus the bivector $\mathbf{i} \wedge \mathbf{j}$ defines the orientation of a surface as anticlockwise, whilst the bivector $\mathbf{j} \wedge \mathbf{i}$ defines the orientation as clockwise. These ideas are shown in Fig. 17.5.

So far, so good. Our rules seem to be leading somewhere. The inner product (17.12) is our old friend the dot product, and does not need explaining. However, the outer product (17.13) does require some further explanation.

The equation

$$\mathbf{ab} = 9 + 12\mathbf{i} \wedge \mathbf{j}$$

simply means that the geometric product of two vectors $\mathbf{a}$ and $\mathbf{b}$ creates a scalar, inner product of 9, and an outer product of 12 on the $\mathbf{ij}$-plane.

For example, given

$$\mathbf{a} = 3\mathbf{i}$$
$$\mathbf{b} = 3\mathbf{i} + 4\mathbf{j}$$

then

$$\mathbf{ab} = 3\mathbf{i} \cdot (3\mathbf{i} + 4\mathbf{j}) + 3\mathbf{i} \wedge (3\mathbf{i} + 4\mathbf{j})$$
$$= 9 + 9\mathbf{i} \wedge \mathbf{i} + 12\mathbf{i} \wedge \mathbf{j}$$
$$\mathbf{ab} = 9 + 12\mathbf{i} \wedge \mathbf{j}.$$

The 9 represents $\|\mathbf{a}\|\|\mathbf{b}\|\cos\theta$, whereas the 12 represents an area $\|\mathbf{a}\|\|\mathbf{b}\|\sin\theta$ on the $\mathbf{ij}$-plane. The angle between the two vectors θ is given by

$$\theta = \cos^{-1}\left(\tfrac{3}{5}\right).$$

However, reversing the product, we obtain

$$\mathbf{ba} = (3\mathbf{i} + 4\mathbf{j}) \cdot 3\mathbf{i} + (3\mathbf{i} + 4\mathbf{j}) \wedge 3\mathbf{i}$$
$$= 9 + 9\mathbf{i} \wedge \mathbf{i} + 12\mathbf{j} \wedge \mathbf{i}$$
$$\mathbf{ab} = 9 - 12\mathbf{i} \wedge \mathbf{j}.$$

The sign of the outer (wedge) product has flipped to reflect the new orientation of the vectors relative to the accepted orientation of the basis bivectors.

So the geometric product combines the scalar and wedge products into a single product, where the scalar product is the symmetric component and the wedge product is the antisymmetric component. Now let's see how these products behave in 3D.

17.8 The Geometric Product in 3D

Before we consider the geometric product in 3D we need to introduce some new notation, which will simplify future algebraic expressions. Rather than use $\mathbf{i}, \mathbf{j}$ and $\mathbf{k}$ to represent the unit basis vectors let's employ $\mathbf{e}_1, \mathbf{e}_2$ and $\mathbf{e}_3$ respectively. This means that (17.15) can be written

$$\mathbf{ab} = \|\mathbf{a}\|\|\mathbf{b}\|\cos\theta + \|\mathbf{a}\|\|\mathbf{b}\|\sin\theta\,\mathbf{e}_1 \wedge \mathbf{e}_2.$$

We begin with two 3D vectors:

$$\mathbf{a} = a_1\mathbf{e}_1 + a_2\mathbf{e}_2 + a_3\mathbf{e}_3$$
$$\mathbf{b} = b_1\mathbf{e}_1 + b_2\mathbf{e}_2 + b_3\mathbf{e}_3$$

therefore, their inner product is

$$\mathbf{a} \cdot \mathbf{b} = (a_1\mathbf{e}_1 + a_2\mathbf{e}_2 + a_3\mathbf{e}_3) \cdot (b_1\mathbf{e}_1 + b_2\mathbf{e}_2 + b_3\mathbf{e}_3)$$
$$= a_1b_1 + a_2b_2 + a_3b_3$$

and their outer product is

$$\mathbf{a} \wedge \mathbf{b} = (a_1\mathbf{e}_1 + a_2\mathbf{e}_2 + a_3\mathbf{e}_3) \wedge (b_1\mathbf{e}_1 + b_2\mathbf{e}_2 + b_3\mathbf{e}_3)$$
$$= a_1b_2\mathbf{e}_1 \wedge \mathbf{e}_2 + a_1b_3\mathbf{e}_1 \wedge \mathbf{e}_3 + a_2b_1\mathbf{e}_2 \wedge \mathbf{e}_1 + a_2b_3\mathbf{e}_2 \wedge \mathbf{e}_3$$
$$+ a_3b_1\mathbf{e}_3 \wedge \mathbf{e}_1 + a_3b_2\mathbf{e}_3 \wedge \mathbf{e}_2$$

Fig. 17.6 The 3D bivectors

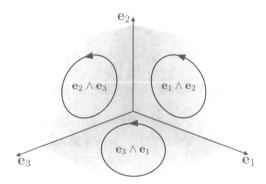

$$\mathbf{a} \wedge \mathbf{b} = (a_1 b_2 - a_2 b_1)\mathbf{e}_1 \wedge \mathbf{e}_2 + (a_2 b_3 - a_3 b_2)\mathbf{e}_2 \wedge \mathbf{e}_3 + (a_3 b_1 - a_1 b_3)\mathbf{e}_3 \wedge \mathbf{e}_1.$$
$$(17.16)$$

This time we have three unit-basis bivectors: $\mathbf{e}_1 \wedge \mathbf{e}_2, \mathbf{e}_2 \wedge \mathbf{e}_3, \mathbf{e}_3 \wedge \mathbf{e}_1$, and three associated scalar multipliers: $(a_1 b_2 - a_2 b_1)$, $(a_2 b_3 - a_3 b_2)$, $(a_3 b_1 - a_1 b_3)$ respectively.

Continuing with the idea described in the previous section, the three bivectors represent the three planes containing the respective vectors as shown in Fig. 17.6, and the scalar multipliers are projections of the area of the vector parallelogram onto the three bivectors as shown in Fig. 17.7. The orientation of the vectors **a** and **b** determine whether the projected areas are positive or negative.

You may think that (17.16) looks familiar. In fact, it looks very similar to the cross product $\mathbf{a} \times \mathbf{b}$:

$$\mathbf{a} \times \mathbf{b} = (a_1 b_2 - a_2 b_1)\mathbf{e}_3 + (a_2 b_3 - a_3 b_2)\mathbf{e}_1 + (a_3 b_1 - a_1 b_3)\mathbf{e}_2. \qquad (17.17)$$

This similarity is no accident. For when Hamilton invented quaternions, he did not recognise the possibility of bivectors, and invented some rules, which eventually

Fig. 17.7 The projections on the three bivectors

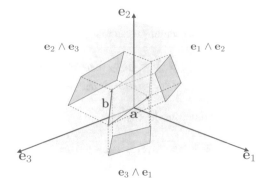

became the cross product! Later in this chapter we discover that quaternions are really bivectors in disguise.

We can see that a simple relationship exists between (17.16) and (17.17):

$$\mathbf{e}_1 \wedge \mathbf{e}_2 \text{ and } \mathbf{e}_3$$
$$\mathbf{e}_2 \wedge \mathbf{e}_3 \text{ and } \mathbf{e}_1$$
$$\mathbf{e}_3 \wedge \mathbf{e}_1 \text{ and } \mathbf{e}_2$$

the wedge product bivectors are perpendicular to the vector components of the cross product. So the wedge product is just another way of representing the cross product. However, the wedge product introduces a very important bonus: it works in space of any dimension, whereas, the cross product is only comfortable in 3D. Not only that, the wedge (outer product) is a product that creates volumes, hypervolumes, and can also be applied to vectors, bivectors, trivectors, etc.

17.9 The Outer Product of Three 3D Vectors

Having seen that the outer product of two 3D vectors is represented by areal projections onto the three basis bivectors, let's explore the outer product of three 3D vectors.

Given

$$\mathbf{a} = a_1\mathbf{e}_1 + a_2\mathbf{e}_2 + a_3\mathbf{e}_3$$
$$\mathbf{b} = b_1\mathbf{e}_1 + b_2\mathbf{e}_2 + b_3\mathbf{e}_3$$
$$\mathbf{c} = c_1\mathbf{e}_1 + c_2\mathbf{e}_2 + c_3\mathbf{e}_3$$

then

$$\mathbf{a} \wedge \mathbf{b} \wedge \mathbf{c} = (a_1\mathbf{e}_1 + a_2\mathbf{e}_2 + a_3\mathbf{e}_3) \wedge (b_1\mathbf{e}_1 + b_2\mathbf{e}_2 + b_3\mathbf{e}_3) \wedge (c_1\mathbf{e}_1 + c_2\mathbf{e}_2 + c_3\mathbf{e}_3)$$
$$= [(a_1b_2 - a_2b_1)\mathbf{e}_1 \wedge \mathbf{e}_2 + (a_2b_3 - a_3b_2)\mathbf{e}_2 \wedge \mathbf{e}_3 + (a_3b_1 - a_1b_3)\mathbf{e}_3 \wedge \mathbf{e}_1]$$
$$\wedge (c_1\mathbf{e}_1 + c_2\mathbf{e}_2 + c_3\mathbf{e}_3).$$

At this stage we introduce another axiom: the outer product is associative. This means that $\mathbf{a} \wedge (\mathbf{b} \wedge \mathbf{c}) = (\mathbf{a} \wedge \mathbf{b}) \wedge \mathbf{c}$. Therefore, knowing that $\mathbf{a} \wedge \mathbf{a} = 0$:

$$\mathbf{a} \wedge \mathbf{b} \wedge \mathbf{c} = c_3(a_1b_2 - a_2b_1)\mathbf{e}_1 \wedge \mathbf{e}_2 \wedge \mathbf{e}_3 + c_1(a_2b_3 - a_3b_2)\mathbf{e}_2 \wedge \mathbf{e}_3 \wedge \mathbf{e}_1$$
$$+ c_2(a_3b_1 - a_1b_3)\mathbf{e}_3 \wedge \mathbf{e}_1 \wedge \mathbf{e}_2.$$

But we are left with the products $e_1 \wedge e_2 \wedge e_3$, $e_2 \wedge e_3 \wedge e_1$ and $e_3 \wedge e_1 \wedge e_2$. Not to worry, because we know that $\mathbf{a} \wedge \mathbf{b} = -\mathbf{b} \wedge \mathbf{a}$. Therefore,

$$e_2 \wedge e_3 \wedge e_1 = -e_2 \wedge e_1 \wedge e_3 = e_1 \wedge e_2 \wedge e_3$$

and

$$e_3 \wedge e_1 \wedge e_2 = -e_1 \wedge e_3 \wedge e_2 = e_1 \wedge e_2 \wedge e_3.$$

Therefore, we can write $\mathbf{a} \wedge \mathbf{b} \wedge \mathbf{c}$ as

$$\mathbf{a} \wedge \mathbf{b} \wedge \mathbf{c} = c_3(a_1b_2 - a_2b_1)e_1 \wedge e_2 \wedge e_3 + c_1(a_2b_3 - a_3b_2)e_1 \wedge e_2 \wedge e_3$$
$$+ c_2(a_3b_1 - a_1b_3)e_1 \wedge e_2 \wedge e_3$$

or

$$\mathbf{a} \wedge \mathbf{b} \wedge \mathbf{c} = [c_3(a_1b_2 - a_2b_1) + c_1(a_2b_3 - a_3b_2) + c_2(a_3b_1 - a_1b_3)]\, e_1 \wedge e_2 \wedge e_3$$

or using a determinant:

$$\mathbf{a} \wedge \mathbf{b} \wedge \mathbf{c} = \begin{vmatrix} a_1 & b_1 & c_1 \\ a_2 & b_2 & c_2 \\ a_3 & b_3 & c_3 \end{vmatrix} e_1 \wedge e_2 \wedge e_3$$

which is the well-known expression for the volume of a parallelpiped formed by three vectors.

The term $e_1 \wedge e_2 \wedge e_3$ is a *trivector* and reminds us that the volume is oriented. If the sign of the determinant is positive, the original three vectors possess the same orientation of the three basis vectors. If the sign of the determinant is negative, the three vectors possess an orientation opposing that of the three basis vectors.

17.10   Axioms

One of the features of geometric algebra is that it behaves very similar to the everyday algebra of scalars:

Axiom 1: The associative rule:

$$\mathbf{a(bc)} = \mathbf{(ab)c}.$$

Axiom 2: The left and right distributive rules:

$$\mathbf{a(b + c)} = \mathbf{ab} + \mathbf{ac}$$
$$\mathbf{(b + c)a} = \mathbf{ba} + \mathbf{ca}.$$

The next four axioms describe how vectors interact with a scalar λ:
Axiom 3:

$$(\lambda\mathbf{a})\mathbf{b} = \lambda(\mathbf{ab}) = \lambda\mathbf{ab}.$$

Axiom 4:

$$\lambda(\phi\mathbf{a}) = (\lambda\phi)\mathbf{a}.$$

Axiom 5:

$$\lambda(\mathbf{a} + \mathbf{b}) = \lambda\mathbf{a} + \lambda\mathbf{b}.$$

Axiom 6:

$$(\lambda + \phi)\mathbf{a} = \lambda\mathbf{a} + \phi\mathbf{a}.$$

The next axiom that is adopted is
Axiom 7:

$$\mathbf{a}^2 = \|\mathbf{a}\|^2$$

which has already emerged as a consequence of the algebra. However, for non-Euclidean geometries, this can be set to $\mathbf{a}^2 = -\|\mathbf{a}\|^2$, which does not concern us here.

17.11 Notation

Having abandoned $\mathbf{i}, \mathbf{j}, \mathbf{k}$ for $\mathbf{e}_1, \mathbf{e}_2, \mathbf{e}_3$, it is convenient to convert geometric products $\mathbf{e}_1\mathbf{e}_2 \ldots \mathbf{e}_n$ to $\mathbf{e}_{12\ldots n}$. For example, $\mathbf{e}_1\mathbf{e}_2\mathbf{e}_3 \equiv \mathbf{e}_{123}$. Furthermore, we must get used to the following substitutions:

$$\mathbf{e}_i\mathbf{e}_i\mathbf{e}_j = \mathbf{e}_j$$
$$\mathbf{e}_{21} = -\mathbf{e}_{12}$$
$$\mathbf{e}_{312} = \mathbf{e}_{123}$$
$$\mathbf{e}_{112} = \mathbf{e}_2$$
$$\mathbf{e}_{121} = -\mathbf{e}_2.$$

17.12 Grades, Pseudoscalars and Multivectors

As geometric algebra embraces such a wide range of objects, it is convenient to *grade* them as follows: scalars are grade 0, vectors are grade 1, bivectors are grade 2, and trivectors are grade 3, and so on for higher dimensions. In such a graded algebra it is traditional to call the highest grade element a *pseudoscalar*. Thus in 2D the pseudoscalar is $\mathbf{e}_{12}$ and in 3D the pseudoscalar is $\mathbf{e}_{123}$.

One very powerful feature of geometric algebra is the idea of a *multivector*, which is a linear combination of a scalar, vector, bivector, trivector or any other higher dimensional object. For example the following are multivectors:

$$\mathbf{A} = 3 + (2\mathbf{e}_1 + 3\mathbf{e}_2 + 4\mathbf{e}_3) + (5\mathbf{e}_{12} + 6\mathbf{e}_{23} + 7\mathbf{e}_{31}) + 8\mathbf{e}_{123}$$
$$\mathbf{B} = 2 + (2\mathbf{e}_1 + 2\mathbf{e}_2 + 3\mathbf{e}_3) + (4\mathbf{e}_{12} + 5\mathbf{e}_{23} + 6\mathbf{e}_{31}) + 7\mathbf{e}_{123}$$

and we can form their sum:

$$\mathbf{A} + \mathbf{B} = 5 + (4\mathbf{e}_1 + 5\mathbf{e}_2 + 7\mathbf{e}_3) + (9\mathbf{e}_{12} + 11\mathbf{e}_{23} + 13\mathbf{e}_{31}) + 15\mathbf{e}_{123}$$

or their difference:

$$\mathbf{A} - \mathbf{B} = 1 + (\mathbf{e}_2 + \mathbf{e}_3) + (\mathbf{e}_{12} + \mathbf{e}_{23} + \mathbf{e}_{31}) + \mathbf{e}_{123}.$$

We can even form their product $\mathbf{AB}$, but at the moment we have not explored the products between all these elements.

We can isolate any grade of a multivector using the following notation:

$$\langle multivector \rangle_g$$

where g identifies a particular grade. For example, say we have the following multivector:

$$2 + 3\mathbf{e}_1 + 2\mathbf{e}_2 - 5\mathbf{e}_{12} + 6\mathbf{e}_{123}$$

we extract the scalar term using:

$$\langle 2 + 3\mathbf{e}_1 + 2\mathbf{e}_2 - 5\mathbf{e}_{12} + 6\mathbf{e}_{123} \rangle_0 = 2$$

the vector term using

$$\langle 2 + 3\mathbf{e}_1 + 2\mathbf{e}_2 - 5\mathbf{e}_{12} + 6\mathbf{e}_{123} \rangle_1 = 3\mathbf{e}_1 + 2\mathbf{e}_2$$

the bivector term using:

$$\langle 2 + 3\mathbf{e}_1 + 2\mathbf{e}_2 - 5\mathbf{e}_{12} + 6\mathbf{e}_{123} \rangle_2 = -5\mathbf{e}_{12}$$

and the trivector term using:

$$\langle 2 + 3\mathbf{e}_1 + 2\mathbf{e}_2 - 5\mathbf{e}_{12} + 6\mathbf{e}_{123} \rangle_3 = 6\mathbf{e}_{123}.$$

It is also worth pointing out that the inner vector product converts two grade 1 elements, i.e. vectors, into a grade 0 element, i.e. a scalar, whereas the outer vector product converts two grade 1 elements into a grade 2 element, i.e. a bivector. Thus the inner product is a grade lowering operation, while the outer product is a grade raising operation. These qualities of the inner and outer products are associated with higher grade elements in the algebra. This is why the scalar product is renamed as the inner product, because the scalar product is synonymous with transforming vectors into scalars. Whereas, the inner product transforms two elements of grade n into a grade $n - 1$ element.

17.13 Redefining the Inner and Outer Products

As the geometric product is defined in terms of the inner and outer products, it seems only natural to expect that similar functions exist relating the inner and outer products in terms of the geometric product. Such functions do exist and emerge when we combine the following two equations:

$$\mathbf{ab} = \mathbf{a} \cdot \mathbf{b} + \mathbf{a} \wedge \mathbf{b} \tag{17.18}$$
$$\mathbf{ba} = \mathbf{a} \cdot \mathbf{b} - \mathbf{a} \wedge \mathbf{b}. \tag{17.19}$$

Adding and subtracting (17.18) and (17.19) we have

$$\mathbf{a} \cdot \mathbf{b} = \tfrac{1}{2}(\mathbf{ab} + \mathbf{ba}) \tag{17.20}$$
$$\mathbf{a} \wedge \mathbf{b} = \tfrac{1}{2}(\mathbf{ab} - \mathbf{ba}). \tag{17.21}$$

Equations (17.20) and (17.21) and used frequently to define the products between different grade elements.

17.14 The Inverse of a Vector

In traditional vector analysis we accept that it is impossible to divide by a vector, but that is not so in geometric algebra. In fact, we don't actually divide a multivector by another vector but find a way of representing the inverse of a vector. For example, we know that a unit vector $\hat{\mathbf{a}}$ is defined as

$$\hat{\mathbf{a}} = \frac{\mathbf{a}}{\|\mathbf{a}\|}$$

and using the geometric product

$$\hat{\mathbf{a}}^2 = \frac{\mathbf{a}^2}{\|\mathbf{a}\|^2} = 1$$

therefore,

$$b = \frac{a^2 b}{\|a\|^2}$$

and exploiting the associative nature of the geometric product we have

$$b = \frac{a(ab)}{\|a\|^2}. \tag{17.22}$$

Equation (17.22) is effectively stating that, given the geometric product ab we can recover the vector b by pre-multiplying by a^{-1}:

$$\frac{a}{\|a\|^2}.$$

Similarly, we can recover the vector a by post-multiplying by b^{-1}:

$$a = \frac{(ab)b}{\|b\|^2}.$$

For example, given two vectors

$$a = e_1 + 2e_2$$
$$b = 3e_1 + 2e_2$$

their geometric product is

$$ab = 7 - 4e_{12}.$$

Therefore, given ab and a, we can recover b as follows:

$$b = \frac{e_1 + 2e_2}{5}(7 - 4e_{12})$$
$$= \tfrac{1}{5}(7e_1 - 4e_{112} + 14e_2 - 8e_{212})$$
$$= \tfrac{1}{5}(7e_1 - 4e_2 + 14e_2 + 8e_1)$$
$$b = 3e_1 + 2e_2.$$

Similarly, give ab and b, a is recovered as follows:

$$a = (7 - 4e_{12})\frac{3e_1 + 2e_2}{13}$$
$$= \tfrac{1}{13}(21e_1 + 14e_2 - 12e_{121} - 8e_{122})$$
$$= \tfrac{1}{13}(21e_1 + 14e_2 + 12e_2 - 8e_1)$$
$$a = e_1 + 2e_2.$$

Note that the inverse of a unit vector is the original vector:

$$\hat{\mathbf{a}}^{-1} = \frac{\hat{\mathbf{a}}}{\|\hat{\mathbf{a}}\|^2} = \hat{\mathbf{a}}.$$

17.15 The Imaginary Properties of the Outer Product

So far we know that the outer product of two vectors is represented by one or more unit basis vectors, such as

$$\mathbf{a} \wedge \mathbf{b} = \lambda_1 \mathbf{e}_{12} + \lambda_2 \mathbf{e}_{23} + \lambda_3 \mathbf{e}_{31}$$

where, in this case, the λ_i terms represent areas projected onto their respective unit basis bivectors. But what has not emerged is that the outer product is an imaginary quantity, which is revealed by expanding $\mathbf{e}_{12}^2$:

$$\mathbf{e}_{12}^2 = \mathbf{e}_{1212}$$

but as

$$\mathbf{e}_{21} = -\mathbf{e}_{12}$$

then

$$\mathbf{e}_{1(21)2} = -\mathbf{e}_{1(12)2}$$
$$= -\mathbf{e}_1^2 \mathbf{e}_2^2$$
$$\mathbf{e}_{12}^2 = -1.$$

Consequently, the geometric product effectively creates a complex number! Thus in a 2D scenario, given two vectors

$$\mathbf{a} = a_1 \mathbf{e}_1 + a_2 \mathbf{e}_2$$
$$\mathbf{b} = b_1 \mathbf{e}_1 + b_2 \mathbf{e}_2$$

their geometric product is

$$\mathbf{ab} = (a_1 b_1 + a_2 b_2) + (a_1 b_2 - a_2 b_1)\mathbf{e}_{12}$$

and knowing that $\mathbf{e}_{12} = i$, then we can write $\mathbf{ab}$ as

$$\mathbf{ab} = (a_1 b_1 + a_2 b_2) + (a_1 b_2 - a_2 b_1)i. \tag{17.23}$$

However, this notation is not generally adopted by the geometric community. The reason being that i is normally only associated with a scalar, with which it commutes. Whereas in 2D, $\mathbf{e}_{12}$ is associated with scalars and vectors, and although scalars present no problem, under some conditions, it anticommutes with vectors. Consequently, an upper-case I is used so that there is no confusion between the two elements. Thus (17.23) is written as

$$\mathbf{ab} = (a_1b_1 + a_2b_2) + (a_1b_2 - a_2b_1)I$$

where
$$I^2 = -1.$$

It goes without saying that the 3D unit basis bivectors are also imaginary quantities, so is $\mathbf{e}_{123}$.

Multiplying a complex number by i rotates it $90°$ on the complex plane. Therefore, it should be no surprise that multiplying a 2D vector by $\mathbf{e}_{12}$ rotates it by $90°$. However, because vectors are sensitive to their product partners, we must remember that pre-multiplying a vector by $\mathbf{e}_{12}$ rotates a vector clockwise and post-multiplying rotates a vector anticlockwise.

Whilst on the subject of rotations, let's consider what happens in 3D. We begin with a 3D vector

$$\mathbf{a} = a_1\mathbf{e}_1 + a_2\mathbf{e}_2 + a_3\mathbf{e}_3$$

and the unit basis bivector $\mathbf{e}_{12}$ as shown in Fig. 17.8. Next we construct their geometric product by pre-multiplying $\mathbf{a}$ by $\mathbf{e}_{12}$:

$$\mathbf{e}_{12}\mathbf{a} = a_1\mathbf{e}_{12}\mathbf{e}_1 + a_2\mathbf{e}_{12}\mathbf{e}_2 + a_3\mathbf{e}_{12}\mathbf{e}_3$$

Fig. 17.8 The effect of pre-multiplying a vector by a bivector

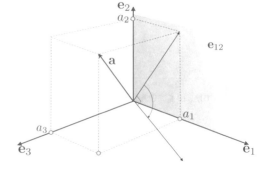

which becomes

$$\mathbf{e}_{12}\mathbf{a} = a_1\mathbf{e}_{121} + a_2\mathbf{e}_{122} + a_3\mathbf{e}_{123}$$
$$= -a_1\mathbf{e}_2 + a_2\mathbf{e}_1 + a_3\mathbf{e}_{123}$$
$$= a_2\mathbf{e}_1 - a_1\mathbf{e}_2 + a_3\mathbf{e}_{123}$$

and contains two parts: a vector $(a_2\mathbf{e}_1 - a_1\mathbf{e}_2)$ and a volume $a_3\mathbf{e}_{123}$.

Figure 17.8 shows how the projection of vector $\mathbf{a}$ is rotated clockwise on the bivector $\mathbf{e}_{12}$. A volume is also created perpendicular to the bivector. This enables us to predict that if the vector is coplanar with the bivector, the entire vector is rotated $-90°$ and the volume component will be zero.

By post-multiplying $\mathbf{a}$ by $\mathbf{e}_{12}$ creates

$$\mathbf{a}\mathbf{e}_{12} = -a_2\mathbf{e}_1 + a_1\mathbf{e}_2 + a_3\mathbf{e}_{123}$$

which shows that while the volumetric element has remained the same, the projected vector is rotated anticlockwise.

You may wish to show that the same happens with the other two bivectors.

17.16 Duality

The ability to exchange pairs of geometric elements such as lines and planes involves a *dual* operation, which in geometric algebra is relatively easy to define. For example, given a multivector $\mathbf{A}$ its dual $\mathbf{A}^*$ is defined as

$$\mathbf{A}^* = I\mathbf{A}$$

where I is the local pseudoscalar. For 2D this is $\mathbf{e}_{12}$ and for 3D it is $\mathbf{e}_{123}$. Therefore, given a 2D vector

$$\mathbf{a} = a_1\mathbf{e}_1 + a_2\mathbf{e}_2$$

its dual is

$$\mathbf{a}^* = \mathbf{e}_{12}(a_1\mathbf{e}_1 + a_2\mathbf{e}_2)$$
$$= a_1\mathbf{e}_{121} + a_2\mathbf{e}_{122}$$
$$= a_2\mathbf{e}_1 - a_1\mathbf{e}_2$$

which is another vector rotated 90° clockwise.

It is easy to show that $(\mathbf{a}^*)^* = -\mathbf{a}$, and two further dual operations return the vector back to $\mathbf{a}$.

In 3D the dual of a vector $\mathbf{e}_1$ is

$$\mathbf{e}_{123}\mathbf{e}_1 = \mathbf{e}_{1231} = \mathbf{e}_{23}$$

which is the perpendicular bivector. Similarly, the dual of $\mathbf{e}_2$ is $\mathbf{e}_{31}$ and the dual of $\mathbf{e}_3$ is $\mathbf{e}_{12}$.

For a general vector $a_1\mathbf{e}_1 + a_2\mathbf{e}_2 + a_3\mathbf{e}_3$ its dual is

$$\mathbf{e}_{123}(a_1\mathbf{e}_1 + a_2\mathbf{e}_2 + a_3\mathbf{e}_3) = a_1\mathbf{e}_{1231} + a_2\mathbf{e}_{1232} + a_3\mathbf{e}_{1233}$$
$$= a_3\mathbf{e}_{12} + a_1\mathbf{e}_{23} + a_2\mathbf{e}_{31}.$$

The duals of the 3D basis bivectors are:

$$\mathbf{e}_{123}\mathbf{e}_{12} = \mathbf{e}_{12312} = -\mathbf{e}_3$$
$$\mathbf{e}_{123}\mathbf{e}_{23} = \mathbf{e}_{12323} = -\mathbf{e}_1$$
$$\mathbf{e}_{123}\mathbf{e}_{31} = \mathbf{e}_{12331} = -\mathbf{e}_2.$$

17.17   The Relationship Between the Vector Product and the Outer Product

We have already discovered that there is a very close relationship between the vector product and the outer product, and just to recap: Given two vectors

$$\mathbf{a} = a_1\mathbf{e}_1 + a_2\mathbf{e}_2 + a_3\mathbf{e}_3$$
$$\mathbf{b} = b_1\mathbf{e}_1 + b_2\mathbf{e}_2 + b_3\mathbf{e}_3$$

then

$$\mathbf{a} \times \mathbf{b} = (a_2b_3 - a_3b_2)\mathbf{e}_1 + (a_3b_1 - a_1b_3)\mathbf{e}_2 + (a_1b_2 - a_2b_1)\mathbf{e}_3 \qquad (17.24)$$

and

$$\mathbf{a} \wedge \mathbf{b} = (a_2b_3 - a_3b_2)\mathbf{e}_2 \wedge \mathbf{e}_3 + (a_3b_1 - a_1b_3)\mathbf{e}_3 \wedge \mathbf{e}_1 + (a_1b_2 - a_2b_1)\mathbf{e}_1 \wedge \mathbf{e}_2$$

or

$$\mathbf{a} \wedge \mathbf{b} = (a_2b_3 - a_3b_2)\mathbf{e}_{23} + (a_3b_1 - a_1b_3)\mathbf{e}_{31} + (a_1b_2 - a_2b_1)\mathbf{e}_{12}. \qquad (17.25)$$

If we multiply (17.25) by I_{123} we obtain

$$I_{123}(\mathbf{a} \wedge \mathbf{b}) = (a_2b_3 - a_3b_2)\mathbf{e}_{123}\mathbf{e}_{23} + (a_3b_1 - a_1b_3)\mathbf{e}_{123}\mathbf{e}_{31} + (a_1b_2 - a_2b_1)\mathbf{e}_{123}\mathbf{e}_{12}$$
$$= -(a_2b_3 - a_3b_2)\mathbf{e}_1 - (a_3b_1 - a_1b_3)\mathbf{e}_2 - (a_1b_2 - a_2b_1)\mathbf{e}_3$$

which is identical to the cross product (17.24) apart from its sign. Therefore, we can state:

$$\mathbf{a} \times \mathbf{b} = -I_{123}(\mathbf{a} \wedge \mathbf{b}).$$

17.18 The Relationship Between Quaternions and Bivectors

Hamilton's rules for the imaginaries i, j and k are shown in Table 17.1, whilst Table 17.2 shows the rules for 3D bivector products.

Although there is some agreement between the table entries, there is a sign reversal in some of them. However, if we switch to a left-handed axial system the bivectors become $\mathbf{e}_{32}$, $\mathbf{e}_{13}$, $\mathbf{e}_{21}$ and their products are as shown in Table 17.3.

If we now create a one-to-one correspondence (isomorphism) between the two systems:

$$i \leftrightarrow \mathbf{e}_{32} \quad j \leftrightarrow \mathbf{e}_{13} \quad k \leftrightarrow \mathbf{e}_{21}$$

there is a true correspondence between quaternions and a left-handed set of bivectors.

Table 17.1 Hamilton's quaternion product rules

	i	j	k
i	-1	k	$-j$
j	$-k$	-1	i
k	j	$-i$	-1

Table 17.2 3D bivector product rules

	$\mathbf{e}_{23}$	$\mathbf{e}_{31}$	$\mathbf{e}_{12}$
$\mathbf{e}_{23}$	-1	$-\mathbf{e}_{12}$	$\mathbf{e}_{31}$
$\mathbf{e}_{31}$	$\mathbf{e}_{12}$	-1	$-\mathbf{e}_{23}$
$\mathbf{e}_{12}$	$-\mathbf{e}_{31}$	$\mathbf{e}_{23}$	-1

Table 17.3 Left-handed 3D bivector product rules

	$\mathbf{e}_{32}$	$\mathbf{e}_{13}$	$\mathbf{e}_{21}$
$\mathbf{e}_{32}$	-1	$\mathbf{e}_{21}$	$-\mathbf{e}_{13}$
$\mathbf{e}_{13}$	$-\mathbf{e}_{21}$	-1	$\mathbf{e}_{32}$
$\mathbf{e}_{21}$	$\mathbf{e}_{13}$	$-\mathbf{e}_{32}$	-1

17.19   Reflections and Rotations

One of geometric algebra's strengths is the elegance it brings to calculating reflections and rotations. Unfortunately, there is insufficient space to examine the derivations of the formulae, but if you are interested, these can be found in the author's books [1, 2]. Let's start with 2D reflections.

17.19.1   2D Reflections

Given a line, whose perpendicular unit vector is $\hat{\mathbf{m}}$ and a vector $\mathbf{a}$ its reflection $\mathbf{a}'$ is given by

$$\mathbf{a}' = \hat{\mathbf{m}}\mathbf{a}\hat{\mathbf{m}}$$

which is rather elegant! For example, Fig. 17.9 shows a scenario where

$$\hat{\mathbf{m}} = \tfrac{1}{\sqrt{2}}(\mathbf{e}_1 + \mathbf{e}_2)$$
$$\mathbf{a} = \mathbf{e}_1$$

therefore,

$$\mathbf{a}' = \tfrac{1}{\sqrt{2}}(\mathbf{e}_1 + \mathbf{e}_2)(\mathbf{e}_1)\tfrac{1}{\sqrt{2}}(\mathbf{e}_1 + \mathbf{e}_2)$$
$$= \tfrac{1}{2}(1 - \mathbf{e}_{12})(\mathbf{e}_1 + \mathbf{e}_2)$$
$$= \tfrac{1}{2}(\mathbf{e}_1 + \mathbf{e}_2 + \mathbf{e}_2 - \mathbf{e}_1)$$
$$\mathbf{a}' = \mathbf{e}_2.$$

Fig. 17.9 The reflection of a 2D vector

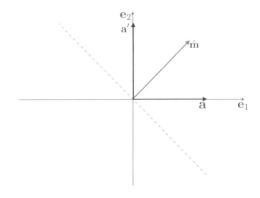

Note that in this scenario a reflection means a mirror image about the perpendicular vector.

17.19.2 3D Reflections

Let's explore the 3D scenario shown in Fig. 17.10 where

$$\mathbf{a} = \mathbf{e}_1 + \mathbf{e}_2 - \mathbf{e}_3$$
$$\hat{\mathbf{m}} = \mathbf{e}_2$$

therefore,

$$\mathbf{a}' = \mathbf{e}_2(\mathbf{e}_1 + \mathbf{e}_2 - \mathbf{e}_3)\mathbf{e}_2$$
$$= \mathbf{e}_{212} + \mathbf{e}_{222} - \mathbf{e}_{232}$$
$$= -\mathbf{e}_1 + \mathbf{e}_2 + \mathbf{e}_3.$$

As one might expect, it is also possible to reflect bivectors, trivectors and higher-dimensional objects, and for reasons of brevity, they are summarised as follows:

Reflecting about a line:

$$
\begin{aligned}
\text{scalars:} \quad & \text{invariant} \\
\text{vectors:} \quad & \mathbf{v}' = \hat{\mathbf{m}}\mathbf{v}\hat{\mathbf{m}} \\
\text{bivectors:} \quad & \mathbf{B}' = \hat{\mathbf{m}}\mathbf{B}\hat{\mathbf{m}} \\
\text{trivectors:} \quad & \mathbf{T}' = \hat{\mathbf{m}}\mathbf{T}\hat{\mathbf{m}}.
\end{aligned}
$$

Fig. 17.10 The reflection of a 3D vector

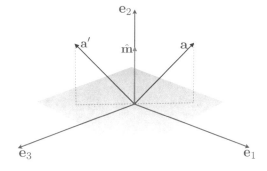

Reflecting about a mirror:

$$\begin{aligned}
\text{scalars:} \quad & \text{invariant} \\
\text{vectors:} \quad & \mathbf{v}' = -\hat{\mathbf{m}}\mathbf{v}\hat{\mathbf{m}} \\
\text{bivectors:} \quad & \mathbf{B}' = \hat{\mathbf{m}}\mathbf{B}\hat{\mathbf{m}} \\
\text{trivectors:} \quad & \mathbf{T}' = -\hat{\mathbf{m}}\mathbf{T}\hat{\mathbf{m}}.
\end{aligned}$$

17.19.3 2D Rotations

Figure 17.11 shows a plan view of two mirrors M and N separated by an angle θ. The point P is in front of mirror M and subtends an angle α, and its reflection P_R exists in the virtual space behind M and also subtends an angle α with the mirror. The angle between P_R and N must be $\theta - \alpha$, and its reflection P' must also lie $\theta - \alpha$ behind N. By inspection, the angle between P and the double reflection P' is 2θ.

   If we apply this double reflection transform to a collection of points, they are effectively all rotated 2θ about the origin where the mirrors intersect. The only slight drawback with this technique is that the angle of rotation is twice the angle between the mirrors.

   Instead of using points, let's employ position vectors and substitute normal unit vectors for the mirrors' orientation. For example, Fig. 17.12 shows the same mirrors with unit normal vectors $\hat{\mathbf{m}}$ and $\hat{\mathbf{n}}$. After two successive reflections, P becomes P',

Fig. 17.11 Rotating a point by a double reflection

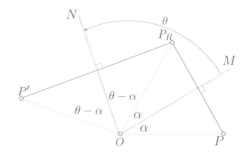

Fig. 17.12 Rotating a point by a double reflection

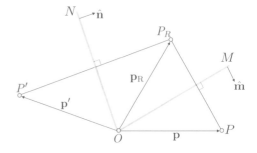

Fig. 17.13 Rotating a point
by 180°

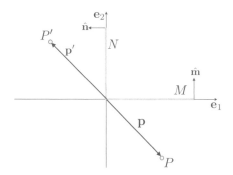

and using the relationship:

$$\mathbf{v}' = -\hat{\mathbf{m}}\mathbf{v}\hat{\mathbf{m}}$$

we compute the reflections as follows:

$$\mathbf{p}_R = -\hat{\mathbf{m}}\mathbf{p}\hat{\mathbf{m}}$$
$$\mathbf{p}' = -\hat{\mathbf{n}}\mathbf{p}_R\hat{\mathbf{n}}$$
$$\mathbf{p}' = \hat{\mathbf{n}}\hat{\mathbf{m}}\mathbf{p}\hat{\mathbf{m}}\hat{\mathbf{n}}$$

which is also rather elegant and compact. However, we must remember that P is rotated twice the angle separating the mirrors, and the rotation is relative to the origin. Let's demonstrate this technique with an example.

Figure 17.13 shows two mirrors M and N with unit normal vectors $\hat{\mathbf{m}}$, $\hat{\mathbf{n}}$ and position vector $\mathbf{p}$:

$$\hat{\mathbf{m}} = \mathbf{e}_2$$
$$\hat{\mathbf{n}} = -\mathbf{e}_1$$
$$P = (1, -1)$$
$$\mathbf{p} = \mathbf{e}_1 - \mathbf{e}_2.$$

As the mirrors are separated by 90° the point P is rotated 180°:

$$\mathbf{p}' = \hat{\mathbf{n}}\hat{\mathbf{m}}\mathbf{p}\hat{\mathbf{m}}\hat{\mathbf{n}}$$
$$= -\mathbf{e}_1\mathbf{e}_2(\mathbf{e}_1 - \mathbf{e}_2)\mathbf{e}_2(-\mathbf{e}_1)$$
$$= \mathbf{e}_{12121} - \mathbf{e}_{12221}$$
$$= -\mathbf{e}_1 + \mathbf{e}_2$$
$$P' = (-1, \ 1).$$

17.20   Rotors

Quaternions are the natural choice for rotating vectors about an arbitrary axis, and although it may not be immediately obvious, we have already started to discover geometric algebra's equivalent.

We begin with

$$\mathbf{p}' = \hat{\mathbf{n}}\hat{\mathbf{m}}\mathbf{p}\hat{\mathbf{m}}\hat{\mathbf{n}}$$

and substitute $\mathbf{R}$ for $\hat{\mathbf{n}}\hat{\mathbf{m}}$ and $\tilde{\mathbf{R}}$ for $\hat{\mathbf{m}}\hat{\mathbf{n}}$, therefore,

$$\mathbf{p}' = \mathbf{R}\mathbf{p}\tilde{\mathbf{R}}$$

where $\mathbf{R}$ and $\tilde{\mathbf{R}}$ are called *rotors* which perform the same function as a quaternion.

Because geometric algebra is non-commutative, the sequence of elements, be they vectors, bivectors, trivectors, etc., is very important. Consequently, it is very useful to include a command that reverses a sequence of elements. The notation generally employed is the tilde (͂) symbol:

$$\mathbf{R} = \hat{\mathbf{n}}\hat{\mathbf{m}}$$
$$\tilde{\mathbf{R}} = \hat{\mathbf{m}}\hat{\mathbf{n}}.$$

Let's unpack a rotor in terms of its angle and bivector as follows:

The bivector defining the plane is $\hat{\mathbf{m}} \wedge \hat{\mathbf{n}}$ and θ is the angle between the vectors. Let

$$\mathbf{R} = \hat{\mathbf{n}}\hat{\mathbf{m}}$$
$$\tilde{\mathbf{R}} = \hat{\mathbf{m}}\hat{\mathbf{n}}$$

where

$$\hat{\mathbf{n}}\hat{\mathbf{m}} = \hat{\mathbf{n}} \cdot \hat{\mathbf{m}} - \hat{\mathbf{m}} \wedge \hat{\mathbf{n}}$$
$$\hat{\mathbf{m}}\hat{\mathbf{n}} = \hat{\mathbf{n}} \cdot \hat{\mathbf{m}} + \hat{\mathbf{m}} \wedge \hat{\mathbf{n}}$$
$$\hat{\mathbf{n}} \cdot \hat{\mathbf{m}} = \cos\theta$$
$$\hat{\mathbf{m}} \wedge \hat{\mathbf{n}} = \hat{\mathbf{B}}\sin\theta.$$

Therefore,

$$\mathbf{R} = \cos\theta - \hat{\mathbf{B}}\sin\theta$$
$$\tilde{\mathbf{R}} = \cos\theta + \hat{\mathbf{B}}\sin\theta.$$

We now have an equation that rotates a vector $\mathbf{p}$ through an angle 2θ about an axis defined by $\hat{\mathbf{B}}$:

Fig. 17.14 Rotating a vector by 90°

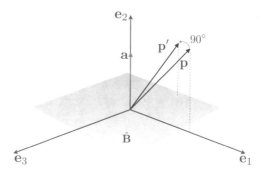

$$\mathbf{p}' = \mathbf{Rp\hat{R}}$$

or

$$\mathbf{p}' = (\cos\theta - \hat{\mathbf{B}}\sin\theta)\mathbf{p}(\cos\theta + \hat{\mathbf{B}}\sin\theta)$$

We can also express this such that it identifies the real angle of rotation α:

$$\mathbf{p}' = (\cos(\alpha/2) - \hat{\mathbf{B}}\sin(\alpha/2))\mathbf{p}\left(\cos(\alpha/2) + \hat{\mathbf{B}}\sin(\alpha/2)\right). \qquad (17.26)$$

Equation (17.26) references a bivector, which may make you feel uncomfortable! But remember, it simply identifies the axis perpendicular to its plane. Let's demonstrate how (17.26) works with two examples.

Figure 17.14 shows a scenario where vector $\mathbf{p}$ is rotated 90° about $\mathbf{e}_2$ which is perpendicular to $\hat{\mathbf{B}}$, where

$$\alpha = 90°$$
$$\mathbf{a} = \mathbf{e}_2$$
$$\mathbf{p} = \mathbf{e}_1 + \mathbf{e}_2$$
$$\hat{\mathbf{B}} = \mathbf{e}_{31}.$$

Therefore,

$$\mathbf{p}' = (\cos 45° - \mathbf{e}_{31}\sin 45°)(\mathbf{e}_1 + \mathbf{e}_2)(\cos 45° + \mathbf{e}_{31}\sin 45°)$$
$$= \left(\tfrac{\sqrt{2}}{2} - \tfrac{\sqrt{2}}{2}\mathbf{e}_{31}\right)(\mathbf{e}_1 + \mathbf{e}_2)\left(\tfrac{\sqrt{2}}{2} + \tfrac{\sqrt{2}}{2}\mathbf{e}_{31}\right)$$
$$= \left(\tfrac{\sqrt{2}}{2}\mathbf{e}_1 + \tfrac{\sqrt{2}}{2}\mathbf{e}_2 - \tfrac{\sqrt{2}}{2}\mathbf{e}_3 - \tfrac{\sqrt{2}}{2}\mathbf{e}_{312}\right)\left(\tfrac{\sqrt{2}}{2} + \tfrac{\sqrt{2}}{2}\mathbf{e}_{31}\right)$$
$$= \tfrac{1}{2}(\mathbf{e}_1 - \mathbf{e}_3 + \mathbf{e}_2 + \mathbf{e}_{231} - \mathbf{e}_3 - \mathbf{e}_1 - \mathbf{e}_{312} - \mathbf{e}_{31231})$$
$$\mathbf{p}' = \mathbf{e}_2 - \mathbf{e}_3.$$

Fig. 17.15 Rotating a vector
by 120°

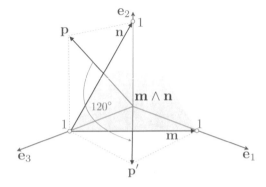

Observe what happens when the bivector's sign is reversed to $-\mathbf{e}_{31}$:

$$\mathbf{p}' = (\cos 45° + \mathbf{e}_{31} \sin 45°)(\mathbf{e}_1 + \mathbf{e}_2)(\cos 45° - \mathbf{e}_{31} \sin 45°)$$

$$= \left(\tfrac{\sqrt{2}}{2} + \tfrac{\sqrt{2}}{2}\mathbf{e}_{31}\right)(\mathbf{e}_1 + \mathbf{e}_2)\left(\tfrac{\sqrt{2}}{2} - \tfrac{\sqrt{2}}{2}\mathbf{e}_{31}\right)$$

$$= \left(\tfrac{\sqrt{2}}{2}\mathbf{e}_1 + \tfrac{\sqrt{2}}{2}\mathbf{e}_2 + \tfrac{\sqrt{2}}{2}\mathbf{e}_3 + \tfrac{\sqrt{2}}{2}\mathbf{e}_{312}\right)\left(\tfrac{\sqrt{2}}{2} - \tfrac{\sqrt{2}}{2}\mathbf{e}_{31}\right)$$

$$= \tfrac{1}{2}(\mathbf{e}_1 + \mathbf{e}_3 + \mathbf{e}_2 + \mathbf{e}_{231} + \mathbf{e}_3 - \mathbf{e}_1 + \mathbf{e}_{312} - \mathbf{e}_{31231})$$

$$\mathbf{p}' = \mathbf{e}_2 + \mathbf{e}_3.$$

the rotation is clockwise about $\mathbf{e}_2$.

Figure 17.15 shows another scenario where vector $\mathbf{p}$ is rotated 120° about the bivector $\mathbf{B}$, where

$$\mathbf{m} = \mathbf{e}_1 - \mathbf{e}_3$$

$$\mathbf{n} = \mathbf{e}_2 - \mathbf{e}_3$$

$$\alpha = 120°$$

$$\mathbf{p} = \mathbf{e}_2 + \mathbf{e}_3$$

$$\mathbf{B} = \mathbf{m} \wedge \mathbf{n}$$

$$= (\mathbf{e}_1 - \mathbf{e}_3) \wedge (\mathbf{e}_2 - \mathbf{e}_3)$$

$$\mathbf{B} = \mathbf{e}_{12} + \mathbf{e}_{31} + \mathbf{e}_{23}.$$

Next, we normalise $\mathbf{B}$ to $\hat{\mathbf{B}}$:

$$\hat{\mathbf{B}} = \frac{1}{\sqrt{3}}(\mathbf{e}_{12} + \mathbf{e}_{23} + \mathbf{e}_{31})$$

therefore,

$$\mathbf{p}' = (\cos 60° - \hat{\mathbf{B}} \sin 60°)\mathbf{p}(\cos 60° + \hat{\mathbf{B}} \sin 60°)$$

$$= \left(\tfrac{1}{2} - \tfrac{1}{\sqrt{3}}(\mathbf{e}_{12} + \mathbf{e}_{23} + \mathbf{e}_{31})\tfrac{\sqrt{3}}{2}\right)(\mathbf{e}_2 + \mathbf{e}_3)\left(\tfrac{1}{2} + \tfrac{1}{\sqrt{3}}(\mathbf{e}_{12} + \mathbf{e}_{23} + \mathbf{e}_{31})\tfrac{\sqrt{3}}{2}\right)$$

$$= \left(\tfrac{1}{2} - \tfrac{\mathbf{e}_{12}}{2} - \tfrac{\mathbf{e}_{23}}{2} - \tfrac{\mathbf{e}_{31}}{2}\right)(\mathbf{e}_2 + \mathbf{e}_3)\left(\tfrac{1}{2} + \tfrac{\mathbf{e}_{12}}{2} + \tfrac{\mathbf{e}_{23}}{2} + \tfrac{\mathbf{e}_{31}}{2}\right)$$

$$= \tfrac{1}{4}(\mathbf{e}_2 + \mathbf{e}_3 - \mathbf{e}_1 - \mathbf{e}_{123} + \mathbf{e}_3 - \mathbf{e}_2 - \mathbf{e}_{312} + \mathbf{e}_1)(1 + \mathbf{e}_{12} + \mathbf{e}_{23} + \mathbf{e}_{31})$$

$$= \tfrac{1}{2}(\mathbf{e}_3 - \mathbf{e}_{123})(1 + \mathbf{e}_{12} + \mathbf{e}_{23} + \mathbf{e}_{31})$$

$$= \tfrac{1}{2}(\mathbf{e}_3 + \mathbf{e}_{312} - \mathbf{e}_2 + \mathbf{e}_1 - \mathbf{e}_{123} - \mathbf{e}_{12312} - \mathbf{e}_{12323} - \mathbf{e}_{12331})$$

$$= \tfrac{1}{2}(\mathbf{e}_3 - \mathbf{e}_2 + \mathbf{e}_1 + \mathbf{e}_3 + \mathbf{e}_1 + \mathbf{e}_2)$$

$$\mathbf{p}' = \mathbf{e}_1 + \mathbf{e}_3.$$

These examples show that rotors behave just like quaternions. Rotors not only rotate vectors, but they can be used to rotate bivectors, and even trivectors, although, as one might expect, a rotated trivector remains unaltered in 3D.

17.21 Applied Geometric Algebra

This has been a very brief introduction to geometric algebra, and it has been impossible to cover all the algebra's features. However, if you have understood the above topics, you will have understood some of the fundamental ideas behind the algebra. Let's now consider some practical applications for geometric algebra.

The sine rule states that for any triangle $\triangle ABC$ with angles α, β and θ, and respective opposite sides a, b and c, then

$$\frac{a}{\sin \alpha} = \frac{b}{\sin \beta} = \frac{c}{\sin \theta}.$$

This rule can be proved using the outer product of two vectors, which we know incorporates the sine of the angle between two vectors:

$$\|\mathbf{a} \wedge \mathbf{b}\| = \|\mathbf{a}\|\|\mathbf{b}\| \sin \alpha.$$

With reference to Fig. 17.16, we can state the triangle's area as

$$\text{area of } \triangle \text{ABC} = \tfrac{1}{2}\| -\mathbf{c} \wedge \mathbf{a}\| = \tfrac{1}{2}\|\mathbf{c}\|\|\mathbf{a}\| \sin \beta$$

$$\text{area of } \triangle \text{BCA} = \tfrac{1}{2}\| -\mathbf{a} \wedge \mathbf{b}\| = \tfrac{1}{2}\|\mathbf{a}\|\|\mathbf{b}\| \sin \theta$$

$$\text{area of } \triangle \text{CAB} = \tfrac{1}{2}\| -\mathbf{b} \wedge \mathbf{c}\| = \tfrac{1}{2}\|\mathbf{b}\|\|\mathbf{c}\| \sin \alpha$$

Fig. 17.16 The sine rule

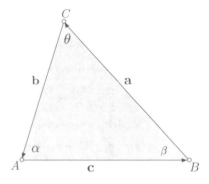

Fig. 17.17 The cosine rule

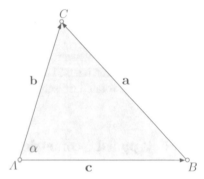

which means that

$$\|\mathbf{c}\|\,\|\mathbf{a}\|\sin\beta = \|\mathbf{a}\|\,\|\mathbf{b}\|\sin\theta = \|\mathbf{b}\|\,\|\mathbf{c}\|\sin\alpha$$

$$\frac{\|\mathbf{a}\|}{\sin\alpha} = \frac{\|\mathbf{b}\|}{\sin\beta} = \frac{\|\mathbf{c}\|}{\sin\theta}.$$

The cosine rule states that for any triangle $\triangle ABC$ with sides a, b and c, then

$$a^2 = b^2 + c^2 - 2bc\cos\alpha$$

where α is the angle between b and c.

Although this is an easy rule to prove using simple trigonometry, the geometric algebra solution is even easier.

Figure 17.17 shows a triangle $\triangle ABC$ constructed from vectors $\mathbf{a}$, $\mathbf{b}$ and $\mathbf{c}$. From Fig. 17.17

$$\mathbf{a} = \mathbf{b} - \mathbf{c}. \tag{17.27}$$

Squaring (17.27) we obtain

$$\mathbf{a}^2 = \mathbf{b}^2 + \mathbf{c}^2 - (\mathbf{b}\mathbf{c} + \mathbf{c}\mathbf{b}).$$

Fig. 17.18 A point P perpendicular to a point T on a line

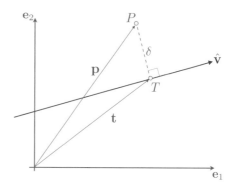

But

$$\mathbf{bc} + \mathbf{cb} = 2\mathbf{b} \cdot \mathbf{c} = 2\|\mathbf{b}\|\|\mathbf{c}\| \cos \alpha$$

therefore,

$$\|\mathbf{a}\|^2 = \|\mathbf{b}\|^2 + \|\mathbf{c}\|^2 - 2\|\mathbf{b}\|\|\mathbf{c}\| \cos \alpha.$$

Figure 17.18 shows a scenario where a line with direction vector $\hat{\mathbf{v}}$ passes through a point T. The objective is to locate another point P perpendicular to $\hat{\mathbf{v}}$ and a distance δ from T. The solution is found by post-multiplying $\hat{\mathbf{v}}$ by the psuedoscalar $\mathbf{e}_{12}$, which rotates $\hat{\mathbf{v}}$ through an angle of $90°$.

As $\hat{\mathbf{v}}$ is a unit vector

$$\overrightarrow{TP} = \delta\hat{\mathbf{v}}\mathbf{e}_{12}$$

therefore,

$$\mathbf{p} = \mathbf{t} + \overrightarrow{TP}$$

and

$$\mathbf{p} = \mathbf{t} + \delta\hat{\mathbf{v}}\mathbf{e}_{12}. \qquad (17.28)$$

For example, Fig. 17.19 shows a 2D scenario where

$$\hat{\mathbf{v}} = \frac{1}{\sqrt{2}}(\mathbf{e}_1 + \mathbf{e}_2)$$
$$T = (4, \ 1)$$
$$\mathbf{t} = 4\mathbf{e}_1 + \mathbf{e}_2$$
$$\delta = \sqrt{32}.$$

Using (17.28)

Fig. 17.19 A point P
perpendicular to a point T on
a line

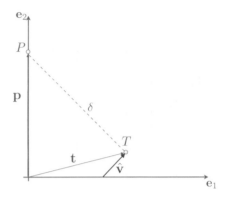

Fig. 17.20 Reflecting a
vector about a vector

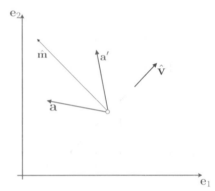

$$\mathbf{p} = \mathbf{t} + \delta\hat{\mathbf{v}}\mathbf{e}_{12}$$
$$- 4\mathbf{e}_1 + \mathbf{e}_2 \mid \sqrt{32} \frac{1}{\sqrt{2}}(\mathbf{e}_1 + \mathbf{e}_2)\mathbf{e}_{12}$$
$$= 4\mathbf{e}_1 + \mathbf{e}_2 + 4\mathbf{e}_2 - 4\mathbf{e}_1$$
$$\mathbf{p} = 5\mathbf{e}_2$$

and

$$P = (0, \ 5).$$

If $\mathbf{p}$ is required on the other side of the line, we pre-multiply $\hat{\mathbf{v}}$ by $\mathbf{e}_{12}$:

$$\mathbf{p} = \mathbf{t} + \delta\mathbf{e}_{12}\hat{\mathbf{v}}$$

which is the same as reversing the sign of δ.

Reflecting a vector about another vector happens to be a rather easy problem for geometric algebra. Figure 17.20 shows the scenario where we see a vector $\mathbf{a}$ reflected about the normal to a line with direction vector $\hat{\mathbf{v}}$.

We begin by calculating $\hat{\mathbf{m}}$:

$$\hat{\mathbf{m}} = \hat{\mathbf{v}}\mathbf{e}_{12} \tag{17.29}$$

Fig. 17.21 Reflecting a
vector about a vector

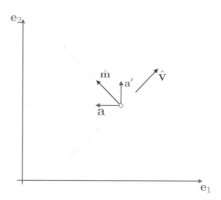

then reflecting **a** about $\hat{\mathbf{m}}$:

$$\mathbf{a}' = \hat{\mathbf{m}}\mathbf{a}\hat{\mathbf{m}}$$

substituting $\hat{\mathbf{m}}$ we have

$$\mathbf{a}' = \hat{\mathbf{v}}\mathbf{e}_{12}\mathbf{a}\hat{\mathbf{v}}\mathbf{e}_{12}. \tag{17.30}$$

As an illustration, consider the scenario shown in Fig. 17.21 where

$$\hat{\mathbf{v}} = \tfrac{1}{\sqrt{2}}(\mathbf{e}_1 + \mathbf{e}_2)$$

$$\mathbf{a} = -\mathbf{e}_1.$$

Therefore, using (17.29)

$$\hat{\mathbf{m}} = \tfrac{1}{\sqrt{2}}(\mathbf{e}_1 + \mathbf{e}_2)\mathbf{e}_{12}$$

$$\hat{\mathbf{m}} = \tfrac{1}{\sqrt{2}}(\mathbf{e}_2 - \mathbf{e}_1)$$

and using (17.30)

$$\mathbf{a}' = \tfrac{1}{\sqrt{2}}(\mathbf{e}_2 - \mathbf{e}_1)(-\mathbf{e}_1)\tfrac{1}{\sqrt{2}}(\mathbf{e}_2 - \mathbf{e}_1)$$

$$= \tfrac{1}{2}(\mathbf{e}_{12} + 1)(\mathbf{e}_2 - \mathbf{e}_1)$$

$$= \tfrac{1}{2}(\mathbf{e}_1 + \mathbf{e}_2 + \mathbf{e}_2 - \mathbf{e}_1)$$

$$\mathbf{a}' = \mathbf{e}_2.$$

In computer graphics we often need to test whether a point is above, below or
on a planar surface. If we already have the plane equation for the surface it is just
a question of substituting the test point in the equation and investigating its signed
value. But here is another way using geometric algebra. For example, if a bivector
is used to represent the orientation of a plane, the outer product of the test point's

Fig. 17.22 Point relative to
a bivector

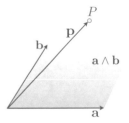

Fig. 17.23 Three points
relative to a bivector

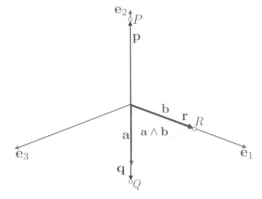

position vector with the bivector computes an oriented volume. Figure 17.22 shows
a bivector $\mathbf{a} \wedge \mathbf{b}$ and a test point P with position vector $\mathbf{p}$ relative to the bivector.
    Let

$\mathbf{a} \wedge \mathbf{b} \wedge \mathbf{p}$ is +ve, then P is 'above' the bivector
$\mathbf{a} \wedge \mathbf{b} \wedge \mathbf{p}$ is -ve, then P is 'below' the bivector
$\mathbf{a} \wedge \mathbf{b} \wedge \mathbf{p}$ is zero, then P is coplanar with the bivector.

    The terms 'above' and 'below' mean in the bivector's positive and negative half-
space respectively.
    As an example, consider the scenario shown in Fig. 17.23 where the plane's ori-
entation is represented by the bivector $\mathbf{a} \wedge \mathbf{b}$, and three test points P, Q and R.
    If $P = (0,\ 1,\ 0)$, $Q = (0,\ -1,\ 0)$, $R = (1,\ 0,\ 0)$

$$\mathbf{a} = \mathbf{e}_1 + \mathbf{e}_3$$
$$\mathbf{b} = \mathbf{e}_1$$

then

$$\mathbf{p} = \mathbf{e}_2$$
$$\mathbf{q} = -\mathbf{e}_2$$
$$\mathbf{r} = \mathbf{e}_1$$

and

$$\mathbf{a} \wedge \mathbf{b} \wedge \mathbf{p} = (\mathbf{e}_1 + \mathbf{e}_3) \wedge \mathbf{e}_1 \wedge \mathbf{e}_2$$
$$= \mathbf{e}_{123}$$
$$\mathbf{a} \wedge \mathbf{b} \wedge \mathbf{q} = (\mathbf{e}_1 + \mathbf{e}_3) \wedge \mathbf{e}_1 \wedge (-\mathbf{e}_2)$$
$$= -\mathbf{e}_{123}$$
$$\mathbf{a} \wedge \mathbf{b} \wedge \mathbf{r} = (\mathbf{e}_1 + \mathbf{e}_3) \wedge \mathbf{e}_1 \wedge \mathbf{e}_1$$
$$= 0.$$

We can see that the signs of the first two volumes show that P is in the positive half-space, Q is in the negative half-space, and R is on the plane.

17.22 Summary

Geometric algebra is a new and exciting subject and is destined to impact upon the way we solve problems in computer games and animation. Hopefully, you have found this chapter interesting, and if you are tempted to take the subject further, then look at the author's books.

References

1. Vince JA (2008) Geometric algebra for computer graphics. Springer
2. Vince JA (2009) Geometric algebra: an algebraic system for computer games and animation. Springer

Chapter 18
Calculus: Derivatives

18.1 Introduction

Calculus is a very large subject, and calculus books have a reputation for being heavy. Therefore, to minimise this book's weight, and provide a gentle introduction to the subject, I have selected specific topics from my book [1], and condensed them into two chapters.

One branch of calculus is concerned with a function's *derivative*, which describes how fast a function changes relative to its independent variable. In this chapter, I show how *limits* are used in this process. We begin with some historical background, and then look at small numerical quantities, and how they can be ignored if they occur in certain products, but remain important in quotients.

18.2 Background

Over a period of 350 years or more, calculus has evolved conceptually and in notation. Up until recently, calculus was described using *infinitesimals*, which are numbers so small, they can be ignored in certain products. However, it was Cauchy and the German mathematician Karl Weierstrass (1815–1897), who showed how limits can replace infinitesimals.

18.3 Small Numerical Quantities

The adjective *small* is a relative term, and requires clarification in the context of numbers. For example, if numbers are in the hundreds, and also contain some decimal component, then it seems reasonable to ignore digits after the 3rd decimal place for any quick calculation. For instance,

© Springer-Verlag London Ltd., part of Springer Nature 2022
J. Vince, *Mathematics for Computer Graphics*, Undergraduate Topics
in Computer Science, https://doi.org/10.1007/978-1-4471-7520-9_18

$$100.000003 \times 200.000006 \approx 20,000$$

and ignoring the decimal part has no significant impact on the general accuracy of the answer, which is measured in tens of thousands.

To develop an algebraic basis for this argument let's divide a number into two parts: a primary part x, and some very small secondary part δx (pronounced *delta x*). In one of the above numbers, $x = 100$ and $\delta x = 0.000003$. Given two such numbers, x_1 and y_1, their product is given by

$$x_1 = x + \delta x$$
$$y_1 = y + \delta y$$
$$x_1 y_1 = (x + \delta x)(y + \delta y)$$
$$= xy + x \cdot \delta y + y \cdot \delta x + \delta x \cdot \delta y.$$

Using $x_1 = 100.000003$ and $y_1 = 200.000006$ we have

$$x_1 y_1 = 100 \times 200 + 100 \times 0.000006 + 200 \times 0.000003 + 0.000003 \times 0.000006$$
$$= 20,000 + 0.0006 + 0.0006 + 0.00000000018$$
$$= 20,000 + 0.0012 + 0.00000000018$$
$$= 20,000.00120000018$$

where it is clear that the products $x \cdot \delta y$, $y \cdot \delta x$ and $\delta x \cdot \delta y$ contribute very little to the result. Furthermore, the smaller we make δx and δy, their contribution becomes even more insignificant. Just imagine if we reduce δx and δy to the level of quantum phenomenon, e.g. 10^{-34}, then their products play no part in every-day numbers. But there is no need to stop there, we can make δx and δy as small as we like, e.g. $10^{-100,000,000,000}$. Later on we employ the device of reducing a number towards zero, such that any products involving them can be dropped from any calculation.

Even though the product of two numbers less than one is an even smaller number, care must be taken with their quotients. For example, in the above scenario, where $\delta y = 0.000006$ and $\delta x = 0.000003$,

$$\frac{\delta y}{\delta x} = \frac{0.000006}{0.000003} = 2$$

so we must watch out for such quotients.

From now on I will employ the term *derivative* to describe a function's rate of change relative to its independent variable. I will now describe two ways of computing a derivative, and provide a graphical interpretation of the process. The first way uses simple algebraic equations, and the second way uses a functional representation. Needless to say, they both give the same result.

18.4 Equations and Limits

18.4.1 Quadratic Function

Here is a simple algebraic approach using limits to compute the derivative of a quadratic function. Starting with the function $y = x^2$, let x change by δx, and let δy be the corresponding change in y. We then have

$$y = x^2$$
$$y + \delta y = (x + \delta x)^2$$
$$= x^2 + 2x \cdot \delta x + (\delta x)^2$$
$$\delta y = 2x \cdot \delta x + (\delta x)^2.$$

Dividing throughout by δx we have

$$\frac{\delta y}{\delta x} = 2x + \delta x.$$

The ratio $\delta y / \delta x$ provides a measure of how fast y changes relative to x, in increments of δx. For example, when $x = 10$

$$\frac{\delta y}{\delta x} = 20 + \delta x,$$

and if $\delta x = 1$, then $\delta y / \delta x = 21$. Equally, if $\delta x = 0.001$, then $\delta y / \delta x = 20.001$. By making δx smaller and smaller, δy becomes equally smaller, and their ratio converges towards a limiting value of 20.

In this case, as δx approaches zero, $\delta y / \delta x$ approaches $2x$, which is written

$$\lim_{\delta x \to 0} \frac{\delta y}{\delta x} = 2x.$$

Thus in the limit, when $\delta x = 0$, we create a condition where δy is divided by zero—which is a meaningless operation. However, if we hold onto the idea of a limit, as $\delta x \to 0$, it is obvious that the quotient $\delta y / \delta x$ is converging towards $2x$. The subterfuge employed to avoid dividing by zero is to substitute another quotient dy/dx to stand for the limiting condition:

$$\frac{dy}{dx} = \lim_{\delta x \to 0} \frac{\delta y}{\delta x} = 2x.$$

dy/dx (pronounced *dee y dee x*) is the derivative of $y = x^2$, i.e. $2x$. For instance, when $x = 0$, $dy/dx = 0$, and when $x = 3$, $dy/dx = 6$. The derivative dy/dx, is the instantaneous rate at which y changes relative to x.

If we had represented this equation as a function:

$$f(x) = x^2$$
$$f'(x) = 2x$$

where $f'(x)$ is another way of expressing dy/dx.

Now let's introduce two constants into the original quadratic equation to see what effect, if any, they have on the derivative. We begin with

$$y = ax^2 + b$$

and increment x and y:

$$y + \delta y = a(x + \delta x)^2 + b$$
$$= a\left(x^2 + 2x \cdot \delta x + (\delta x)^2\right) + b$$
$$\delta y = a\left(2x \cdot \delta x + (\delta x)^2\right).$$

Dividing throughout by δx:

$$\frac{\delta y}{\delta x} = a(2x + \delta x)$$

and the derivative is

$$\frac{dy}{dx} = \lim_{\delta x \to 0} \frac{\delta y}{\delta x} = 2ax.$$

Thus we see the added constant b disappears (i.e. because it does not change), whilst the multiplied constant a is transmitted through to the derivative.

18.4.2   Cubic Equation

Now let's repeat the above analysis for $y = x^3$:

$$y = x^3$$
$$y + \delta y = (x + \delta x)^3$$
$$= x^3 + 3x^2 \cdot \delta x + 3x(\delta x)^2 + (\delta x)^3$$
$$\delta y = 3x^2 \cdot \delta x + 3x(\delta x)^2 + (\delta x)^3.$$

Dividing throughout by δx:

$$\frac{\delta y}{\delta x} = 3x^2 + 3x \cdot \delta x + (\delta x)^2.$$

Employing the idea of infinitesimals, one would argue that any term involving δx can be ignored, because its numerical value is too small to make any contribution to the result. Similarly, using the idea of limits, one would argue that as δx is made increasingly smaller, towards zero, any term involving δx rapidly disappears.

Using limits, we have

$$\lim_{\delta x \to 0} \frac{\delta y}{\delta x} = 3x^2$$

or

$$\frac{dy}{dx} = \lim_{\delta x \to 0} \frac{\delta y}{\delta x} = 3x^2.$$

We could also show that if $y = ax^3 + b$ then

$$\frac{dy}{dx} = 3ax^2.$$

This incremental technique can be used to compute the derivative of all sorts of functions.

If we continue computing the derivatives of higher-order polynomials, we discover the following pattern:

$$y = x^2, \quad \frac{dy}{dx} = 2x$$

$$y = x^3, \quad \frac{dy}{dx} = 3x^2$$

$$y = x^4, \quad \frac{dy}{dx} = 4x^3$$

$$y = x^5, \quad \frac{dy}{dx} = 5x^4.$$

Clearly, the rule is

$$y = x^n, \quad \frac{dy}{dx} = nx^{n-1}$$

but we need to prove why this is so. The solution is found in the binomial expansion for $(x + \delta x)^n$, which can be divided into three components:

1. Decreasing terms of x.
2. Increasing terms of δx.
3. The terms of Pascal's triangle.

For example, the individual terms of $(x + \delta x)^4$ are:

Decreasing terms of x:	x^4	x^3	x^2	x^1	x^0
Increasing terms of δx:	$(\delta x)^0$	$(\delta x)^1$	$(\delta x)^2$	$(\delta x)^3$	$(\delta x)^4$
The terms of Pascal's triangle:	1	4	6	4	1

which when combined produce

$$x^4 + 4x^3(\delta x) + 6x^2(\delta x)^2 + 4x(\delta x)^3 + (\delta x)^4.$$

Thus when we begin an incremental analysis:

$$y = x^4$$
$$y + \delta y = (x + \delta x)^4$$
$$= x^4 + 4x^3(\delta x) + 6x^2(\delta x)^2 + 4x(\delta x)^3 + (\delta x)^4$$
$$\delta y = 4x^3(\delta x) + 6x^2(\delta x)^2 + 4x(\delta x)^3 + (\delta x)^4.$$

Dividing throughout by δx:

$$\frac{\delta y}{\delta x} = 4x^3 + 6x^2(\delta x)^1 + 4x(\delta x)^2 + (\delta x)^3.$$

In the limit, as δx slides to zero, only the second term of the original binomial expansion remains:

$$4x^3.$$

The second term of the binomial expansion $(x + \delta x)^n$ is always of the form

$$nx^{n-1}$$

which is the proof we require.

18.4.3   Functions and Limits

In order to generalise the above findings, let's approach the above analysis using a function of the form $y = f(x)$. We begin by noting some arbitrary value of its independent variable and note the function's value. In general terms, this is x and $f(x)$ respectively. We then increase x by a small amount δx, to give $x + \delta x$, and measure the function's value again: $f(x + \delta x)$. The function's change in value is $f(x + \delta x) - f(x)$, whilst the change in the independent variable is δx. The quotient of these two quantities approximates to the function's rate of change at x:

$$\frac{f(x + \delta x) - f(x)}{\delta x}. \tag{18.1}$$

By making δx smaller and smaller towards zero, (18.1) converges towards a limiting value expressed as

$$\frac{dy}{dx} = \lim_{\delta x \to 0} \frac{f(x + \delta x) - f(x)}{\delta x} \tag{18.2}$$

which can be used to compute all sorts of functions. For example, to compute the derivative of $\sin x$ we proceed as follows:

$$y = \sin x$$
$$y + \delta y = \sin(x + \delta x).$$

Using the identity $\sin(A + B) = \sin A \cos B + \cos A \sin B$, we have

$$y + \delta y = \sin x \cos(\delta x) + \cos x \sin(\delta x)$$
$$\delta y = \sin x \cos(\delta x) + \cos x \sin(\delta x) - \sin x$$
$$= \sin x (\cos(\delta x) - 1) + \cos x \sin(\delta x).$$

Dividing throughout by δx we have

$$\frac{\delta y}{\delta x} = \frac{\sin x}{\delta x} (\cos(\delta x) - 1) + \frac{\sin(\delta x)}{\delta x} \cos x.$$

In the limit as $\delta x \rightarrow 0$, $(\cos(\delta x) - 1) \rightarrow 0$ and $\sin(\delta x)/\delta x = 1$, and

$$\frac{dy}{dx} = \frac{d(\sin x)}{dx} = \cos x.$$

Before moving on, let's compute the derivative of $\cos x$.

$$y = \cos x$$
$$y + \delta y = \cos(x + \delta x).$$

Using the identity $\cos(A + B) = \cos A \cos B - \sin A \sin B$, we have

$$y + \delta y = \cos x \cos(\delta x) - \sin x \sin(\delta x)$$
$$\delta y = \cos x \cos(\delta x) - \sin x \sin(\delta x) - \cos x$$
$$= \cos x (\cos(\delta x) - 1) - \sin x \sin(\delta x).$$

Dividing throughout by δx we have

$$\frac{\delta y}{\delta x} = \frac{\cos x}{\delta x} (\cos(\delta x) - 1) - \frac{\sin(\delta x)}{\delta x} \sin x.$$

In the limit as $\delta x \rightarrow 0$, $(\cos(\delta x) - 1) \rightarrow 0$ and $\sin(\delta x)/\delta x = 1$ (see Appendix A) and

$$\frac{dy}{dx} = \frac{d(\cos x)}{dx} = - \sin x.$$

We will continue to employ this strategy to compute the derivatives of other functions later on.

18.4.4   *Graphical Interpretation of the Derivative*

To illustrate this limiting process graphically, consider the scenario in Fig. 18.1 where the sample point is P. In this case the function is $f(x) = x^2$ and P's coordinates are $(x,\ x^2)$. We identify another point R, displaced δx to the right of P, with coordinates $(x + \delta x,\ x^2)$. The point Q on the curve, vertically above R, has coordinates $(x + \delta x,\ (x + \delta x)^2)$. When δx is relatively small, the slope of the line PQ approximates to the function's rate of change at P, which is the graph's slope. This is given by

$$\text{slope} = \frac{QR}{PR} = \frac{(x + \delta x)^2 - x^2}{\delta x}$$

$$= \frac{x^2 + 2x(\delta x) + (\delta x)^2 - x^2}{\delta x}$$

$$= \frac{2x(\delta x) + (\delta x)^2}{\delta x}$$

$$= 2x + \delta x.$$

We can now reason that as δx is made smaller and smaller, Q approaches P, and *slope* becomes the graph's slope at P. This is the *limiting* condition:

$$\frac{dy}{dx} = \lim_{\delta x \to 0} (2x + \delta x) = 2x.$$

Thus, for any point with coordinates $(x,\ x^2)$, the slope is given by $2x$. For example, when $x = 0$, the slope is 0, and when $x = 4$, the slope is 8, etc.

Fig. 18.1 Sketch of $f(x) = x^2$

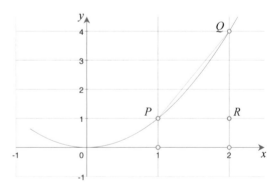

18.4.5 Derivatives and Differentials

Given a function $f(x)$, the ratio df/dx represents the instantaneous change of f for some x, and is called the *first derivative* of $f(x)$. For linear functions, this is constant, for other functions, the derivative's value changes with x and is represented by a function.

The elements df, dy and dx are called *differentials*, and historically, the derivative used to be called the *differential coefficient*, but has now been dropped in favour of *derivative*. One can see how the idea of a differential coefficient arose if we write, for example:

$$\frac{dy}{dx} = 3x$$

as

$$dy = 3x\ dx.$$

In this case, $3x$ acts like a coefficient of dx, nevertheless, we will use the word *derivative*. It is worth noting that if $y = x$, then $dy/dx = 1$, or $dy = dx$. The two differentials are individual algebraic quantities, which permits us to write statements such as

$$\frac{dy}{dx} = 3x, \qquad dy = 3x\ dx, \qquad dx = \frac{dy}{3x}.$$

Now let's find dy/dx, for

$$y = 6x^3 - 4x^2 + 8x + 6.$$

Differentiating y:

$$\frac{dy}{dx} = 18x^2 - 8x + 8$$

which is the instantaneous change of y relative to x. When $x = 1$, $dy/dx = 18 - 8 + 8 = 18$, which means that y is changing 18 times faster than x. Consequently, $dx/dy = 1/18$.

18.4.6 Integration and Antiderivatives

If it is possible to differentiate a function, it seems reasonable to assume the existence of an inverse process to convert a derivative back to its associated function. Fortunately, this is the case, but there are some limitations. This inverse process is called *integration* and reveals the *antiderivative* of a function. Many functions can be paired together in the form of a derivative and an antiderivative, such as $2x$ with x^2, and $\cos x$ with $\sin x$. However, there are many functions where it is impossible to derive its antiderivative in a precise form. For example, there is no simple, finite functional antiderivative for $\sin x^2$ or $(\sin x)/x$. To understand integration, let's begin with a simple derivative.

If we are given

$$\frac{dy}{dx} = 18x^2 - 8x + 8$$

it is not too difficult to reason that the original function could have been

$$y = 6x^3 - 4x^2 + 8x.$$

However, it could have also been

$$y = 6x^3 - 4x^2 + 8x + 2$$

or

$$y = 6x^3 - 4x^2 + 8x + 20$$

or with any other constant. Consequently, when integrating the original function, the integration process has to include a constant:

$$y = 6x^3 - 4x^2 + 8x + C.$$

The value of C is not always required, but it can be determined if we are given some extra information, such as $y = 10$ when $x = 0$, then $C = 10$.

The notation for integration employs a curly 'S' symbol $\int$, which may seem strange, but is short for *sum* and will be explained later. So, starting with

$$\frac{dy}{dx} = 18x^2 - 8x + 8$$

we rewrite this as

$$dy = (18x^2 - 8x + 8)dx$$

and integrate both sides, where dy becomes y and the right-hand-side becomes

$$\int \left(18x^2 - 8x + 8\right) dx$$

although brackets are not always used:

$$y = \int 18x^2 - 8x + 8 \, dx.$$

This equation reads: "*y is the integral of* $18x^2 - 8x + 8$ *dee x.*" The dx reminds us that x is the independent variable. In this case we can write the answer:

$$dy = 18x^2 - 8x + 8 \ dx$$

$$y = \int 18x^2 - 8x + 8 \ dx$$

$$= 6x^3 - 4x^2 + 8x + C$$

where C is some constant.

For example, let's find y, given

$$dy = 6x^2 + 10x \ dx.$$

Integrating:

$$y = \int 6x^2 + 10x \ dx$$

$$= 2x^3 + 5x^2 + C.$$

Now let's find y, given

$$dy = dx.$$

Integrating:

$$y = \int 1 \ dx$$

$$= x + C.$$

The antiderivatives for the sine and cosine functions are written:

$$\int \sin x \ dx = -\cos x + C$$

$$\int \cos x \ dx = \sin x + C$$

which you may think obvious, as we have just computed their derivatives. However, the reason for introducing integration alongside differentiation, is to make you familiar with the notation, and memorise the two distinct processes, as well as lay the foundations for the next chapter.

18.5 Function Types

Mathematical functions come in all sorts of shapes and sizes. Sometimes they are described explicitly where y equals some function of its independent variable(s), such as

$$y = x \sin x$$

or implicitly where y, and its independent variable(s) are part of an equation, such as

$$x^2 + y^2 = 10.$$

A function may reference other functions, such as

$$y = \sin\left(\cos^2 x\right)$$

or

$$y = x^{\sin x}.$$

There is no limit to the way functions can be combined, which makes it impossible to cover every eventuality. Nevertheless, we will explore some useful combinations that prepare us for any future surprises.

First, we examine how to differentiate different types of functions, that include sums, products and quotients, which are employed later on to differentiate specific functions such as trigonometric, logarithmic and hyperbolic. Where relevant, I include the appropriate antiderivative to complement its derivative.

18.6 Differentiating Groups of Functions

So far we have only considered simple individual functions, which, unfortunately, do not represent the equations found in mathematics, science, physics or even computer graphics. In general, the functions we have to differentiate include sums of functions, functions of functions, function products and function quotients. Let's explore these four scenarios.

18.6.1 Sums of Functions

A function normally computes a numerical value from its independent variable(s), and if it can be differentiated, its derivative generates another function with the same independent variable. Consequently, if a function contains two functions of x, such as u and v, where

$$y = u(x) + v(x)$$

which can be abbreviated to

$$y = u + v$$

then

Fig. 18.2 Graph of
$y = 2x^6 + \sin x + \cos x$ and
its derivative,
$\frac{dy}{dx} = 12x^5 + \cos x - \sin x$
(dashed)

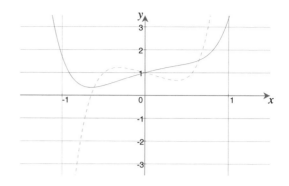

$$\frac{dy}{dx} = \frac{du}{dx} + \frac{dv}{dx}$$

where we just sum their individual derivatives.

As an example, find dy/dx, given

$$u = 2x^6$$
$$v = 3x^5$$
$$y = u + v$$
$$y = 2x^6 + 3x^5.$$

Differentiating y:

$$\frac{dy}{dx} = 12x^5 + 15x^4.$$

Similarly, find dy/dx, given

$$u = 2x^6$$
$$v = \sin x$$
$$w = \cos x$$
$$y = u + v + w$$
$$y = 2x^6 + \sin x + \cos x.$$

Differentiating y:

$$\frac{dy}{dx} = 12x^5 + \cos x - \sin x.$$

Figure 18.2 shows a graph of $y = 2x^6 + \sin x + \cos x$ and its derivative $y = 12x^5 + \cos x - \sin x$. Differentiating such functions is relatively easy, so too, is integrating. Given

$$\frac{dy}{dx} = \frac{du}{dx} + \frac{dv}{dx}$$

then

$$y = \int \frac{du}{dx} \, dx + \int \frac{dv}{dx} \, dx$$
$$= \int \left(\frac{du}{dx} + \frac{dv}{dx} \right) dx.$$

For example, let's find y, given

$$\frac{dy}{dx} = 12x^5 + \cos x - \sin x.$$

Integrating:

$$dy = \left(12x^5 + \cos x - \sin x \right) dx$$
$$y = \int 12x^5 \, dx + \int \cos x \, dx - \int \sin x \, dx$$
$$= 2x^6 + \sin x + \cos x + C.$$

18.6.2 *Function of a Function*

One of the advantages of modern mathematical notation is that it lends itself to unlimited elaboration without introducing any new symbols. For example, the polynomial $3x^2 + 2x$ is easily raised to some power by adding brackets and an appropriate index: $(3x^2 + 2x)^2$. Such an object is a *function of a function*, because the function $3x^2 + 2x$ is subjected to a further squaring function. The question now is: how are such functions differentiated? Well, the answer is relatively easy, but does introduce some new ideas.

Imagine that Heidi swims twice as fast as John, who in turn, swims three times as fast as his dog, Monty. It should be obvious that Heidi swims six (2 × 3) times faster than Monty. This product rule, also applies to derivatives, because if y changes twice as fast as u, i.e. $dy/du = 2$, and u changes three times as fast as x, i.e. $du/dx = 3$, then y changes six times as fast as x:

$$\frac{dy}{dx} = \frac{dy}{du} \cdot \frac{du}{dx}.$$

To differentiate

$$y = \left(3x^2 + 2x \right)^2$$

we substitute

$$u = 3x^2 + 2x$$

then

$$y = u^2$$

and

$$\frac{dy}{du} = 2u$$
$$= 2\left(3x^2 + 2x\right)$$
$$= 6x^2 + 4x.$$

Next, we require du/dx:

$$u = 3x^2 + 2x$$
$$\frac{du}{dx} = 6x + 2$$

therefore, we can write

$$\frac{dy}{dx} = \frac{dy}{du} \cdot \frac{du}{dx}$$
$$= \left(6x^2 + 4x\right)(6x + 2)$$
$$= 36x^3 + 36x^2 + 8x.$$

This result is easily verified by expanding the original polynomial and differentiating:

$$y = \left(3x^2 + 2x\right)^2$$
$$= \left(3x^2 + 2x\right)\left(3x^2 + 2x\right)$$
$$= 9x^4 + 12x^3 + 4x^2$$
$$\frac{dy}{dx} = 36x^3 + 36x^2 + 8x.$$

Figure 18.3 shows a graph of $y = \left(3x^2 + 2x\right)^2$ and its derivative $y = 36x^3 + 36x^2 + 8x$.

Now let's differentiate $y = \sin(ax)$, which is a function of a function.

Substitute u for ax:

Fig. 18.3 Graph of
$y = (3x^2 + 2x)^2$ and its
derivative,
$\frac{dy}{dx} = 36x^3 + 36x^2 + 8x$
(dashed)

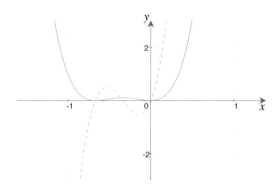

$$y = \sin u$$

$$\frac{dy}{du} = \cos u$$

$$= \cos(ax).$$

Next, we require du/dx:

$$u = ax$$

$$\frac{du}{dx} = a$$

therefore, we can write

$$\frac{dy}{dx} = \frac{dy}{du} \cdot \frac{du}{dx}$$

$$= \cos(ax) \cdot a$$

$$= a \cos(ax).$$

Consequently, given

$$\frac{dy}{dx} = \cos(ax)$$

then

$$dy = \cos(ax)\, dx$$

$$y = \int \cos(ax)\, dx$$

$$= \frac{1}{a} \sin(ax) + C.$$

Similarly, given

$$\frac{dy}{dx} = \sin(ax)$$

then

$$dy = \sin(ax)\,dx$$

$$y = \int \sin(ax)\,dx$$

$$= -\frac{1}{a}\cos(ax) + C.$$

To differentiate $y = \sin\left(x^2\right)$, which is also a function of a function, we substitute u for x^2:

$$y = \sin u$$

$$\frac{dy}{du} = \cos u$$

$$= \cos\left(x^2\right).$$

Next, we require du/dx:

$$u = x^2$$

$$\frac{du}{dx} = 2x$$

therefore, we can write

$$\frac{dy}{dx} = \frac{dy}{du} \cdot \frac{du}{dx}$$

$$= \cos\left(x^2\right) \cdot 2x$$

$$= 2x\cos\left(x^2\right).$$

Figure 18.4 shows a graph of $y = \sin\left(x^2\right)$ and its derivative $y = 2x\cos\left(x^2\right)$. In general, there can be any depth of functions within a function, which permits us to write the *chain rule* for derivatives:

$$\frac{dy}{dx} = \frac{dy}{du} \cdot \frac{du}{dv} \cdot \frac{dv}{dw} \cdot \frac{dw}{dx}.$$

Fig. 18.4 Graph of
$y = \sin\left(x^2\right)$ and its
derivative, $\frac{dy}{dx} = 2x \cos\left(x^2\right)$
(dashed)

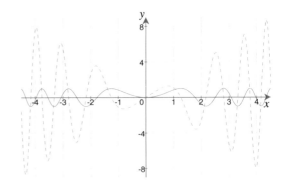

18.6.3   *Function Products*

Function products occur frequently in every-day mathematics, and involve the product of two, or more functions. Here are three simple examples:

$$y = \left(3x^2 + 2x\right)\left(2x^2 + 3x\right)$$
$$y = \sin x \cos x$$
$$y = x^2 \sin x.$$

When it comes to differentiating function products of the form

$$y = uv$$

it seems natural to assume that

$$\frac{dy}{dx} = \frac{du}{dx} \cdot \frac{dv}{dx} \tag{18.3}$$

which unfortunately, is incorrect. For example, in the case of

$$y = \left(3x^2 + 2x\right)\left(2x^2 + 3x\right)$$

differentiating using the above rule (18.3) produces

$$\frac{dy}{dx} = (6x + 2)(4x + 3)$$
$$= 24x^2 + 26x + 6.$$

However, if we expand the original product and then differentiate, we obtain

$$y = (3x^2 + 2x)(2x^2 + 3x)$$
$$= 6x^4 + 13x^3 + 6x^2$$
$$\frac{dy}{dx} = 24x^3 + 39x^2 + 12x$$

which is correct, but differs from the first result. Obviously, (18.3) must be wrong. So let's return to first principles and discover the correct rule.

So far we have incremented the independent variable—normally x—by δx to discover the change in y—normally δy. Next, we see how the same notation can be used to increment functions.

Given the following functions of x, u and v, where

$$y = uv$$

if x increases by δx, then there will be corresponding changes of δu, δv and δy, in u, v and y respectively. Therefore,

$$y + \delta y = (u + \delta u)(v + \delta v)$$
$$= uv + u\delta v + v\delta u + \delta u\delta v$$
$$\delta y = u\delta v + v\delta u + \delta u\delta v.$$

Dividing throughout by δx we have

$$\frac{\delta y}{\delta x} = u\frac{\delta v}{\delta x} + v\frac{\delta u}{\delta x} + \delta u\frac{\delta v}{\delta x}.$$

In the limiting condition:

$$\frac{dy}{dx} = \lim_{\delta x \to 0}\left(u\frac{\delta v}{\delta x}\right) + \lim_{\delta x \to 0}\left(v\frac{\delta u}{\delta x}\right) + \lim_{\delta x \to 0}\left(\delta u\frac{\delta v}{\delta x}\right).$$

As $\delta x \to 0$, then $\delta u \to 0$ and $\left(\delta u\frac{\delta v}{\delta x}\right) \to 0$. Therefore,

$$\frac{dy}{dx} = u\frac{dv}{dx} + v\frac{du}{dx}. \tag{18.4}$$

Applying (18.4) to the original function product:

$$u = 3x^2 + 2x$$
$$v = 2x^2 + 3x$$
$$y = uv$$
$$\frac{du}{dx} = 6x + 2$$

Fig. 18.5 Graph of
$y = (3x^2 + 2x)(2x^2 + 3x)$
and its derivative,
$\frac{dy}{dx} = 24x^3 + 39x^2 + 12x$
(dashed)

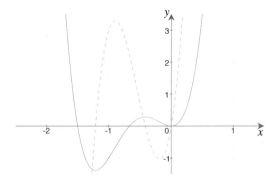

$$\frac{dv}{dx} = 4x + 3$$

$$\frac{dy}{dx} = u\frac{dv}{dx} + v\frac{du}{dx}$$

$$= (3x^2 + 2x)(4x + 3) + (2x^2 + 3x)(6x + 2)$$

$$= (12x^3 + 17x^2 + 6x) + (12x^3 + 22x^2 + 6x)$$

$$= 24x^3 + 39x^2 + 12x$$

which agrees with our previous prediction. Figure 18.5 shows a graph of $y = (3x^2 + 2x)(2x^2 + 3x)$ and its derivative $y = 24x^3 + 39x^2 + 12x$.

Now let's differentiate $y = \sin x \cos x$ using (18.4).

$$y = \sin x \cos x$$

$$u = \sin x$$

$$\frac{du}{dx} = \cos x$$

$$v = \cos x$$

$$\frac{dv}{dx} = -\sin x$$

$$\frac{dy}{dx} = u\frac{dv}{dx} + v\frac{du}{dx}$$

$$= \sin x(-\sin x) + \cos x \cos x$$

$$= \cos^2 x - \sin^2 x$$

$$= \cos(2x).$$

Using the identity $\sin(2x) = 2\sin x \cos x$, we can rewrite the original function as

$$y = \sin x \cos x$$

$$= \tfrac{1}{2}\sin(2x)$$

Fig. 18.6 Graph of
$y = \sin x \cos x$ and its
derivative, $\frac{dy}{dx} = \cos(2x)$
(dashed)

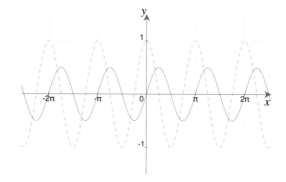

$$\frac{dy}{dx} = \cos(2x)$$

which confirms the above derivative. Now let's consider the antiderivative of $\cos 2x$.
Given

$$\frac{dy}{dx} = \cos(2x)$$

then

$$dy = \cos(2x)\, dx$$

$$y = \int \cos(2x)\, dx$$

$$= \tfrac{1}{2} \sin(2x) + C$$

$$= \sin x \cos x + C.$$

Figure 18.6 shows a graph of $y = \sin x \cos$ and its derivative $y = \cos(2x)$.

Let's differentiate $y = x^2 \sin x$, using (18.4):

$$y = x^2 \sin x$$

$$u = x^2$$

$$\frac{du}{dx} = 2x$$

$$v = \sin x$$

$$\frac{dv}{dx} = \cos x$$

$$\frac{dy}{dx} = u\frac{dv}{dx} + v\frac{du}{dx}$$

$$= x^2 \cos x + 2x \sin x.$$

Fig. 18.7 Graph of
$y = x^2 \sin x$ and its
derivative
$y = x^2 \cos x + 2x \sin x$
(dashed)

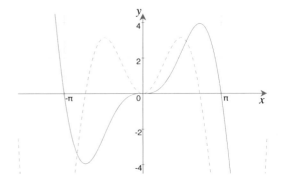

Figure 18.7 shows a graph of $y = x^2 \sin x$ and its derivative $x^2 \cos x + 2x \sin x$.

18.6.4   Function Quotients

Next, we investigate how to differentiate the quotient of two functions. We begin
with two functions of x, u and v, where

$$y = \frac{u}{v}$$

which makes y also a function of x.

We now increment x by δx and measure the change in u as δu, and the change in
v as δv. Consequently, the change in y is δy:

$$y + \delta y = \frac{u + \delta u}{v + \delta v}$$

$$\delta y = \frac{u + \delta u}{v + \delta v} - \frac{u}{v}$$

$$= \frac{v(u + \delta u) - u(v + \delta v)}{v(v + \delta v)}$$

$$= \frac{v\delta u - u\delta v}{v(v + \delta v)}.$$

Dividing throughout by δx we have

$$\frac{\delta y}{\delta x} = \frac{v\dfrac{\delta u}{\delta x} - u\dfrac{\delta v}{\delta x}}{v(v + \delta v)}.$$

As $\delta x \to 0$, δu, δv and δy also tend towards zero, and the limiting conditions are

$$\frac{dy}{dx} = \lim_{\delta x \to 0} \frac{\delta y}{\delta x}$$

$$v\frac{du}{dx} = \lim_{\delta x \to 0} v\frac{\delta u}{\delta x}$$

$$u\frac{dv}{dx} = \lim_{\delta x \to 0} u\frac{\delta v}{\delta x}$$

$$v^2 = \lim_{\delta x \to 0} v(v + \delta v)$$

therefore,

$$\frac{dy}{dx} = \frac{v\dfrac{du}{dx} - u\dfrac{dv}{dx}}{v^2}.$$

To illustrate this, let's differentiate y, given

$$y = \frac{x^3 + 2x^2 + 3x + 6}{x^2 + 3}.$$

Substitute $u = x^3 + 2x^2 + 3x + 6$ and $v = x^2 + 3$, then

$$\frac{du}{dx} = 3x^2 + 4x + 3$$

$$\frac{dv}{dx} = 2x$$

$$\frac{dy}{dx} = \frac{(x^2 + 3)(3x^2 + 4x + 3) - (x^3 + 2x^2 + 3x + 6)2x}{(x^2 + 3)^2}$$

$$= \frac{(3x^4 + 4x^3 + 3x^2 + 9x^2 + 12x + 9) - (2x^4 + 4x^3 + 6x^2 + 12x)}{x^4 + 6x^2 + 9}$$

$$= \frac{x^4 + 6x^2 + 9}{x^4 + 6x^2 + 9}$$

$$= 1$$

which is not a surprising result when one sees that the original function has the factors

$$y = \frac{(x^2 + 3)(x + 2)}{x^2 + 3} = x + 2$$

whose derivative is 1. Figure 18.8 shows a graph of $y = (x^2 + 3)(x + 2)/(x^2 + 3)$ and its derivative $y = 1$.

Fig. 18.8 Graph of $y = (x^2 + 3)(x + 2)/(x^2 + 3)$ and its derivative, $\frac{dy}{dx} = 1$ (dashed)

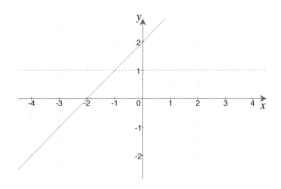

18.7   Differentiating Implicit Functions

Functions conveniently fall into two types: explicit and implicit. An explicit function, describes a function in terms of its independent variable(s), such as

$$y = a \sin x + b \cos x$$

where the value of y is determined by the values of a, b and x. On the other hand, an implicit function, such as

$$x^2 + y^2 = 25$$

combines the function's name with its definition. In this case, it is easy to untangle the explicit form:

$$y = \sqrt{25 - x^2}.$$

So far, we have only considered differentiating explicit functions, so now let's examine how to differentiate implicit functions. Let's begin with a simple explicit function and differentiate it as it is converted into its implicit form.
     Let

$$y = 2x^2 + 3x + 4$$

then

$$\frac{dy}{dx} = 4x + 3.$$

Now let's start the conversion into the implicit form by bringing the constant 4 over to the left-hand side:

$$y - 4 = 2x^2 + 3x$$

differentiating both sides:

$$\frac{dy}{dx} = 4x + 3.$$

Bringing 4 and $3x$ across to the left-hand side:

$$y - 3x - 4 = 2x^2$$

differentiating both sides:

$$\frac{dy}{dx} - 3 = 4x$$

$$\frac{dy}{dx} = 4x + 3.$$

Finally, we have

$$y - 2x^2 - 3x - 4 = 0$$

differentiating both sides:

$$\frac{dy}{dx} - 4x - 3 = 0$$

$$\frac{dy}{dx} = 4x + 3$$

which seems straight forward. The reason for working through this example is to remind us that when y is differentiated we get dy/dx.

Let's find dy/dx, given

$$y + \sin x + 4x = 0.$$

Differentiating the individual terms:

$$y + \sin x + 4x = 0$$

$$\frac{dy}{dx} + \cos x + 4 = 0$$

$$\frac{dy}{dx} = -\cos x - 4.$$

$$y + x^2 - \cos x = 0$$

$$\frac{dy}{dx} + 2x + \sin x = 0$$

$$\frac{dy}{dx} = -2x - \sin x.$$

But how do we differentiate $y^2 + x^2 = r^2$? Well, the important difference between this implicit function and previous functions, is that it involves a function of a function. y is not only a function of x, but is squared, which means that we must employ the chain rule described earlier:

$$\frac{dy}{dx} = \frac{dy}{du} \cdot \frac{du}{dx}.$$

Therefore, given

$$y^2 + x^2 = r^2$$

$$2y\frac{dy}{dx} + 2x = 0$$

$$\frac{dy}{dx} = \frac{-2x}{2y}$$

$$= \frac{-x}{\sqrt{r^2 - x^2}}.$$

This is readily confirmed by expressing the original function in its explicit form and differentiating:

$$y = \left(r^2 - x^2\right)^{\frac{1}{2}}$$

which is a function of a function.

Let $u = r^2 - x^2$, then

$$\frac{du}{dx} = -2x.$$

As $y = u^{\frac{1}{2}}$, then

$$\frac{dy}{du} = \frac{1}{2}u^{-\frac{1}{2}}$$

$$= \frac{1}{2u^{\frac{1}{2}}}$$

$$= \frac{1}{2\sqrt{r^2 - x^2}}.$$

However,

$$\frac{dy}{dx} = \frac{dy}{du} \cdot \frac{du}{dx}$$

$$= \frac{-2x}{2\sqrt{r^2 - x^2}}$$

$$= \frac{-x}{\sqrt{r^2 - x^2}}$$

which agrees with the implicit differentiated form.

As another example, let's find dy/dx, given

$$x^2 - y^2 + 4x = 6y.$$

Differentiating the individual terms:

$$2x - 2y\frac{dy}{dx} + 4 = 6\frac{dy}{dx}.$$

Rearranging the terms, we have

$$2x + 4 = 6\frac{dy}{dx} + 2y\frac{dy}{dx}$$
$$= \frac{dy}{dx}(6 + 2y)$$
$$\frac{dy}{dx} = \frac{2x + 4}{6 + 2y}.$$

If, for example, we have to find the slope of $x^2 - y^2 + 4x = 6y$ at the point $(4, 3)$, then we simply substitute $x = 4$ and $y = 3$ in dy/dx to obtain the answer 1.

Finally, let's find dy/dx, given

$$x^n + y^n = a^n$$
$$nx^{n-1} + ny^{n-1}\frac{dy}{dx} = 0$$
$$\frac{dy}{dx} = -\frac{nx^{n-1}}{ny^{n-1}}$$
$$\frac{dy}{dx} = -\frac{x^{n-1}}{y^{n-1}}.$$

18.8 Differentiating Exponential and Logarithmic Functions

18.8.1 Exponential Functions

Exponential functions have the form $y = a^x$, where the independent variable is the exponent. Such functions are used to describe various forms of growth or decay, from the compound interest law, to the rate at which a cup of tea cools down. One special value of a is 2.718282.., called e, where

$$e = \lim_{n \to \infty} \left(1 + \frac{1}{n}\right)^n.$$

Raising e to the power x:

$$e^x = \lim_{n \to \infty} \left(1 + \frac{1}{n}\right)^{nx}$$

Fig. 18.9 Graphs of $y = e^x$ and $y = e^{-x}$

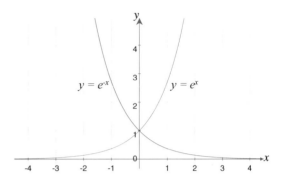

which, using the Binomial Theorem, is

$$e^x = 1 + x + \frac{x^2}{2!} + \frac{x^3}{3!} + \frac{x^4}{4!} + \cdots .$$

If we let

$$y = e^x$$

$$\frac{dy}{dx} = 1 + x + \frac{x^2}{2!} + \frac{x^3}{3!} + \frac{x^4}{4!} + \cdots$$

$$= e^x .$$

which is itself. Figure 18.9 shows graphs of $y = e^x$ and $y = e^{-x}$.

Now let's differentiate $y = a^x$. We know from the rules of logarithms that

$$\log x^n = n \log x$$

therefore, given

$$y = a^x$$

then

$$\ln y = \ln a^x = x \ln a$$

therefore

$$y = e^{x \ln a}$$

which means that

$$a^x = e^{x \ln a} .$$

Consequently,

$$\frac{d}{dx}a^x = \frac{d}{dx}e^{x\ln a}$$
$$= \ln a\ e^{x\ln a}$$
$$= a^x \ln a.$$

Similarly, it can be shown that

$$y = e^{-x}, \quad \frac{dy}{dx} = -e^{-x}$$

$$y = e^{ax}, \quad \frac{dy}{dx} = ae^{ax}$$

$$y = e^{-ax}, \quad \frac{dy}{dx} = -ae^{-ax}$$

$$y = a^x, \quad \frac{dy}{dx} = \ln a\ a^x$$

$$y = a^{-x}, \quad \frac{dy}{dx} = -\ln a\ a^{-x}.$$

The exponential antiderivatives are written:

$$\int e^x\ dx = e^x + C$$

$$\int e^{-x}\ dx = -e^{-x} + C$$

$$\int e^{ax}\ dx = \frac{1}{a}e^{ax} + C$$

$$\int e^{-ax}\ dx = -\frac{1}{a}e^{-ax} + C$$

$$\int a^x\ dx = \frac{1}{\ln a}a^x + C$$

$$\int a^{-x}\ dx = -\frac{1}{\ln a}a^{-x} + C.$$

18.8.2 *Logarithmic Functions*

Given a function of the form

$$y = \ln x$$

then

$$x = e^y.$$

Fig. 18.10 Graph of
$y = \ln x$ and its derivative,
$\frac{dy}{dx} = \frac{1}{x}$ (dashed)

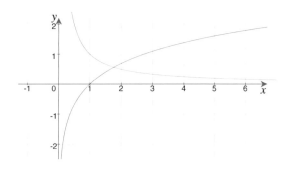

Therefore,

$$\frac{dx}{dy} = e^y$$

$$= x$$

$$\frac{dy}{dx} = \frac{1}{x}.$$

Thus

$$\frac{d}{dx} \ln x = \frac{1}{x}.$$

Figure 18.10 shows the graph of $y = \ln x$ and its derivative $y = 1/x$. Conversely,

$$\int \frac{1}{x}\, dx = \ln |x| + C.$$

When differentiating logarithms to a base a, we employ the conversion formula:

$$y = \log_a x$$

$$= (\ln x)(\log_a e)$$

whose derivative is

$$\frac{dy}{dx} = \frac{1}{x} \log_a e.$$

When $a = 10$, then $\log_{10} e = 0.4343...$ and

$$\frac{d}{dx} \log_{10} x \approx \frac{0.4343}{x}$$

Figure 18.11 shows the graph of $y = \log_{10} x$ and its derivative $y \approx 0.4343/x$.

Fig. 18.11 Graph of $y = \log_{10} x$ and its derivative, $\frac{dy}{dx} \approx \frac{0.4343}{x}$ (dashed)

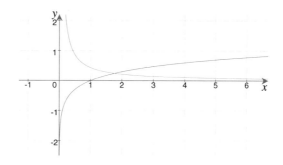

18.9 Differentiating Trigonometric Functions

We have only differentiated two trigonometric functions: $\sin x$ and $\cos x$, so let's add $\tan x$, $\csc x$, $\sec x$ and $\cot x$ to the list, as well as their inverse forms.

18.9.1 Differentiating tan

Rather than return to first principles and start incrementing x by δx, we can employ the rules for differentiating different function combinations and various trigonometric identities. In the case of $\tan(ax)$, this can be written as

$$\tan(ax) = \frac{\sin(ax)}{\cos(ax)}$$

and employ the quotient rule:

$$\frac{dy}{dx} = \frac{v\dfrac{du}{dx} - u\dfrac{dv}{dx}}{v^2}.$$

Therefore, let $u = \sin(ax)$ and $v = \cos(ax)$, and

$$\frac{dy}{dx} = \frac{a\cos(ax)\cos(ax) + a\sin(ax)\sin(ax)}{\cos^2(ax)}$$

$$= \frac{a\left(\cos^2(ax) + \sin^2(ax)\right)}{\cos^2(ax)}$$

$$= \frac{a}{\cos^2(ax)}$$

$$= a\sec^2(ax)$$

$$= a\left(1 + \tan^2(ax)\right).$$

Fig. 18.12 Graph of
$y = \tan x$ and its derivative,
$\frac{dy}{dx} = 1 + \tan^2 x$ (dashed)

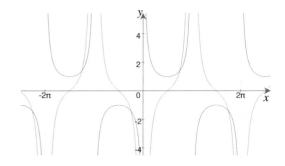

Figure 18.12 shows the graph of $y = \tan x$ and its derivative $y = 1 + \tan^2 x$.

It follows that

$$\int \sec^2(ax)\, dx = \frac{1}{a}\tan(ax) + C.$$

18.9.2   Differentiating csc

Using the quotient rule:

$$y = \csc(ax)$$

$$= \frac{1}{\sin(ax)}$$

$$\frac{dy}{dx} = \frac{0 - a\cos(ax)}{\sin^2(ax)}$$

$$= \frac{-a\cos(ax)}{\sin^2(ax)}$$

$$= -\frac{a}{\sin(ax)} \cdot \frac{\cos(ax)}{\sin(ax)}$$

$$= -a\csc(ax) \cdot \cot(ax).$$

Figure 18.13 shows the graph of $y = \csc x$ and its derivative $y = -\csc x \cot x$.

It follows that

$$\int \csc(ax) \cdot \cot(ax)\, dx = -\frac{1}{a}\csc(ax) + C.$$

Fig. 18.13 Graph of
$y = \csc x$ and its derivative,
$\frac{dy}{dx} = -\csc x \cot x$ (dashed)

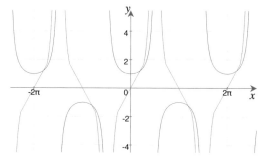

Fig. 18.14 Graph of
$y = \sec x$ and its derivative,
$\frac{dy}{dx} = \sec x \tan x$ (dashed)

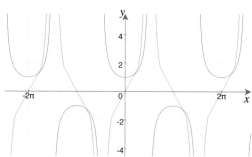

18.9.3 Differentiating sec

Using the quotient rule:

$$y = \sec(ax)$$
$$= \frac{1}{\cos(ax)}$$
$$\frac{dy}{dx} = \frac{-(-a\sin(ax))}{\cos^2(ax)}$$
$$= \frac{a\sin(ax)}{\cos^2(ax)}$$
$$= \frac{a}{\cos(ax)} \cdot \frac{\sin(ax)}{\cos(ax)}$$
$$= a\sec(ax) \cdot \tan(ax).$$

Figure 18.14 shows the graph of $y = \sec x$ and its derivative $y = \sec x \tan x$.

It follows that

$$\int \sec(ax) \cdot \tan(ax)\, dx = \frac{1}{a}\sec(ax) + C.$$

18.9.4 Differentiating cot

Using the quotient rule:

$$y = \cot(ax)$$

$$= \frac{1}{\tan(ax)}$$

$$\frac{dy}{dx} = \frac{-a\sec^2(ax)}{\tan^2(ax)}$$

$$= -\frac{a}{\cos^2(ax)} \cdot \frac{\cos^2(ax)}{\sin^2(ax)}$$

$$= -\frac{a}{\sin^2(ax)}$$

$$= -a\csc^2(ax)$$

$$= -a\left(1 + \cot^2(ax)\right).$$

Figure 18.15 shows the graph of $y = \cot x$ and its derivative $y = -(1 + \cot^2 x)$.

It follows that

$$\int \csc^2(ax)\, dx = -\frac{1}{a}\cot(ax) + C.$$

18.9.5 Differentiating arcsin, arccos and arctan

These inverse functions are solved using a clever strategy.
  Let

$$x = \sin y$$

Fig. 18.15 Graph of
$y = \cot x$ and its derivative,
$\frac{dy}{dx} = -(1 + \cot^2 x)$
(dashed)

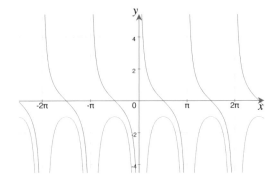

then
$$y = \arcsin x.$$

Differentiating the first expression, we have

$$\frac{dx}{dy} = \cos y$$

$$\frac{dy}{dx} = \frac{1}{\cos y}$$

and as $\sin^2 y + \cos^2 y = 1$, then

$$\cos y = \sqrt{1 - \sin^2 y} = \sqrt{1 - x^2}$$

and

$$\frac{d}{dx} \arcsin x = \frac{1}{\sqrt{1 - x^2}}.$$

Using a similar technique, it can be shown that

$$\frac{d}{dx} \arccos x = -\frac{1}{\sqrt{1 - x^2}}$$

$$\frac{d}{dx} \arctan x = \frac{1}{1 + x^2}.$$

It follows that

$$\int \frac{dx}{\sqrt{1 - x^2}} = \arcsin x + C$$

$$\int \frac{dx}{1 + x^2} = \arctan x + C.$$

18.9.6 Differentiating arccsc, arcsec and arccot

Let
$$y = \text{arccsc } x$$

then

$$x = \csc y$$

$$= \frac{1}{\sin y}$$

$$\frac{dx}{dy} = \frac{-\cos y}{\sin^2 y}$$

$$\frac{dy}{dx} = \frac{-\sin^2 y}{\cos y}$$

$$= -\frac{1}{x^2}\frac{x}{\sqrt{x^2-1}}$$

$$\frac{d}{dx}\operatorname{arccsc} x = -\frac{1}{x\sqrt{x^2-1}}.$$

Similarly,

$$\frac{d}{dx}\operatorname{arcsec} x = \frac{1}{x\sqrt{x^2-1}}$$

$$\frac{d}{dx}\operatorname{arccot} x = -\frac{1}{x^2+1}.$$

It follows:

$$\int \frac{dx}{x\sqrt{x^2-1}} = \operatorname{arcsec}|x| + C$$

$$\int \frac{dx}{x^2+1} = -\operatorname{arccot} x + C.$$

18.10   Differentiating Hyperbolic Functions

Trigonometric functions are useful for parametric, circular motion, whereas hyperbolic functions arise in equations for the absorption of light, mechanics and in integral calculus. Figure 18.16 shows graphs of the unit circle and a hyperbola whose respective equations are

Fig. 18.16 Graphs of the unit circle $x^2 + y^2 = 1$ and the hyperbola $x^2 - y^2 = 1$

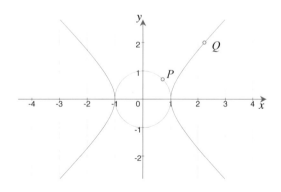

$$x^2 + y^2 = 1$$
$$x^2 - y^2 = 1$$

where the only difference between them is a sign. The parametric form for the trigonometric, or circular functions and the hyperbolic functions are respectively:

$$\sin^2 \theta + \cos^2 \theta = 1$$
$$\cosh^2 x - \sinh^2 x = 1.$$

The three hyperbolic functions have the following definitions:

$$\sinh x = \frac{e^x - e^{-x}}{2}$$
$$\cosh x = \frac{e^x + e^{-x}}{2}$$
$$\tanh x = \frac{\sinh x}{\cosh x} = \frac{e^{2x} - 1}{e^{2x} + 1}$$

and their reciprocals are:

$$\operatorname{cosech} x = \frac{1}{\sinh x} = \frac{2}{e^x - e^{-x}}$$
$$\operatorname{sech} x = \frac{1}{\cosh x} = \frac{2}{e^x + e^{-x}}$$
$$\coth x = \frac{1}{\tanh x} = \frac{e^{2x} + 1}{e^{2x} - 1}.$$

Other useful identities include:

$$\operatorname{sech}^2 x = 1 - \tanh^2 x$$
$$\operatorname{cosech}^2 x = \coth^2 x - 1.$$

The coordinates of P and Q in Fig. 18.16 are given by $P(\cos \theta, \sin \theta)$ and $Q(\cosh x, \sinh x)$. Table 18.1 shows the names of the three hyperbolic functions, their reciprocals and inverse forms. As these functions are based upon e^x and e^{-x}, they are relatively easy to differentiate.

Table 18.1 Hyperbolic function names

Function	Reciprocal	Inverse Function	Inverse Reciprocal
sinh	cosech	arsinh	arcsch
cosh	sech	arcosh	arsech
tanh	coth	artanh	arcoth

18.10.1   *Differentiating sinh, cosh and tanh*

Here are the rules for differentiating hyperbolic functions:

y	dy/dx
$\sinh x$	$\cosh x$
$\cosh x$	$\sinh x$
$\tanh x$	$\text{sech}^2 x$
$\text{cosech}\, x$	$-\text{cosech}\, x \;\coth x$
$\text{sech}\, x$	$-\text{sech}\, x \;\tanh x$
$\coth x$	$-\text{cosech}^2 x$

and the inverse, hyperbolic functions:

y	dy/dx
$\text{arsinh}\, x$	$\dfrac{1}{\sqrt{1+x^2}}$
$\text{arcosh}\, x$	$\dfrac{1}{\sqrt{x^2-1}}$
$\text{artanh}\, x$	$\dfrac{1}{1-x^2}$
$\text{arcsch}\, x$	$-\dfrac{1}{x\sqrt{1+x^2}}$
$\text{arsech}\, x$	$-\dfrac{1}{x\sqrt{1-x^2}}$
$\text{arcoth}\, x$	$-\dfrac{1}{x^2-1}$

Here are the rules for integrating hyperbolic functions:

$f(x)$	$\int f(x)\, dx$
$\sinh x$	$\cosh x + C$
$\cosh x$	$\sinh x + C$
$\text{sech}^2 x$	$\tanh x + C$

and the inverse, hyperbolic functions:

$f(x)$	$\int f(x)\, dx$
$\dfrac{1}{\sqrt{1+x^2}}$	$\text{arsinh}\, x + C$
$\dfrac{1}{\sqrt{x^2-1}}$	$\text{arcosh}\, x + C$
$\dfrac{1}{1-x^2}$	$\text{artanh}\, x + C.$

18.11 Higher Derivatives

There are three parts to this section: The first part shows what happens when a function is repeatedly differentiated; the second shows how these higher derivatives resolve local minimum and maximum conditions; and the third section provides a physical interpretation for these derivatives. Let's begin by finding the higher derivatives of simple polynomials.

18.12 Higher Derivatives of a Polynomial

We have previously seen that polynomials of the form

$$y = ax^r + bx^s + cx^t \dots$$

are differentiated as follows:

$$\frac{dy}{dx} = rax^{r-1} + sbx^{s-1} + tcx^{t-1} \dots.$$

For example, given

$$y = 3x^3 + 2x^2 - 5x$$

then

$$\frac{dy}{dx} = 9x^2 + 4x - 5$$

which describes how the slope of the original function changes with x.

Figure 18.17 shows the graph of $y = 3x^3 + 2x^2 - 5x$ and its derivative $y = 9x^2 + 4x - 5$, and we can see that when $x = -1$ there is a local maximum, where the function reaches a value of 4, then begins a downward journey to 0, where the slope is -5. Similarly, when $x \simeq 0.55$, there is a point where the function reaches a local

Fig. 18.17 Graph of
$y = 3x^3 + 2x^2 - 5x$
and its derivative
$\frac{dy}{dx} = 9x^2 + 4x - 5$ (dashed)

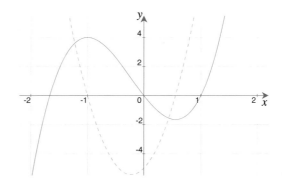

minimum with a value of approximately -1.65. The slope is zero at both points, which is reflected in the graph of the derivative.

Having differentiated the function once, there is nothing to prevent us differentiating a second time, but first we require a way to annotate the process, which is performed as follows. At a general level, let y be some function of x, then the first derivative is

$$\frac{d}{dx}(y).$$

The second derivative is found by differentiating the first derivative:

$$\frac{d}{dx}\left(\frac{dy}{dx}\right)$$

and is written:

$$\frac{d^2y}{dx^2}.$$

Similarly, the third derivative is

$$\frac{d^3y}{dx^3}$$

and the nth derivative:

$$\frac{d^ny}{dx^n}.$$

When a function is expressed as $f(x)$, its derivative is written $f'(x)$. The second derivative is written $f''(x)$, and so on for higher derivatives.

Returning to the original function, the first and second derivatives are

$$\frac{dy}{dx} = 9x^2 + 4x - 5$$

$$\frac{d^2y}{dx^2} = 18x + 4$$

and the third and fourth derivatives are

$$\frac{d^3y}{dx^3} = 18$$

$$\frac{d^4y}{dx^4} = 0.$$

Figure 18.18 shows the original function and the first two derivatives. The graph of the first derivative shows the slope of the original function, whereas the graph of the second derivative shows the slope of the first derivative. These graphs help us identify a local maximum and minimum. By inspection of Fig. 18.18, when the first derivative equals zero, there is a local maximum or a local minimum. Algebraically,

Fig. 18.18 Graph of
$y = 3x^3 + 2x^2 - 5x$, its first
derivative
$\frac{dy}{dx} = 9x^2 + 4x - 5$ (short
dashes) and its second
derivative $\frac{d^2 y}{dx^2} = 18x + 4$
(long dashes)

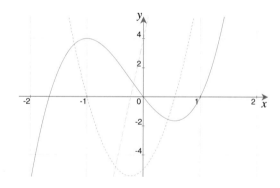

this is when

$$\frac{dy}{dx} = 0$$
$$9x^2 + 4x - 5 = 0.$$

Solving this quadratic in x we have

$$x = \frac{-b \pm \sqrt{b^2 - 4ac}}{2a}$$

where $a = 9, \quad b = 4, \quad c = -5$:

$$x = \frac{-4 \pm \sqrt{16 + 180}}{18}$$
$$x_1 = -1, \quad x_2 \approx 0.555$$

which confirms our earlier analysis. However, what we don't know, without referring
to the graphs, whether it is a minimum, or a maximum.

18.13   Identifying a Local Maximum or Minimum

Figure 18.19 shows a function containing a local maximum of 5 when $x = -1$. Note
that as the independent variable x, increases from -2 towards 0, the slope of the
graph changes from positive to negative, passing through zero at $x = -1$. This is
shown in the function's first derivative, which is the straight line passing through the
points $(-2, 6)$, $(-1, 0)$ and $(0, -6)$. A natural consequence of these conditions
implies that the slope of the first derivative must be negative:

Fig. 18.19 A function
containing a local maximum,
and its first derivative
(dashed)

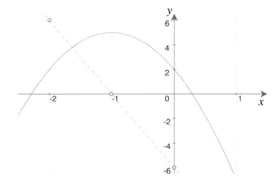

Fig. 18.20 A function
containing a local minimum,
and its first derivative
(dashed)

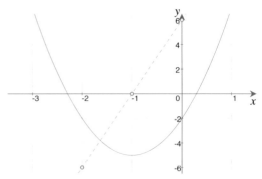

$$\frac{d^2 y}{dx^2} = -\text{ve.}$$

Figure 18.20 shows another function containing a local minimum of -5 when $x = -1$. Note that as the independent variable x, increases from -2 towards 0, the slope of the graph changes from negative to positive, passing through zero at $x = -1$. This is shown in the function's first derivative, which is the straight line passing through the points $(-2, -6)$, $(-1, 0)$ and $(0, 6)$. A natural consequence of these conditions implies that the slope of the first derivative must be positive:

$$\frac{d^2 y}{dx^2} = +\text{ve.}$$

We can now apply this observation to the original function for the two values of x, $x_1 = -1$, $x_2 \approx 0.555$:

$$\frac{dy}{dx} = 9x^2 + 4x - 5$$

$$\frac{d^2 y}{dx^2} = 18x + 4.$$

Fig. 18.21 Graph of $y = -3x^3 + 9x$, its first derivative $y = -9x^2 + 9$ (short dashes) and its second derivative $y = -18x$ (long dashes)

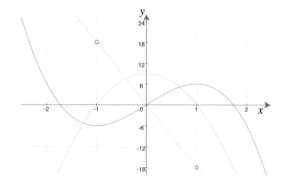

$$x = -1, \quad \frac{d^2 y}{dx^2} = 18 \times (-1) + 4 = -14$$

$$x = 0.555, \quad \frac{d^2 y}{dx^2} = 18 \times (0.555) + 4 = 14.$$

Which confirms that when $x = -1$ there is a local maximum, and when $x \approx 0.555$, there is a local minimum, as shown in Fig. 18.17.

Now let's find the local minimum and maximum for y, given

$$y = -3x^3 + 9x.$$

The first derivative is

$$\frac{dy}{dx} = -9x^2 + 9$$

and second derivative

$$\frac{d^2 y}{dx^2} = -18x$$

as shown in Fig. 18.21. For a local maximum or minimum, the first derivative equals zero:

$$-9x^2 + 9 = 0$$

which implies that $x = \pm 1$.
The sign of the second derivative determines whether there is a local minimum or maximum.

$$\frac{d^2 y}{dx^2} = -18x$$
$$= -18 \times (-1) = +\text{ve}$$
$$= -18 \times (+1) = -\text{ve}$$

therefore, when $x = -1$ there is a local minimum, and when $x = +1$ there is a local maximum, as confirmed by Fig. 18.21.

18.14   Partial Derivatives

Up to this point we have used functions with one independent variable, such as $y = f(x)$. However, we must be able to compute derivatives of functions with more than one independent variable, such as $y = f(u, v, w)$. The technique employed is to assume that only one variable changes, whilst the other variables are held constant. This means that a function can possess several derivatives—one for each independent variable. Such derivatives are called *partial derivatives* and employ a new symbol ∂, which can be read as '*partial dee*'.

Given a function $f(u, v, w)$, the three partial derivatives are defined as

$$\frac{\partial f}{\partial u} = \lim_{h \to 0} \frac{f(u + h, v, w) - f(u, v, w)}{h}$$

$$\frac{\partial f}{\partial v} = \lim_{h \to 0} \frac{f(u, v + h, w) - f(u, v, w)}{h}$$

$$\frac{\partial f}{\partial w} = \lim_{h \to 0} \frac{f(u, v, w + h) - f(u, v, w)}{h}.$$

For example, a function for the volume of a cylinder is

$$V(r, h) = \pi r^2 h$$

where r is the radius, and h is the height. Say we wish to compute the function's partial derivative with respect to r. First, the partial derivative is written

$$\frac{\partial V}{\partial r}.$$

Second, we hold h constant, whilst allowing r to change. This means that the function becomes

$$V(r, h) = kr^2 \tag{18.5}$$

where $k = \pi h$. Thus the partial derivative of (18.5) with respect to r is

$$\frac{\partial V}{\partial r} = 2kr$$

$$= 2\pi h r.$$

Next, by holding r constant, and allowing h to change, we have

$$\frac{\partial V}{\partial h} = \pi r^2.$$

Sometimes, for purposes of clarification, the partial derivatives identify the constant variable(s):

$$\left(\frac{\partial V}{\partial r}\right)_h = 2\pi h r$$

$$\left(\frac{\partial V}{\partial h}\right)_r = \pi r^2.$$

Partial differentiation is subject to the same rules for ordinary differentiation—we just to have to remember which independent variable changes, and those held constant. As with ordinary derivatives, we can compute higher-order partial derivatives.

As an example, let's find the second-order partial derivatives of f, given

$$f(u, v) = u^4 + 2u^3v^2 - 4v^3.$$

The first partial derivatives are

$$\frac{\partial f}{\partial u} = 4u^3 + 6u^2v^2$$

$$\frac{\partial f}{\partial v} = 4u^3v - 12v^2$$

and the second-order partial derivatives are

$$\frac{\partial^2 f}{\partial u^2} = 12u^2 + 12uv^2$$

$$\frac{\partial^2 f}{\partial v^2} = 4u^3 - 24v.$$

Now let's find the second-order partial derivatives of f, given

$$f(u, v) = \sin(4u)\cos(5v)$$

the first partial derivatives are

$$\frac{\partial f}{\partial u} = 4\cos(4u)\cos(5v)$$

$$\frac{\partial f}{\partial v} = -5\sin(4u)\sin(5v)$$

and the second-order partial derivatives are

$$\frac{\partial^2 f}{\partial u^2} = -16 \sin(4u) \cos(5v)$$

$$\frac{\partial^2 f}{\partial v^2} = -25 \sin(4u) \cos(5v).$$

In general, given $f(u, v) = uv$, then

$$\frac{\partial f}{\partial u} = v$$

$$\frac{\partial f}{\partial v} = u$$

and the second-order partial derivatives are

$$\frac{\partial^2 f}{\partial u^2} = 0$$

$$\frac{\partial^2 f}{\partial v^2} = 0.$$

Similarly, given $f(u, v) = u/v$, then

$$\frac{\partial f}{\partial u} = \frac{1}{v}$$

$$\frac{\partial f}{\partial v} = -\frac{u}{v^2}$$

and the second-order partial derivatives are

$$\frac{\partial^2 f}{\partial u^2} = 0$$

$$\frac{\partial^2 f}{\partial v^2} = \frac{2u}{v^3}.$$

Finally, given $f(u, v) = u^v$, then

$$\frac{\partial f}{\partial u} = vu^{v-1}$$

whereas, $\partial f/\partial v$ requires some explaining. First, given

$$f(u, v) = u^v$$

taking natural logs of both sides, we have

$$\ln f(u, v) = v \ln u$$

and

$$f(u, v) = e^{v \ln u}.$$

Therefore,

$$\frac{\partial f}{\partial v} = e^{v \ln u} \ln u$$

$$= u^v \ln u.$$

The second-order partial derivatives are

$$\frac{\partial^2 f}{\partial u^2} = v(v-1)u^{v-2}$$

$$\frac{\partial^2 f}{\partial v^2} = u^v \ln^2 u.$$

18.14.1 Visualising Partial Derivatives

Functions of the form $y = f(x)$ are represented by a 2D graph, and the function's derivative $f'(x)$ represents the graph's slope at any point x. Functions of the form $z = f(x, y)$ can be represented by a 3D surface, like the one shown in Fig. 18.22, which is $z(x, y) = 2.5x^2 - 2.5y^2$. The two partial derivatives are

$$\frac{\partial z}{\partial x} = 8x$$

$$\frac{\partial z}{\partial y} = -4y$$

Fig. 18.22 Surface of $z = 2.5x^2 - 2.5y^2$ using a right-handed axial system with a vertical z-axis

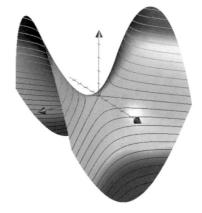

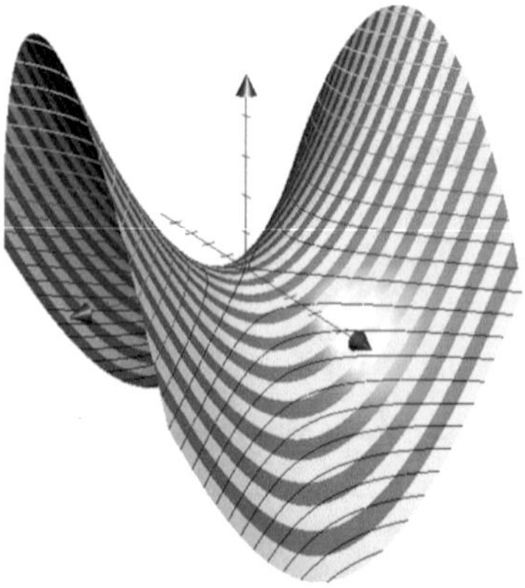

Fig. 18.23 $\frac{\partial z}{\partial x}$ describes the
slopes of these contour lines

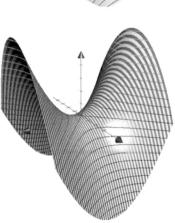

Fig. 18.24 $\frac{\partial z}{\partial y}$ describes the
slopes of these contour lines

where $\partial z/\partial x$ is the slope of the surface in the x-direction, as shown in Fig. 18.23,
and $\partial z/\partial y$ is the slope of the surface in the y-direction, as shown in Fig. 18.24.

The second-order partial derivatives are

$$\frac{\partial^2 z}{\partial x^2} = 8 = +\text{ve}$$

$$\frac{\partial^2 z}{\partial y^2} = -4 = -\text{ve}.$$

As $\partial^2 z/\partial x^2$ is positive, there is a local minimum in the x-direction, and as $\partial^2 z/\partial y^2$
is negative, there is a local maximum in the y-direction, as confirmed by Fig. 18.23.

18.14.2 *Mixed Partial Derivatives*

We have seen that, given a function of the form $f(u, v)$, the partial derivatives $\partial f/\partial u$ and $\partial f/\partial v$ provide the relative instantaneous changes in f and u, and f and v, respectively, whilst the second independent variable remains fixed. However, nothing prevents us from differentiating $\partial f/\partial u$ with respect to v, whilst keeping u constant:

$$\frac{\partial}{\partial v}\left(\frac{\partial f}{\partial u}\right)$$

which is also written as

$$\frac{\partial^2 f}{\partial v \partial u}$$

and is a *mixed partial derivative*.

As an example, let's find the mixed partial derivative of f, given

$$f(u, v) = u^3 v^4.$$

Therefore,

$$\frac{\partial f}{\partial u} = 3u^2 v^4$$

and

$$\frac{\partial^2 f}{\partial v \partial u} = 12u^2 v^3.$$

It should be no surprise that reversing the differentiation gives the same result. Let

$$f(u, v) = u^3 v^4$$

then

$$\frac{\partial f}{\partial v} = 4u^3 v^3$$

and

$$\frac{\partial^2 f}{\partial u \partial v} = 12u^2 v^3.$$

Generally, for continuous functions, we can write

$$\frac{\partial^2 f}{\partial u \partial v} = \frac{\partial^2 f}{\partial v \partial u}.$$

Let's look at two examples. The formula for the volume of a cylinder is given by $V(r, h) = \pi r^2 h$, where r and h are the cylinder's radius and height, respectively. The mixed partial derivative is computed as follows.

$$V(r, h) = \pi r^2 h$$

$$\frac{\partial V}{\partial r} = 2\pi h r$$

$$\frac{\partial^2 V}{\partial h \partial r} = 2\pi r$$

or

$$V(r, h) = \pi r^2 h$$

$$\frac{\partial V}{\partial h} = \pi r^2$$

$$\frac{\partial^2 V}{\partial r \partial h} = 2\pi r.$$

Given

$$f(u, v) = \sin(4u)\cos(3v)$$

then

$$\frac{\partial f}{\partial u} = 4\cos(4u)\cos(3v)$$

$$\frac{\partial^2 f}{\partial v \partial u} = -12\cos(4u)\sin(3v)$$

or

$$\frac{\partial f}{\partial v} = -3\sin(4u)\sin(3v)$$

$$\frac{\partial^2 f}{\partial u \partial v} = -12\cos(4u)\sin(3v).$$

18.15   Chain Rule

Earlier, we came across the chain rule for computing the derivatives of functions of functions. For example, to compute the derivative of $y = \sin(x^2)$ we substitute $u = x^2$, then

$$y = \sin u$$

$$\frac{dy}{du} = \cos u$$

$$= \cos(x^2).$$

Next, we compute du/dx:

$$u = x^2$$

$$\frac{du}{dx} = 2x$$

and dy/dx is the product of the two derivatives using the chain rule:

$$\frac{dy}{dx} = \frac{dy}{du} \cdot \frac{du}{dx}$$
$$= \cos\left(x^2\right)2x$$
$$= 2x \cos\left(x^2\right).$$

But say we have a function where w is a function of two variables x and y, which in turn, are a function of u and v. Then we have

$$w = f(x, y)$$
$$x = r(u, v)$$
$$y = s(u, v).$$

With such a scenario, we have the following partial derivatives:

$$\frac{\partial w}{\partial x}, \quad \frac{\partial w}{\partial y}$$
$$\frac{\partial w}{\partial u}, \quad \frac{\partial w}{\partial v}$$
$$\frac{\partial x}{\partial u}, \quad \frac{\partial x}{\partial v}$$
$$\frac{\partial y}{\partial u}, \quad \frac{\partial y}{\partial v}.$$

These are chained together as follows

$$\frac{\partial w}{\partial u} = \frac{\partial w}{\partial x} \cdot \frac{\partial x}{\partial u} + \frac{\partial w}{\partial y} \cdot \frac{\partial y}{\partial u} \tag{18.6}$$

$$\frac{\partial w}{\partial v} = \frac{\partial w}{\partial x} \cdot \frac{\partial x}{\partial v} + \frac{\partial w}{\partial y} \cdot \frac{\partial y}{\partial v}. \tag{18.7}$$

Here is an example of the chain rule. Find $\partial w/\partial u$ and $\partial w/\partial v$, given

$$w = f(2x + 3y)$$
$$x = r\left(u^2 + v^2\right)$$
$$y = s\left(u^2 - v^2\right).$$

Therefore

$$\frac{\partial w}{\partial x} = 2, \quad \frac{\partial w}{\partial y} = 3$$

$$\frac{\partial x}{\partial u} = 2u, \quad \frac{\partial x}{\partial v} = 2v$$

$$\frac{\partial y}{\partial u} = 2u, \quad \frac{\partial y}{\partial v} = -2v$$

and plugging these into (18.6) and (18.7) we have

$$\frac{\partial w}{\partial u} = \frac{\partial w}{\partial x}\frac{\partial x}{\partial u} + \frac{\partial w}{\partial y}\frac{\partial y}{\partial u}$$
$$= 2 \times 2u + 3 \times 2u$$
$$= 10u$$
$$\frac{\partial w}{\partial v} = \frac{\partial w}{\partial x}\frac{\partial x}{\partial v} + \frac{\partial w}{\partial y}\frac{\partial y}{\partial v}$$
$$= 2 \times 2v + 3 \times (-2v)$$
$$= -2v.$$

Thus, when $u = 2$ and $v = 1$

$$\frac{\partial w}{\partial u} = 20, \quad \text{and} \quad \frac{\partial w}{\partial v} = -2.$$

18.16   Total Derivative

Given a function with three independent variables, such as $w = f(x, y, t)$, where $x = g(t)$ and $y = h(t)$, there are three primary partial derivatives:

$$\frac{\partial w}{\partial x}, \quad \frac{\partial w}{\partial y}, \quad \frac{\partial w}{\partial t}$$

which show the differential change of w with x, y and t respectively. There are also three derivatives:

$$\frac{dx}{dt}, \quad \frac{dy}{dt}, \quad \frac{dt}{dt}$$

where $dt/dt = 1$. The partial and ordinary derivatives can be combined to create the *total derivative* which is written

$$\frac{dw}{dt} = \frac{\partial w}{\partial x}\frac{dx}{dt} + \frac{\partial w}{\partial y}\frac{dy}{dt} + \frac{\partial w}{\partial t}.$$

dw/dt measures the instantaneous change of w relative to t, when all three independent variables change.

Let's find dw/dt, given

$$w = x^2 + xy + y^3 + t^2$$
$$x = 2t$$
$$y = t - 1.$$

Therefore,

$$\frac{dx}{dt} = 2$$

$$\frac{dy}{dt} = 1$$

$$\frac{\partial w}{\partial x} = 2x + y = 4t + t - 1 = 5t - 1$$

$$\frac{\partial w}{\partial y} = x + 3y^2 = 2t + 3(t-1)^2 = 3t^2 - 4t + 3$$

$$\frac{\partial w}{\partial t} = 2t$$

$$\frac{dw}{dt} = \frac{\partial w}{\partial x}\frac{dx}{dt} + \frac{\partial w}{\partial y}\frac{dy}{dt} + \frac{\partial w}{\partial t}$$

$$= (5t - 1)2 + (3t^2 - 4t + 3) + 2t = 3t^2 + 8t + 1$$

and the total derivative equals

$$\frac{dw}{dt} = 3t^2 + 8t + 1$$

and when $t = 1$, $dw/dt = 12$.

18.17 Summary

This chapter has shown how limits provide a useful tool for computing a function's derivative. Basically, the function's independent variable is disturbed by a very small quantity, typically δx, which alters the function's value. The quotient

$$\frac{f(x + \delta x) - f(x)}{\delta x}$$

is a measure of the function's rate of change relative to its independent variable. By making δx smaller and smaller towards zero, we converge towards a limiting value called the function's derivative. Unfortunately, not all functions possess a derivative, therefore we can only work with functions that can be differentiated.

We have seen how to differentiate generic functions such as sums, products, quotients and a function of a function, and we have also seen how to address explicit and implicit forms. These techniques were then used to differentiate exponential, logarithmic, trigonometric and hyperbolic functions, which will be employed in later chapters to solve various problems. Where relevant, integrals of certain functions have been included to show the intimate relationship between derivatives and antiderivatives.

Hopefully, it is now clear that differentiation is like an operator—in that it describes how fast a function changes relative to its independent variable in the form of another function.

Reference

1. Vince JA (2013) Calculus for computer graphics, 2nd edn. Springer

Chapter 19
Calculus: Integration

19.1 Introduction

In this chapter I develop the idea that integration is the inverse of differentiation, and examine standard algebraic strategies for integrating functions, where the derivative is unknown; these include simple algebraic manipulation, trigonometric identities, integration by parts, integration by substitution and integration using partial fractions.

19.2 Indefinite Integral

In the previous chapter we have seen that given a simple function, such as

$$y = \sin x + 23$$

$$\frac{dy}{dx} = \cos x$$

and the constant term 23 disappears. Inverting the process, we begin with

$$dy = \cos x \, dx$$

and integrating:

$$y = \int \cos x \, dx$$

$$= \sin x + C.$$

An integral of the form

$$\int f(x) \, dx$$

© Springer-Verlag London Ltd., part of Springer Nature 2022
J. Vince, *Mathematics for Computer Graphics*, Undergraduate Topics
in Computer Science, https://doi.org/10.1007/978-1-4471-7520-9_19

is known as an *indefinite integral*; and as we don't know whether the original function contains a constant term, a constant C has to be included. Its value remains undetermined unless we are told something about the original function. In this example, if we are told that when $x = \pi/2$, $y = 24$, then

$$24 = \sin\left(\pi/2\right) + C$$
$$= 1 + C$$
$$C = 23.$$

19.3   Integration Techniques

19.3.1   *Continuous Functions*

Functions come in all sorts of shapes and sizes, which is why we have to be very careful before they are differentiated or integrated. If a function contains any form of discontinuity, then it cannot be differentiated or integrated. For example, the square-wave function shown in Fig. 19.1 cannot be differentiated as it contains discontinuities. Consequently, to be very precise, we identify an *interval* $[a, \ b]$, over which a function is analysed, and stipulate that it must be continuous over this interval. For example, a and b define the upper and lower bounds of the interval such that

$$a \leq x \leq b$$

then we can say that for $f(x)$ to be continuous

$$\lim_{h \to 0} f(x + h) = f(x).$$

Even this needs further clarification as h must not take x outside of the permitted interval. So, from now on, we assume that all functions are continuous and can be integrated without fear of singularities.

Fig. 19.1 A discontinuous square-wave function

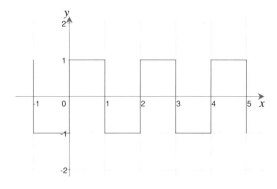

19.3.2 Difficult Functions

There are many functions that cannot be differentiated and represented by a finite collection of elementary functions. For example, the derivative $f'(x) = (\sin x)/x$ does not exist, which precludes the possibility of its integration. Figure 19.2 shows this function, and even though it is continuous, its derivative and integral can only be approximated. Similarly, the derivative $f'(x) = \sqrt{x} \sin x$ does not exist, and also precludes the possibility of its integration. Figure 19.3 shows this continuous function. So now let's examine how most functions have to be rearranged to secure their integration.

19.3.3 Trigonometric Identities

Sometimes it is possible to simplify the integrand by substituting a trigonometric identity. To illustrate this, let's evaluate $\int \sin^2 x \, dx, \int \cos^2 x \, dx, \int \tan^2 x \, dx$ and $\int \sin(3x) \cos x \, dx$.

Fig. 19.2 Graph of $y = (\sin x)/x$

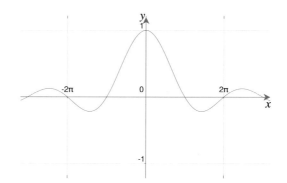

Fig. 19.3 Graph of $y = \sqrt{x} \sin x$

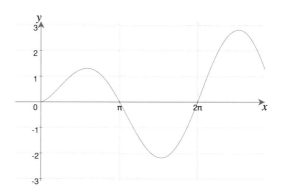

The identity $\sin^2 x = \frac{1}{2}(1 - \cos(2x))$ converts $\sin^2 x$ into a double-angle form:

$$\int \sin^2 x \, dx = \frac{1}{2} \int 1 - \cos(2x) \, dx$$

$$= \frac{1}{2} \int dx - \frac{1}{2} \int \cos(2x) \, dx$$

$$= \frac{1}{2}x - \frac{1}{4} \sin(2x) + C.$$

Figure 19.4 shows the graphs of $y = \sin^2 x$ and $y = \frac{1}{2}x - \frac{1}{4}\sin(2x)$.

The identity $\cos^2 x = \frac{1}{2}(\cos(2x) + 1)$ converts $\cos^2 x$ into a double-angle form:

$$\int \cos^2 x \, dx = \frac{1}{2} \int \cos(2x) + 1 \, dx$$

$$= \frac{1}{2} \int \cos(2x) \, dx + \frac{1}{2} \int dx$$

$$= \frac{1}{4} \sin(2x) + \frac{1}{2}x + C.$$

Figure 19.5 shows the graphs of $y = \cos^2 x$ and $y = \frac{1}{4}\sin(2x) + \frac{1}{2}x$.

Fig. 19.4 The graphs of $y = \sin^2 x$ (dashed) and $y = \frac{1}{2}x - \frac{1}{4}\sin(2x)$

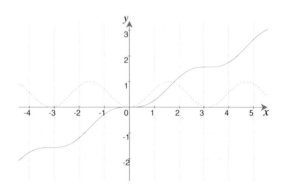

Fig. 19.5 The graphs of $y = \cos^2 x$ (dashed) and $y = \frac{1}{4}\sin(2x) + \frac{1}{2}x$

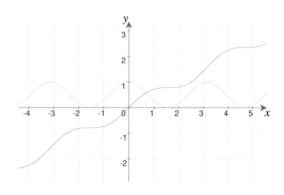

Fig. 19.6 The graphs of $y = \tan^2 x$ (dashed) and $y = \tan x - x$

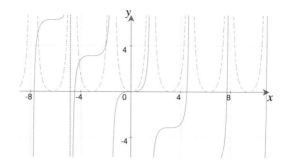

The identity $\sec^2 x = 1 + \tan^2 x$, permits us to write

$$\int \tan^2 x \; dx = \int \sec^2 x - 1 \; dx$$

$$= \int \sec^2 x \; dx - \int dx$$

$$= \tan x - x + C.$$

Figure 19.6 shows the graphs of $y = \tan^2 x$ and $y = \tan x - x$.

Finally, to evaluate $\int \sin(3x) \cos x \; dx$ we use the identity

$$2 \sin a \cos b = \sin(a + b) + \sin(a - b)$$

which converts the integrand's product into the sum and difference of two angles:

$$\sin(3x) \cos x = \tfrac{1}{2}(\sin(4x) + \sin(2x))$$

$$\int \sin(3x) \cos x \; dx = \tfrac{1}{2} \int \sin(4x) + \sin(2x) \; dx$$

$$= \tfrac{1}{2} \int \sin(4x) \; dx + \tfrac{1}{2} \int \sin(2x) \; dx$$

$$= -\tfrac{1}{8} \cos(4x) - \tfrac{1}{4} \cos(2x) + C.$$

Figure 19.7 shows the graphs of $y = \sin(3x) \cos x$ and $y = -\tfrac{1}{8} \cos(4x) - \tfrac{1}{4} \cos(2x)$.

19.3.4 Exponent Notation

Radicals are best replaced by their equivalent exponent notation. For example, to evaluate

Fig. 19.7 The graphs of
$y = \sin(3x)\cos x$ (dashed)
and $y =$
$-\frac{1}{8}\cos(4x) - \frac{1}{4}\cos(2x)$

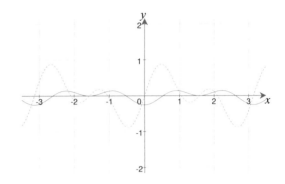

$$\int \frac{2}{\sqrt[4]{x}}\, dx$$

we proceed as follows:

The constant 2 is moved outside the integral, and the integrand is converted into an exponent form:

$$2\int \frac{1}{\sqrt[4]{x}}\, dx = 2\int x^{-\frac{1}{4}}$$

$$= 2\left(\frac{x^{\frac{3}{4}}}{\frac{3}{4}}\right) + C$$

$$= 2\left(\frac{4}{3}x^{\frac{3}{4}}\right) + C$$

$$= \frac{8}{3}x^{\frac{3}{4}} + C.$$

Figure 19.8 shows the graphs of $y = 2/\sqrt[4]{x}$ and $y = 8x^{\frac{3}{4}}/3$.

Fig. 19.8 The graphs of
$y = 2/\sqrt[4]{x}$ (dashed) and
$y = 8x^{\frac{3}{4}}/3$

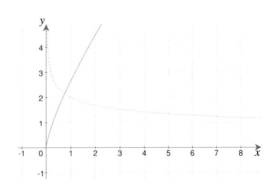

19.3.5 Completing the Square

Where possible, see if an integrand can be simplified by completing the square. For example, to evaluate

$$\int \frac{1}{x^2 - 4x + 8} \, dx$$

we proceed as follows:
We have already seen that

$$\int \frac{1}{1 + x^2} \, dx = \arctan x + C$$

and it's not too difficult to prove that

$$\int \frac{1}{a^2 + x^2} \, dx = \frac{1}{a} \arctan \left(\frac{x}{a} \right) + C.$$

Therefore, if we can manipulate an integrand into this form, then the integral will reduce to an arctan result. The following needs no manipulation:

$$\int \frac{1}{4 + x^2} \, dx = \tfrac{1}{2} \arctan \left(\frac{x}{2} \right) + C.$$

However, the original integrand has $x^2 - 4x + 8$ as the denominator, which is resolved by completing the square:

$$x^2 - 4x + 8 = 4 + (x - 2)^2.$$

Therefore,

$$\int \frac{1}{x^2 - 4x + 8} \, dx = \int \frac{1}{2^2 + (x - 2)^2} \, dx$$

$$= \tfrac{1}{2} \arctan \left(\frac{x - 2}{2} \right) + C.$$

Figure 19.9 shows the graphs of $y = 1/(x^2 - 4x + 8)$ and $y = \tfrac{1}{2} \arctan \left(\frac{x-2}{2} \right)$.

To evaluate

$$\int \frac{1}{x^2 + 6x + 10} \, dx.$$

we factorize the denominator:

Fig. 19.9 The graphs of
$y = 1/(x^2 - 4x + 8)$
(dashed) and
$y = \frac{1}{2} \arctan\left(\frac{x-2}{2}\right)$

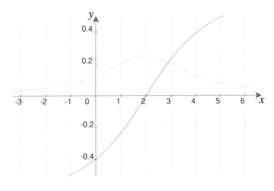

Fig. 19.10 The graphs of
$y = 1/(x^2 + 6x + 10)$
(dashed) and
$y = \arctan(x + 3)$

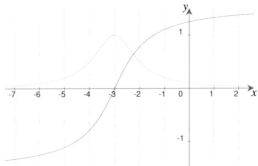

$$\int \frac{1}{x^2 + 6x + 10} \, dx = \int \frac{1}{1^2 + (x + 3)^2} \, dx$$
$$= \arctan(x + 3) + C.$$

Figure 19.10 shows the graphs of $y = 1/(x^2 + 6x + 10)$ and $y = \arctan(x + 3)$.

19.3.6   The Integrand Contains a Derivative

An integral of the form

$$\int f(x) f'(x) \, dx$$

is relatively easy to integrate. For example, let's evaluate

$$\int \frac{\arctan x}{1 + x^2} \, dx.$$

Knowing that

$$\frac{d}{dx} \arctan x = \frac{1}{1 + x^2}$$

let $u = \arctan x$, then

$$\frac{du}{dx} = \frac{1}{1 + x^2}$$

and

$$\int \frac{\arctan x}{1 + x^2} \, dx = \int u \, du$$
$$= \tfrac{1}{2} u^2 + C$$
$$= \tfrac{1}{2} (\arctan x)^2 + C.$$

Figure 19.11 shows the graphs of $y = \frac{\arctan x}{1+x^2}$ and $y = \tfrac{1}{2}(\arctan x)^2$.
An integral of the form

$$\int \frac{f'(x)}{f(x)} \, dx$$

is also relatively easy to integrate. For example, let's evaluate

$$\int \frac{\cos x}{\sin x} \, dx.$$

Knowing that

$$\frac{d}{dx} \sin x = \cos x$$

let $u = \sin x$, then

$$\frac{du}{dx} = \cos x$$

Fig. 19.11 The graphs of
$y = \frac{\arctan x}{1+x^2}$ (dashed) and
$y = \tfrac{1}{2}(\arctan x)^2$

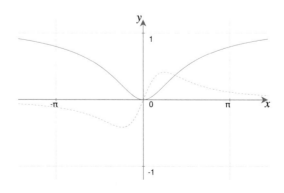

Fig. 19.12 The graphs of
$y = \cos x / \sin x$ (dashed)
and $y = \ln |\sin x|$

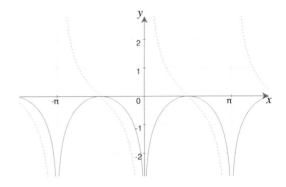

and

$$\int \frac{\cos x}{\sin x}\, dx = \int \frac{1}{u}\, du$$
$$= \ln |u| + C$$
$$= \ln |\sin x| + C.$$

Figure 19.12 shows the graphs of $y = \cos x / \sin x$ and $y = \ln |\sin x|$.

19.3.7   Converting the Integrand into a Series of Fractions

Integration is often made easier by converting an integrand into a series of fractions.
For example, to integrate

$$\int \frac{4x^3 + x^2 - 8 + 12x \cos x}{4x}\, dx$$

we divide the numerator by $4x$:

$$\int \frac{4x^3 + x^2 - 8 + 12x \cos x}{4x}\, dx = \int x^2\, dx + \int \frac{x}{4}\, dx - \int \frac{2}{x}\, dx + \int 3 \cos x\, dx$$
$$= \tfrac{1}{3}x^3 + \tfrac{1}{8}x^2 - 2 \ln |x| + 3 \sin x + C.$$

Figure 19.13 shows the graphs of $y = \left(4x^3 + x^2 - 8 + 12x \cos x\right)/4x$ and $y = \tfrac{1}{3}x^3 + \tfrac{1}{8}x^2 - 2 \ln |x| + 3 \sin x$.

Fig. 19.13 The graphs of
$y = (4x^3 + x^2 - 8 + 12x \cos x)/4x$ (dashed) and
$y = \frac{1}{3}x^3 + \frac{1}{8}x^2 - 2 \ln |x| + 3 \sin x$

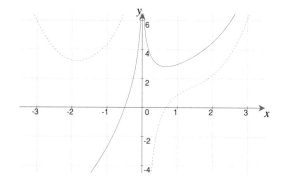

19.3.8 *Integration by Parts*

Integration by parts is based upon the rule for differentiating function products where

$$\frac{d}{dx} uv = u \frac{dv}{dx} + v \frac{du}{dx}$$

therefore,

$$uv = \int uv' \, dx + \int vu' \, dx$$

which rearranged, gives

$$\int uv' \, dx = uv - \int vu' \, dx.$$

Thus, if an integrand contains a product of two functions, we can attempt to integrate it by parts. For example, let's evaluate

$$\int x \sin x \, dx.$$

In this case, we try the following:

$$u = x \quad \text{and} \quad v' = \sin x$$

therefore

$$u' = 1 \quad \text{and} \quad v = C_1 - \cos x.$$

Integrating by parts:

Fig. 19.14 The graphs of
$y = x \sin x$ (dashed) and
$y = -x \cos x + \sin x$

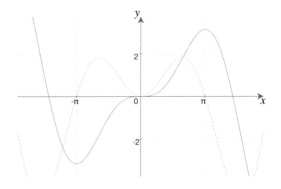

$$\int uv' \, dx = uv - \int vu' \, dx$$

$$\int x \sin x \, dx = x(C_1 - \cos x) - \int (C_1 - \cos x)(1) \, dx$$

$$= C_1 x - x \cos x - C_1 x + \sin x + C$$

$$= -x \cos x + \sin x + C.$$

Figure 19.14 shows the graphs of $y = x \sin x$ and $y = -x \cos x + \sin x$.
Note the problems that arise if we make the wrong substitution:

$$u = \sin x \quad \text{and} \quad v' = x$$

therefore

$$u' = \cos x \quad \text{and} \quad v = \tfrac{1}{2}x^2 + C_1$$

Integrating by parts:

$$\int uv' \, dx = uv - \int vu' \, dx$$

$$\int x \sin x \, dx = \sin x \left(\tfrac{1}{2}x^2 + C_1 \right) - \int \left(\tfrac{1}{2}x^2 + C_1 \right) \cos x \, dx$$

which requires to be integrated by parts, and is even more difficult, which suggests
the substitution was not useful.
    Now let's evaluate

$$\int x^2 \cos x \, dx.$$

In this case, we try the following:

$$u = x^2 \quad \text{and} \quad v' = \cos x$$

therefore

$$u' = 2x \quad \text{and} \quad v = \sin x + C_1.$$

Integrating by parts:

$$\int uv' \, dx = uv - \int vu' \, dx$$

$$\int x^2 \cos x \, dx = x^2(\sin x + C_1) - 2 \int (\sin x + C_1)(x) \, dx$$

$$= x^2 \sin x + C_1 x^2 - 2C_1 \int x \, dx - 2 \int x \sin x \, dx$$

$$= x^2 \sin x + C_1 x^2 - 2C_1 \left(\tfrac{1}{2}x^2 + C_2 \right) - 2 \int x \sin x \, dx$$

$$= x^2 \sin x - C_3 - 2 \int x \sin x \, dx.$$

At this point we come across $\int x \sin x \, dx$, which we have already solved:

$$\int x^2 \cos x \, dx = x^2 \sin x - C_3 - 2(-x \cos x + \sin x + C_4)$$

$$= x^2 \sin x - C_3 + 2x \cos x - 2 \sin x - C_5$$

$$= x^2 \sin x + 2x \cos x - 2 \sin x + C$$

Figure 19.15 shows the graphs of $y = x^2 \cos x$ and $y = x^2 \sin x + 2x \cos x - 2 \sin x$. Now let's evaluate

$$\int x \ln x \, dx.$$

In this case, we try the following:

$$u = \ln x \quad \text{and} \quad v' = x$$

Fig. 19.15 The graphs of $y = x^2 \cos x$ (dashed) and $y = x^2 \sin x + 2x \cos x - 2 \sin x$

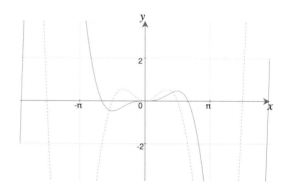

therefore

$$u' = \frac{1}{x} \quad \text{and} \quad v = \tfrac{1}{2}x^2.$$

Integrating by parts:

$$\int uv' \, dx = uv - \int vu' \, dx$$

$$\int x \ln x \, dx = \tfrac{1}{2}x^2 \ln x - \int \left(\tfrac{1}{2}x^2 \right) \frac{1}{x} \, dx$$

$$= \tfrac{1}{2}x^2 \ln x - \tfrac{1}{2} \int x \, dx$$

$$= \tfrac{1}{2}x^2 \ln x - \tfrac{1}{4}x^2 + C.$$

Figure 19.16 shows the graphs of $y = x \ln x$ and $y = \tfrac{1}{2}x^2 \ln x - \tfrac{1}{4}x^2$.
    Finally, let's evaluate

$$\int \sqrt{1 + x^2} \, dx.$$

Although this integrand does not look as though it can be integrated by parts, if we rewrite it as

$$\int \sqrt{1 + x^2}(1) \, dx.$$

then we can use the formula.
    Let

$$u = \sqrt{1 + x^2} \quad \text{and} \quad v' = 1$$

therefore

$$u' = \frac{x}{\sqrt{1 + x^2}} \quad \text{and} \quad v = x.$$

Fig. 19.16 The graphs of
$y = x \ln x$ (dashed) and
$y = \tfrac{1}{2}x^2 \ln x - \tfrac{1}{4}x^2$

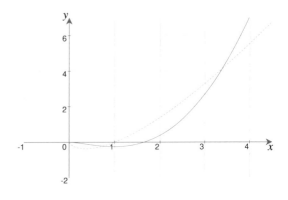

Fig. 19.17 The graphs of $y = \sqrt{1+x^2}$ (dashed) and $y = \frac{1}{2}x\sqrt{1+x^2} + \frac{1}{2}\operatorname{arsinh} x$

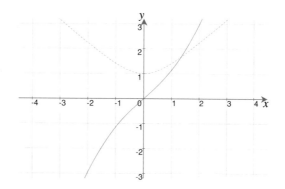

Integrating by parts:

$$\int uv' \, dx = uv - \int vu' \, dx$$

$$\int \sqrt{1+x^2} \, dx = x\sqrt{1+x^2} - \int \frac{x^2}{\sqrt{1+x^2}} \, dx.$$

Now we simplify the right-hand integrand:

$$\int \sqrt{1+x^2} \, dx = x\sqrt{1+x^2} - \int \frac{(1+x^2) - 1}{\sqrt{1+x^2}} \, dx$$

$$= x\sqrt{1+x^2} - \int \frac{1+x^2}{\sqrt{1+x^2}} \, dx + \int \frac{1}{\sqrt{1+x^2}} \, dx$$

$$= x\sqrt{1+x^2} - \int \sqrt{1+x^2} \, dx + \operatorname{arsinh} x + C_1.$$

Now we have the original integrand on the right-hand side, therefore

$$2\int \sqrt{1+x^2} \, dx = x\sqrt{1+x^2} + \operatorname{arsinh} x + C_1$$

$$\int \sqrt{1+x^2} \, dx = \frac{1}{2}x\sqrt{1+x^2} + \frac{1}{2}\operatorname{arsinh} x + C.$$

Figure 19.17 shows the graphs of $y = \sqrt{1+x^2}$ and $y = \frac{1}{2}x\sqrt{1+x^2} + \frac{1}{2}\operatorname{arsinh} x$.

19.3.9   Integration by Substitution

Integration by substitution is based upon the chain rule for differentiating a function of a function, which states that if y is a function of u, which in turn is a function of x, then

$$\frac{dy}{dx} = \frac{dy}{du}\frac{du}{dx}.$$

For example, let's evaluate

$$\int x^2\sqrt{x^3}\,dx.$$

This is easily solved by rewriting the integrand:

$$\int x^2\sqrt{x^3}\,dx = \int x^{\frac{7}{2}}\,dx$$

$$= \tfrac{2}{9}x^{\frac{9}{2}} + C.$$

However, introducing a constant term within the square-root requires integration by substitution. For example,

$$\text{evaluate}\quad \int x^2\sqrt{x^3+1}\,dx.$$

First, we let $u = x^3 + 1$, then

$$\frac{du}{dx} = 3x^2 \quad\text{or}\quad dx = \frac{du}{3x^2}.$$

Substituting u and dx in the integrand gives

$$\int x^2\sqrt{x^3+1}\,dx = \int x^2\sqrt{u}\,\frac{du}{3x^2}$$

$$= \tfrac{1}{3}\int \sqrt{u}\,du$$

$$= \tfrac{1}{3}\int u^{\frac{1}{2}}\,du$$

$$= \tfrac{1}{3}\cdot\tfrac{2}{3}u^{\frac{3}{2}} + C$$

$$= \tfrac{2}{9}\left(x^3+1\right)^{\frac{3}{2}} + C.$$

Figure 19.18 shows the graphs of $y = x^2\sqrt{x^3+1}$ and $y = \tfrac{2}{9}\left(x^3+1\right)^{\frac{3}{2}}$.
Now let's evaluate

$$\int 2\sin x \cdot \cos x\,dx.$$

Integrating by substitution we let $u = \sin x$, then

$$\frac{du}{dx} = \cos x \quad\text{or}\quad dx = \frac{du}{\cos x}.$$

Fig. 19.18 The graphs of $y = x^2\sqrt{x^3 + 1}$ (dashed) and $y = \frac{2}{9}(x^3 + 1)^{\frac{3}{2}}$

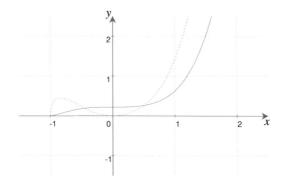

Fig. 19.19 The graphs of $y = 2\sin x \cdot \cos x$ (dashed) and $y = \sin^2 x$

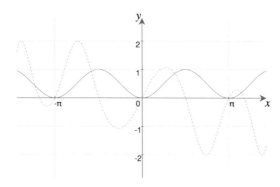

Substituting u and dx in the integrand gives

$$\int 2\sin x \cdot \cos x \, dx = 2\int u \cos x \, \frac{du}{\cos x}$$

$$= 2\int u \, du$$

$$= u^2 + C_1$$

$$= \sin^2 x + C.$$

Figure 19.19 shows the graphs of $y = 2\sin x \cdot \cos x$ and $y = \sin^2 x$.
To evaluate

$$\int 2e^{\cos 2x} \sin x \cdot \cos x \, dx.$$

we integrate by substitution, and let $u = \cos(2x)$, then

$$\frac{du}{dx} = -2\sin(2x) \quad \text{or} \quad dx = -\frac{du}{2\sin(2x)}.$$

Fig. 19.20 The graphs of
$y = 2e^{\cos(2x)} \sin x \cdot \cos x$
(dashed) and $y = -\frac{1}{2}e^{\cos(2x)}$

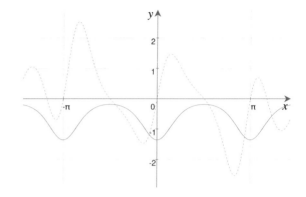

Substituting a double-angle identity, u and du:

$$\int 2e^{\cos 2x} \sin x \cdot \cos x \, dx = -\int e^u \sin(2x) \frac{du}{2\sin(2x)}$$

$$= -\frac{1}{2}\int e^u \, du$$

$$= -\frac{1}{2}e^u + C$$

$$= -\frac{1}{2}e^{\cos(2x)} + C.$$

Figure 19.20 shows the graphs of $y = 2e^{\cos(2x)} \sin x \cdot \cos x$ and $y = -\frac{1}{2}e^{\cos(2x)}$.
To evaluate

$$\int \frac{\cos x}{(1 + \sin x)^3} \, dx.$$

we integrate by substitution, and let $u = 1 + \sin x$, then

$$\frac{du}{dx} = \cos x \quad \text{or} \quad dx = \frac{du}{\cos x}.$$

$$\int \frac{\cos x}{(1 + \sin x)^3} \, dx = \int \frac{\cos x}{u^3} \frac{du}{\cos x}$$

$$= \int u^{-3} \, du$$

$$= -\frac{1}{2}u^{-2} + C$$

$$= -\frac{1}{2}(1 + \sin x)^{-2} + C$$

$$= -\frac{1}{2(1 + \sin x)^2} + C.$$

Figure 19.21 shows the graphs of $y = \cos x/(1 + \sin x)^3$ and $y = -\frac{1}{2}(1 + \sin x)^{-2}$.

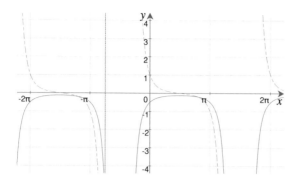

Fig. 19.21 The graphs of $y = \cos x / (1 + \sin x)^3$ (dashed) and $y = -\frac{1}{2}(1 + \sin x)^{-2}$

To evaluate

$$\int \sin(2x) \, dx.$$

we integrate by substitution, and let $u = 2x$, then

$$\frac{du}{dx} = 2 \quad \text{or} \quad dx = \frac{du}{2}.$$

$$\int \sin(2x) \, dx = \frac{1}{2} \int \sin u \, du$$
$$= -\frac{1}{2} \cos u + C$$
$$= -\frac{1}{2} \cos(2x) + C$$

Figure 19.22 shows the graphs of $y = \sin(2x)$ and $y = -\frac{1}{2} \cos(2x)$.

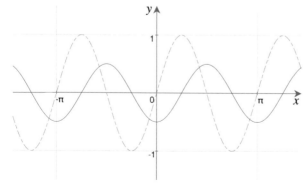

Fig. 19.22 The graphs of $y = \sin(2x)$ (dashed) and $y = -\frac{1}{2} \cos(2x)$

19.3.10 Partial Fractions

Integration by *partial fractions* is used when an integrand's denominator contains a product that can be split into two fractions. For example, it should be possible to convert

$$\int \frac{3x + 4}{(x + 1)(x + 2)}\, dx$$

into

$$\int \frac{A}{x + 1}\, dx + \int \frac{B}{x + 2}\, dx$$

which individually, are easy to integrate. Let's compute A and B:

$$\frac{3x + 4}{(x + 1)(x + 2)} = \frac{A}{x + 1} + \frac{B}{x + 2}$$
$$3x + 4 = A(x + 2) + B(x + 1)$$
$$= Ax + 2A + Bx + B.$$

Equating constants and terms in x:

$$4 = 2A + B \tag{19.1}$$
$$3 = A + B \tag{19.2}$$

Subtracting (19.2) from (19.1), gives $A = 1$ and $B = 2$. Therefore,

$$\int \frac{3x + 4}{(x + 1)(x + 2)}\, dx = \int \frac{1}{x + 1}\, dx + \int \frac{2}{x + 2}\, dx$$
$$= \ln(x + 1) + 2 \ln(x + 2) + C.$$

Figure 19.23 shows the graphs of $y = \frac{3x+4}{(x+1)(x+2)}$ and $y = \ln(x + 1) + 2 \ln(x + 2)$.

Fig. 19.23 The graphs of $y = \frac{3x+4}{(x+1)(x+2)}$ (dashed) and $y = \ln(x + 1) + 2 \ln(x + 2)$

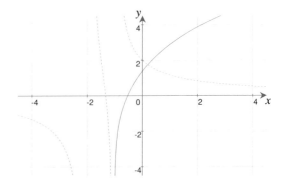

Fig. 19.24 The graphs of $y = \frac{5x-7}{(x-1)(x-2)}$ (dashed) and $y = 2\ln(x-1) + 3\ln(x-2)$

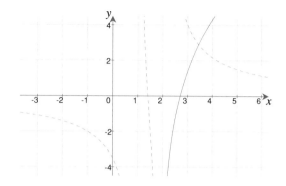

Now let's evaluate

$$\int \frac{5x-7}{(x-1)(x-2)}\,dx.$$

Integrating by partial fractions:

$$\frac{5x-7}{(x-1)(x-2)} = \frac{A}{x-1} + \frac{B}{x-2}$$
$$5x - 7 = A(x-2) + B(x-1)$$
$$= Ax + Bx - 2A - B.$$

Equating constants and terms in x:

$$-7 = -2A - B \tag{19.3}$$
$$5 = A + B \tag{19.4}$$

Subtracting (19.3) from (19.4), gives $A = 2$ and $B = 3$. Therefore,

$$\int \frac{5x-7}{(x-1)(x-2)}\,dx = \int \frac{2}{x-1}\,dx + \int \frac{3}{x-2}\,dx$$
$$= 2\ln(x-1) + 3\ln(x-2) + C.$$

Figure 19.24 shows the graphs of $y = \frac{5x-7}{(x-1)(x-2)}$ and $y = 2\ln(x-1) + 3\ln(x-2)$.
  Finally, let's evaluate

$$\int \frac{6x^2 + 5x - 2}{x^3 + x^2 - 2x}\,dx$$

using partial fractions:

Fig. 19.25 The graphs of $y = \frac{6x^2+5x-2}{x^3+x^2-2x}$ (dashed) and $y = \ln x + 2\ln(x+2) + 3\ln(x-1)$

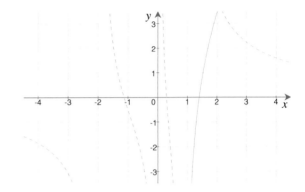

$$\frac{6x^2 + 5x - 2}{x^3 + x^2 - 2x} = \frac{A}{x} + \frac{B}{x+2} + \frac{C}{x-1}$$

$$6x^2 + 5x - 2 = A(x+2)(x-1) + Bx(x-1) + Cx(x+2)$$

$$= Ax^2 + Ax - 2A + Bx^2 - Bx + Cx^2 + 2Cx.$$

Equating constants, terms in x and x^2:

$$-2 = -2A \qquad\qquad\qquad (19.5)$$

$$5 = A - B + 2C \qquad\qquad (19.6)$$

$$6 = A + B + C \qquad\qquad (19.7)$$

Manipulating (19.5), (19.6) and (19.7): $A = 1$, $B = 2$ and $C = 3$, therefore,

$$\int \frac{6x^2 + 5x - 2}{x^3 + x^2 - 2x} \, dx = \int \frac{1}{x} \, dx + \int \frac{2}{x+2} \, dx + \int \frac{3}{x-1} \, dx$$

$$= \ln x + 2\ln(x+2) + 3\ln(x-1) + C.$$

Figure 19.25 shows the graphs of $y = \frac{6x^2+5x-2}{x^3+x^2-2x}$ and $y = \ln x + 2\ln(x+2) + 3\ln(x-1)$.

19.4 Area Under a Graph

The ability to calculate the area under a graph is one of the most important discoveries of integral calculus. Prior to calculus, area was computed by dividing a zone into very small strips and summing the individual areas. The accuracy of the result is improved simply by making the strips smaller and smaller, taking the result towards some limiting value. In this section, I show how integral calculus provides a way to compute the area between a function's graph and the x- and y-axis.

19.5 Calculating Areas

Before considering the relationship between area and integration, let's see how area
is calculated using functions and simple geometry.

Figure 19.26 shows the graph of $y = 1$, where the area A of the shaded zone is

$$A = x, \quad x > 0.$$

For example, when $x = 4$, $A = 4$, and when $x = 10$, $A = 10$. An interesting observation is that the original function is the derivative of A:

$$\frac{dA}{dx} = 1 = y.$$

Figure 19.27 shows the graph of $y = 2x$. The area A of the shaded triangle is

$$A = \tfrac{1}{2}\text{base} \times \text{height}$$
$$= \tfrac{1}{2}x \times 2x$$
$$= x^2.$$

Fig. 19.26 Area of the
shaded zone is $A = x$

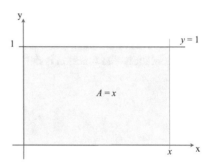

Fig. 19.27 Area of the
shaded zone is $A = x^2$

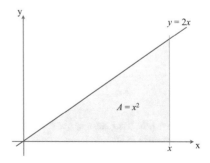

Fig. 19.28 Graph of
$y = \sqrt{r^2 - x^2}$

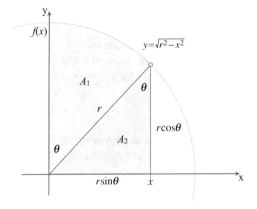

Thus, when $x = 4$, $A = 16$. Once again, the original function is the derivative of A:

$$\frac{dA}{dx} = 2x = y$$

which is no coincidence.

Finally, Fig. 19.28 shows a circle where $x^2 + y^2 = r^2$, and the curve of the first quadrant is described by the function

$$y = \sqrt{r^2 - x^2}, \quad 1 \le x \le r.$$

The total area of the shaded zones is the sum of the two parts A_1 and A_2. To simplify the calculations the function is defined in terms of the angle θ, such that

$$x = r \sin \theta$$

and

$$y = r \cos \theta.$$

Therefore,

$$A_1 = \tfrac{1}{2}r^2\theta$$
$$A_2 = \tfrac{1}{2}(r \cos \theta)(r \sin \theta) = \tfrac{1}{4}r^2 \sin(2\theta)$$
$$A = A_1 + A_2$$
$$= \tfrac{1}{2}r^2 \left(\theta + \tfrac{1}{2} \sin(2\theta)\right).$$

To show that the total area is related to the function's derivative, let's differentiate A with respect to θ:

$$\frac{dA}{d\theta} = \tfrac{1}{2}r^2 \left(1 + \cos(2\theta)\right) = r^2 \cos^2 \theta.$$

But we want the derivative $\frac{dA}{dx}$, which requires the chain rule

$$\frac{dA}{dx} = \frac{dA}{d\theta}\frac{d\theta}{dx}$$

where

$$\frac{dx}{d\theta} = r\cos\theta$$

or

$$\frac{d\theta}{dx} = \frac{1}{r\cos\theta}$$

therefore,

$$\frac{dA}{dx} = \frac{r^2\cos^2\theta}{r\cos\theta} = r\cos\theta = y$$

which is the equation for the quadrant.

Hopefully, these three examples provide strong evidence that the derivative of the function for the area under a graph, equals the graph's function:

$$\frac{dA}{dx} = f(x)$$

which implies that

$$A = \int f(x)\,dx.$$

Now let's prove this observation using Fig. 19.29, which shows a continuous function $y = f(x)$. Next, we define a function $A(x)$ to represent the area under the graph over the interval $[a,\ x]$. δA is the area increment between x and $x + \delta x$, and

$$\delta A \approx f(x) \cdot \delta x.$$

Fig. 19.29 Relationship between $y = f(x)$ and $A(x)$

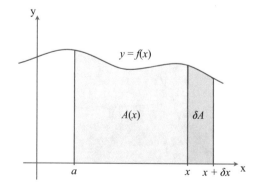

We can also reason that

$$\delta A = A(x + \delta x) - A(x) \approx f(x) \cdot \delta x$$

and the derivative $\frac{dA}{dx}$ is the limiting condition:

$$\frac{dA}{dx} = \lim_{\delta x \to 0} \frac{A(x + \delta x) - A(x)}{\delta x} = \lim_{\delta x \to 0} \frac{f(x) \cdot \delta x}{\delta x} = f(x)$$

thus,

$$\frac{dA}{dx} = f(x),$$

whose antiderivative is

$$A(x) = \int f(x) \, dx.$$

The function $A(x)$ computes the area over the interval $[a, \; b]$ and is represented by

$$A(x) = \int_a^b f(x) \, dx$$

which is called *the integral* or *definite integral*.

Let's assume that $A(b)$ is the area under the graph of $f(x)$ over the interval $[0, \; b]$, as shown in Fig. 19.30, and is written

$$A(b) = \int_0^b f(x) \, dx.$$

Similarly, let $A(a)$ be the area under the graph of $f(x)$ over the interval $[0, \; a]$, as shown in Fig. 19.31, and is written

Fig. 19.30 $A(b)$ is the area under the graph $y = f(x)$, $0 \leq x \leq b$

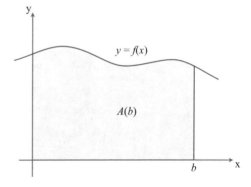

Fig. 19.31 $A(a)$ is the area under the graph $y = f(x)$, $0 \le x \le a$

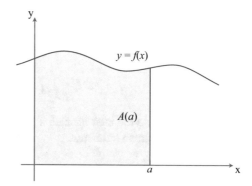

Fig. 19.32 $A(b) - A(a)$ is the area under the graph $y = f(x)$, $a \le x \le b$

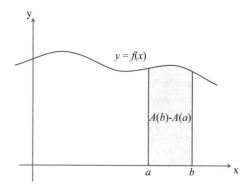

$$A(a) = \int_0^a f(x)\, dx.$$

Figure 19.32 shows that the area of the shaded zone over the interval $[a, b]$ is calculated by

$$A = A(b) - A(a)$$

which is written

$$A = \int_0^b f(x)\, dx - \int_0^a f(x)\, dx$$

and is contracted to

$$A = \int_a^b f(x)\, dx. \tag{19.8}$$

The *fundamental theorem of calculus* states that the definite integral

$$\int_a^b f(x)\, dx = F(b) - F(a)$$

where

$$F(a) = \int f(x)\, dx, \quad x = a$$

$$F(b) = \int f(x)\, dx, \quad x = b.$$

In order to compute the area beneath a graph of $f(x)$ over the interval $[a,\ b]$, we first integrate the graph's function

$$F(x) = \int f(x)\, dx$$

and then calculate the area, which is the difference

$$A = F(b) - F(a).$$

To illustrate how (19.8) is used in the context of the earlier three examples, let's calculate the area over the interval $[1,\ 4]$ for $y = 1$, as shown in Fig. 19.33. We begin with

$$A = \int_1^4 1\, dx.$$

Next, we integrate the function, and transfer the interval bounds employing the *substitution symbol* $\Big|_1^4$, or square brackets $\Big[\ \Big]_1^4$. Using $\Big|_1^4$, we have

Fig. 19.33 Area under the graph is $\int_1^4 1\, dx$

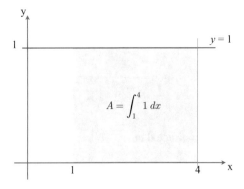

$$A = \int_1^4 1\, dx$$

Fig. 19.34 Area under the
graph is $\int_1^4 2x\,dx$

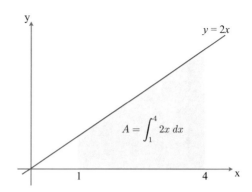

$$A = \Big|_1^4 \; x$$
$$= 4 - 1$$
$$= 3$$

or using $\left[\;\right]_1^4$, we have

$$A = \Big[\, x \,\Big]_1^4$$
$$= 4 - 1$$
$$= 3.$$

I will continue with square brackets.

Now let's calculate the area over the interval $[1,\ 4]$ for $y = 2x$, as shown in Fig. 19.34. We begin with

$$A = \int_1^4 2x\,dx.$$

Next, we integrate the function and evaluate the area

$$A = \Big[\, x^2 \,\Big]_1^4$$
$$= 16 - 1$$
$$= 15.$$

Finally, let's calculate the area over the interval $[0,\ r]$ for $y = \sqrt{r^2 - x^2}$, which is the equation for a circle, as shown in Fig. 19.35. We begin with

Fig. 19.35 Area under the
graph is $\int_0^r \sqrt{r^2 - x^2}\, dx$

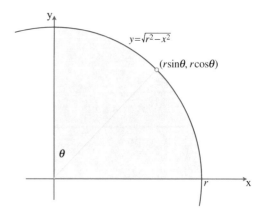

$$A = \int_0^r \sqrt{r^2 - x^2}\, dx. \qquad (19.9)$$

Unfortunately, (19.9) contains a function of a function, which is resolved by substituting another independent variable. In this case, the geometry of the circle suggests

$$x = r \sin \theta$$

therefore,

$$\sqrt{r^2 - x^2} = r \cos \theta$$

and

$$\frac{dx}{d\theta} = r \cos \theta. \qquad (19.10)$$

However, changing the independent variable requires changing the interval for the integral. In this case, changing $0 \le x \le r$ into $\theta_1 \le \theta \le \theta_2$:

When $x = 0$, $r \sin \theta_1 = 0$, therefore $\theta_1 = 0$.
When $x = r$, $r \sin \theta_2 = r$, therefore $\theta_2 = \pi/2$.
Thus, the new interval is $[0, \pi/2]$.
Finally, the dx in (19.9) has to be changed into $d\theta$, which using (19.10) makes

$$dx = r \cos \theta \, d\theta.$$

Now we are in a position to rewrite the original integral using θ as the independent variable:

$$A = \int_0^{\frac{\pi}{2}} (r \cos \theta)(r \cos \theta) \, d\theta$$

$$= r^2 \int_0^{\frac{\pi}{2}} \cos^2 \theta \, d\theta$$

$$= \frac{r^2}{2} \int_0^{\frac{\pi}{2}} 1 + \cos(2\theta) \, d\theta$$

$$= \frac{r^2}{2} \left[\theta + \tfrac{1}{2} \sin(2\theta) \right]_0^{\frac{\pi}{2}}$$

$$= \frac{r^2}{2} \left(\frac{\pi}{2} \right)$$

$$= \frac{\pi r^2}{4}$$

which makes the area of a full circle πr^2.

19.6 Positive and Negative Areas

Area in the real world is always regarded as a positive quantity—no matter how it is measured. In mathematics, however, area is often a signed quantity, and is determined by the clockwise or anticlockwise direction of vertices. As we generally use a left-handed Cartesian axial system in calculus, areas above the x-axis are positive, whilst areas below the x-axis are negative. This can be illustrated by computing the area of the positive and negative parts of a sine wave.

Figure 19.36 shows a sketch of a sine wave over one cycle, where the area above the x-axis is labelled A_1, and the area below the x-axis is labelled A_2. These areas are computed as follows.

Fig. 19.36 The two areas associated with a sine wave

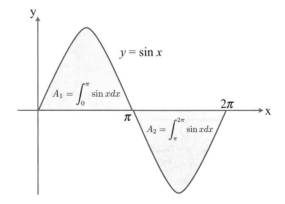

Fig. 19.37 The accumulated
area of a sine wave

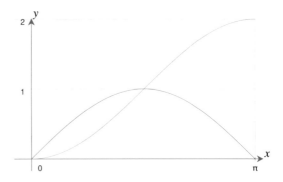

$$A_1 = \int_0^\pi \sin x \, dx$$

$$= \left[-\cos x \right]_0^\pi$$

$$= 1 + 1$$

$$= 2.$$

However, A_2 gives a negative result:

$$A_2 = \int_\pi^{2\pi} \sin x \, dx$$

$$= \left[-\cos x \right]_\pi^{2\pi}$$

$$= -1 - 1$$

$$= -2.$$

This means that the area is zero over the bounds 0 to 2π, .

$$A_2 = \int_0^{2\pi} \sin x \, dx$$

$$= \left[-\cos x \right]_0^{2\pi}$$

$$= -1 + 1$$

$$= 0.$$

Consequently, one must be very careful using this technique for functions that are negative in the interval under investigation. Figure 19.37 shows a sine wave over the interval $[0, \ \pi]$ and its accumulated area.

19.7 Area Between Two Functions

Figure 19.38 shows the graphs of $y = x^2$ and $y = x^3$, with two areas labelled A_1 and A_2. A_1 is the area trapped between the two graphs over the interval $[-1, \ 0]$ and A_2 is the area trapped between the two graphs over the interval $[0, \ 1]$. These areas are calculated very easily: in the case of A_1 we sum the individual areas under the two graphs, remembering to reverse the sign for the area associated with $y = x^3$. For A_2 we subtract the individual areas under the two graphs.

$$
\begin{aligned}
A_1 &= \int_{-1}^{0} x^2 \, dx - \int_{-1}^{0} x^3 \, dx \\
&= \left[\frac{x^3}{3} \right]_{-1}^{0} - \left[\frac{x^4}{4} \right]_{-1}^{0} \\
&= \tfrac{1}{3} + \tfrac{1}{4} \\
&= \tfrac{7}{12}.
\end{aligned}
$$

$$
\begin{aligned}
A_2 &= \int_{0}^{1} x^2 \, dx - \int_{0}^{1} x^3 \, dx \\
&= \left[\frac{x^3}{3} \right]_{0}^{1} - \left[\frac{x^4}{4} \right]_{0}^{1} \\
&= \tfrac{1}{3} - \tfrac{1}{4} \\
&= \tfrac{1}{12}.
\end{aligned}
$$

Note, that in both cases the calculation is the same, which implies that when we employ

$$
A = \int_{a}^{b} [f(x) - g(x)] \, dx
$$

A is always the area trapped between $f(x)$ and $g(x)$ over the interval $[a, \ b]$.

Fig. 19.38 Two areas between $y = x^2$ and $y = x^3$

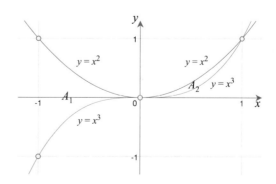

Fig. 19.39 The area
between $y = \sin x$ and
$y = 0.5$

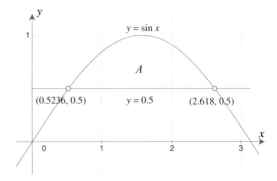

Let's take another example, by computing the area A between $y = \sin x$ and the
line $y = 0.5$, as shown in Fig. 19.39. The horizontal line intersects the sine curve at
$x = 30°$ and $x = 150°$, marked in radians as 0.5236 and 2.618 respectively.

$$A = \int_{30°}^{150°} \sin x \, dx - \int_{\pi/6}^{5\pi/6} 0.5 \, dx$$

$$= \left[-\cos x \right]_{30°}^{150°} - \frac{1}{2} \left[x \right]_{\pi/6}^{5\pi/6}$$

$$= \left(\frac{\sqrt{3}}{2} + \frac{\sqrt{3}}{2} \right) - \frac{1}{2} \left(\frac{5\pi}{6} - \frac{\pi}{6} \right)$$

$$= \sqrt{3} - \frac{\pi}{3}$$

$$\approx 0.685.$$

19.8   Areas with the y-Axis

So far we have only calculated areas between a function and the x-axis. So let's
compute the area between a function and the y-axis. Figure 19.40 shows the function
$y = x^2$ over the interval $[0, 4]$, where A_1 is the area between the curve and the x-
axis, and A_2 is the area between the curve and y-axis. The sum $A_1 + A_2$ must equal
$4 \times 16 = 64$, which is a useful control. Let's compute A_1.

Fig. 19.40 The areas between the x-axis and the y-axis

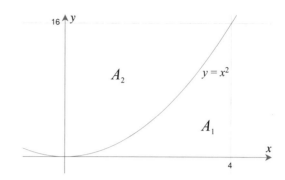

$$A_1 = \int_0^4 x^2 \, dx$$

$$= \left[\frac{x^3}{3} \right]_0^4$$

$$= \frac{64}{3}$$

$$\approx 21.333$$

which means that $A_2 \approx 42.666$. To compute A_2 we construct an integral relative to dy with a corresponding interval. If $y = x^2$ then $x = y^{\frac{1}{2}}$, and the interval is $[0, \ 16]$:

$$A_2 = \int_0^{16} y^{\frac{1}{2}} \, dy$$

$$= \left[\frac{2}{3} y^{\frac{3}{2}} \right]_0^{16}$$

$$= \frac{2}{3} 64$$

$$\approx 42.666.$$

19.9   Area with Parametric Functions

When working with functions of the form $y = f(x)$, the area under its curve and the x-axis over the interval $[a, \ b]$ is

$$A = \int_a^b f(x) \, dx.$$

However, if the curve has a parametric form where

$$x = f_x(t) \quad \text{and} \quad y = f_y(t)$$

then we can derive an equivalent integral as follows.
First: We need to establish equivalent limits $[\alpha, \ \beta]$ for t, such that

$$a = f_x(\alpha) \quad \text{and} \quad b = f_x(\beta).$$

Second: Any point on the curve has corresponding Cartesian and parametric coordinates:

$$x \quad \text{and} \quad f_x(t)$$

$$y = f(x) \quad \text{and} \quad f_y(t).$$

Third:

$$x = f_x(t)$$
$$dx = f_x'(t)dt$$
$$A = \int_a^b f(x) \, dx$$
$$= \int_\alpha^\beta f_y(t) f_x'(t) \, dt$$

therefore

$$A = \int_\alpha^\beta f_y(t) f_x'(t) \, dt. \qquad\qquad (19.11)$$

Let's apply (19.11) using the parametric equations for a circle

$$x = -r \cos t$$
$$y = r \sin t.$$

as shown in Fig. 19.41. Remember that the Cartesian interval is $[a, b]$ left to right, and the polar interval $[\alpha, \ \beta]$, must also be left to right, which is why $x = -r \cos t$. Therefore,

$$f_x't = r \sin t$$
$$f_y(t) = r \sin t$$

Fig. 19.41 The parametric
functions for a circle

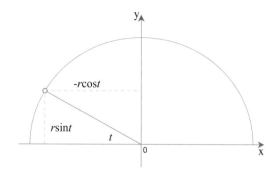

$$A = \int_\alpha^\beta f_y(t) f_x'(t)\ dt$$

$$= \int_0^\pi r \sin t \cdot r \sin t\ dt$$

$$= r^2 \int_0^\pi \sin^2 t\ dt$$

$$= \frac{r^2}{2} \int_0^\pi 1 - \cos(2t)\ dt$$

$$= \frac{r^2}{2} \Big[t + \tfrac{1}{2} \sin(2t) \Big]_0^\pi$$

$$= \frac{\pi r^2}{2}$$

which makes the area of a full circle πr^2.

19.10 The Riemann Sum

The German mathematician Bernhard Riemann (1826–1866) (pronounced 'Ree-man') made major contributions to various areas of mathematics, including integral calculus, where his name is associated with a formal method for summing areas and volumes. Through the *Riemann Sum*, Riemann provides an elegant and consistent notation for describing single, double and triple integrals when calculating area and volume. Let's see how the Riemann sum explains why the area under a curve is the function's integral.

Figure 19.42 shows a function $f(x)$ divided into eight equal sub-intervals where

$$\Delta x = \frac{b - a}{8}$$

Fig. 19.42 The graph of
function $f(x)$ over the
interval $[a, b]$

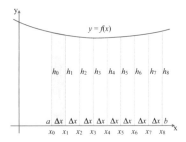

and

$$a = x_0 < x_1 < x_2 < \cdots < x_7 < x_8 = b.$$

In order to compute the area under the curve over the interval $[a, b]$, the interval
is divided into some large number of sub-intervals. In this case, eight, which is not
very large, but convenient to illustrate. Each sub-interval becomes a rectangle with
a common width Δx and a different height. The area of the first rectangular sub-
interval shown shaded, can be calculated in various ways. We can take the left-most
height $f(x_0)$ and form the product $f(x_0)\Delta x$, or we can take the right-most height
$f(x_1)$ and form the product $f(x_1)\Delta x$. On the other hand, we could take the mean
of the two heights $(f(x_0) + f(x_1))/2$ and form the product $(f(x_0) + f(x_1))\Delta x/2$.
A solution that shows no bias towards either left, right or centre, is to let $f(x_i^*)$ be
anywhere in a specific sub-interval Δx_i, then the area of the rectangle associated
with the sub-interval is $f(x_i^*)\Delta x_i$, and the sum of the rectangular areas is given by

$$A = \sum_{i=1}^{8} f(x_i^*)\Delta x_i.$$

Dividing the interval into eight equal sub-intervals will not generate a very accurate
result for the area under the graph. But increasing it to eight-thousand or eight-
million, will take us towards some limiting value. Rather than specify some specific
large number, it is common practice to employ n, and let n tend towards infinity,
which is written

$$A = \sum_{i=1}^{n} f(x_i^*)\Delta x_i. \tag{19.12}$$

The right-hand side of (19.12) is called a Riemann sum, of which there are many.
For the above description, I have assumed that the sub-intervals are equal, which is
not a necessary requirement.

   If the number of sub-intervals is n, then

$$\Delta x = \frac{b - a}{n}$$

and the definite integral is defined as

$$\int_a^b f(x)\,dx = \lim_{n\to\infty} \sum_{i=1}^n f(x_i^*)\Delta x_i.$$

19.11 Summary

In this chapter we have discovered the double role of integration. Integrating a function reveals another function, whose derivative is the function under investigation. Simultaneously, integrating a function computes the area between the function's graph and the x- or y-axis. Although the concept of area in every-day life is an unsigned quantity, within mathematics, and in particular calculus, area *is* a signed quality, and one must be careful when making such calculations.

Chapter 20
Worked Examples

20.1 Introduction

This chapter examines a variety of problems encountered in computer graphics and develops mathematical strategies for their solution. Such strategies may not be the most efficient, however, they will provide the reader with a starting point, which may be improved upon.

20.2 Area of Regular Polygon

Given a regular polygon with n sides, side length s, and radius r of the circumscribed circle, its area can be computed by dividing it into n isosceles triangles and summing their total area.

Figure 20.1 shows one of the isosceles triangles OAB formed by an edge s and the centre O of the polygon. From Fig. 20.1 we observe that

$$\frac{s}{2h} = \tan\left(\frac{\pi}{n}\right)$$

therefore,

$$h = \frac{s}{2}\cot\left(\frac{\pi}{n}\right)$$

$$\text{area}(\Delta OAB) = \frac{sh}{2} = \frac{s^2}{4}\cot\left(\frac{\pi}{n}\right)$$

but there are n such triangles, therefore,

© Springer-Verlag London Ltd., part of Springer Nature 2022
J. Vince, *Mathematics for Computer Graphics*, Undergraduate Topics
in Computer Science, https://doi.org/10.1007/978-1-4471-7520-9_20

Fig. 20.1 One of the isosceles triangles forming a regular polygon

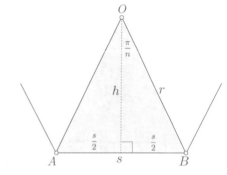

Table 20.1 Area of first 6 regular polygons

n	Area
3	0.433
4	1
5	1.72
6	2.598
7	3.634
8	4.828

$$\text{area} = \frac{ns^2}{4}\cot\left(\frac{\pi}{n}\right).$$

Table 20.1 shows the area for the first six regular polygons with $s = 1$.

20.3 Area of Any Polygon

Figure 20.2 shows a polygon with the following vertices in anticlockwise sequence, and by inspection, the area is 9.5.

x	0	2	5	4	2
y	2	0	1	3	3

The area of a polygon is given by

$$area = \frac{1}{2}\sum_{i=0}^{n-1}(x_i\,y_{i+1(\mathrm{mod}\ n)} - y_i\,x_{i+1(\mathrm{mod}\ n)})$$

$$= \frac{1}{2}(0 \times 0 + 2 \times 1 + 5 \times 3 + 4 \times 3 + 2 \times 2 - 2 \times 2$$
$$- 0 \times 5 - 1 \times 4 - 3 \times 2 - 3 \times 0)$$

$$area = \frac{1}{2}(33 - 14) = 9.5.$$

Fig. 20.2 A five-sided
irregular polygon

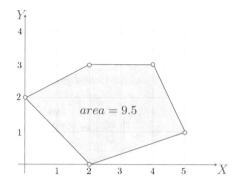

$area = 9.5$

20.4   Dihedral Angle of a Dodecahedron

The dodecahedron is a member of the five Platonic solids, which are constructed from regular polygons. The dihedral angle is the internal angle between two touching faces. Figure 20.3 shows a dodecahedron with one of its pentagonal sides.

Figure 20.4 illustrates the geometry required to fold two pentagonal sides through the dihedral angle γ.

The point P has coordinates

$$P(x, \ y, \ z) = (\sin 72°, \ 0, \ -\cos 72°)$$

Fig. 20.3 A dodecahedron
with one of its pentagonal
sides

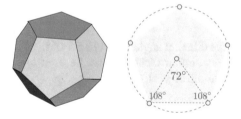

Fig. 20.4 The dihedral
angle γ between two
pentagonal sides

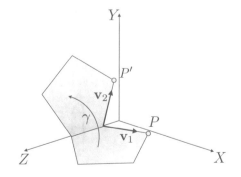

and for simplicity, we will use a unit vector to represent an edge, therefore

$$\|\mathbf{v}_1\| = \|\mathbf{v}_2\| = 1.$$

The coordinates of the rotated point P' are given by the following transform:

$$\begin{bmatrix} x' \\ y' \\ z' \end{bmatrix} = \begin{bmatrix} \cos\gamma & -\sin\gamma & 0 \\ \sin\gamma & \cos\gamma & 0 \\ 0 & 0 & 1 \end{bmatrix} \begin{bmatrix} \sin 72° \\ 0 \\ -\cos 72° \end{bmatrix}$$

where

$$x' = \cos\gamma \sin 72°$$
$$y' = \sin\gamma \sin 72°$$
$$z' = -\cos 72°.$$

But

$$\mathbf{v}_1 \cdot \mathbf{v}_2 = \|\mathbf{v}_1\|\|\mathbf{v}_2\| \cos\theta = xx' + yy' + zz'$$

therefore,

$$\cos\theta = \cos\gamma \sin^2 72° + \cos^2 72°$$

but $\theta = 108°$ (internal angle of a regular pentagon), therefore,

$$\cos\gamma = \frac{\cos 108° - \cos^2 72°}{\sin^2 72°} = \frac{\cos 72°}{\cos 72° - 1}.$$

The dihedral angle $\gamma \approx 116.56505°$.

A similar technique can be used to calculate the dihedral angles of the other Platonic objects.

20.5 Vector Normal to a Triangle

Very often in computer graphics we have to calculate a vector normal to a plane containing three points. The most effective tool to achieve this is the vector product. For example, given three points $P_1(5, 0, 0)$, $P_2(0, 0, 5)$ and $P_3(10, 0, 5)$, we can create two vectors $\mathbf{a}$ and $\mathbf{b}$ as follows:

$$\mathbf{a} = \begin{bmatrix} x_2 - x_1 \\ y_2 - y_1 \\ z_2 - z_1 \end{bmatrix}, \quad \mathbf{b} = \begin{bmatrix} x_3 - x_1 \\ y_3 - y_1 \\ z_3 - z_1 \end{bmatrix},$$

therefore,

$$a = -5i + 5k, \qquad b = 5i + 5k.$$

The normal vector $\mathbf{n}$ is given by

$$\mathbf{n} = \mathbf{a} \times \mathbf{b} = \begin{vmatrix} i & j & k \\ -5 & 0 & 5 \\ 5 & 0 & 5 \end{vmatrix} = 50\mathbf{j}.$$

20.6 Area of a Triangle Using Vectors

The vector product is also useful in calculating the area of a triangle using two of its sides as vectors. For example, using the same points and vectors in the previous example:

$$area = \tfrac{1}{2}\|\mathbf{a} \times \mathbf{b}\| = \tfrac{1}{2}\begin{vmatrix} i & j & k \\ -5 & 0 & 5 \\ 5 & 0 & 5 \end{vmatrix} = \tfrac{1}{2}\|50\mathbf{j}\| = 25.$$

20.7 General Form of the Line Equation from Two Points

The general form of the line equation is given by

$$ax + by + c = 0$$

and it may be required to compute this equation from two known points. For example, Fig. 20.5 shows two points $P_1(x_1, \ y_1)$ and $P_2(x_2, \ y_2)$, from which it is possible to determine $P(x, \ y)$.

Fig. 20.5 A line formed from two points P_1 and P_2

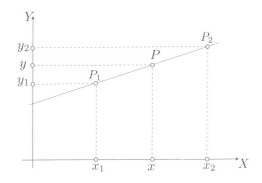

From Fig. 20.5

$$\frac{y_2 - y_1}{x_2 - x_1} = \frac{y - y_1}{x - x_1}$$
$$(y_2 - y_1)(x - x_1) = (x_2 - x_1)(y - y_1)$$
$$(y_2 - y_1)x - (y_2 - y_1)x_1 = (x_2 - x_1)y - (x_2 - x_1)y_1$$
$$(y_2 - y_1)x + (x_1 - x_2)y = x_1 y_2 - x_2 y_1$$

therefore,

$$a = y_2 - y_1 \quad b = x_1 - x_2 \quad c = -(x_1 y_2 - x_2 y_1).$$

If the two points are $P_1(1,\ 0)$ and $P_2(3,\ 4)$, then

$$(4 - 0)x + (1 - 3)y - (1 \times 4 - 3 \times 0) = 0$$

and

$$4x - 2y - 4 = 0.$$

20.8   Angle Between Two Straight Lines

Given two line equations it is possible to compute the angle between the lines using the scalar product. For example, if the line equations are

$$a_1 x + b_1 y + c_1 = 0$$
$$a_2 x + b_2 y + c_2 = 0$$

their normal vectors are $\mathbf{n} = a_1 \mathbf{i} + b_1 \mathbf{j}$ and $\mathbf{m} = a_2 \mathbf{i} + b_2 \mathbf{j}$ respectively, therefore,

$$\mathbf{n} \cdot \mathbf{m} = \|\mathbf{n}\| \|\mathbf{m}\| \cos \alpha$$

and the angle between the lines α is given by

$$\alpha = \cos^{-1} \left(\frac{\mathbf{n} \cdot \mathbf{m}}{\|\mathbf{n}\| \|\mathbf{m}\|} \right).$$

Figure 20.6 shows two lines with equations

$$2x + 2y - 4 = 0$$
$$2x + 4y - 4 = 0$$

Fig. 20.6 Two lines
intersecting at an angle α

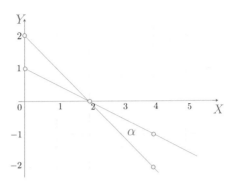

therefore,

$$\alpha = \cos^{-1}\left(\frac{2 \times 2 + 2 \times 4}{\sqrt{2^2 + 2^2}\sqrt{2^2 + 4^2}}\right) \approx 18.435°.$$

20.9   Test if Three Points Lie on a Straight Line

Figure 20.7 shows three points P_1, P_2 and P_3 which lie on a straight line. There are all sorts of ways to detect such a condition. For example, we could assume that the points are the vertices of a triangle, and if the triangle's area is zero, then the points lie on a line. Here is another approach.

Given $P_1(x_1, y_1)$, $P_2(x_2, y_2)$, $P_3(x_3, y_3)$ and $\mathbf{r} = \overrightarrow{P_1P_2}$ and $\mathbf{s} = \overrightarrow{P_1P_3}$, the three points lie on a straight line when $\mathbf{s} = \lambda\mathbf{r}$ where λ is a scalar.

Let the points be

$$P_1(0, -2), \quad P_2(1, -1), \quad P_3(4, 2)$$

Fig. 20.7 Three points on a
common line

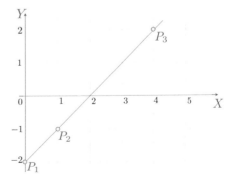

then

$$\mathbf{r} = \mathbf{i} + \mathbf{j}, \quad \text{and} \quad \mathbf{s} = 4\mathbf{i} + 4\mathbf{j}$$

and

$$\mathbf{s} = 4\mathbf{r}$$

therefore, the points lie on a straight line as confirmed by the diagram.
    Another way is to compute

$$\begin{vmatrix} x_1 & y_1 & 1 \\ x_2 & y_2 & 1 \\ x_3 & y_3 & 1 \end{vmatrix} = \begin{vmatrix} 0 & -2 & 1 \\ 1 & -1 & 1 \\ 4 & 2 & 1 \end{vmatrix} = 0$$

which is twice the area of $\Delta P_1 P_2 P_3$, and as this equals zero, the points must be
co-linear.

20.10   Position and Distance of the Nearest Point on a Line to a Point

Suppose we have a line and some arbitrary point P, and we require to find the
nearest point on the line to P. Vector analysis provides a very elegant way to solve
such problems. Figure 20.8 shows a line and a point P and the nearest point Q
on the line. The nature of the geometry is such that the line connecting P to Q is
perpendicular to the reference line, which is exploited in the analysis. The objective
is to determine the position vector $\mathbf{q}$.
    We start with the line equation

$$ax + by + c = 0$$

and declare $Q(x, y)$ as the nearest point on the line to P.

Fig. 20.8 Q is the nearest
point on the line to P

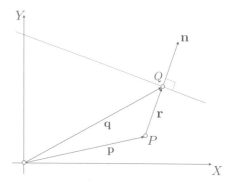

The normal to the line must be

$$\mathbf{n} = a\mathbf{i} + b\mathbf{j}$$

and the position vector for Q is

$$\mathbf{q} = x\mathbf{i} + y\mathbf{j}.$$

Therefore,

$$\mathbf{n} \cdot \mathbf{q} = -c. \qquad (20.1)$$

$\mathbf{r}$ is parallel to $\mathbf{n}$, therefore,

$$\mathbf{r} = \lambda \mathbf{n} \qquad (20.2)$$

where λ is some scalar.
    Taking the scalar product of (20.2)

$$\mathbf{n} \cdot \mathbf{r} = \lambda \mathbf{n} \cdot \mathbf{n} \qquad (20.3)$$

but as

$$\mathbf{r} = \mathbf{q} - \mathbf{p} \qquad (20.4)$$
$$\mathbf{n} \cdot \mathbf{r} = \mathbf{n} \cdot \mathbf{q} - \mathbf{n} \cdot \mathbf{p}. \qquad (20.5)$$

Substituting (20.1) and (20.3) in (20.5) we obtain

$$\lambda \mathbf{n} \cdot \mathbf{n} = -c - \mathbf{n} \cdot \mathbf{p}$$

therefore,

$$\lambda = \frac{-(\mathbf{n} \cdot \mathbf{p} + c)}{\mathbf{n} \cdot \mathbf{n}}.$$

From (20.4) we get

$$\mathbf{q} = \mathbf{p} + \mathbf{r}. \qquad (20.6)$$

Substituting (20.2) in (20.6) we obtain the position vector for Q:

$$\mathbf{q} = \mathbf{p} + \lambda \mathbf{n}.$$

The distance PQ must be the magnitude of $\mathbf{r}$:

$$PQ = \|\mathbf{r}\| = \lambda \|\mathbf{n}\|.$$

Let's test this result with an example where the answer can be predicted.

Fig. 20.9 Q is the nearest point on the line to P

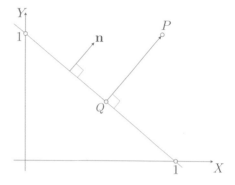

Figure 20.9 shows a line whose equation is $x + y - 1 = 0$, and the associated point is $P(1, \ 1)$. By inspection, the nearest point is $Q\left(\frac{1}{2}, \ \frac{1}{2}\right)$ and the distance $PQ \approx 0.7071$.

From the line equation

$$a = 1, \quad b = 1, \quad c = -1$$

therefore,

$$\lambda = -\frac{2 - 1}{2} = -\frac{1}{2}$$

and

$$x_Q = x_P + \lambda x_n = 1 - \frac{1}{2} \times 1 = \frac{1}{2}$$
$$y_Q = y_P + \lambda y_n = 1 - \frac{1}{2} \times 1 = \frac{1}{2}.$$

The nearest point is $Q\left(\frac{1}{2}, \ \frac{1}{2}\right)$ and the distance is

$$PQ = \|\lambda\mathbf{n}\| = \frac{1}{2}\|\mathbf{i} + \mathbf{j}\| \approx 0.7071.$$

20.11   Position of a Point Reflected in a Line

Suppose that instead of finding the nearest point on a line we require the reflection Q of P in the line. Once more, we set out to discover the position vector for Q. Figure 20.10 shows the vectors used in the analysis. We start with the line equation

$$ax + by + c = 0$$

Fig. 20.10 The vectors
required to find the reflection
of *P* in the line

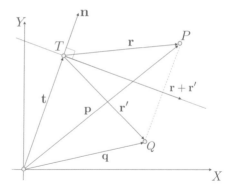

and declare $T(x, y)$ as the nearest point on the line to O with $\mathbf{t} = x\mathbf{i} + y\mathbf{j}$ as its
position vector.

From the line equation

$$\mathbf{n} = a\mathbf{i} + b\mathbf{j}$$

therefore,

$$\mathbf{n} \cdot \mathbf{t} = -c. \tag{20.7}$$

We note that $\mathbf{r} + \mathbf{r}'$ is orthogonal to $\mathbf{n}$, therefore,

$$\mathbf{n} \cdot (\mathbf{r} + \mathbf{r}') = 0$$

and

$$\mathbf{n} \cdot \mathbf{r} + \mathbf{n} \cdot \mathbf{r}' = 0. \tag{20.8}$$

We also note that $\mathbf{p} - \mathbf{q}$ is parallel to $\mathbf{n}$, therefore,

$$\mathbf{p} - \mathbf{q} = \mathbf{r} - \mathbf{r}' = \lambda \mathbf{n}$$

where λ is some scalar, therefore,

$$\lambda = \frac{\mathbf{r} - \mathbf{r}'}{\mathbf{n}}. \tag{20.9}$$

From the figure we note that

$$\mathbf{r} = \mathbf{p} - \mathbf{t}. \tag{20.10}$$

Substituting (20.7) in (20.10)

$$\mathbf{n} \cdot \mathbf{r} = \mathbf{n} \cdot \mathbf{p} - \mathbf{n} \cdot \mathbf{t} = \mathbf{n} \cdot \mathbf{p} + c. \tag{20.11}$$

Fig. 20.11 Q is the
reflection of P in the line

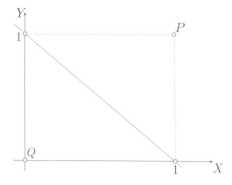

Substituting (20.8) and (20.11) in (20.9)

$$\lambda = \frac{\mathbf{n} \cdot \mathbf{r} - \mathbf{n} \cdot \mathbf{r}'}{\mathbf{n} \cdot \mathbf{n}} = \frac{2\mathbf{n} \cdot \mathbf{r}}{\mathbf{n} \cdot \mathbf{n}}$$

$$\lambda = \frac{2(\mathbf{n} \cdot \mathbf{p} + c)}{\mathbf{n} \cdot \mathbf{n}}$$

and the position vector is

$$\mathbf{q} = \mathbf{p} - \lambda\mathbf{n}.$$

Let's again test this formula with a scenario that can be predicted in advance.
Given the line equation

$$x + y - 1 = 0$$

and the point $P(1, \ 1)$, the reflection must be the origin, as shown in Fig. 20.11.
Now let's confirm this prediction. From the line equation

$$a = 1, \quad b = 1, \quad c = -1$$

and

$$x_P = 1$$
$$y_P = 1$$
$$\lambda = \frac{2 \times (2 - 1)}{2} = 1$$

therefore,

$$x_Q = x_P - \lambda x_n = 1 - 1 \times 1 = 0$$
$$y_Q = y_P - \lambda y_n = 1 - 1 \times 1 = 0$$

and the reflection point is $Q(0, \ 0)$.

20.12 Intersection of a Line and a Sphere

In ray tracing and ray casting it is necessary to detect whether a ray (line) intersects objects within a scene. Such objects may be polygonal, constructed from patches, or defined by equations. In this example, we explore the intersection between a line and a sphere.

There are three possible scenarios: the line intersects, touches or misses the sphere. It just so happens, that the cosine rule proves very useful in setting up a geometric condition that identifies the above scenarios, which are readily solved using vector analysis.

Figure 20.12 shows a sphere with radius r located at C. The line is represented parametrically, which lends itself to this analysis. The objective is to discover whether there are points in space that satisfy both the line equation and the sphere equation. If there is a point, a position vector will locate it.

The position vector for C is

$$\mathbf{c} = x_c\mathbf{i} + y_c\mathbf{j} + z_c\mathbf{k}$$

and the equation of the line is

$$\mathbf{p} = \mathbf{t} + \lambda\mathbf{v}$$

where λ is a scalar, and

$$\|\mathbf{v}\| = 1. \tag{20.12}$$

For an intersection at P

$$\|\mathbf{q}\| = r$$
$$\|\mathbf{q}\|^2 = r^2$$
$$\|\mathbf{q}\|^2 - r^2 = 0.$$

Fig. 20.12 The vectors required to locate a possible intersection

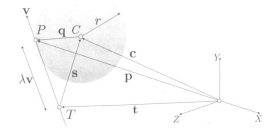

Using the cosine rule

$$\|\mathbf{q}\|^2 = \|\lambda\mathbf{v}\|^2 + \|\mathbf{s}\|^2 - 2\|\lambda\mathbf{v}\|\|\mathbf{s}\|\cos\theta \qquad (20.13)$$

$$\|\mathbf{q}\|^2 = \lambda^2\|\mathbf{v}\|^2 + \|\mathbf{s}\|^2 - 2\|\mathbf{v}\|\|\mathbf{s}\|\lambda\cos\theta. \qquad (20.14)$$

Substituting (20.12) in (20.14)

$$\|\mathbf{q}\|^2 = \lambda^2 + \|\mathbf{s}\|^2 - 2\|\mathbf{s}\|\lambda\cos\theta. \qquad (20.15)$$

Now let's identify $\cos\theta$:

$$\mathbf{s}\cdot\mathbf{v} = \|\mathbf{s}\|\|\mathbf{v}\|\cos\theta$$

therefore,

$$\cos\theta = \frac{\mathbf{s}\cdot\mathbf{v}}{\|\mathbf{s}\|}. \qquad (20.16)$$

Substituting (20.16) in (20.15)

$$\|\mathbf{q}\|^2 = \lambda^2 - 2\mathbf{s}\cdot\mathbf{v}\lambda + \|\mathbf{s}\|^2$$

therefore,

$$\|\mathbf{q}\|^2 - r^2 = \lambda^2 - 2\mathbf{s}\cdot\mathbf{v}\lambda + \|\mathbf{s}\|^2 - r^2 = 0. \qquad (20.17)$$

Equation (20.17) is a quadratic in λ where

$$\lambda = \mathbf{s}\cdot\mathbf{v} \pm \sqrt{(\mathbf{s}\cdot\mathbf{v})^2 - \|\mathbf{s}\|^2 + r^2} \qquad (20.18)$$

and

$$\mathbf{s} = \mathbf{c} - \mathbf{t}.$$

The discriminant of (20.18) determines whether the line intersects, touches or misses the sphere.

The position vector for P is given by

$$\mathbf{p} = \mathbf{t} + \lambda\mathbf{v}$$

where

$$\lambda = \mathbf{s}\cdot\mathbf{v} \pm \sqrt{(\mathbf{s}\cdot\mathbf{v})^2 - \|\mathbf{s}\|^2 + r^2}$$

and

$$\mathbf{s} = \mathbf{c} - \mathbf{t}.$$

For a miss condition

$$(\mathbf{s}\cdot\mathbf{v})^2 - \|\mathbf{s}\|^2 + r^2 < 0.$$

Fig. 20.13 Three lines that miss, touch and intersect the sphere

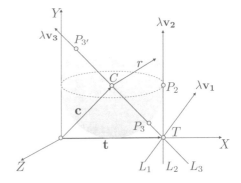

For a touch condition

$$(\mathbf{s} \cdot \mathbf{v})^2 - \|\mathbf{s}\|^2 + r^2 = 0.$$

For an intersect condition

$$(\mathbf{s} \cdot \mathbf{v})^2 - \|\mathbf{s}\|^2 + r^2 > 0.$$

To test these formulae we will create all three scenarios and show that the equations are well behaved.

Figure 20.13 shows a sphere with three lines represented by their direction vectors $\lambda \mathbf{v}_1$, $\lambda \mathbf{v}_2$ and $\lambda \mathbf{v}_3$. The sphere has radius $r = 1$ and is located at C with position vector

$$\mathbf{c} = \mathbf{i} + \mathbf{j}$$

whilst the three lines L_1, L_2 and L_3 miss, touch and intersect the sphere respectively.

The lines are of the form

$$\mathbf{p} = \mathbf{t} + \lambda \mathbf{v}$$

therefore,

$$\mathbf{p}_1 = \mathbf{t}_1 + \lambda \mathbf{v}_1$$
$$\mathbf{p}_2 = \mathbf{t}_2 + \lambda \mathbf{v}_2$$
$$\mathbf{p}_3 = \mathbf{t}_3 + \lambda \mathbf{v}_3$$

where,

$$\mathbf{t}_1 = 2\mathbf{i}, \quad \mathbf{v}_1 = \tfrac{1}{\sqrt{2}}\mathbf{i} + \tfrac{1}{\sqrt{2}}\mathbf{j}$$
$$\mathbf{t}_2 = 2\mathbf{i}, \quad \mathbf{v}_2 = \mathbf{j}$$
$$\mathbf{t}_3 = 2\mathbf{i}, \quad \mathbf{v}_3 = -\tfrac{1}{\sqrt{2}}\mathbf{i} + \tfrac{1}{\sqrt{2}}\mathbf{j}$$

and
$$c = i + j.$$

Let's substitute the lines in the original equations:
L_1:

$$s = -i + j$$
$$(s \cdot v)^2 - \|s\|^2 + r^2 = 0 - 2 + 1 = -1$$

the negative discriminant confirms a miss condition.
L_2:

$$s = -i + j$$
$$(s \cdot v)^2 - \|s\|^2 + r^2 = 1 - 2 + 1 = 0$$

the zero discriminant confirms a touch condition, therefore $\lambda = 1$ and the touch point is $P_2(2, \ 1, \ 0)$ which is correct.
L_3:

$$s = -i + j$$
$$(s \cdot v)^2 - \|s\|^2 + r^2 = 2 - 2 + 1 = 1$$

the positive discriminant confirms an intersect condition, therefore,

$$\lambda = \tfrac{2}{\sqrt{2}} \pm 1 = 1 + \sqrt{2} \quad \text{or} \quad \sqrt{2} - 1.$$

The intersection points are given by the two values of λ:
When $\lambda = 1 + \sqrt{2}$

$$x_P = 2 + \left(1 + \sqrt{2}\right)\left(-\tfrac{1}{\sqrt{2}}\right) = 1 - \tfrac{1}{\sqrt{2}}$$
$$y_P = 0 + \left(1 + \sqrt{2}\right)\tfrac{1}{\sqrt{2}} = 1 + \tfrac{1}{\sqrt{2}}$$
$$z_P = 0.$$

When $\lambda = \sqrt{2} - 1$

$$x_P = 1 + \left(\sqrt{2} - 1\right)\left(-\tfrac{1}{\sqrt{2}}\right) = 1 + \tfrac{1}{\sqrt{2}}$$
$$y_P = 0 + \left(\sqrt{2} - 1\right)\tfrac{1}{\sqrt{2}} = 1 - \tfrac{1}{\sqrt{2}}$$
$$z_P = 0.$$

The intersection points are

$$P_{3'}\left(1 - \tfrac{1}{\sqrt{2}},\ 1 + \tfrac{1}{\sqrt{2}},\ 0\right)$$
$$P_3\left(1 + \tfrac{1}{\sqrt{2}},\ 1 - \tfrac{1}{\sqrt{2}},\ 0\right)$$

which are correct.

20.13 Sphere Touching a Plane

A sphere will touch a plane if the perpendicular distance from its centre to the plane equals its radius. The geometry describing this condition is identical to finding the position and distance of the nearest point on a plane to a point.

Figure 20.14 shows a sphere located at P with position vector $\mathbf{p}$. A potential touch condition occurs at Q, and the objective of the analysis is to discover its position vector $\mathbf{q}$. Given the following plane equation

$$ax + by + cz + d = 0$$

its surface normal is

$$\mathbf{n} = a\mathbf{i} + b\mathbf{j} + c\mathbf{k}.$$

The nearest point Q on the plane to a point P is given by the position vector

$$\mathbf{q} = \mathbf{p} + \lambda\mathbf{n} \tag{20.19}$$

where

$$\lambda = -\frac{\mathbf{n} \cdot \mathbf{p} + d}{\mathbf{n} \cdot \mathbf{n}}$$

Fig. 20.14 The vectors used to detect when a sphere touches a plane

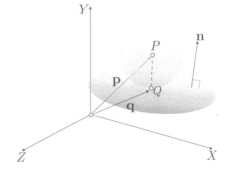

Fig. 20.15 A sphere touching a plane

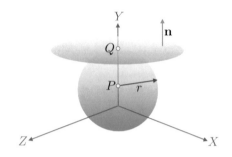

the distance

$$PQ = \|\lambda\mathbf{n}\|.$$

If P is the centre of the sphere with radius r, and position vector $\mathbf{p}$, the touch point is also given by (20.19) when

$$PQ = \|\lambda\mathbf{n}\| = r.$$

Let's test the above equations with a simple example, as shown in Fig. 20.15, which shows a sphere with radius $r = 1$ and centred at $P(1, \ 1, \ 1)$.

The plane equation is

$$y - 2 = 0$$

therefore,

$$\mathbf{n} = \mathbf{j}$$

and

$$\mathbf{p} = \mathbf{i} + \mathbf{j} + \mathbf{k}$$

therefore,

$$\lambda = -(1 - 2) = 1$$

which equals the sphere's radius and therefore the sphere and plane touch. The touch point is

$$x_Q = 1 + 1 \times 0 = 1$$
$$y_Q = 1 + 1 \times 1 = 2$$
$$z_Q = 1 + 1 \times 0 = 1$$
$$Q = (1, \ 2, \ 1).$$

20.14 Summary

Unfortunately, problem solving is not always obvious, and it is possible to waste hours of analysis simply because the objective of the solution has not been well formulated. Hopefully, though, the reader has discovered some of the strategies used in solving the above geometric problems, and will be able to implement them in other scenarios. At the end of the day, practice makes perfect!

Appendix A
Limit of $(\sin \theta)/\theta$

This appendix proves that

$$\lim_{\theta \to 0} \frac{\sin \theta}{\theta} = 1, \quad \text{where } \theta \text{ is in radians.}$$

From high-school mathematics we know that $\sin \theta \approx \theta$, for small values of θ. For example:

$$\sin 0.1 = 0.099833$$
$$\sin 0.05 = 0.04998$$
$$\sin 0.01 = 0.0099998$$

and

$$\frac{\sin 0.1}{0.1} = 0.99833$$
$$\frac{\sin 0.05}{0.05} = 0.99958$$
$$\frac{\sin 0.01}{0.01} = 0.99998.$$

Therefore, we can reason that in the limit, as $\theta \to 0$:

$$\lim_{\theta \to 0} \frac{\sin \theta}{\theta} = 1.$$

Figure A.1 shows a graph of $(\sin \theta)/\theta$, which confirms this result. However, this is an observation, rather than a proof. So, let's pursue a geometric line of reasoning.

From Fig. A.2 we see as the circle's radius is unity, $OA = OB = 1$, and $AC = \tan \theta$. As part of the strategy, we need to calculate the area of the triangle $\triangle OAB$, the sector OAB and the $\triangle OAC$:

© Springer-Verlag London Ltd., part of Springer Nature 2022
J. Vince, *Mathematics for Computer Graphics*, Undergraduate Topics
in Computer Science, https://doi.org/10.1007/978-1-4471-7520-9

Fig. A.1 Graph of $(\sin\theta)/\theta$

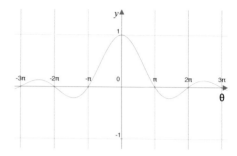

Fig. A.2 Unit radius circle
with trigonometric ratios

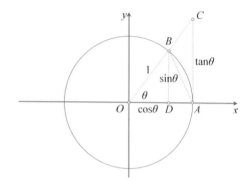

$$\text{area}(\triangle OAB) = \text{area}(\triangle ODB) + \text{area}(\triangle DAB)$$
$$= \tfrac{1}{2}\cos\theta\sin\theta + \tfrac{1}{2}(1-\cos\theta)\sin\theta$$
$$= \tfrac{1}{2}\cos\theta\sin\theta + \tfrac{1}{2}\sin\theta - \tfrac{1}{2}\cos\theta\sin\theta$$
$$= \tfrac{1}{2}\sin\theta.$$
$$\text{area of sector } OAB = \frac{\theta}{2\pi}\pi(1)^2 = \tfrac{1}{2}\theta.$$
$$\text{area}(\triangle OAC) = \tfrac{1}{2}(1)\tan\theta = \tfrac{1}{2}\tan\theta.$$

From the geometry of a circle, we know that

$$\tfrac{1}{2}\sin\theta < \tfrac{1}{2}\theta < \tfrac{1}{2}\tan\theta$$
$$\sin\theta < \theta < \frac{\sin\theta}{\cos\theta}$$
$$1 < \frac{\theta}{\sin\theta} < \frac{1}{\cos\theta}$$
$$1 > \frac{\sin\theta}{\theta} > \cos\theta$$

and as $\theta \to 0$, $\cos \theta \to 1$ and $\dfrac{\sin \theta}{\theta} \to 1$. This holds, even for negative values of θ, because

$$\frac{\sin(-\theta)}{-\theta} = \frac{-\sin \theta}{-\theta} = \frac{\sin \theta}{\theta}.$$

Therefore,

$$\lim_{\theta \to 0} \frac{\sin \theta}{\theta} = 1.$$

Appendix B
Integrating $\cos^n \theta$

We start with

$$\int \cos^n x \, dx = \int \cos x \, \cos^{n-1} x \, dx.$$

Let $u = \cos^{n-1} x$ and $v' = \cos x$, then

$$u' = -(n-1)\cos^{n-2} x \, \sin x$$

and

$$v = \sin x.$$

Integrating by parts:

$$\int uv' \, dx = uv - \int v \, u' \, dx + C$$

$$\int \cos^{n-1} x \, \cos x \, dx = \cos^{n-1} x \, \sin x + \int \sin x \, (n-1)\cos^{n-2} x \, \sin x \, dx + C$$

$$= \sin x \, \cos^{n-1} x + (n-1)\int \sin^2 x \, \cos^{n-2} x \, dx + C$$

$$= \sin x \, \cos^{n-1} x + (n-1)\int (1 - \cos^2 x) \, \cos^{n-2} x \, dx + C$$

$$= \sin x \, \cos^{n-1} x + (n-1)\int \cos^{n-2} dx - (n-1)\int \cos^n x \, dx + C$$

$$n \int \cos^n x \, dx = \sin x \, \cos^{n-1} x + (n-1)\int \cos^{n-2} dx + C$$

$$\int \cos^n x \, dx = \frac{\sin x \, \cos^{n-1} x}{n} + \frac{n-1}{n}\int \cos^{n-2} dx + C$$

where n is an integer, $\neq 0$.
Similarly,

$$\int \sin^n x \, dx = -\frac{\cos x \, \sin^{n-1} x}{n} + \frac{n-1}{n}\int \sin^{n-2} dx + C.$$

© Springer-Verlag London Ltd., part of Springer Nature 2022
J. Vince, *Mathematics for Computer Graphics*, Undergraduate Topics
in Computer Science, https://doi.org/10.1007/978-1-4471-7520-9

For example,

$$\int \cos^3 x\, dx = \frac{\sin x\ \cos^2 x}{3} + \tfrac{2}{3} \sin x + C.$$

Index

A

Adding
  complex numbers, 152
  matrices, 165
  ordered pairs, 158
  quaternions, 241, 256
Additive form of a quaternion, 245
Aitken, Alexander, 314
Aleph-zero, 27
Algebra, 31
Algebraic number, 20, 26
Altmann, Simon, 238, 243, 269
Analytic geometry, 325
Angle
  compound, 60
Angle between
  a line and plane, 366
  two planes, 365
  two straight lines, 536
Annulus, 333
Anticlockwise, 67
Antiderivative, 445
Antisymmetric
  functions, 402
  matrix, 128
Area, 113
  between two functions, 523
  circle, 514
  negative, 521
  of any polygon, 532
  of a regular polygon, 531
  of a shape, 67
  of a triangle, 341, 387, 535
  parametric function, 525
  positive, 521
  under a graph, 512
  with the y-axis, 524

Areal coordinates, 185
Argand, Jean-Robert, 95
Associative law, 10
Atan2, 57
Axial vector, 243
Axioms, 10, 412

B

Babylonians, 77
Back-face detection, 106
Barycentric coordinates, 185, 371
Base, 11
Bernstein polynomials, 303
Bézier
  curves, 301, 398
  matrix, 311
Bézier, Pierre, 301
Binary
  addition, 16
  negative number, 18
  number, 13
  operation, 10
  subtraction, 18
Binary form of a quaternion, 245
Binomial expansion, 304, 442
Bivector, 408
Blending
  curve, 308
  function, 312
Brahmagupta, 7
B-splines, 315
Bürgi, Joost, 38

C

Calculus, 437

© Springer-Verlag London Ltd., part of Springer Nature 2022
J. Vince, *Mathematics for Computer Graphics*, Undergraduate Topics
in Computer Science, https://doi.org/10.1007/978-1-4471-7520-9

Camera space, 206
Cantor, Georg, 2, 27
Cardano, Girolamo, 77
Cardinality, 27
Cartesian
    coordinates, 65, 66
    plane, 66
    vector, 103
Casteljau, Paul de, 301
Cauchy, Augustin-Louis, 77, 79, 437
Cayley, Arthur, 77, 119, 261, 269
Cayley numbers, 255, 261
Centre of gravity, 328
Centroid, 374
Ceva, Giovanni, 372
Ceva's Theorem, 372
Chain rule, 486
Change of axes, 202
    2D, 202
    3D, 205
Circle, 332
    equation, 302
Clifford, William Kingdon, 96, 401
Clockwise, 67
Closed interval, 42
Cofactor expansion, 140
Column vector, 96, 120, 123, 131, 184
Commutative law, 10
Complex
    number, 24, 151
    plane, 22
Complex conjugate, 25
    complex number, 154
    quaternion, 246
Complex number
    absolute value, 154
    adding matrix, 165
    addition, 152
    complex conjugate, 154
    conjugate matrix, 167
    definition, 151
    inverse, 156
    inverse matrix, 167
    matrix, 164
    modulus, 154
    norm, 154, 166
    ordered pair, 158
    product, 153
    product matrix, 166
    quotient, 155
    quotient matrix, 168
    square, 154
    subtracting matrix, 165

    subtraction, 152
Compound angles, 60
Continuity, 317, 437
Continuous functions, 492
Control
    point, 307, 315
    vertex, 307
Convex hull, 307, 387
Coordinates
    barycentric, 371
    Cartesian, 65
    cylindrical, 72
    polar, 70
    spherical polar, 71
Cosecant, 54
Cosine, 54
    rule, 59, 430
Cotangent, 54
Counting, 5
Cross product, 107
Cubic
    Bernstein polynomials, 307
    Bézier surface patch, 322
    equation, 440
    function, 66
    interpolant, 308
    interpolation, 289
Curves and patches, 301
Cylindrical coordinates, 72

D
Decimal
    number, 12
    system, 7
Definite integral, 516
Degree, 51
De Moivre's Formula, 62
Dependent variable, 41
Derivative, 437, 445
    graphical interpretation, 444
    partial, 480
    total, 488
Descartes, René, 21, 32, 65
Determinant, 77, 125, 220
    complex, 92
    expansion, 92
    order, 89
    property, 91
    second-order, 79
    third-order, 82, 88
    value, 89
Diagonal matrix, 142

Differential, 445
Differentiating, 448
    arccos function, 470
    arccot function, 471
    arccsc function, 471
    arcsec function, 471
    arcsin function, 470
    arctan function, 470
    cosh function, 474
    cot function, 470
    csc function, 468
    exponential functions, 463
    function of a function, 450
    function products, 454
    function quotients, 458
    hyperbolic functions, 472
    implicit functions, 460
    logarithmic functions, 465
    partial, 481
    sec function, 469
    sine function, 451
    sinh function, 474
    sums of functions, 448
    tan function, 467
    tanh function, 474
    trigonometric functions, 467
Dihedral angle of a dodecahedron, 533
Dirac, Paul, 3
Direction cosines
    2D, 204
    3D, 206
Distance
    between two 2D points, 68
    between two 3D points, 69
Distributive law, 11
Division algebra, 255
Dodecahedron, 533
Domain, 43, 55
Dot product, 104
Double angle, 268, 269
Double-angle identities, 61
Duality, 419

E
Element, 6
Ellipse equation, 302
Equation
    explicit, 40
    implicit, 40
    linear, 78, 81
Equilateral triangle, 329
Euler

angles, 208
    rotations, 196
Euler, Leonhard, 40, 62
Even function, 44
Explicit equation, 40
Exterior angle, 327

F
Fermat, Pierre de, 65
Feynman, Richard, 243
Frobenius, Ferdinand Georg, 255
Function, 40, 442
    continuous, 492
    cubic, 66, 440
    domain, 43
    even, 44
    graph, 66
    linear, 66
    notation, 41
    odd, 44
    power, 46
    quadratic, 66, 439
    range, 43
    second derivative, 479
    trigonometric, 66
Function of a function
    differentiating, 450
Fundamental theorem of calculus, 517

G
Gauss, Johann, 77, 79, 119
General form of a line equation, 535
Geometric
    algebra, 401
    continuity, 317
    product in 2D, 407
    product in 3D, 409
    transform, 181
Gibbs, Josiah Willard, 96, 236
Gimbal lock, 199
Gödel, Kurt, 32
Golden section, 327
Grades, 413
Grassmann, Hermann Günther, 96, 236
Graves, John Thomas, 261

H
Half-angle identities, 63
Half-open interval, 42
Hamilton's rules, 233, 421
Hamilton, William Rowan, 95, 108, 269

Hermite, Charles, 293
Hermite interpolation, 293
Hessian normal form, 335, 343
Hestenes, David, 401
Hexadecimal number, 13
Higher derivatives, 475
Homogeneous coordinates, 184

I

Identity matrix, 191
Image space, 206
Imaginary number, 21
Implicit equation, 40
Indefinite integral, 491
Independent variable, 41
Indeterminate form, 8
Indices, 37
    laws of, 38
Infinitesimals, 437
Infinity, 27
Inner product, 405, 415
Integer, 19
    number, 6
Integral
    definite, 516
    indefinite, 491
Integrating, 445
    arccos function, 470
    arccot function, 471
    arccsc function, 471
    arcsec function, 471
    arcsin function, 470
    arctan function, 470
    by parts, 501
    by substitution, 505
    completing the square, 497
    cot function, 470
    csc function, 468
    difficult functions, 493
    exponential function, 465
    integrand contains a derivative, 498
    logarithmic function, 466
    partial fractions, 510
    radicals, 495
    sec function, 469
    tan function, 467
    techniques, 492
    trigonometric identities, 493
Intercept theorems, 326
Interior angle, 327
Interpolating
    quaternions, 297

vectors, 294
Interpolation, 285
    cubic, 289
    linear, 286, 380
    non-linear, 288
    trigonometric, 288
Intersecting
    circle and line, 345
    line and sphere, 543
    line segments, 339
    planes, 359
    straight lines, 338, 348
Interval, 42
    closed, 42
    half-open, 42
    open, 42
Inverse
    complex number, 156
    matrix, 134
    of a vector, 415
    quaternion, 252, 258
    trigonometric function, 56
Irrational number, 20
Isosceles triangle, 328

K
Kronecker, Leopold, 2, 19

L
Lambert's law, 105
Laplace expansion, 89
Laplace, Pierre-Simon, 77, 89, 140
Laplacian expansion, 140
Left-handed axes, 69
Leibniz, Gottfried von, 40, 77
Lerp, 381
L'Hôpital, Guillaume de, 77
Lighting calculations, 105
Limits, 437, 442
Linear
    equation, 78, 81
    function, 66
    interpolation, 286, 305, 312, 380
Linearly independent, 78
Lobachevsky, Nikolai, 236, 325
Local coordinates, 371
Logarithm, 38
    base, 39

M
Mass points, 374, 391

Matrix, 88, 119, 184, 253
   addition, 130
   algebra, 119
   antisymmetric, 128
   determinant, 125
   diagonal, 142
   dimension, 122
   inverse, 134
   multiplication, 121
   notation, 122, 184
   null, 123
   order, 122
   orthogonal, 141
   products, 130
   rectangular, 134
   scalar multiplication, 130
   singular, 135
   skew-symmetric, 128
   square, 122, 133
   subtraction, 130
   symmetric, 126
   trace, 124
   transpose, 125
   unit, 123
Maxima, 477
Median, 328
Member, 6
Minima, 477
Mixed partial derivative, 485
Möbius, August, 185, 371
Moivre, Abraham de, 62
Multiple-angle identities, 62
Multivectors, 413

N
Napier, John, 38
Natural number, 19
Nearest point to a line, 538
Negative number, 8
Non-linear interpolation, 288
Non-rational B-splines, 315
Non-uniform
   B-splines, 318
   rational B-splines, 319
Non-Uniform Rational B-Splines (NURBS), 319
Norm
   complex number, 154, 166
   ordered pair, 161
   quaternion, 247, 257
   quaternion product, 251
Normalised quaternion, 248

Notation, 3
Null matrix, 123
Number
   algebraic, 20, 26
   arithmetic, 9
   binary, 13
   complex, 24
   hexadecimal, 13
   imaginary, 21
   integer, 6, 19
   line, 8
   natural, 19
   negative, 8
   octal, 12
   positive, 8
   rational, 6, 19, 20
   real, 6, 20
   transcendental, 20, 26

O
Object space, 206
Octal number, 12
Octaves, 261
Odd function, 44
One-to-one correspondence, 27
Open interval, 42
Ordered pair, 158, 238, 241
   absolute value, 161
   addition, 158
   complex conjugate, 161
   inverse, 162
   modulus, 161
   multiplying by a scalar, 159
   norm, 161
   product, 158, 159
   quotient, 161
   square, 160
   subtraction, 158
Oriented axes, 66
Origin, 66
Orthogonal
   matrix, 141, 208, 254
Outer product, 405, 415
   3D, 411
   imaginary properties, 417

P
Parallelogram, 331
Partial derivative, 480
   chain rule, 486
   first, 481

mixed, 485
second, 482
visualising, 483
Pascal, Blaise, 65
Pascal's triangle, 303, 442
Peirce, Benjamin, 119
Peirce, Charles, 119
Perimeter relationships, 63
Perspective projection, 222
Pitch, 197, 208
Placeholder, 7
Planar surface patch, 319
Plane equation, 351
    Cartesian form, 351
    from three points, 357
    general form, 353
    parametric form, 354
Plücker, Julius, 185
Poincaré, Henri, 2
Point inside a triangle, 341, 385
Point reflected in a line, 540
Polar
    coordinates, 70
    vector, 243
Polynomial equation, 20
Position vector, 102
Power
    functions, 46
    series, 52
Product
    complex number, 153
    pure quaternion, 249
    quaternion, 248, 257
    unit-norm quaternion, 249
Pseudoscalars, 413
Pseudovector, 243
Pure quaternion, 243
    product, 249

Q
Quadratic
    Bézier curve, 306
    Bézier surface patch, 320
    equation, 33
    function, 66, 439
Quadrilateral, 330
Quaternion, 233, 261, 421
    addition, 241, 256
    additive form, 245
    algebra, 254
    binary form, 245
    conjugate, 246

interpolating, 297
inverse, 252, 258
matrix, 253, 271
norm, 247, 257
normalised, 248
product, 239, 241, 248, 249, 257, 261
pure, 243
square, 250, 258
subtraction, 241
unit-norm, 248, 257
units, 238, 239, 244
Quotient
    complex number, 155
    quaternion, 252

R
Radian, 51, 52, 325
Radius of the inscribed circle, 390
Range, 43, 55
Rational
    B-splines, 315
    coefficients, 20
    number, 6, 19
Ratios, 373
Real
    number, 6, 20
    quaternion, 242
Rectangular matrix, 134
Recursive Bézier curve, 310
Reflecting a vector, 432
Reflections, 422
Regular polygon, 332
Rhombus, 331
Riemann, Bernhard, 325, 527
Riemann sum, 527
Right-handed axes, 69
Right-hand rule, 112
Right triangle, 329
Rodrigues, Benjamin Olinde, 269
Rodrigues, Olinde, 236
Roll, 197, 208
Rotating about an axis, 200, 211
Rotation, 422
    matrix, 164
Rotors, 426
Row vector, 96, 123, 131, 184
Russell, Bertrand, 31

S
Sarrus, Pierre, 88
Sarrus's rule, 88, 92

Scalar product, 103, 104
Secant, 54
Second derivative, 479
Seki, Takakazu, 77
Series
    cosine, 52
    power, 52
    sine, 52
Servois, François-Joseph, 236
Set, 6
    member, 6
Simultaneous equations, 93
Sine, 54
    differentiating, 451
    rule, 58, 429
Singular matrix, 135
Skew-symmetric matrix, 128
Space partitioning, 337
Sphere touching a plane, 547
Spherical polar coordinates, 71
Square
    complex number, 154
    matrix, 122, 133
    quaternion, 250, 258
Square-root of i
    complex number, 156
    matrix, 169
    ordered pair, 163
Straight line equation, 347
Subtracting
    complex numbers, 152
    matrices, 165
    ordered pairs, 158
    quaternions, 241
Surface patch, 319
Symmetric
    functions, 402
    matrix, 126

T
Tait, Peter Guthrie, 234
Tangent, 54
Thales, 326
Theorem of
    Pythagoras, 68, 69, 329
    Thales, 329
3D
    complex numbers, 108
    coordinates, 69
    reflections, 423
    rotation transform, 196
    transforms, 194

vector, 99
Three intersecting planes, 362
Total derivative, 488
Trace, 124
Transcendental number, 20, 26
Transform
    2D, 182
    2D reflection, 183, 187, 193
    2D rotation about a point, 194
    2D scaling, 182, 186
    2D shearing, 189
    2D translation, 182, 186
    3D reflection, 202
    3D scaling, 195
    3D translation, 195
    affine, 192
Transforming vectors, 218
Transpose matrix, 125
Trapezoid, 330
Triangle
    centre of gravity, 328
    equilateral, 329
    isosceles, 328
    right, 329
Trigonometric
    function, 53, 66
    identities, 58
    interpolation, 288
    inverse function, 56
    ratios, 53
Trigonometry, 51
Trivector, 412, 413
2D
    analytic geometry, 334
    polygon, 67
    reflections, 422
    rotations, 424
    scaling transform, 192
    vector, 96
Two's complement, 18

U
Uniform B-splines, 315
Unit
    matrix, 123
    normal vector, 112
    quaternion, 244
    vector, 102
Unit-norm
    quaternion, 248, 249, 257

V

Vector
    2D, 96
    3D, 99
    addition, 101
    Cartesian, 103
    column, 96, 120, 123, 131, 184
    interpolating, 294
    magnitude, 98
    normalising, 102
    normal to a triangle, 534
    position, 102
    product, 103, 107
    row, 96, 123, 131, 184
    scaling, 100
    subtraction, 101
    transforming, 218
    unit, 102
Vertices, 67
Virtual camera, 205
Volume of a tetrahedron, 396

W

Warren, John, 95
Weierstrass, Karl, 437
Wessel, Caspar, 95
Whitehead, Alfred North, 31
Wilson, Edwin Bidwell, 96, 236
Wittgenstein, Ludwig, 2
Witt, Jan de, 77
World space, 206

X

Xy-plane, 65

Y

Yaw, 197, 208

Z

Zero, 7

tes

ablisher Services